Small Organic Molecules-Based Fluorescent Biosensors and their Applications

This book provides an overview of organic molecule-based fluorescent compounds and their applications as sensors and biosensors. The initial chapter introduces fundamental fluorescence concepts and their significance in biosensing. The book, in turn, details the synthesis of various scaffolds including xanthene, BODIPY, julolidine, cyanine, quinoline, phenanthiridine, acridine, rhodamine, benzothiazole, coumarin, perylene, and carbazole. The subsequent section covers the use of these organic fluorescent molecules in sensing proteins and DNA through selective binding, ion indicators for real-time tracking, and receptor-specific ligands for interaction studies. It also explores cellular component visualization using organelle markers and membrane probes. Additionally, the book delves into the application of fluorescent organic molecules for sensing lipids, carbohydrates, and other biological molecules, fostering interdisciplinary understanding. Addressing environmental concerns, the book highlights the use of fluorescent probes for analyte analysis, providing insights into pollution monitoring and water quality assessment. This book is useful for researchers, students, and professionals seeking to understand and harness the potential of these innovative biosensing technologies.

Small Organic Molecules-Based Fluorescent Biosensors and their Applications

Edited by

Abha Sharma

CRC Press
Taylor & Francis Group
Boca Raton London New York

CRC Press is an imprint of the
Taylor & Francis Group, an **informa** business

Designed cover image: Abha Sharma

First edition published 2026
by CRC Press
2385 NW Executive Center Drive, Suite 320, Boca Raton FL 33431

and by CRC Press
4 Park Square, Milton Park, Abingdon, Oxon, OX14 4RN

CRC Press is an imprint of Taylor & Francis Group, LLC

ISBN: 978-1-032-70616-0 (hbk)
ISBN: 978-1-032-70621-4 (pbk)
ISBN: 978-1-032-70624-5 (ebk)

DOI: 10.1201/9781032706245

Typeset in Times
by Deanta Global Publishing Services, Chennai, India

Contents

About the Editor

Abha Sharma earned her PhD from the Defence Research and Development Establishment (awarded by Jiwaji University), Gwalior, India, in 2006. She has worked at the Indian Institute of Toxicology Research (IITR), Lucknow, as a Research Associate for more than three and a half years. Since 2010, she has been working as a faculty member in the Department of Medicinal Chemistry at the National Institute of Pharmaceutical Education and Research, Raebareli, India. Currently, she is serving as an Associate Professor in the same department. Her research area includes the design and synthesis of molecules for the treatment of Alzheimer's disease (AD) and the development of sensors/probes for the detection of analytes/biomarkers of various central nervous system diseases like AD. Her publications include nearly 60 high-impact journal papers and book chapters, and she has been granted several patents. She has also served as a referee for some international journals, and she has guided several students through the completion of their research projects.

1 Photoluminescence and Fluorescence Mechanisms

Preeti Ashokkumar Chaudhran and Abha Sharma
National Institute of Pharmaceutical Education and Research, Raebareli, Lucknow, UP, India

LIST OF ABBREVIATIONS

UV	Ultraviolet
ISC	Intersystem crossing
PET	Photoinduced electron transfer
FRET	Fluorescence resonance energy transfer
ICT	Intramolecular charge transfer
EET	Electronic energy transfer
MLCT	Metal-ligand charge transfer
LLCT	Ligand-ligand charge transfer
TICT	Twisted intramolecular charge transfer
AIE	Aggregation-induced emission
ESIPT	Excited-state intramolecular proton transfer
ET	Electron transfer
CT	Charge transfer
HOMO	Highest occupied molecular orbital
LUMO	Lowest unoccupied molecular orbital
ES	Excited state
GS	Ground state
D-A	Donor-acceptor
D-A-D	Donor-acceptor-donor
DFT	Density functional theory
HBO	2-(2'-hydroxyphenyl)benzoxazole
HBT	2-(2'-hydroxyphenyl)benzothiazole
HBI	2-(2'-hydroxyphenyl)benzimidazole
RPT	Reverse proton transfer
AB	Azobenzene
E-SB	Trans Stibene
DPCB	1,2-diphenylcyclobutene
NP	Naphthalene

1.1 INTRODUCTION

The term "luminescence" is a Latin word where lumen means "light" and was first introduced by German physicist and science historian Eilhard Wiedemann as *luminescenz* in 1888 for all the phenomena of light without being conditioned by high temperature. Let's look into the historical evolution of the "luminescence" phenomenon. There are various types of luminescence categorized upon the mode of excitation, in particular photoluminescence where the emission of light is the result of photoexcitation of the emitting species. Two well-known types of photoluminescence are fluorescence and phosphorescence, whereas others differing by the excitation source are radioluminescence, bioluminescence, chemiluminescence, electroluminescence, thermoluminescence, etc.

1.2 A BRIEF HISTORY

In photoluminescence, the term phosphorescence belongs to Greek origin and means "bearing light." The word phosphor was formerly assigned to the materials glowing even in the dark upon removing the source of light; the most popular one was Bolognian phosphor founded by a cobbler, Vincenzo Cascariolo, in 1602 in Bologna. Next, a physicist and a professor of mathematics, Sir George Gabriel Stokes, at Cambridge introduced the term fluorescence. Nevertheless, the report by Spanish physician Nicolas Monardes in 1565 stated the unusual blue color of an infusion of wood (later called *Lignum nephriticum*) used to treat urinary and kidney diseases. Aztecs already knew the diuretic properties and peculiar color effect of the wood but it was an expensive and scarce medicine (Muyskens, 2006). Thus, it was of great interest to explore more about that wood and its blue color (Acuña and Amat-Guerri, 2007). This was further examined by Newton, Boyle, and others; however, it remains misunderstood (Harvey, 2005). Recently, the intense blue color of the wood was found as the result of the chemical species: a four-ring tetrahydromethanobenzofuro[2,3-d]oxacine (matlaline) (Acuña et al., 2009). Though this chemical is absent in the plant, it is produced as a result of the oxidation of the tree's flavonoids. Later, David Brewster, a Scottish preacher in 1833, performed an experiment with the alcoholic extract of leaves (chlorophyll) (Brewster, 1834), and John Herschel, in 1845, performed an experiment with the solution of quinine sulfate where at the surface of the solution blue color was observed; similar to *Lignum nephriticum*, both the cases of homogeneous liquid were superficially colored whereas colorless internally. He named this phenomenon epiploic dispersion (it is of Greek origin meaning surface) (John Frederick William, 1845). In 1852, the phenomenon was reexamined by Stokes, and a renowned article entitled "On the Refrangibility of Light" (Stokes, 1852) was published. He revealed that this phenomenon was an example of emission of light after the absorption of light by the emitting species (both organic and inorganic samples). One of his experiments was remarkable, where he performed a solar spectrum using a prism, where a solution of quinine sulfate

DOI: 10.1201/9781032706245-1

filled in a tube was moved through visible light of the spectrum and appeared colorless. However, when the tube was exposed to the light beyond the violet portion, i.e., nonvisible zone corresponding to the UV radiation, surprisingly blue light was emitted by the solution. From this experiment, Stokes stated that the emitted light is always of longer wavelength than the excitation light, later known as Stokes law. In his first paper, Stokes called the phenomenon "dispersive reflection," but in the footnote he wrote: "I confess I do not like this term. I am almost inclined to coin the word and named the observed phenomenon as 'Fluorescence' from fluorspar, as the analogous term 'opalescence' is derived from the name of a mineral." In his later articles, he used the word fluorescence (Stokes, 1853).

1.3 DISTINCTION BETWEEN FLUORESCENCE AND PHOSPHORESCENCE

In the 19th century, based on experimental results, the difference between fluorescence and phosphorescence was made where the disappearance of the emission of light after removing the source of excitation was considered as fluorescence, while the phosphorescence persists even after the end of excitation. In both cases, the emission lasts longer than the excitation; however, the difference based on the duration of emission is not correct as we know there are examples of short-lived phosphorescence (e.g., ZnS: violet luminescence) and long-lived fluorescence (e.g., uranyl salts). The first theoretical difference was stated by Francis Perrin (Jean Perrin's son) in his thesis: "If the molecule passes through an intermediate state between absorption and emission, there is phosphorescence" (Perrin, 1929). Further, to describe the processes following the absorption by molecule, an energy diagram was presented where electronic states and possible transitions between them were indicated using arrows. A simplified diagram is shown in Figure 1.1. Since the 1970s, this diagram had often been called Jablonski's diagram (after Aleksander Jablonski, Polish physicist). However, the diagram should be named as "Perrin-Jablonski diagram" as the energy level diagram showing absorption and emission was first used by Jean Perrin that appropriately describes thermally activated delayed fluorescence containing metastable intermediate state. Surprisingly, G. N. Lewis credited Jablonski for this outcome and not Perrin, which led to a misnomer in the discovery. However, Jean and Francis Perrin's diagram was lacking as the intermediate state cannot revert radiatively or otherwise to the ground state (GS), whereas Jablonski's work allows such transition translating the possibility of second emission at even longer wavelengths (phosphorescence) (Jabłoński, 1935). That was the only fact differing from the Perrin and Jablonski diagram (Berberan-Santos, 2001).

In 1943, A. Teren recognized the behavior of intermediate state; further, G. Lewis and M. Kasha improvised in 1944 and stated that the intermediate state can be the triplet excited state in contrast to the singlet excited state. Here,

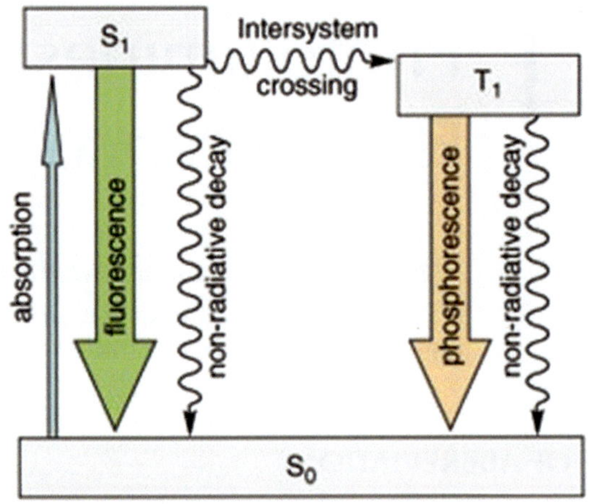

FIGURE 1.1 Simplified Perrin-Jablonski diagram depicting fluorescence and phosphorescence. Reprinted with permission from Ref. Bernard Valeur & Berberan-Santos, Copyright (2011) American Chemical Society.

the singlet and triplet excited states describe the symmetry/asymmetry (spin state) of the wavefunction, where the symmetric wavefunction has three forms, i.e., multiplicity of three, called as triplet state, and the asymmetric wavefunction has one form, i.e., multiplicity of one, called as singlet state. Thus, fluorescence is an emission process from the singlet excited state, whereas radiative emission from the triplet excited state is as phosphorescence. Jablonski's diagram provides a suitable representation of the excited-state structure and possible transitions between intermediate states, separated by the energies in the order of 10,000 cm^{-1}. Each electronic state is further split into multiple sublevels demonstrating the vibrational modes of the molecule. The vibrational levels are separated by the energy of 100 cm^{-1}. According to thermodynamics, molecules reside in electronic GS as the energy gap between the excited state and GS is considerably larger. The transitions may occur between electronic and vibrational states of polyatomic fluorescent molecules. Photoirradiation leads to the excitation of molecules from the lowest vibrational level in the electronic GS to an accessible vibrational level of the excited electronic state. Here, as the ground electronic state is singlet, the destined electronic state should also be singlet. After excitation, the molecule immediately relaxes to the lowest vibrational level of the excited electronic state on the timescale of femtosecond to picosecond, which results in a Stokes shift that describes the photons emitted are of longer wavelength in comparison to excitation radiation. The molecule remains in the lowest vibrational level of the excited electronic state for a timescale of nanoseconds, called the fluorescence lifetime. Another phenomenon is occurring simultaneously, intersystem crossing (ISC), where molecules relax from singlet excited state to the lower-energy triplet excited state that further needs to change the multiplicity from triplet to singlet when getting relaxed back to the GS. Phosphorescence lifetime lies in the order of microseconds to seconds, whereas the rate is much lower for this process when compared with the nonradiative decay processes such as collision quenching. Hence, phosphorescence is hardly detectable at room temperature in aqueous system. Studies involving protein conformation at cryogenic conditions utilize phosphorescence (So and Dong, 2001).

There are certain photophysical characteristics should be focused for a molecule to serve as a fluorophore, such as wavelength of excitation and emission, Stokes shift, brightness, quantum yield and fluorescence lifetime, anisotropy, and photobleaching. Fluorophores absorb the light to get excited and emit light of higher wavelength; however, the excitation light should be compatible. As if the fluorophore proposed to be used in vivo, then UV light can cause tissue damage if exposed directly (Janssen et al., 1993), whereas the infrared light can cause tissue heating (Smith, 1974). For surface imaging applications, blue and/or green light color are compatible, though such wavelengths have poor tissue penetration hence only appropriate for small animal imaging or superficial structures imaging. The optimal wavelength for excitation of a fluorophore is near-infrared or deep red range as it offers low autofluorescence and good tissue penetration (Ntziachristos et al., 2003). The asymmetry between excitation and emission wavelength known as the Stokes shift is a vital parameter in designing a fluorescent probe. Emission wavelengths should lie in the 650–900 nm range, which defines the coverage of near-infrared and deep red regions. Here, the maximal light penetration is obtained with minimal autofluorescence. Another consideration is the brightness of a fluorophore, where the brighter the fluorophore, the higher signal-to-noise ratio is obtained, leading to deeper penetration. Also, a higher quantum yield requires less excitation light for fluorescence to take place (Kobayashi et al., 2010).

1.4 APPLICATION OF RADIATIVE EMISSION: FLUORESCENCE

In 1857, Edmond Becquerel gave the idea of coating the inner surface of the tubes and lamps with luminescent material; however, the tubes were not efficient and the lifetime was inadequate for practical application. Later, the first commercial tubes were available in the 1930s based on the discharge of low-pressure mercury vapor, producing UV radiation that in turn irradiates the fluorescent inner coating of the tube. In the beginning, the coating consists of zinc orthosilicate and magnesium tungstate, later substituted by doped calcium halophosphates. However, in the present compact bulbs and tubes, lanthanides such as Tb(III), Eu(I), and Eu(III) are employed, producing green, blue, and red lights, respectively. Further, fluorescence evolved as an analytical tool (O'Haver, 1978), and in 1862 the first research article by Victor Pierre, a professor at Prague, was published on the same, where he studied the effect of solvents and acidity/alkalinity (Valeur and Berberan-Santos, 2011). In 1868, F. Goppelsroder used the term *fluorescence analysis* for the first time, where he explained the complexation of a hydroxyflavone derivative with aluminum resulting in the drastic enhancement of fluorescence intensity, serving as a detection tool for the metal ion. Later, fluorescence analysis was broadly used, and a book by Radley and Grant was published in 1933 listing the reported applications of the same (Radley and Grant, 1939). At present, fluorescence sensing is a field accelerating with a high pace in research and has become the most preferred topic for researchers. Fluorescent chemosensors capable of selectively identifying the species of interest are more considerable. A fluorescent chemosensor consists of two components: a fluorophore responsible for signaling and a receptor that possesses a recognition potential for the guest analyte, and both units are connected by a spacer to generate a fluorophore-spacer-receptor system (Lodeiro et al., 2010). When the guest analyte binds to the receptor, the fluorescence intensity, fluorescence lifetime, and emission wavelength change through different mechanisms. The mechanism of interaction between the fluorophore and recognition units is of continuous interest to the researchers, and exploration of new mechanisms is always inevitable (Gunnlaugsson et al., 2006). Conventional mechanisms of signaling have been used broadly for optical detection including photoinduced electron transfer (PET) (Gunnlaugsson et al., 2005; Xu et al., 2010), fluorescence resonance energy transfer (FRET) (Sapsford et al., 2006), intramolecular charge transfer (ICT) (Callan et al., 2005), electronic energy transfer (EET) (Martinez-Manez and Sancenon, 2003), metal-ligand charge transfer (MLCT) (Zhao et al., 2010), twisted intramolecular charge transfer (TICT) (Rettig and Lapouyade, 2002), and excimer/exciplex formation (Lodeiro and Pina, 2009). The development of supramolecular chemistry has emerged as a new signaling mechanism for fluorescent sensing. Another approach attracting the researchers' attention is the conformational restriction of the small molecules as this phenomenon can cause considerable change in the fluorescence of the molecule depending upon the provided environment. The mechanisms based on conformational restriction include aggregation-induced emission (AIE) (Wang et al., 2010), C=N isomerization (Wu *et al.*, 2007), and excited-state intramolecular proton transfer (ESIPT) (Berezin and Achilefu, 2010).

1.5 CONVENTIONAL SIGNAL MECHANISMS

Conventional signaling mechanisms involve some basic photophysical processes: electron transfer (ET), charge transfer (CT), energy transfer, and excimer/exciplex formation. A brief of the above processes: electron transfer includes PET where electrons are transferred between donor and acceptor that are inhibited/activated by the attachment of the analyte, CT includes intramolecular, twisted intramolecular, and metal-ligand types of charge transfer, whereas energy transfer includes electronic and fluorescence resonance types of energy transfer. Excimer/exciplex involves the short-lived homodimeric/heterodimeric complex formation in the excited state.

1.5.1 Electron Transfer

1.5.1.1 Photoinduced Electron Transfer

The classical electron transfer process named photoinduced electron transfer (PET) is generally employed in the designing of fluorescent probes (Balzani et al., 2001; Grabchev et al., 2003). In 1985, de Silva et al. (de Silva and Rupasinghe, 1985) developed the first PET-based fluorescent sensor. Later, PET-based fluorescent probes became a point of interest for biological, medical, and chemical researchers. A classic PET-based fluorescent probe consists of three basic elements: recognition/activating groups, appropriate linkers, and fluorophores (Escudero, 2016). Some important factors that must be considered for designing PET-based probes are the following: (1) The Rehm-Weller equation governs Gibbs free energy (distance dependent) that is influenced by local environment, such as the structures that possess ICT excited state. (2) PET is directional, e.g., photosynthetic reaction center. PET is further categorized into two types depending upon the direction of electron transfer between the recognition element and fluorophore: donor-excited PET (d-PET) and acceptor-excited PET (a-PET) (De Silva et al., 2009). For a-PET, electron transfer occurs from the highest occupied molecular orbital (HOMO) of the recognition group to the lowest unoccupied molecular orbital (LUMO), and HOMO of the recognition group lies between HOMO and LUMO of the fluorophore. As a result, the electron in the HOMO of the recognition group transfers to the HOMO of the fluorophore and the electron in LUMO of the fluorophore could not revert back to the GS (HOMO of the fluorophore), which results in the fluorescence quenching.

For d-PET, LUMO of the recognition group lies between HOMO and LUMO of the fluorophore. The fluorophore excites and the electron moves from HOMO to LUMO; however, when it relaxes, the electron via nonradiative transition reaches to LUMO of a fluorophore, which leads to fluorescence quenching. The fluorescent probes having an "on-off" or "off-on" response are usually developed based on the PET phenomenon (Dong et al., 2016). At the time of response, the recognition group binds to the analyte, resulting in either restriction or activation of the PET phenomenon leading to turn on or off, respectively. Also, PET is affected by the chemical properties of the analyte as it is responsible for the change in energy level of HOMO and LUMO of the recognition group (Yang et al., 2017).

The ability of electron donating and receiving can be expressed in the form of oxidation/reduction potential. Depending upon the change in oxidation/reduction potential, the designing of the fluorescent probes is categorized into four kinds (Figure 1.2): (1) After binding to the analyte, oxidation potential of the recognition unit increases and it becomes an ineffective electron donor, which results in inhibition of a-PET (Xia et al., 2017). (2) A decrease in the oxidation potential of the recognition group results in a decrease of the electron-accepting efficiency resulting in the forbidden d-PET (Wang et al., 2012). (3) After binding to the analyte, the recognition unit detaches and a-PET deactivates (Xiao et al., 2021). (4) Also, d-PET can be deactivated after the attachment of the analyte, resulting in the detachment of the recognition group (Xiao et al., 2021). In all the above-discussed cases, the fluorophore can be linked to the more electron-rich/deficient groups that can act as electron donor/acceptor, permitting both a-PET/d-PET processes, respectively. Hence, a fluorophore can be employed for dual functioning, as "off-on" and "on-off" type PET-based probes (Yang et al., 2017). Such probes have found application in recognition (detection), disease therapy, and microscopic imaging including cell microenvironment (i.e., pH, polarity), bioactive small molecules (biothiols, gaseous molecules, and reactive oxygen species), and biological small molecules (Fan et al., 2012; Lee et al., 2014; Zhang et al., 2013), etc.

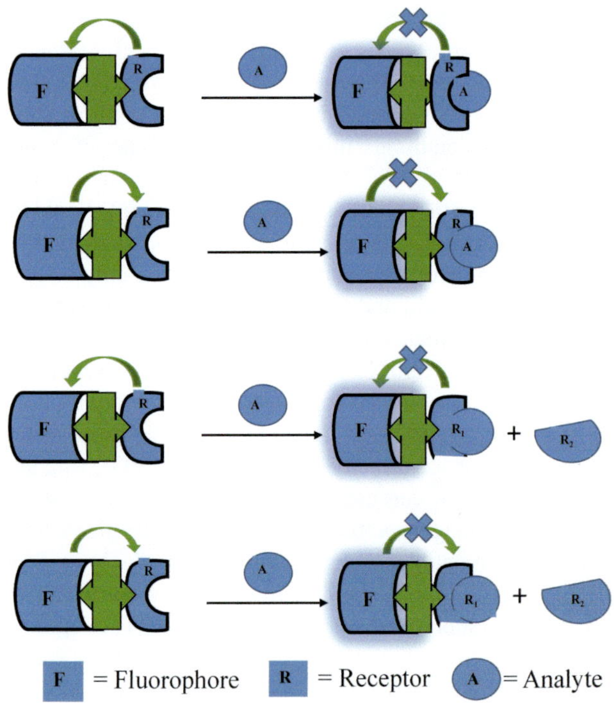

FIGURE 1.2 Four design types of PET-based fluorescent probes.

1.5.2 Charge Transfer

In contrast to the PET system, the charge transfer does not require any linker/spacer, hence the fluorophore could be directly attached to the receptor and their orbitals overlap upon each other. Here, one part of the molecule acts as electron rich and the other part as electron deficient. In this case, excitation leads to the redistribution of the electron density, and a dipole is generated. When such a system binds to the analyte, the interaction occurs with the excited-state dipole (Valeur and Berberan-Santos, 2002).

1.5.2.1 Intramolecular Charge Transfer

Intramolecular charge transfer (ICT) comprises electronic charge transfer from donor to acceptor, i.e., electron rich/deficient, respectively, present within the same molecule. The charge transfer is an essential process in biology and chemistry, hence becoming a reason to gain attention from the scientific community (Bulheller et al., 2008; Sedghi et al., 2008; Zhao et al., 2007). The fundamental process involves many biological processes, such as photosynthesis and metabolism. If the acceptor (A) and donor (D) are involved from two different atoms, clusters, and molecules, then the process is named Intermolecular CT, whereas if both the donor and acceptor are within the same molecule, then it becomes Intramolecular CT. Another one involves "through-bond charge transfer," where the electrons pass from donor to acceptor via π–electron bridge. Interestingly, CT can occur via through-space CT, if the donor and acceptor are not even connected through the π–electron bridge. Though "through-space" CT process is comparatively less recognized, it could help unfold many of the CT-driven processes (Tsiperman et al., 1999). When such ultrafast processes occur in the molecule, the photophysics of the electron transfer becomes somewhat more complicated than what has been seen in Jablonski's diagram (Arzhantsev et al., 2006; Druzhinin et al., 2010; Misra et al., 2011).

For ICT, the emission competes with the nonradiative pathways, and dual fluorescence is usually seen in the high-polarity solvents (Figure 1.3). The dipole moments (μ) of a molecule can drastically increase in ES from GS as there is a change in electronic state distribution over the molecule in ES. This increase in μ is quite obvious for D-A-based ES CT molecules that generally exist in two forms in GS and ES. Two different forms involve normal and quinonoid forms (Figure 1.4) (Misra and Bhattacharyya, 2018; Paul et al., 2010).

The reactivity and the acid-base character of the molecule change with the alteration in the electronic charge distribution. When the donor and acceptor are part of a molecule, no interactions occur in the GS however CT takes place in ES, provided with the energy feasibility (Haidekker et al., 2005). In D-π-A system, the molecules can be excited in two ways: First, a donor in ES transfers an electron to

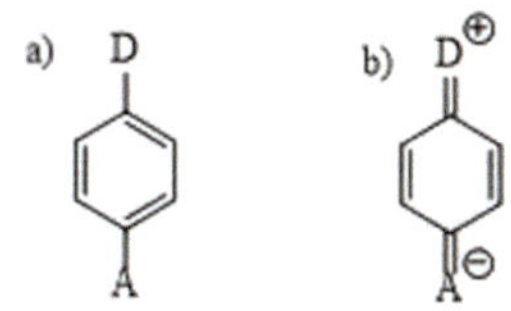

FIGURE 1.4 The normal (a) and quinonoid (b) forms of D-π-A system with benzene as the spacer.

the acceptor, whereas in another case, the electron transfers from a donor to an excited acceptor. Hence, either of acceptor or donor excites using appropriate wavelength, which results in the charge-separated state (D^+-π-A^-). To facilitate ES ICT process, a molecule absorbs suitable energy that excites it to the first/higher ES. After reaching the ES, electron density rearranges to have higher μ and lower energy (Cornelissen-Gude and Rettig, 1998). The normal ES is designated as LE, whereas the new state generated due to electronic state rearrangement is called ICT state. Two different states (LE and ICT) lead to dual fluorescence of the molecule. The generation of these states is not arbitrary as it is governed by some rule, given in Equations 1.1 and 1.2:

$$E_{LE} - E_{ICT} > 0 \quad (1.1)$$

where E_{LE} and E_{ICT} are the energies of the LE and ICT states.

$$E_{ICT} = IP(D) - EA(A) + C + E_{solv} \quad (1.2)$$

Here, EA(A) and IP(D) are electron affinity of the acceptor and ionization potential of the donor, respectively. "C" represents the mutual coulombic attraction between acceptor and donor, E_{solv} defines solvation energy. From the above equations, it is anticipated that the energy of the LE state must be higher in comparison to the ICT state. The formation of ICT from the LE state is also affected by medium properties. Some molecules show dual emission, whereas

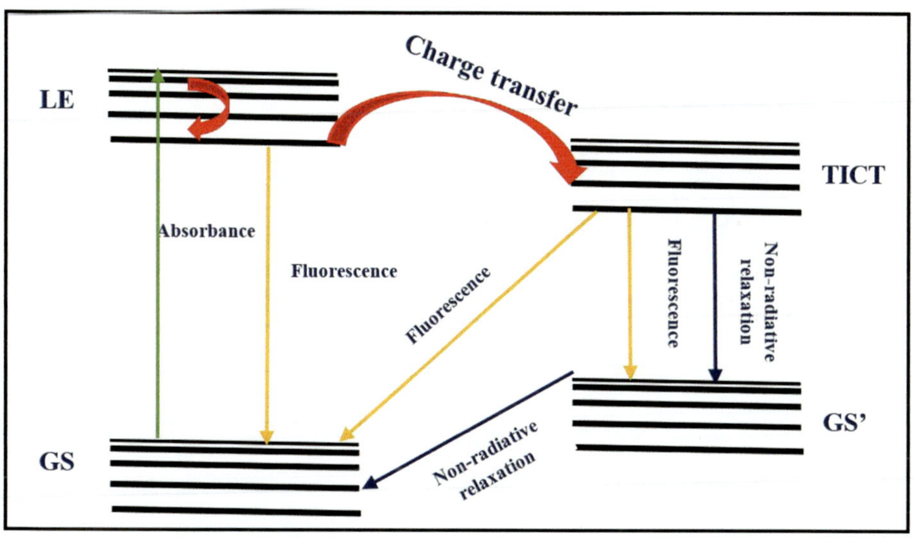

FIGURE 1.3 Jablonski's diagram shows ICT process after photoexcitation, where GS: ground state, ES: excited state, ISC: intersystem crossing, ICT: intramolecular charge transfer.

some ICT-based molecules show emission from either of the states (LE/ICT) (Mallick et al., 2007).

Many reports are available where ICT mechanisms in a variety of molecules have been discussed. Wolf and coworkers reported CT process in D-A-D and D-A types of molecules possessing a sulfur-bearing acceptor unit and an anthracene moiety as a donor unit (Figure 1.5i). They state that the oxidation state of a sulfur atom affects CT process in these molecules (Pahlavanlu et al., 2016). Jana et al. studied the molecule 5-(4-dimethylaminophenyl)-penta-2,4-dienoic acid ethyl ester (DMAPPDE) and found dual emission behavior in polar solvents that were recognized due to the LE and ICT states existing in the ES (Figure 1.5ii). The absorption and the emission spectra of the molecules are affected by the variation in the acceptor's strength (Jana et al., 2013). Another group, Gopidas et al., studied the planar nature of tetrahydropyrene derivatives and revealed them as CT species (Figure 1.5iii and iv). Absorption and the emission properties of the molecules are pH-dependent, making them potential candidates for pH-sensing applications. The same group studied vinyltetrahydropyrene derivatives for ICT process experimentally (Sumalekshmy and Gopidas, 2005). Kim and coworkers evaluated bulky meso-substitution effect on meso-diarylamino subporphyrins (Lee et al., 2016) (Figure 1.5v). Carlotti et al. reported ICT process in the D-π-A system containing three cationic sites, where the methylpyridinium/quinolinium group acts as acceptor, the dimethyl amino group acts as donor, and ethylene and butadiene groups are present as spacers. Time-resolved studies revealed the formation of ICT state in such molecules (Carlotti et al., 2015) (Figure 1.5vi and vii).

Bai et al. reported an interesting finding where ICT process was studied for two types of molecules: (1) asymmetrical DA structure and (ii) symmetrical DAD structure.

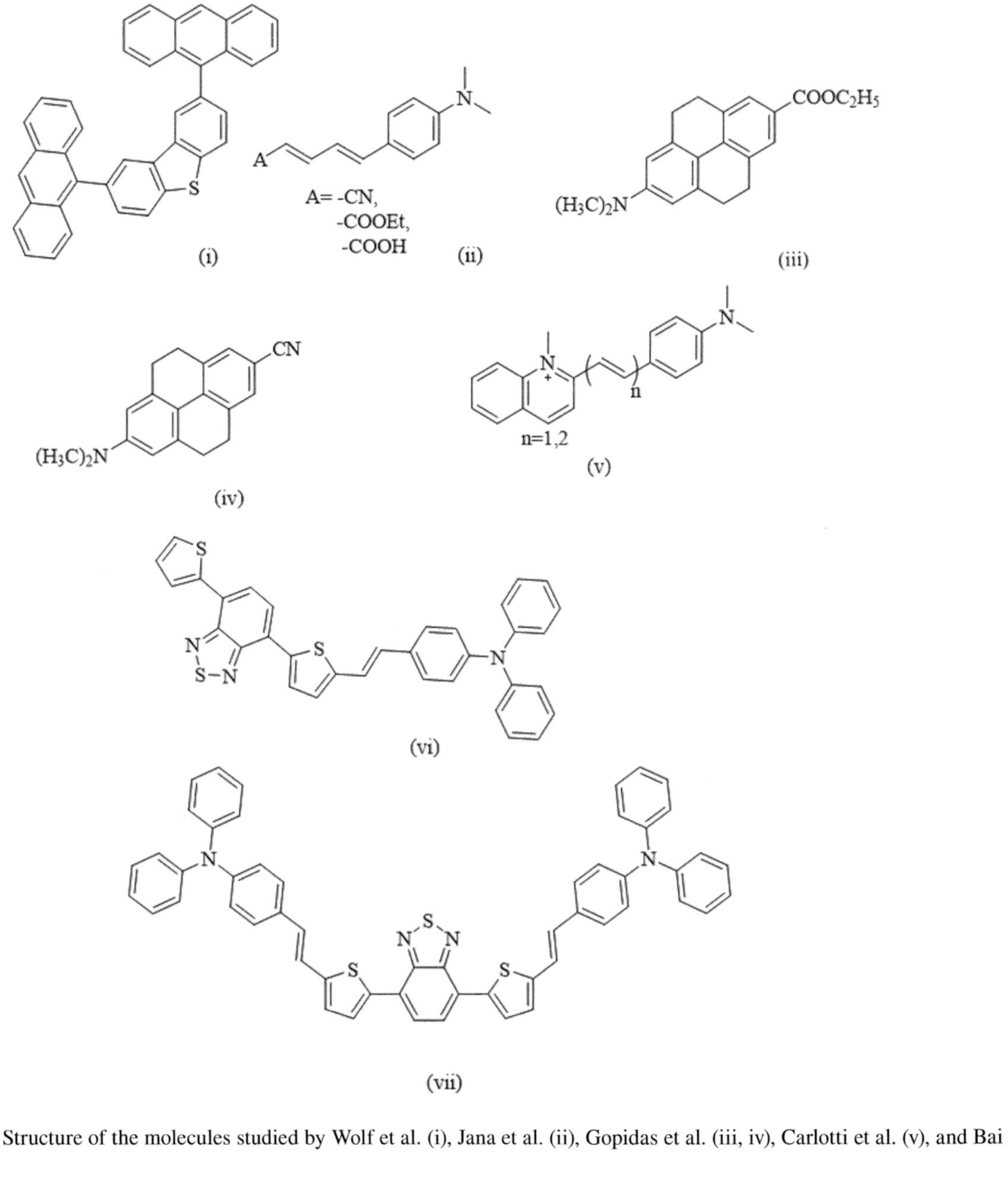

FIGURE 1.5 Structure of the molecules studied by Wolf et al. (i), Jana et al. (ii), Gopidas et al. (iii, iv), Carlotti et al. (v), and Bai et al. (vi, vii).

The results showed increased dipole moment as excitation is higher in the DA system in comparison to the DAD symmetrical system. Also, the change in the dipole between ES and GS was estimated theoretically through the Lippert-Mataga analysis, and little shift in absorption maxima was found with an increase in solvent polarity while the emission maxima were affected to a larger extent. From both the above results, they concluded that the change in the dipole moment of ES is much larger than that in the GS, which indicates the presence of ICT process in the molecule. They also performed femtosecond fluorescence depletion measurements to comprehend the solvation process and dynamics of the process. The results concluded that DA molecules solvate faster than DAD molecules, which relates with a relaxed ICT state. All the above studies suggest a larger charge distribution for DA molecules in the ES than DAD molecules. The efficiency of photovoltaic devices relies on the push-pull ability of the chromophores, where DA chromophores could serve to function better than symmetric DAD chromophores in those devices (Gong et al., 2007).

There is another upgradation of the ICT phenomenon called TICT, which basically occurs in the molecules where donor and acceptor units are connected through a single bond; subsequent intramolecular twisting leads the molecules to return to GS either by the nonradiative relaxation or the red-shifted emission. Saha et al. studied the theoretical calculations of the trans-4-[4'-(N,N'-dimethy lamino)styryl]pyridine molecule and predicted the twisting of donor with respect to acceptor group resulting in TICT states (Sowmiya et al. 2011). This particular class of active fluorophores, also known as "molecular rotors," shows enhanced fluorescence intensity upon exposure to the sterically restricted environments such as in highly viscous media (Haidekker and Theodorakis, 2007). Viscous environment makes TICT-active fluorophores incapable of overcoming potential barriers against LE-TICT interconversions, or in some cases, viscosity hinders internal conversion from TICT state. The example of a molecular rotor includes 9-(2-carboxy-2-cyanovinyl)julolidine and its analogs (Haidekker and Theodorakis, 2010). The versatility is exhibited not only in molecular biology, such as in peptide-protein interaction studies (Goh et al., 2014), imaging intracellular microviscosity (Suhling et al., 2012), and amyloid deposit research (Cao et al., 2012), but also in polymer science (Priestley et al., 2005), contact mechanics (Sasaki et al., 2016), and fluid dynamics. Another well-known TICT-based molecular rotor, Thioflavin T, is extensively used for in vitro biomedical assays and biological cell imaging (Amdursky et al., 2012). Many molecular rotors are emerging, such as α-cyanostilbene derivatives (Kuimova, 2012a, 2012b; Zhu and Zhao, 2013).

1.5.2.2 Charge Transfer in Metal-Ligand Complexes

The photophysical characteristics of the metal complexes that recognize the fluorescent emission are: emission lifetimes in nanosecond scale, intense emission under aerated conditions, and a small Stokes shift. The fluorescence in metal-ligand complexes is the result of ligand-based charge transfer where prominent emission corresponds to poor metal-ligand interaction showing small metal contribution in the excited state or, when compared to the ISC rate constant, could be a competitive fluorescence.

Usually, the singlet ES of such complexes is short-lived, e.g., about 15 ± 10 fs for $[Ru(bpy)_3]^{2+}$ and 100 fs for $[Ir(ppy)_3]$, as the metal ion takes part in transforming the singlet ES to the triplet ES with extremely rapid ISC rate of 10^{12} s^{-1}. Hence, in metal-ligand complexes, the fluorescence is generally challenging to observe, whereas quantum yield of $[Ru(bpy)_3]^{2+}$ complex was calculated as 9 x 10^{-5}, which is insignificant. However, in the early 2000s, certain complexes with intense fluorescence were reported with a lifetime in nanosecond scale. Another interesting luminescence was fluorescence–phosphorescence dual emission, discussed below with relevant examples. The complexes can exist in different geometry, i.e., octahedral, tetrahedral, square planar, and linear.

(A) Octahedral complexes

In the 1980s and 1990s, the photophysical properties of the rhodium complexes, $[Ru(bpy)_3]^{2+}$ and $[Ru(bpy)_3]^3$, were broadly studied, and they were found to have oxidation-dependent photophysical behavior. Various bands of high intensity were observed below 350 nm attributed to LC π-π* transition, whereas no MLCT transition band was seen. When observed at low temperature (77 K), the emission band at 448 nm with a lifetime of 2.2 ms could be seen in the rigid glass. In 2001, a metallocycle complex, 2,5-bis(ditolylethynyl)rhodium cyclopentadienes, was reported by a group (Figure 1.6i). The absorption wavelength was within the range of 453–517 nm, whereas emission lies in the range of 496–590 nm. The lifetime lies in the nanosecond scale range of 0.45–1.21 ns, and the small Stokes shift of 1,870–2,390 cm^{-1} indicates the metallocycle of rhodium (rhodacycles) displays fluorescence emission with the quantum yield of 0.01–0.18. The slow ISC of 10^8 s^{-1} results in the long-lived emission from the singlet state. A cadmium (II) complex $[Cd_3(4\text{-}PTZ)_2(OH)_2Cl_2]$ in which 4-PTZ = 5-(4-pyridyl)tetrazolate emits at 390 nm with a fluorescence lifetime of 1.5 ns and the quantum yield of 0.31 (Figure 1.6ii). The fluorescence was stated to be the result of LLCT transition also; a weak phosphorescence was at 470 nm with a lifetime of 2.0 s and quantum yield of 0.15.

(B) Square planar complexes

In addition to the octahedral complexes, some square planar and linear complexes exhibiting fluorescence emission are also known. Platinum-containing square planar complexes are well explored due to the anti-cancer properties of cis- and trans-$[PtCl_2(NH_3)_2]$ and the photophysical properties of platinum terpyridyl complexes (Williams, 2007). Hudson et al. in 2009 reported an impressive triarylboron-platinum (II) complex that displayed dual emission at 77 K. The Pt(II) complex, $[Pt(N,N\text{-}Si\text{-}BNPA)Ph_2)]$ (Figure 1.7i), has an unstable six-membered nonplanar N,N-chelate ring;

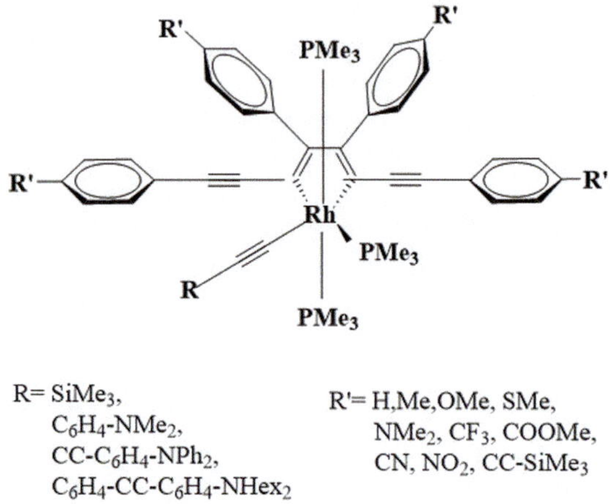

(i)

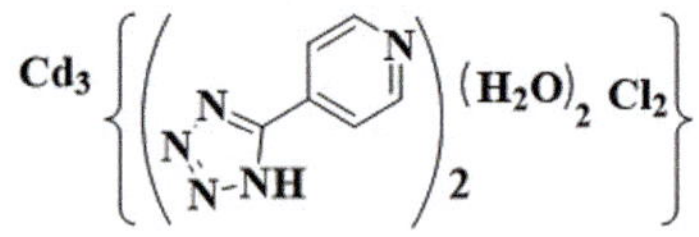

(ii)

FIGURE 1.6 Structure of a metallocycle complex, 2,5-bis(ditolylethynyl)rhodium cyclipenta-dienes (i), cadmium (II) complex $[Cd_3(4\text{-}PTZ)_2(OH)_2Cl_2]$ in which 4-PTZ = 5-(4-pyridyl)tetrazolate (ii).

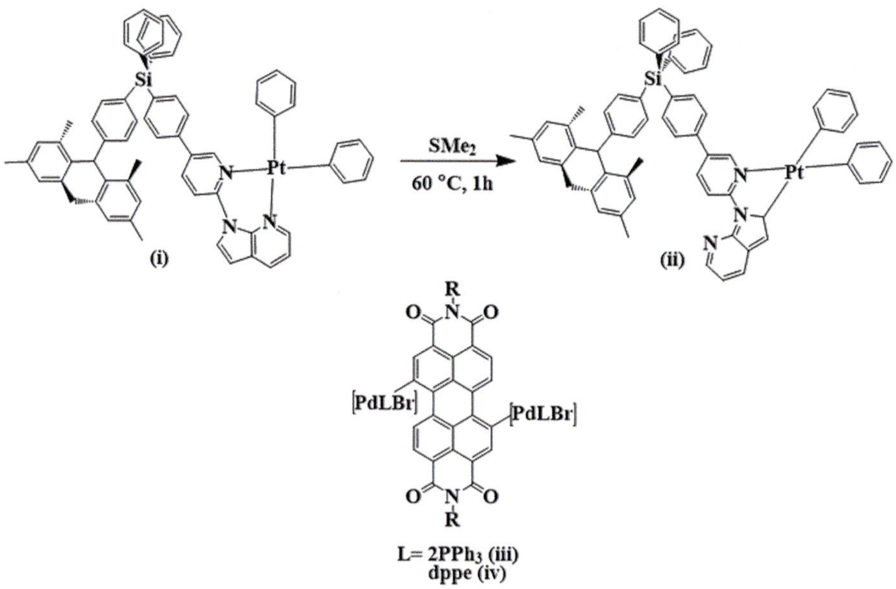

FIGURE 1.7 The structure of Pt(II) complex, $[Pt(N,N\text{-}Si\text{-}BNPA)Ph_2)]$ (i), five-membered N,C-chelate complex, $[Pt(N,C\text{-}Si\text{-}BNPA)(SMe_2)Ph)]$ (ii), and the structure studied by Weissman and group (iii, iv).

further intramolecular C-H activation takes place to form a more stable five-membered N,C-chelate complex, $[Pt(N,C\text{-}Si\text{-}BNPA)(SMe_2)Ph)]$ (Figure 1.7ii). The complexes 7i and 7ii show singlet emission at 392 and 399 nm, respectively, which was stated as the result of Si-BNPA ligand transition of mesityl-boron charge transfer, whereas the triplet emission at 494 and 495 nm, observed for 7i and 7ii, respectively, was the result of NPA-centered π-π* transition. As discussed earlier, the phosphorescence intensity increases at lower temperatures as the quenching process is eliminated. An interesting singlet-triplet dual emissions could be the result of (i) the heavy metal effect of Pt(II) that facilitates ISC to generate phosphorescence. (ii) N,N- and N,C- chelation mode affects the intensity of the singlet and the triplet emission (Williams, 2007). Apart from Pt complex, Pd complexes have also been found by Weissman and group, exhibiting strong fluorescence. Two Pd(II) complexes, 7iii and 7iv, show intense emission at 584 nm (Figure 1.7iii,iv). The quantum yield of 0.65 and 0.22 and the lifetime of 7.5 ns and 2.8 ns for complexes 7iii and 7iv, respectively, were observed. Such nanoscale lifetime suggests it as fluorescence emission. Further, femtosecond and nanosecond transient absorption experiments were performed to confirm the fluorescence process of both complexes (Weissman et al., 2007).

(C) Tetrahedral complexes

About a decade ago, fluorescence tetrahedral complexes of Cu(I) were reported. In 2003, Abedin-Siddique and his group reported three fluorescent Cu(I) complexes (Figure 1.8i, ii, and iii). The fluorescence lifetime of these complexes lies

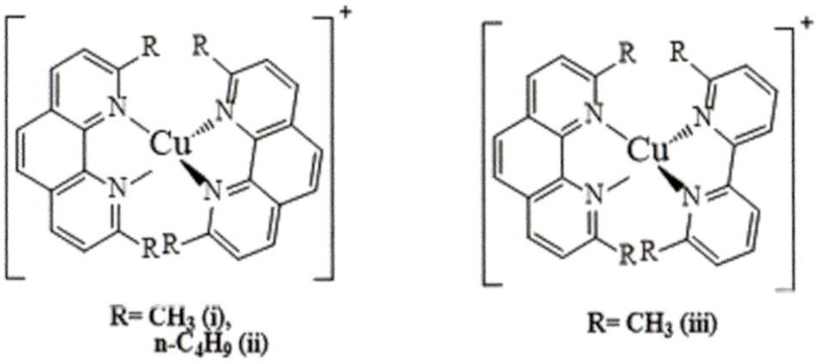

FIGURE 1.8 Three tetrahedral Cu complexes reported by Abedin-Siddique and group (i, ii, iii).

in the range of 13–16 ps, and the emissions are short-lived that were assigned due to MLCT. For complex (Figure 1.8i) the structure was flattened due to the Jahn-Teller effect in 3d^9 electronic configuration. In 2010, Deaton et al. reported a Cu(I) complex, [Cu(PNP-^{t}Bu)]$_2$, exhibiting a quantum yield of 0.57, the radiative decay of 7.24 x 106 s^{-1}, and decay time of 11.5 μs. The singlet state is assigned based on the extinction coefficients that lie in the normal range for singlet MLCT transitions (Kirchhoff et al., 1983).

(D) Linear complexes

Lu et al., in 2003, reported Au(I) complex 1.9i that shows strong fluorescence emission at 413 and 428 nm with a quantum yield of 0.22 and lifetime of ˂50 ns. The small Stokes shift of 1,040 cm^{-1} confirms the fluorescence emission, whereas no phosphorescence was observed at 77 K as the T1 state must be close to the energy of GS (Lu et al., 2003). Another, Au (I) complex 9ii displayed three emission bands at 384, 490, and 524 nm in the solid state (Figure 1.9ii). The emission at 384 nm was stated due to the intra-ligand π-π* transition with a short lifetime of 10 μs, and it was further noted that it was a singlet emission as it has a short lifetime and relatively small Stokes shift (3,028 cm^{-1}) compared to the triplet excited state (4,000 cm^{-1}) (Bayón et al., 2005).

1.5.3 Energy Transfer

1.5.3.1 Fluorescence Resonance Energy Transfer

The term fluorescence resonance energy transfer (FRET) was named after Theodor Forster, a theoretical physicist who proposed an equation in 1948 to quantify the electronic excitation transfer efficiency from an energy donor to acceptor (Förster, 1948). FRET occurs between two molecules and serves as a crucial physical phenomenon for

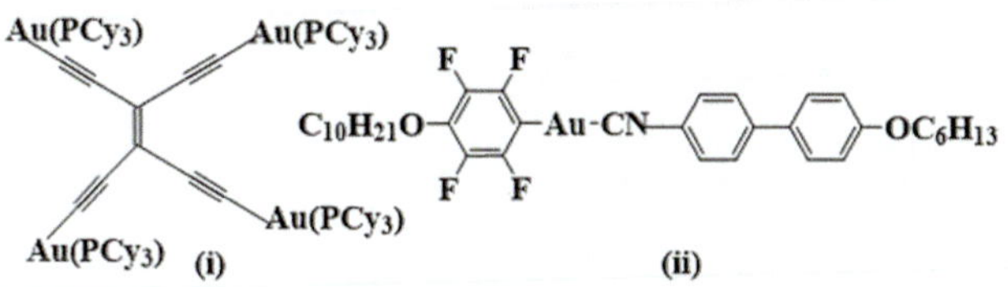

FIGURE 1.9 Linear Au(I) complex.

the understanding of some biological systems (Kozlov et al., 1997). It was discovered nearly 50 years ago and developed with time in biomedical research and drug discovery. Basically, FRET is a radiationless and distance-dependent energy transfer from an excited state donor to the acceptor fluorophore. The mechanism of the FRET involves an excited donor and a suitable acceptor that can excite by the energy transferred by the donor in a radiationless manner through a long-range dipole-dipole interaction (Figure 1.10) (Wu et al., 2020; Yuan et al., 2013).

The theory supporting energy transfer is based on the idea of treating an excited fluorophore as an oscillating dipole that exchanges the energy with the second dipole having a similar frequency. FRET is distance dependent and at the same time found insensitive to the surrounding solvent, which can be utilized to produce information such as excited-state reactions, solvent relaxation, fluorescence quenching, anisotropic measurements, etc. Usually, the surrounding solvents have a profound impact on the spectral properties of the donor and acceptor. FRET can transfer energy over a longer distance in comparison to the short-range solvents effect, and the dielectric nature of constituents positioned between the fluorophores has less effect on the ability of the FRET mechanism. A pair of molecules in which energy transfers through FRET is referred to as a donor-acceptor pair.

$$D + h\nu \rightarrow D^*$$

$$D^* + A \rightarrow D + A^* \text{ (D: donor, A: acceptor)}$$

$$A^* \rightarrow A + h\nu'$$

Certain criteria must be fulfilled for FRET to take place in a molecule. These are: (1) The fluorescence emission of the donor molecule must overlap with the absorption spectra of the acceptor. The degree of overlap describes the spectral overlap integral (J). (2) Both the donor and acceptor should be in close proximity to each other, typically 1–10 nm. (3) The orientation of the dipole moment of donor and acceptor must be parallel to each other. (4) The fluorescence lifetime duration of the donor molecule should be appropriate to allow FRET to take place. The rate of FRET relies on the extent of spectral overlap, the quantum yield, the orientation of the dipole moment, and of course the distance between the donor and acceptor. Hence, any factor affecting the distance between the donor and the acceptor will affect the rate of FRET. For detection and quantification of FRET, different ways are available: Simply, FRET can be observed by exciting the molecule containing both the donor and acceptor moieties with the wavelength centered near the emission maxima of the acceptor. FRET results in the decrease of the donor fluorescence and enhancement of the acceptor fluorescence, hence allowing the ratiometric determination of the signals. The alternative method is to measure the fluorescence lifetime of donor fluorophore

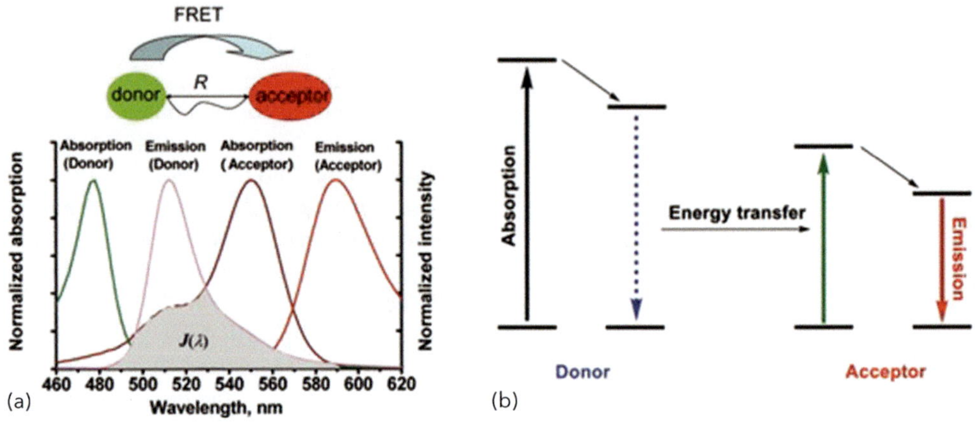

FIGURE 1.10 Jablonski's diagram showing FRET process (i), absorption and emission spectra of an ideal donor-acceptor pair and spectral overlap (J) in gray color (ii). Reprinted with permission from ref. Wu et al., Copyright (2020) Royal Society of Chemistry & Yuan, et al., Copyright (2013) American Chemical Society.

in the absence and presence of the acceptor, where FRET will decrease the excited lifetime of the donor fluorophore (Arshad Hussain, 2009). This energy transfer mechanism has wide utilization in determining the structure and the dynamics of proteins and nucleic acid, in the visualization and the detection of intermolecular association, and in the development of intermolecular binding assays (Willard et al., 2001). A research group, Goswami et al., developed a probe containing a quinoline-benzothiazole-based donor and a rhodamine-based acceptor and demonstrated its application in Cd^{2+} detection (Figure 1.11). The molecule was excited at 360 nm and displayed an emission at 470 nm that corresponds to the quinoline-benzothiazole moiety whereas no emission in the range of 585 nm was found that describes the presence of rhodamine moiety predominantly in spirolactum form. Upon addition of Cd^{2+} ion, a new absorption peak at 565 nm was observed indicating the ring opening of the spirolactum of the rhodamine moiety, which leads to the emission shift to 585 nm (Aich et al., 2015; Yuan et al., 2013).

1.5.4 New Emerging Signaling Mechanism

1.5.4.1 Excited-State Intramolecular Proton Transfer

The excited-state intramolecular proton transfer (ESIPT) process was discovered in the 1950s by Weller for salicylic acid (Weller, 1955). Later, the photophysics of salicylic acid was explored extensively (Padalkar and Seki, 2016). Typically, it is defined as the molecule containing both proton donor and acceptor groups that can undergo proton transfer in the excited state due to the increased basicity/acidity. In the excited state of salicylic acid, the hydroxyl hydrogen tends to be more electropositive and the carboxyl oxygen attains a more electronegative character resulting in the transfer of proton from the hydroxyl group to the carboxylic group (Figure 1.12) (Joshi and Antonov, 2021).

Many systems are known where the proton donor and acceptor groups are not in close proximity to transfer the proton; in such cases, either solvent or concentration assists the transfer of proton over a long distance within the molecule, termed as pseudo-intramolecular ESIPT

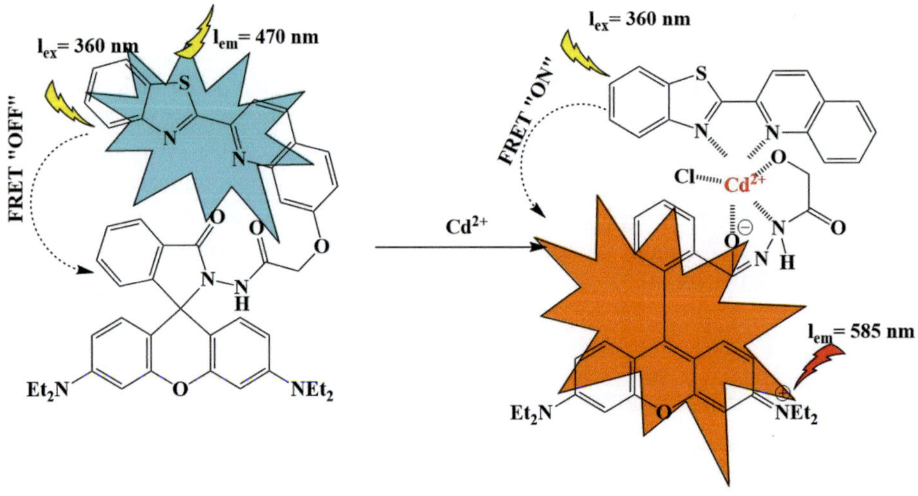

FIGURE 1.11 Quinoline-benzothiazole/rhodamine-based FRET probe for Cd^{2+} detection.

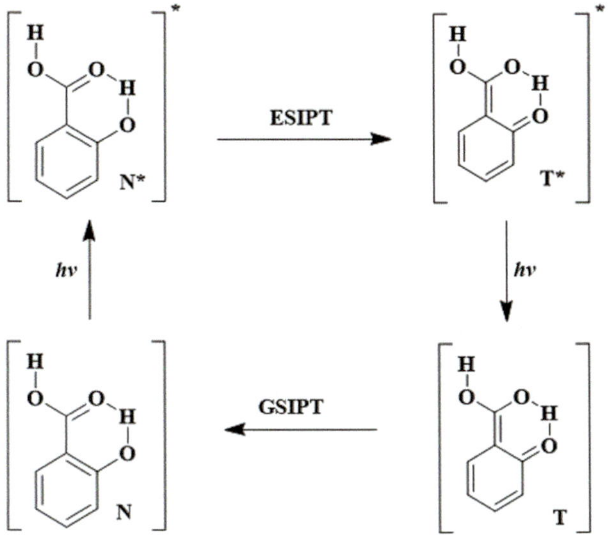

FIGURE 1.12 Photocycle of the ESIPT process in salicylic acid.

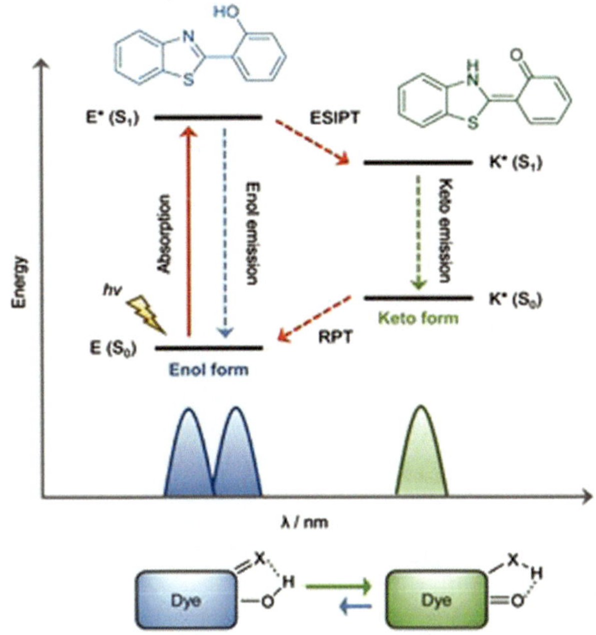

FIGURE 1.13 A diagrammatical representation of ESIPT process. Reprinted with permission from ref. Sedgwick et al., Copyright (2018) Royal Society of Chemistry.

(Konijnenberg et al., 1989). Most common fluorophores serving as analogues for ESIPT are 2-(2'-hydroxyphenyl) benzoxazole (HBO), 2-(2'-hydroxyphenyl)benzimidazole (HBI), and 2-(2'-hydroxyphenyl)benzothiazole (HBT). Other fluorophores reported for ESIPT include benzophenones, benzotriazoles, N-salicylideneaniline, quinoxalines, quinolone, anthraquinones, and flavones. Basically, ESIPT is a distinctive four-level process, where the GS of the fluorophore exists in the enol form that upon photoexcitation redistributes its electronic charge leading to the enhanced acidity for the hydrogen bond of the donor and the increased basicity of the hydrogen bond of the acceptor within enol form. This results in the rapid enol-to-keto phototautomerization, where the excited-state enol gets converted into the excited keto form. Then, the excited keto form radiatively decays to the GS through a reverse proton transfer (RPT) and generates the original E form (Figure 1.13) (Das et al., 1994; Sarkar et al., 1995; Sedgwick et al., 2018).

Unusually, a large Stokes shift (~200 nm) was observed for the ESIPT fluorophore, which helps to avoid self-reabsorption and inner filter effects. These fast four-level photochemical processes are characterized by such attractive features that found their application as imaging agents and fluorescent probes. This ESIPT emission is highly sensitive to the local surroundings, where the polar protic solvents can result in the inhibition of ESIPT process. This shows the requirement of a large ratio of organic solvents or cetyltrimethyammonium bromide to generate a hydrophobic micellar environment to produce a fluorescent response. Many ESIPT fluorophores show dual emission arising from the excited enol and keto form that can find application in the ratiometric sensing (Lee et al., 2015). General strategy considered for designing and developing of ESIPT-based probe includes blocking of hydrogen donor of ESIPT fluorophore with a reactive unit that inhibits the ESIPT process. This results in the emission from only enol form of the fluorophore as no exchangeable protons are available. However, when exposed to the analyte, the keto form generated revives the ESIPT process again, consequently serving as a detection system for the analyte.

The excitation wavelength dependence on the ESIPT was first reported by Hetherington et al. for 7-azaindole (Parada et al., 2015). They found that excited-state double proton transfer takes place at room temperature in less than 5 ps and also exhibits the excitation wavelength-dependent ESIPT at 77 K. Pant et al. found that the excitation wavelength dependent-ESIPT was also observed in the solid state, where two emission bands were observed at 370 nm (U band) and 450 nm (B band). Here, U-band was considered to be the result of the dimer molecules, where the ESIPT is not taking place, whereas the B band has a comparatively larger Stokes shift, suggesting ESIPT process, as at 77 K and with longer excitation wavelength, B band shifts drastically. The decay from normal and tautomeric states results in emission, which is found exponential with the decay times at higher temperatures and shorter excitation wavelength. This shows that a fast excited-state equilibrium between normal and the proton transferred form exists (Pant et al., 1994). However, the dependence on temperature and excitation wavelength requires a comprehensive theoretical study. ESIPT systems have found application in luminescent solar cells (Misra and Mishra, 2008), photo stabilizers (Catalan et al., 1990), white light generator (Moraes et al., 2020), organic light-emitting diodes (Zhang et al., 2016), thermally activated delayed fluorescence (Long et al., 2020), sensing of anion and cation (Y. Wu et al., 2007), etc.

1.5.4.2 Aggregation-Induced Emission

Around 20 years ago, Tang et al. had described a system, where the term "aggregation-induced emission (AIE)" was introduced contrary to the concept of fluorescence quenching occurring in the solid-state fluorophore as a result of ACQ (Luo et al., 2001). The same group has reported the emission in the solid state that includes enhancement of phosphorescence in xanthene dyes in the solvent system of glycerol:water. However, in the first half of the 20th century, new fluorescent bands associated with intramolecular interaction were reported. In 1936, unusual behavior of pseudoisocyanine chloride in the aqueous solution was studied by Scheibe (Scheibe, 1937) and Jelley (Jelley, 1937) independently where the absorption maximum shifts to the lower energy (red shift) when compared to the spectrum in ethanol whereas upon increasing the concentration of dye in water, the band became more intense and sharp. The molecules, upon aggregation, shift the band to a longer wavelength (red shift) with characteristic increased fluorescence intensity and smaller Stokes shift than monomer are named Scheibe aggregates or J-aggregates (J denotes Jelley, the pioneer) (Kasha et al., 1965). In contrast, when the absorption band shifts to the shorter wavelength (hypsochromic shift) for the aggregates in comparison to the monomer are described as H-aggregates (H represents hypsochromic), in such cases fluorescence quenching was observed (Brixner et al., 2017). In several systems, emission enhancement by the J-aggregates can be seen (Figure 1.14).

The working mechanism of the AIE has been explored where a proposed mechanism involves the restriction of intramolecular rotation (RIR) (Hong et al., 2009), evaluated through viscosity, temperature, and intramolecular steric hindrance-dependent luminescence quantum efficiency. Another proposed mechanism involves a restriction of intramolecular vibration (RIV) (Mei et al., 2014) that plays a role in the solid-state luminescence. Further, comprehensive model involves the combination of both above-mentioned mechanisms, i.e., RIR and RIV, and is termed a principle of restriction of motion (RIM). Later, other hypothesis were proposed including photocyclization (Gao et al., 2017), restriction for the formation of twisted intramolecular charge transfer (Liu *et al.,* 2017), and Z/E isomerization (Garg et al., 2015), whereas RIM was described as the main mechanistic cause of the AIE phenomenon usually occurring in the pure aromatic molecular rotors.

In Figure 1.15, the structural comparison in the presence of various similar and versatile effects was carried out as follows to examine the effect of rigidification. Azobenzene (AB) with $\phi_F \approx 0$, by simply restriction of the twist around a single bond, results in the 9,10-diazaphenanthrene (DPA) structure where a lifetime of 4.27 ns and $\phi_F \approx 0.02$ in ethanol was obtained (Takezaki et al., 1996). The E/Z form of stilbene are poorly fluorescent at room temperature with $\phi_F \approx 0.04$–0.05, however found temperature dependency as at T = 83.2 K, the molecular structure can be rigidified with $\phi_F \approx 0.95$ (Charlton and Saltiel, 1977). Also, the introduction of a central double bond generates the structure of indeno[2,1-a]indene (Ind-Ind) and 1,2-diphenylcyclobutene (DPCB), having restricted twisting of the molecule makes them highly fluorescent with unitary ϕ_F value (Saltiel *et al.,* 1968). Another example of structural comparison is naphthalene (NP) and retinol pair, both containing the same number of C-C double bonds, with the difference of ϕ_F, where ϕ_F for naphthalene is 0.21 in ethanol (Valeur and Berberan-Santos, 2012) whereas $\phi_F \approx 0.029$ for retinol in

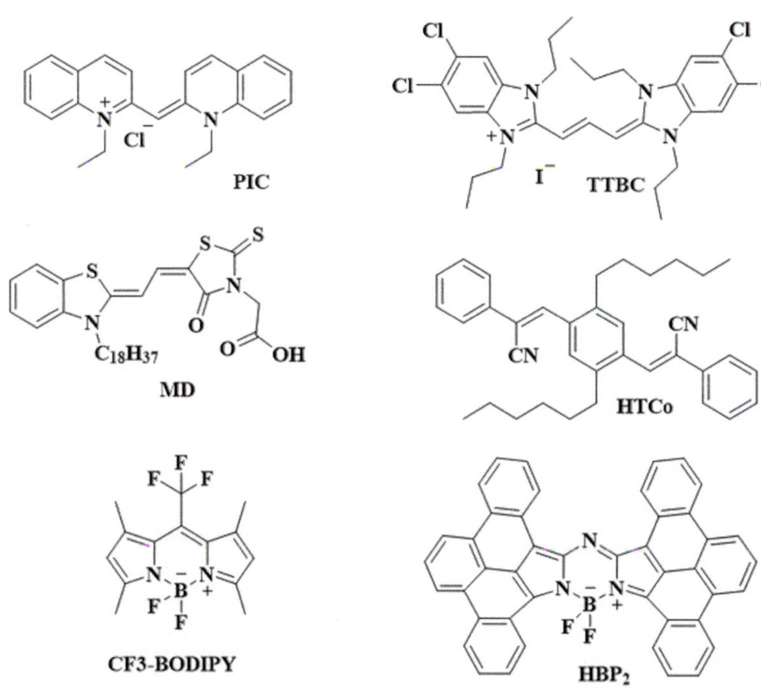

FIGURE 1.14 Molecules showing enhancement of emission due to formation of J-aggregates.

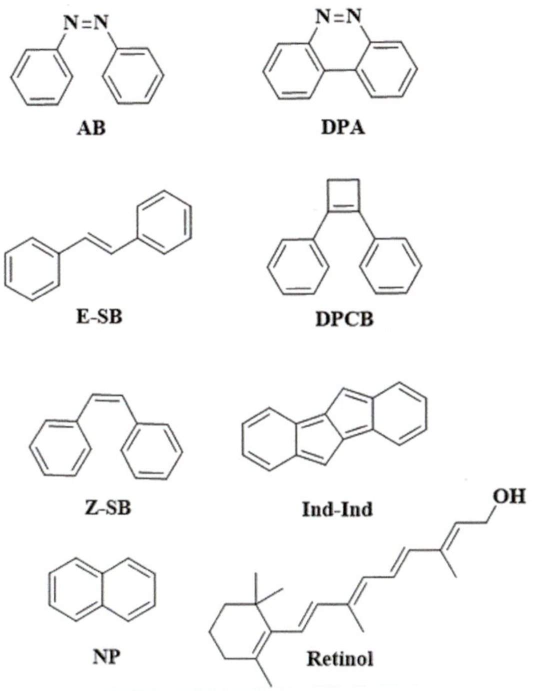

FIGURE 1.15 Compounds with similar structures but differ in the flexibility imparted by C=C bonds and the molecular rigidity.

hexane (Tsin *et al.,* 1988), at a temperature of T = 293 K, due to the rigidity in the structure of naphthalene, which is absent in retinol.

Tang et al. have discussed a molecule, 1,1,2,2-tetraphenylethylene (TPE); although the emission of TPE was known 50 years ago, it was explored for further insights into the interesting characteristics of the molecule. In 1980s and at the end of 1990s, the reports for the Π-twist of the central double bond in the excited state of TPE and viscosity-dependent fluorescence enhancement of the structure were published (Klingenberg et al., 1973). In TPE, the AIE phenomenon was likely due to free rotor or loose bolt effect. This results in the increase or decrease of the coupling modes of the vibronic transitions. The phenomenon of changing of C-C double bond to partial single bond upon excitation is known as loose bolt effect leading to the AIE effect. However, the topic is still under deliberation for the exact molecular mechanism, as recent studies claim the interpretation favors RIM and E/Z isomerization, which upon inhibiting or blocking results into the AIE phenomenon (Rodrigues and de Melo, 2022).

CONCLUSION

Photoluminescence is a fascinating topic that has been explored by researchers in the best possible way. Photoluminescence is a broad term involving processes: fluorescence and phosphorescence. Fluorescence has evolved incredibly over the years and assisted in understanding the mechanism of the processes occurring miraculously in nature, since the creation of mother Earth. Till date, so many mechanisms of fluorescence have been revealed, surprising the researchers every time. Fluorescence was found to be the result of electron transfer through the PET process, charge transfer via ICT, modified ICT: TICT, and MLCT, and energy transfer including FRET, and many new signaling mechanisms are emerging such as ESIPT, AIE, etc. Further, the applications of various types of fluorophores and their fluorescence mechanisms are covered in the following chapters.

REFERENCES

Acuña, A. U., & Amat-Guerri, F. (2007). Early history of solution fluorescence: The *Lignum nephriticum* of Nicolás Monardes. Berberan-Santos M.N. (Eds.) In *Fluorescence of supermolecules, polymers, and nanosystems* (pp. 3–20). Springer.

Acuña, A. U., Amat-Guerri, F., Morcillo, P., Liras, M., & Rodríguez, B. (2009). Structure and formation of the fluorescent compound of *lignum nephriticum. Organic Letters, 11*(14), 3020–3023.

Aich, K., Goswami, S., Das, S., Mukhopadhyay, C. D., Quah, C. K., & Fun, H. K. (2015). Cd(2+) triggered the FRET "ON": A new molecular switch for the ratiometric detection of Cd(2+) with live-cell imaging and bound X-ray structure. *Inorganic Chemistry, 54*(15), 7309–7315.

Amdursky, N., Erez, Y., & Huppert, D. (2012). Molecular rotors: What lies behind the high sensitivity of the thioflavin-T fluorescent marker. *Accounts of Chemical Research, 45*(9), 1548–1557.

Arshad Hussain, S. (2009). *An introduction to fluorescence resonance energy transfer (FRET).* doi:10.48550/arXiv.0908.1815

Arzhantsev, S., Zachariasse, K. A., & Maroncelli, M. (2006). Photophysics of *trans*-4-(Dimethylamino)-4 '-cyanostilbene and its use as a solvation probe. *The Journal of Physical Chemistry A, 110*(10), 3454–3470.

Balzani, V., Piotrowiak, P., Rodgers, M., Mattay, J., & Astruc, D. (2001). *Electron transfer in chemistry* (Vol. 1). Wiley-VCh Weinheim.

Bayón, R., Coco, S., & Espinet, P. (2005). Gold liquid crystals displaying luminescence in the mesophase and short F· · ·F interactions in the solid state. *Chemistry – A European Journal, 11*(4), 1079–1085.

Berberan-Santos, M. N. (2001). Pioneering contributions of Jean and Francis Perrin to molecular luminescence. Valeur, B. & Brochon JC. (Eds.) In *New trends in fluorescence spectroscopy: Applications to chemical and life sciences* (pp. 7–33). Springer.

Berezin, M. Y., & Achilefu, S. (2010). Fluorescence lifetime measurements and biological imaging. *Chemical Reviews, 110*(5), 2641–2684.

Brewster, D. (1834). On the colours of natural bodies *(Vol. 12). Earth and Environmental Science Transactions of The Royal Society of Edinburgh, 12*(2), 538–545.

Brixner, T., Hildner, R., Köhler, J., Lambert, C., & Würthner, F. (2017). Exciton transport in molecular aggregates – from natural antennas to synthetic chromophore systems. *Advanced Energy Materials, 7*(16), 1700236.

Bulheller, B. M., Miles, A. J., Wallace, B. A., & Hirst, J. D. (2008). Charge-transfer transitions in the vacuum-ultraviolet of protein circular dichroism spectra. *The Journal of Physical Chemistry B, 112*(6), 1866–1874.

Callan, J. F., De Silva, A. P., & Magri, D. C. T. (2005). Luminescent sensors and switches in the early 21st century. *Tetrahedron, 61*(36), 8551–8588.

Cao, K., Farahi, M., Dakanali, M., Chang, W. M., Sigurdson, C. J., Theodorakis, E. A., & Yang, J. (2012). Aminonaphthalene 2-cyanoacrylate (ANCA) probes fluorescently discriminate between amyloid-β and prion plaques in brain. *Journal of the American Chemical Society, 134*(42), 17338–17341.

Carlotti, B., Benassi, E., Barone, V., Consiglio, G., Elisei, F., Mazzoli, A., & Spalletti, A. (2015). Effect of the π bridge and acceptor on intramolecular charge transfer in push–pull cationic chromophores: An ultrafast spectroscopic and TD-DFT computational study. *ChemPhysChem, 16*(7), 1440–1450.

Catalan, J., Fabero, F., Soledad Guijarro, M., Claramunt, R. M., Santa Maria, M. D., Foces-Foces, M. D. l. C., Hernandez Cano, F., Elguero, J., & Sastre, R. (1990). Photoinduced intramolecular proton transfer as the mechanism of ultraviolet stabilizers: A reappraisal. *Journal of the American Chemical Society, 112*(2), 747–759.

Charlton, J. L., & Saltiel, J. (1977). An analysis of *trans*-stilbene fluorescence quantum yields and lifetimes. *The Journal of Physical Chemistry, 81*(20), 1940–1944.

Cornelissen-Gude, C., & Rettig, W. (1998). Dual fluorescence and multiple charge transfer nature in derivatives of N-pyrrolobenzonitrile. *The Journal of Physical Chemistry A, 102*(40), 7754–7760.

Das, K., Sarkar, N., Ghosh, A. K., Majumdar, D., Nath, D. N., & Bhattacharyya, K. (1994). Excited-state intramolecular proton transfer in 2-(2-Hydroxyphenyl)benzimidazole and -benzoxazole: Effect of rotamerism and hydrogen bonding. *The Journal of Physical Chemistry, 98*(37), 9126–9132.

De Silva, A. P., Moody, T. S., & Wright, G. D. (2009). Fluorescent PET (Photoinduced Electron Transfer) sensors as potent analytical tools. *Analyst, 134*(12), 2385–2393.

de Silva, A. P., & Rupasinghe, R. D. (1985). A new class of fluorescent pH indicators based on photo-induced electron transfer. *Journal of the Chemical Society, Chemical Communications, 23*, 1669–1670.

Dong, B., Song, X., Wang, C., Kong, X., Tang, Y., & Lin, W. (2016). Dual site-controlled and lysosome-targeted intramolecular charge transfer–photoinduced electron transfer–fluorescence resonance energy transfer fluorescent probe for monitoring pH changes in living cells. *Analytical Chemistry, 88*(7), 4085–4091.

Druzhinin, S. I., Mayer, P., Stalke, D., von Bu'low, R., Noltemeyer, M., & Zachariasse, K. A. (2010). Intramolecular charge transfer with 1-tert-butyl-6-cyano-1, 2, 3, 4-tetrahydroquinoline (ntc6) and other aminobenzonitriles: A comparison of experimental vapor phase spectra and crystal structures with calculations. *Journal of the American Chemical Society, 132*(22), 7730–7744.

Escudero, D. (2016). Revising intramolecular photoinduced electron transfer (PET) from first-principles. *Accounts of Chemical Research, 49*(9), 1816–1824.

Fan, J., Liu, X., Hu, M., Zhu, H., Song, F., & Peng, X. (2012). Development of an oxidative dehydrogenation-based fluorescent probe for $Cu^{2+}$and its biological imaging in living cells. *Analytica Chimica Acta, 735*, 107–113.

Förster, T. (1948). Zwischenmolekulare Energiewanderung und Fluoreszenz. *Annalen der physik, 437*(1–2), 55–75.

Gao, Y. J., Chang, X. P., Liu, X. Y., Li, Q. S., Cui, G., & Thiel, W. (2017). Excited-state decay paths in tetraphenylethene derivatives. *The Journal of Physical Chemistry A, 121*(13), 2572–2579.

Garg, K., Ganapathi, E., Rajakannu, P., & Ravikanth, M. (2015). Stereochemical modulation of emission behaviour in E/Z isomers of diphenyldipyrroethene from aggregation induced emission to crystallization induced emission. *Physical Chemistry Chemical Physics, 17*(29), 19465–19473.

Goh, W. L., Lee, M. Y., Joseph, T. L., Quah, S. T., Brown, C. J., Verma, C., Brenner, S., Ghadessy, F. J., & Teo, Y. N. (2014). Molecular rotors as conditionally fluorescent labels for rapid detection of biomolecular interactions. *Journal of American Chemical Society, 136*(17), 6159–6162.

Gong, Y., Guo, X., Wang, S., Su, H., Xia, A., He, Q., & Bai, F. (2007). Photophysical properties of photoactive molecules with conjugated push– pull structures. *The Journal of Physical Chemistry A, 111*(26), 5806–5812.

Grabchev, I., Chovelon, J.-M., & Qian, X. (2003). A copolymer of 4-N, N-dimethylaminoethylene-N-allyl-1, 8-naphthalimide with methylmethacrylate as a selective fluorescent chemosensor in homogeneous systems for metal cations. *Journal of Photochemistry Photobiology A: Chemistry, 158*(1), 37–43.

Gunnlaugsson, T., Ali, H. D. P., Glynn, M., Kruger, P. E., Hussey, G. M., Pfeffer, F. M., dos Santos, C. M., & Tierney, J. J. (2005). Fluorescent photoinduced electron transfer (PET) sensors for anions; from design to potential application. *Journal of Fluorescence, 15*, 287–299.

Gunnlaugsson, T., Glynn, M., Tocci, G. M., Kruger, P. E., & Pfeffer, F. M. (2006). Anion recognition and sensing in organic and aqueous media using luminescent and colorimetric sensors. *Coordination Chemistry Reviews, 250*(23), 3094–3117.

Haidekker, M. A., Brady, T., Lichlyter, D., & Theodorakis, E. (2005). Effects of solvent polarity and solvent viscosity on the fluorescent properties of molecular rotors and related probes. *Bioorganic Chemistry, 33*(6), 415–425.

Haidekker, M. A., & Theodorakis, E. A. (2007). Molecular rotors—fluorescent biosensors for viscosity and flow. *Organic Biomolecular Chemistry, 5*(11), 1669–1678.

Haidekker, M. A., & Theodorakis, E. A. (2010). Environment-sensitive behavior of fluorescent molecular rotors. *Journal of Biological Engineering, 4*(1), 11.

Harvey, E. N. (2005). *A history of luminescence: From the earliest times until 1900*. Dover Publications.

Hong, Y., Lam, J. W. Y., & Tang, B. Z. (2009). Aggregation-induced emission: phenomenon, mechanism and applications. *Chemical Communications, 29*, 4332–4353.

Jabłoński, A. J. Z. f. P. (1935). Über den mechanismus der photolumineszenz von farbstoffphosphoren. *Zeitschrift für Physik, 94*(1–2), 38–46.

Jana, S., Dalapati, S., Ghosh, S., & Guchhait, N. (2013). Excited state intramolecular charge transfer process in 5-(4-dimethylamino-phenyl)-penta-2, 4-dienoic acid ethyl ester and effect of acceptor functional groups. *Journal of Photochemistry Photobiology A: Chemistry, 261*, 31–40.

Janssen, Y. M., Van Houten, B., Borm, P. J., & Mossman, B. T. (1993). Cell and tissue responses to oxidative damage. *Laboratory Investigation; A Journal of Technical Methods and Pathology, 69*(3), 261–274.

Jelley, E. E. (1937). Molecular, nematic and crystal states of I: I-Diethyl--cyanine chloride. *Nature, 139*(3519), 631–631.

John Frederick William, H. (1845). *On a case of superficial colour presented by a homogeneous liquid internally colourless* (Vol. 135). The Royal Society.

Joshi, H. C., & Antonov, L. (2021). Excited-state intramolecular proton transfer: A short introductory review. *Molecules, 26*(5), 1475.

Kasha, M., Rawls, H. R., & El-Bayoumi, M. A. (1965). The exciton model in molecular spectroscopy. *Pure and Applied Chemistry, 11*(3–4), 371–392.

Kirchhoff, J. R., Gamache Jr, R. E., Blaskie, M. W., Del Paggio, A. A., Lengel, R. K., & McMillin, D. R. (1983). Temperature dependence of luminescence from Cu (NN) $^{2+}$ systems in fluid solution: Evidence for the participation of two excited states. *Inorganic Chemistry, 22*(17), 2380–2384.

Klingenberg, H. H., Lippert, E., & Rapp, W. (1973). The influence of an intramolecular relaxation process on the frequency-time behaviour or the fluorescence of tetraphenylethylene in highly viscous solutions. *Chemical Physics Letters, 18*(3), 417–419.

Kobayashi, H., Ogawa, M., Alford, R., Choyke, P. L., & Urano, Y. (2010). New strategies for fluorescent probe design in medical diagnostic imaging. *Chemical Reviews, 110*(5), 2620–2640.

Konijnenberg, J., Ekelmans, G. B., Huizer, A. H., & Varma, C. A. G. O. (1989). Mechanism and solvent dependence of the solvent-catalysed pseudo-intramolecular proton transfer of 7-hydroxyquinoline in the first electronically excited singlet state and in the ground state of its tautomer. *Journal of the Chemical Society, Faraday Transactions 2: Molecular and Chemical Physics, 85*(1), 39–51.

Kozlov, V., Bulović, V., Burrows, P., & Forrest, S. (1997). Laser action in organic semiconductor waveguide and double-heterostructure devices. *Nature, 389*(6649), 362–364.

Kuimova, M. K. (2012a). Mapping viscosity in cells using molecular rotors. *Physical Chemistry Chemical Physics, 14*(37), 12671–12686.

Kuimova, M. K. (2012b). Molecular rotors image intracellular viscosity. *Chimia, 66*(4), 159–159.

Lee, M. H., Kim, J. S., & Sessler, J. L. (2015). Small molecule-based ratiometric fluorescence probes for cations, anions, and biomolecules. *Chemical Society Reviews, 44*(13), 4185–4191.

Lee, M. H., Park, N., Yi, C., Han, J. H., Hong, J. H., Kim, K. P., Kang, D. H., Sessler, J. L., Kang, C., & Kim, J. S. (2014). Mitochondria-immobilized pH-sensitive off–on fluorescent probe. *Journal of the American Chemical Society, 136*(40), 14136–14142.

Lee, S. K., Kim, J. O., Shimizu, D., Osuka, A., & Kim, D. (2016). Effect of bulky *meso*-substituents on photoinduced twisted intramolecular charge transfer processes in *meso*-diarylamino subporphyrins. *Journal of Porphyrins Phthalocyanines, 20*(06), 663–669.

Liu, J., Zhang, C., Dong, J., Zhu, J., Shen, C., Yang, G., & Zhang, X. (2017). Endowing a triarylboron compound showing ACQ with AIE characteristics by transforming its emissive TICT state to be dark. *RSC Advances, 7*(24), 14511–14515.

Lodeiro, C., Capelo, J. L., Mejuto, J. C., Oliveira, E., Santos, H. M., Pedras, B., & Nuñez, C. (2010). Light and colour as analytical detection tools: A journey into the periodic table using polyamines to bio-inspired systems as chemosensors. *Chemical Society Reviews, 39*(8), 2948–2976.

Lodeiro, C., & Pina, F. (2009). Luminescent and chromogenic molecular probes based on polyamines and related compounds. *Coordination Chemistry Reviews, 253*(9–10), 1353–1383.

Long, Y., Mamada, M., Li, C., dos Santos, P. L., Colella, M., Danos, A., Adachi, C., & Monkman, A. (2020). Excited state dynamics of thermally activated delayed fluorescence from an excited state intramolecular proton transfer system. *The Journal of Physical Chemistry Letters, 11*(9), 3305–3312.

Lu, W., Zhu, N., & Che, C. M. (2003). Luminescent properties of carbon-rich starburst gold(I) acetylide complexes. Crystal structure of [TEE][Au(PCy3)]4 ([TEE] H4=tetraethynylethene). *Journal of Organometallic Chemistry, 670*(1), 11–16.

Luo, J., Xie, Z., Lam, J. W. Y., Cheng, L., Chen, H., Qiu, C., Kwok, H. S., Zhan, X., Liu, Y., Zhu, D., & Tang, B. Z. (2001). Aggregation-induced emission of 1-methyl-1,2,3,4,5-pentaphenylsilole. *Chemical Communications, 18*, 1740–1741.

Mallick, A., Purkayastha, P., & Chattopadhyay, N. (2007). Photoprocesses of excited molecules in confined liquid environments: An overview. *Journal of Photochemistry Photobiology C: Photochemistry Reviews, 8*(3), 109–127.

Martinez-Manez, R., & Sancenon, F. (2003). Fluorogenic and chromogenic chemosensors and reagents for anions. *Chemical Reviews, 103*(11), 4419–4476.

Mei, J., Hong, Y., Lam, J. W. Y., Qin, A., Tang, Y., & Tang, B. Z. (2014). Aggregation-induced emission: The whole is more brilliant than the parts. *Advanced Materials, 26*(31), 5429–5479.

Misra, R., & Bhattacharyya, S. P. (2018). *Intramolecular charge transfer: Theory and applications*. John Wiley & Sons.

Misra, R., Bhattacharyya, S. P., & Mandal, A. (2011). Spectral response of 4-methyl-2, 6-dicarbomethoxyphenol, an excited-state intramolecular proton-transfer probe in cyclohexane–ethanol mixtures: Signatures of medium microheterogeneity. *The Journal of Physical Chemistry B, 115*(41), 11840–11851.

Misra, V., & Mishra, H. (2008). Photoinduced proton transfer coupled with energy transfer: Mechanism of sensitized luminescence of terbium ion by salicylic acid doped in polymer. *The Journal of Chemical Physics, 128*(24), 244701.

Moraes, E. S., Teixeira Alves Duarte, L. G., Germino, J. C., & Atvars, T. D. Z. (2020). Near attack conformation as strategy for ESIPT modulation for white-light generation. *The Journal of Physical Chemistry C, 124*(41), 22406–22415.

Muyskens, M. (2006). The fluorescence of *Lignum nephriticum*: A flash back to the past and a simple demonstration of natural substance fluorescence. *Journal of Chemical Education, 83*(5), 765.

Ntziachristos, V., Bremer, C., & Weissleder, R. (2003). Fluorescence imaging with near-infrared light: New technological advances that enable *in vivo* molecular imaging. *European Radiology, 13*, 195–208.

O'Haver, T. (1978). Development of luminescence spectrometry as an analytical tool. *Journal of Chemical Education, 55*(7), 423.

Padalkar, V. S., & Seki, S. (2016). Excited-state intramolecular proton-transfer (ESIPT)-inspired solid state emitters. *Chemical Society Reviews, 45*(1), 169–202.

Pahlavanlu, P., Christensen, P. R., Therrien, J. A., & Wolf, M. O. (2016). Controlled intramolecular charge transfer using a sulfur-containing acceptor group. *The Journal of Physical Chemistry C, 120*(1), 70–77.

Pant, D. D., Joshi, H. C., Bisht, P. B., & Tripathi, H. B. (1994). Dual emission and double proton transfer in salicylic acid. *Chemical Physics, 185*(1), 137–144.

Parada, G. A., Markle, T. F., Glover, S. D., Hammarström, L., Ott, S., & Zietz, B. (2015). Control over excited state intramolecular proton transfer and photoinduced tautomerization: Influence of the hydrogen-bond geometry. *Chemistry, 21*(17), 6362–6366.

Paul, B. K., Samanta, A., & Guchhait, N. (2010). Exploring hydrophobic subdomain IIA of the protein bovine serum albumin in the native, intermediate, unfolded, and refolded states by a small fluorescence molecular reporter. *The Journal of Physical Chemistry B, 114*(18), 6183–6196.

Perrin, F. (1929). *La fluorescence des solutions-Induction moléculaire.–Polarisation et durée d'émission.–Photochimie.* Paper presented at the Annales de physique.

Priestley, R. D., Ellison, C. J., Broadbelt, L. J., & Torkelson, J. M. (2005). Structural relaxation of polymer glasses at surfaces, interfaces, and in between. *Science, 309*(5733), 456–459.

Radley, J. A., & Grant, J. (Eds.) (1939). Fluorescence analysis in ultra-violet light. In *Fluorescence analysis in ultra-violet light* (3rd ed.). Chapman & Hall.

Rettig, W., & Lapouyade, R. (2002). Fluorescence probes based on twisted intramolecular charge transfer (TICT) states and other adiabatic photoreactions. In Lakowicz, J. (Eds.), *Topics in fluorescence spectroscopy: probe design and chemical sensing* (pp. 109–149), Springer.

Rodrigues, A. C. B., & de Melo, J. S. S. (2022). Aggregation-induced emission: From small molecules to polymers—historical background, mechanisms and photophysics. In Y. Tang & B. Z. Tang (Eds.), *Aggregation-induced emission* (pp. 209–246). Springer International Publishing.

Saltiel, J., Zafiriou, O. C., Megarity, E. D., & Lamola, A. A. (1968). Tests of the singlet mechanism for *cis-trans* photoisomerization of the stilbenes. *Journal of the American Chemical Society, 90*(17), 4759–4760.

Sapsford, K. E., Berti, L., & Medintz, I. L. (2006). Materials for fluorescence resonance energy transfer analysis: beyond traditional donor–acceptor combinations. *Angewandte Chemie International Edition, 45*(28), 4562–4589.

Sarkar, N., Das, K., Das, S., Datta, A., Nath, D., & Bhattacharyya, K. (1995). Excited-state intramolecular proton transfer of 2-(2'-Hydroxyphenyl)benzimidazole in micelles. *The Journal of Physical Chemistry, 99*(50), 17711–17714.

Sasaki, S., Drummen, G. P., & Konishi, G. I. (2016). Recent advances in twisted intramolecular charge transfer (TICT) fluorescence and related phenomena in materials chemistry. *Journal of Materials Chemistry C, 4*(14), 2731–2743.

Scheibe, G. (1937). Über die Veränderlichkeit der Absorptionsspektren in Lösungen und die Nebenvalenzen als ihre Ursache. *Angewandte Chemie International Edition, 50*(11), 212–219.

Sedghi, G., Sawada, K., Esdaile, L. J., Hoffmann, M., Anderson, H. L., Bethell, D., Haiss, W., Higgins, S. J., & Nichols, R. (2008). Single molecule conductance of porphyrin wires with ultralow attenuation. *Journal of the American Chemical Society, 130*(27), 8582–8583.

Sedgwick, A. C., Wu, L., Han, H. H., Bull, S. D., He, X. P., James, T. D., Sessler, J. L., Tang, B. Z., Tian, H., & Yoon, J. (2018). Excited-state intramolecular proton-transfer (ESIPT) based fluorescence sensors and imaging agents. *Chemical Society Reviews, 47*(23), 8842–8880.

Smith, K. C. (1974). *The science of photobiology*. Springer US.

So, P. T., & Dong, C. Y. (2002). Fluorescence spectrophotometry. *Encyclopedia of Life Sciences*. Macmillan Publishers Ltd, Nature Publishing Group/www.els.net

Sowmiya, M., Tiwari, A. K., & Saha, S. K. (2011). Study on intramolecular charge transfer fluorescence properties of *trans*-4-[4′-(N, N′-dimethylamino) styryl] pyridine: Effect of solvent and pH. *Journal of Photochemistry Photobiology A: Chemistry 218*(1), 76–86.

Stokes, G. G. (1852). On the change of refrangibility of light. *Philosophical Transactions of the Royal Society, 142*, 463–562.

Stokes, G. G. (1853). On the change of refrangibility of light.—— No. II. *Philosophical Transactions of the Royal Society of London, 143*, 385–396.

Suhling, K., Levitt, J. A., Chung, P. H., Kuimova, M. K., & Yahioglu, G. (2012). Fluorescence lifetime imaging of molecular rotors in living cells. *Journal of Visualized Experiments, 60*, 2925.

Sumalekshmy, S., & Gopidas, K. (2005). Intramolecular charge transfer processes in donor-acceptor substituted vinyltetrahydropyrenes. *Photochemical Photobiological Sciences, 4*, 539–546.

Takezaki, M., Hirota, N., & Terazima, M. (1996). Excited state dynamics of 9,10-diazaphenanthrene studied by the time-resolved transient grating method. *The Journal of Physical Chemistry, 100*(24), 10015–10020.

Tsin, A. T. C., Pedrozo-Fernandez, H. A., Gallas, J. M., & Chambers, J. P. (1988). The fluorescence quantum yield of vitamin A2. *Life Sciences, 43*(17), 1379–1384.

Tsiperman, E., Regev, T., Becker, J. Y., Bernstein, J., Ellern, A., Khodorkovsky, V., Shames, A., & Shapiro, L. (1999). Synthesis of a novel rigid tetrathiafulvalene-σ-p-benzoquinone diad (TTF-σ-Q) with inherent structural configuration suitable for intramolecular charge-transfer. *Chemical Communications, 12*, 1125–1126.

Valeur, B., & Berberan-Santos, M. N. (2002). *Molecular fluorescence Wiley*. Wiley.

Valeur, B., & Berberan-Santos, M. N. (2011). A brief history of fluorescence and phosphorescence before the emergence of quantum theory. *Journal of Chemical Education, 88*(6), 731–738.

Valeur, B., & Berberan-Santos, M. N. (2012). *Molecular fluorescence: Principles and applications*. John Wiley & Sons.

Wang, M., Zhang, G., Zhang, D., Zhu, D., & Tang, B. Z. (2010). Fluorescent bio/chemosensors based on silole and tetraphenylethene luminogens with aggregation-induced emission feature. *Journal of Materials Chemistry, 20*(10), 1858–1867.

Wang, R., Yu, F., Chen, L., Chen, H., Wang, L., & Zhang, W. (2012). A highly selective turn-on near-infrared fluorescent probe for hydrogen sulfide detection and imaging in living cells. *Chemical Communications, 48*(96), 11757–11759.

Weissman, H., Shirman, E., Ben-Moshe, T., Cohen, R., Leitus, G., Shimon, L. J., & Rybtchinski, B. (2007). Palladium complexes of perylene diimides: Strong fluorescence despite direct attachment of late transition metals to organic dyes. *Inorganic Chemistry, 46*(12), 4790–4792.

Weller, A. (1955). Über die Fluoreszenz der Salizylsäure und verwandter Verbindungen. *Naturwissenschaften, 42*, 175–176.

Willard, D. M., Carillo, L. L., Jung, J., & Van Orden, A. (2001). CdSe- ZnS quantum dots as resonance energy transfer donors in a model protein- protein binding assay. *Nano Letters, 1*(9), 469–474.

Williams, J. A. G. (2007). Photochemistry and photophysics of coordination compounds: Platinum. In V. Balzani & S. Campagna (Eds.), *Photochemistry and photophysics of coordination compounds II* (pp. 205–268). Springer Berlin Heidelberg.

Wu, J. S., Liu, W. M., Zhuang, X. Q., Wang, F., Wang, P. F., Tao, S. L., Zhang, X. H., Wu, S. K., & Lee, S. T. (2007). Fluorescence turn on of coumarin derivatives by metal cations: a new signaling mechanism based on C= N isomerization. *Organic Letters, 9*(1), 33–36.

Wu, L., Huang, C., Emery, B. P., Sedgwick, A. C., Bull, S. D., He, X. P., Tian, H., Yoon, J., Sessler, J. L., & James, T. D. (2020). Förster resonance energy transfer (FRET)-based small-molecule sensors and imaging agents. *Chemical Society Reviews, 49*(15), 5110–5139.

Wu, Y., Peng, X., Fan, J., Gao, S., Tian, M., Zhao, J., & Sun, S. (2007). Fluorescence Sensing of Anions Based on Inhibition of Excited-State Intramolecular Proton Transfer. *The Journal of Organic Chemistry, 72*(1), 62–70.

Xia, H. C., Xu, X. H., & Song, Q. H. (2017). BODIPY-based fluorescent sensor for the recognization of phosgene in solutions and in gas phase. *Analytical Chemistry, 89*(7), 4192–4197.

Xiao, P., Liu, J., Wang, Z., Tao, F., Yang, L., Yuan, G., Sun, W., & Zhang, X. (2021). A color turn-on fluorescent probe for real-time detection of hydrogen sulfide and identification of food spoilage. *Chemical Communications, 57*(41), 5012–5015.

Xu, Z., Yoon, J., & Spring, D. R. (2010). Fluorescent chemosensors for Zn^{2+}. *Chemical Society Reviews, 39*(6), 1996–2006.

Yang, L., Niu, J. Y., Sun, R., Xu, Y. J., & Ge, J. F. (2017). The pH-influenced PET processes between pyronine and different heterocycles. *Organic Biomolecular Chemistry, 15*(39), 8402–8409.

Yuan, L., Lin, W., Zheng, K., & Zhu, S. (2013). FRET-based small-molecule fluorescent probes: Rational design and bioimaging applications. *Accounts of Chemical Research, 46*(7), 1462–1473.

Zhang, W., Li, P., Yang, F., Hu, X., Sun, C., Zhang, W., Chen, D., & Tang, B. (2013). Dynamic and reversible fluorescence imaging of superoxide anion fluctuations in live cells and *in vivo*. *Journal of the American Chemical Society, 135*(40), 14956–14959.

Zhang, Z., Chen, Y. A., Hung, W. Y., Tang, W. F., Hsu, Y. H., Chen, C. L., Meng, F. Y., & Chou, P. T. (2016). Control of the reversibility of excited-state intramolecular proton transfer (ESIPT) reaction: Host-polarity tuning white organic light emitting diode on a new Thiazolo[5,4-d]thiazole ESIPT system. *Chemistry of Materials, 28*(23), 8815–8824.

Zhao, G. J., Liu, J. Y., Zhou, L. C., & Han, K. L. (2007). Site-selective photoinduced electron transfer from alcoholic solvents to the chromophore facilitated by hydrogen bonding: A new fluorescence quenching mechanism. *The Journal of Physical Chemistry B, 111*(30), 8940–8945.

Zhao, Q., Li, F., & Huang, C. (2010). Phosphorescent chemosensors based on heavy-metal complexes. *Chemical Society Reviews, 39*(8), 3007–3030.

Zhu, L., & Zhao, Y. (2013). Cyanostilbene-based intelligent organic optoelectronic materials. *Journal of Materials Chemistry C, 1*(6), 1059–1065.

2 Acridine-Based Fluorescent Probes
Synthesis, Biological, and Nonbiological Applications

Dheeraj Pandey and Abha Sharma
National Institute of Pharmaceutical Education and Research,
Raebareli, Lucknow, UP, India

LIST OF ABBREVIATIONS

AO	Acridine orange
UV	Ultraviolet
DNA	Deoxyribonucleic acid
RNA	Ribonucleic acid
PPA	Polyphosphoric acid
FMT	Fluorescence molecular tomography
LOD	Limit of detection
LOQ	Limit of quantification
DFT	Density functional theory
TD-DFT	Time-dependent density functional theory
NMR	Nuclear magnetic resonance
HRMS	High-resolution mass spectroscopy
FTIR	Fourier transform infrared spectroscopy
ROS	Reactive oxygen species
Cys	Cysteine
Hcy	Homocysteine

2.1 INTRODUCTION

Fluorescent probes play a crucial role in both biological and chemical research by providing a flexible method to visualize and identify different molecules and processes with a high level of accuracy (Yao et al., 2014). These probes are specifically designed molecules that emit light when excited by an external light source, typically ultraviolet or visible light. They are engineered to interact selectively with particular targets, such as biomolecules, ions, or cellular components, allowing researchers to monitor their location, behavior, and activity in real time. The necessity for fluorescent probes arises from the limitations of traditional techniques used in biochemical and cellular studies. Traditional methods often involve time-consuming sample preparation, lack specificity, or require invasive labeling techniques (Rizvi et al., 2010). Fluorescent probes address these issues by offering numerous benefits. Fluorescent probes can detect even minute concentrations of target molecules, enabling researchers to study biological processes at the molecular level. These probes can be customized to bind exclusively to specific targets, enabling researchers to study individual components within complex biological systems. Fluorescent probes enable the real-time imaging of dynamic processes in living cells and tissues, offering valuable insights into cellular behavior. Fluorescent probes can be adapted for various applications, including monitoring pH changes, detecting ions, tracking biomolecular interactions, and visualizing cellular structures (Zhu et al., 2016). Overall, fluorescent probes have transformed biological and chemical imaging, empowering researchers to explore complex biological phenomena with unprecedented precision and depth. Their adaptability and sensitivity drive advancements across numerous scientific disciplines, from basic research to clinical diagnostics and drug development (Tsien, 1989).

The fluorescent molecules consist of a fluorescent core that is responsible for the phenomenon of fluorescence. The fluorescent core is the part of a molecule responsible for emitting light when energized by an external source like UV or visible light. It typically consists of a system of π electrons that are conjugated, allowing for efficient absorption and subsequent release of light (Mahesh et al., 2019). Compared to nonfluorescent molecules, the excited state of the fluorescent core tends to be more stable, resulting in the emission of photons as the molecule returns to its resting state. The specific chemical makeup of the fluorescent core dictates the wavelength of emitted light, offering a way to customize the molecule's fluorescence for various uses, such as imaging, sensing, or detection applications. Acridine is one of the widely used fluorescent cores that emerged after the discovery of acridine-based dyes that were extracted from coal tar (Herole et al., 2022).

Acridines are a group of heterocyclic compounds distinguished by three interconnected six-membered rings, forming a flat aromatic structure (Pirota et al., 2020). Acridines are found in both natural products and synthetic dyes, serving various purposes such as pharmaceuticals, optical materials, and sensors (Khandelwal et al., 2020). Acridine, classified as an alkaloid derived from anthracene, is alternatively referred to as dibenzopyridine, 2,3,5,6-dibenzopyridine, or 10-azaanthracene. It emits a pungent odor and typically forms colorless to pale yellow needle-like crystals. Its melting point is 110°C, while it boils at 346°C.

 DOI: 10.1201/9781032706245-2

Notably, acridine is recognized for its skin-irritating properties and the characteristic blue fluorescence displayed by its salt solutions. Acridine and its related compounds are known for their stability and exhibit weakly basic properties. Acridine, in particular, exhibits a pKa value of 5.6, which is comparable to that of pyridine, another aromatic compound (Kumar et al., 2012). Apart from the fluorescent properties of acridine it also possesses a wide range of biological activities including antibacterial, antimalarial, antifungal, antitumor, anticancer, cytotoxic, and anti-multidrug-resistant properties (Khandelwal et al., 2020). The versatile synthesis of acridine enables its easy derivatization during the initial stages of synthesis and can be further utilized as a fluorescent probe for various applications (Pordel et al., 2024).

Acridine displays fluorescence because of its distinct chemical composition (Prasher et al., 2018). When it absorbs light energy, like UV or visible light, it becomes excited and moves to a higher energy state. As it returns to its original state, acridine releases the absorbed light, resulting in fluorescence. Moreover, environmental factors can influence acridine fluorescence, enhancing its adaptability for various fluorescence-based applications and imaging methods. Acridine orange (AO) is one of the widely studied dyes based on acridine core. The AO is a fluorescent dye widely employed in biological research and diagnostics and is a positively charged dye that selectively attaches to nucleic acids like DNA and RNA. Upon exposure to light, it emits orange-red fluorescence (Jahanmehr et al., 1987). This versatile dye finds applications in various fields, such as staining cells for microscopy, detecting nucleic acids in gel electrophoresis, and evaluating cell viability in experiments. Its specificity for nucleic acids enables researchers to visualize cellular components, explore DNA/RNA behaviors, and identify tissue abnormalities. Notably, AO exhibits dual fluorescence: it appears green when bound to single-stranded nucleic acids and orange-red when bound to double-stranded ones, offering insights into DNA/RNA structures (Byvaltsev et al., 2019).

2.2 SYNTHETIC METHODS OF ACRIDINE

Although multiple natural sources consisting of the heterocyclic nucleus have been already reported, still a considerable amount of attention is drawn by the researchers on the synthesis of acridine and its derivatives. In 1870, Carl Grabe and Heinrich Caro achieved the first isolation of acridine, a significant accomplishment in the realm of organic chemistry in Germany. The acridine was extracted from high boiling fraction of coal tar (Herole et al., 2022). The extraction process involved several steps. Initially, acridine was separated from coal tar by shaking it with dilute sulfuric acid. Then, it was precipitated from the resulting sulfuric acid solution using potassium dichromate. The formed acridine dichromate was subsequently decomposed using ammonia in the final step of the process. This pioneering work not only led to the isolation of acridine but also contributed to the understanding of its chemical properties, paving the way for further research and applications in organic chemistry (Kumar et al., 2012). The most common synthetic method employed for the synthesis of acridine and its derivatives is the Ullmann acridine synthesis method. Ullmann's discovery in 1902 marked the inception of a significant reaction method for synthesizing acridine derivatives. This process entails treating N-aryl anthranilic acid with an acylating agent, commonly polyphosphoric acid (PPA). Subsequent steps involve reduction and dehydration. Essentially, the acylating agent initiates the reaction by adding an acyl group to the N-aryl anthranilic acid molecule. Then, through reduction and dehydration, the desired acridine derivatives are formed. This method has become pivotal in organic synthesis, enabling the production of various substituted acridines with wide-ranging applications. Various acylating reagents, including $POCl_3$, H_2SO_4, $(CF_3CO)_2O$, HCl, TFA, or PPA, can be employed in this reaction. Notably, acridine-bearing substituents at position 3 cannot be synthesized from 2-chlorobenzoic acid and 3-aminobenzoic acid due to the formation of a mixture of 1- and 3-substituted products during cyclization, making separation challenging. Acridones, on the other hand, can be transformed into corresponding acridines via reduction with Na/BuOH or Al/Hg/NaOH followed by dehydration using $FeCl_3$/HCl (Wang, 2010). A typical cited example of Ullmann method is the conversion of 2-[(4-ethylphenyl)amino]isophthalic acid (**1**) to 7-ethyl-9-oxoacridan-4-carboxylic acid (**2**), followed by further transformations to obtain 7-ethylacridine-4-carboxylic acid (**3**). Initially, **1** reacted with polyphosphoric acid to form a yellow precipitate. The precipitate was dissolved in a mixture of methanol and sodium hydroxide, filtered, and acidified to yield **2**. This acid underwent a series of reactions, was treated with triethylamine in ethanol, then reacted with aluminum foil amalgamated with mercuric chloride in ethanol. The resulting product was further treated with iron(III) chloride under reflux, leading to the formation of **3** (Figure 2.1A) (Spicer et al., 1997).

Another widely used method for the synthesis of acridine is Bernthsen acridine synthesis, which was first reported by Heinrich August Bernthsen in 1884 (Bernthsen, 1884). The Bernthsen synthesis is a chemical process where diphenylamine (**4**) reacts with a carboxylic acid in the presence of zinc chloride heated at 200–270°C, leading to the production of 9-substituted acridine (**5**) (Figure 2.1B). This reaction starts with the formation of an intermediate called acridinium salt, which subsequently undergoes a ring-closing step catalyzed by zinc chloride to yield acridine. In essence, it's a method for making acridine by combining diphenylamine, a carboxylic acid, and catalytic amount of zinc chloride (Goni et al., 2022). A modification was made in the Bernthsen method by George B. for the green synthesis of acridine. In this approach, acridine derivatives were produced using an environmentally friendly approach, involving a condensation reaction between **4** and either aliphatic or aromatic carboxylic acids in a medium where the components are not uniformly distributed. This method yielded high amounts of product 9-ethyl acridine

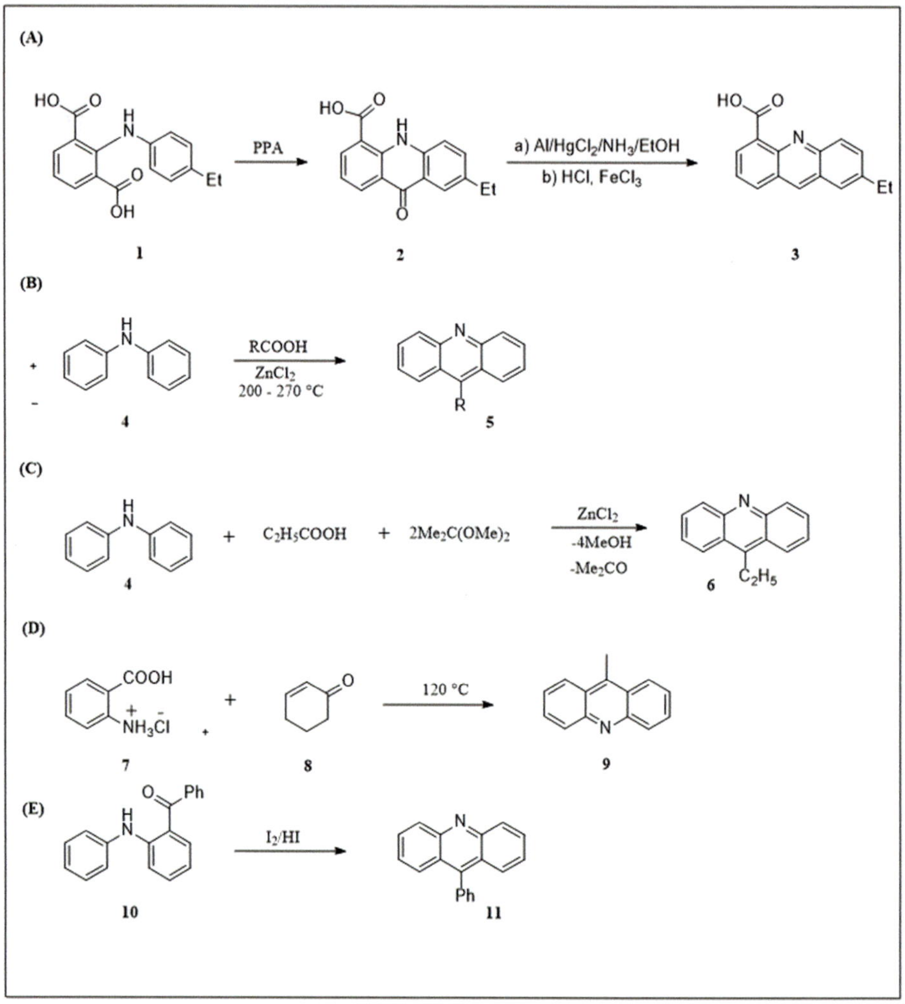

FIGURE 2.1 Synthetic methods of acridine including (A) Ullman method, (B) Bernthsen method, (C) modified Bernthsen method, (D) Friedlander and (E) synthesis from C-acylated diphenylamine.

(**6**) when a substance that promotes the reaction, like zinc chloride, and 2,2-dimethoxypropane, which helps remove water, were present. The resulting products could be easily separated from the reaction medium, leading to a reduction in energy consumption (Figure 2.1C) (Bratulescu, 2016). Another method for the synthesis of acridine is Friedlander synthesis. This synthesis entails the reaction between the anthranilic acid salt (**7**) and cyclohex-2-enone (**8**) at 120°C to yield 9-methylacridine (**9**) (Figure 2.1D). The reaction mechanism likely involves a condensation reaction, where a water molecule is eliminated, facilitating the bond formation between cyclohex-2-enone and the anthranilic acid derivative. As a result, 9-methylacridine, a significant compound with various applications, is obtained. This synthesis provides a practical route for synthesizing 9-methylacridine from easily accessible starting materials under controlled conditions. Acridine can also be synthesized from C-acylated diphenyl amines (**10**) by heating the C-acylated diphenyl amines in the presence of iodine (I_2) and hydrogen iodide (HI) to produce 9-phenylacridine (Figure 2.1E) (Kumar et al., 2012).

2.3 ACRIDINE-BASED FLUORESCENT PROBES FOR BIOLOGICAL APPLICATIONS

Fluorescent molecules have diverse applications in biological sciences due to their ability to emit light upon excitation. They are commonly employed for visualizing cellular structures, organelles, and biomolecules, facilitating studies on cellular processes, protein localization, and dynamic cellular behaviors. Flow cytometry relies on fluorescent dyes to sort and analyze cells based on their fluorescent characteristics, playing a crucial role in identifying cell populations and studying cell surface markers (Adan et al., 2017). In vivo imaging techniques like fluorescence microscopy and fluorescence molecular tomography (FMT) (Stuker et al., 2011) utilize fluorescent probes to noninvasively observe biological processes, enabling the visualization of disease progression, therapy monitoring, and physiological studies (Gao et al., 2017). Additionally, fluorescent tags are utilized for labeling and tracking proteins within cells and tissues, aiding research on protein dynamics, turnover, and localization. These fluorescent molecules can also be utilized for disease diagnosis employing the biomarkers associated with particular disease (Drummen, 2012).

2.3.1 Acridine-Based Fluorescent Probes for Disease Diagnosis

Fluorescent probes play a vital role in disease diagnosis by selectively binding to and illuminating abnormal tissues or biomarkers. They facilitate noninvasive imaging methods like fluorescence microscopy and molecular imaging, enabling the early detection, precise localization, and monitoring of conditions such as cancer, cardiovascular issues, and infectious diseases. This focused visualization enhances diagnostic precision, assists in treatment planning, and evaluates treatment effectiveness. Ravichandiran et al. synthesized a fluorescent probe for the selective detection of cancer cells. A new low-cost, easy-to-prepare dual-channel fluorescence chemosensor, composed of acridine-diphenylacetyl moieties (**12**), was developed to enable remarkable detection of Sn^{4+} in water and biological medium (Figure 2.3). The resulting **12**–Sn^{4+} complex was utilized for the distinguished identification of $Cr_2O_7^{2-}$ ions from other anions and biomolecules. These investigations involved absorption, fluorescence, and electrochemical methods for detecting Sn^{4+} and $Cr_2O_7^{2-}$ ions in pure water. The mechanism for probe-mediated Sn^{4+} detection was experimentally determined by FT-IR, NMR titrations, mass (ESI) analyses, and density functional theory (DFT) calculations. The obtained results indicated that the chemosensor **12** possessed excellent performance characteristics, including good water solubility and compatibility, quick response time (less than 10 seconds), high sensitivity (Sn^{4+} = 0.268 μM and $Cr_2O_7^{2-}$ = 0.160 μM), and selectivity against coexisting metals, anions, amino acids, and peptides. The chemosensor **12** induced negligible toxicity in live cells and was successfully utilized as a biomarker for tracking Sn^{4+} in human normal and cancer cells. More importantly, **12** demonstrated distinguished recognition of Sn^{4+} in human cancer cells rather than in normal live cells. The cytotoxicity of the probe was assessed by the cell viability assay. Following subsequent treatment with 0.781 μM Sn^{4+} or **12** for 20 minutes, both normal (HaCaT, L929) and cancer (FaDu, SNU-1041) cells exhibited minimal fluorescence under green and red filters individually. However, when treated with a combination of 0.781 μM Sn^{4+} and **12** for the same duration, significant and intense fluorescence emissions were observed under both filters. Notably, FaDu and SNU-1041 cells displayed strong fluorescence, whereas HaCaT and L929 cells showed negligible emission. Furthermore, FaDu cells exhibited a 1.49-fold increase in green emission compared to SNU-1041 cells, while SNU-1041 cells demonstrated a 1.30-fold higher red emission than FaDu cells. Most of the fluorescence originated from the cell cytoplasm, with some emission also detected from the nucleus. Notably, the cancer cells displayed higher fluorescence emission due to **12**-Sn^{4+} complexing compared to the normal cells (Ravichandiran et al., 2021).

2.3.2 Acridine-Based Fluorescent Probes for Cell Imaging

Cellular imaging involves using specialized techniques to observe and analyze cellular structures and activities at a very small scale. Scientists achieve this by tagging specific components within cells, like proteins, DNA, or organelles, with fluorescent molecules. When these tagged components are exposed to a particular type of light, they emit light of a specific color. By using a microscope to detect this emitted light, researchers can visualize the labeled cellular components. This allows them to track their distribution, movements, and interactions over time, either by observing them in real time or by capturing still images. Essentially, cellular imaging with fluorescent molecules provides a powerful tool for studying the inner workings of cells with high precision and detail. Cellular damage and death are common aspects of various diseases, but existing methods to detect them are restricted and often unable to identify subtle pathological alterations. Fluorescent probes can be utilized for the visualization of apoptotic cells. Plemel et al. reported the use of AO as a fluorescent probe to study and distinguish the involvement of apoptosis and necrosis in cell death. In this study the author used AO as a nucleic acid dye known for its distinct spectral properties. AO facilitated the real-time assessment of RNA and DNA levels as indicators of cell viability during exposure to different harmful stimuli. This method illuminated spectral patterns specific to different types of cell death, distinguishing between apoptosis and necrosis/necroptosis. The innovative approach demonstrated that cellular RNA decreases in cases of necrotic, necroptotic, and apoptotic cell deaths resulting from demyelinating, ischemic, and traumatic injuries, suggesting its involvement in a broad spectrum of tissue pathologies. The study involved the use of several cell death stimuli, such as H_2O_2 triggers cell death through a mechanism that does not involve caspase-3, resulting in a combination of apoptosis and necrosis, and staurosporine, a potent inhibitor of a wide range of protein kinases, effectively triggers apoptosis in oligodendrocytes, etc. In the study, MO3 cells underwent pulsing with either 1 mM H_2O_2 or 50 nM staurosporine, chosen to represent distinct modes of cell death, under uniform conditions. Following exposure to the toxins, cells were fixed at various time intervals and stained with AO, with RNA extraction carried out simultaneously in parallel cultures. Crucially, there was a significant correlation between RNA concentration and the intensity of red AO emission. The result suggested that an apoptotic cell death induced by staurosporine led to a decrease in RNA levels without any observed decline in RNA quality or cellular energy levels, indicating active RNA degradation during apoptosis. The rise in necrotic cell death induced by H_2O_2 exposure led to an initial metabolic decrease, followed by a reduction in RNA integrity and concentration after 2 hours. This sequence of events suggests that metabolic deterioration could be a contributing factor to RNA loss during necrosis. The methodology underscored AO-labeled cytoplasmic RNA as a crucial early indicator of cellular injury and a highly responsive indicator of various modes of cell death across diverse experimental models (Plemel et al., 2017).

The surrounding conditions, including viscosity, polarity, pH, and oxygen levels, are crucial for metabolism, and deviations in viscosity and polarity are strongly linked to the onset and progression of conditions like diabetes, neurodegenerative disorders, and inflammation. Peng et al. synthesized a series of four fluorescent probes by reaction of 9-acridine carboxaldehyde with cyano compounds to afford viscosity and polarity-sensitive probes. Among the synthesized probes, probes **13a** and **13b** were found to be polarity sensitive while probes **13c** and **13d** were observed as viscosity-sensitive probes. As solvent polarity increased, the maximum emission wavelength of probe **13a**, utilizing acridine-dicyanoisophorone, transitioned from 553 nm to 594 nm. Consequently, the fluorescence quantum yield surged from 0.5% to 35.6%, and the fluorescence intensity experienced a 38-fold increase. Likewise, the polarity response of probe **13b**, based on acridine-cyanofuranone, closely resembled that of **13a**. However, with the escalation of solution viscosity from 0.89 cP (100% water) to 856 cP (1% water), the fluorescence intensity of probe **13c**, employing acridine-tricyanodihydrofuran, increased by 5.6 times at 430 nm. The viscosity response of probe **13d**, utilizing acridine-cyanobenzothiazole, was akin to that of **13c**. Additionally, probes **13a–b** were chosen for HeLa cell imaging experiments due to their outstanding photostability. The results indicated that probe **13a** could specifically target lipid droplets, while probes **13b–c** effectively stained lysosomes. Furthermore, probes **13a–b** facilitated the real-time monitoring of intracellular polarity changes (Peng et al., 2022). Enbanathan et al. developed an acridine-based fluorescent probe (**14**) for selective detection of BF_3. Followed by the interaction of probe **14** with BF_3, the fluorescence intensity rose significantly, increasing by a factor of 10, and the color of the probe changed rapidly, within just one second. The probe **14** demonstrated a low detection limit of 7.39 nM and exhibited high sensitivity toward BF_3. To elucidate the sensing mechanism, mass spectrometry, DFT studies, and NMR titration were conducted. The sensor displayed consistent and reproducible performance. To elucidate the practical use of the probe, test strips were formed and tested and it was observed that test strips quickly and effectively detect trace amounts of BF_3 in the air. Furthermore, its capacity to sense dynamic changes of BF_3 in live cells was demonstrated, and the experiments using fluorescent bioimaging confirmed that the probe could detect BF_3 within active HeLa cells (Enbanathan et al., 2024). In 2023, tetrahydro-acridine salt hybrid coumarin dyes were designed and developed as a near-infrared (NIR) fluorescent probe (**15**) by An et al. for the detection and fluorescent bioimaging of two important bio-thiols, cysteine (Cys) and homocysteine (Hcy). The probe **15** containing activated chlorine atom enabled a more efficient nucleophilic substitution reaction, leading to a rapid response time (3 minutes toward three equivalents of Cys) and high sensitivity to Cys, with a detection limit of 46 nM. Notably, the probe exhibited "turn-on" NIR fluorescence emission (Em_{max}: 674 nm) with a large Stokes shift (approximately 200 nm) in response to Cys/Hcy. Additionally, the probe served as an effective tool for imaging and detecting Cys/Hcy within the mitochondria of living cells and *in vivo*. This work introduced a novel design strategy for NIR fluorescent probe platforms for the detection and fluorescence imaging of bio-thiols (An et al., 2023). An et al. developed a novel NIR fluorescent probe (**16**), based on a xanthene hybrid tetrahydro-acridine salt dye, specifically designed to track Cys. In this design, a chlorine-substituted tetrahydro-acridine served as a highly Cys-reactive site and provided water solubility. Probe **16** demonstrated a significant "turn-on" NIR emission at 830 nm with an exceptionally large Stokes shift of 305 nm, enabling efficient monitoring of Cys. The probe **16** also showed high selectivity, a fast response time of 6 minutes, and high sensitivity, with a limit of detection (LOD) of 0.5 μM. The sensing mechanism of probe toward Cys was thoroughly characterized and analyzed using HPLC, MS spectra, and quantum theory calculations. The strong NIR fluorescent detection capabilities of this novel probe enabled successful monitoring of both exogenous and endogenous fluctuations in Cys concentrations in living cells and in vivo. The dynamics of Cys in the cell lines was easily monitored and visualized using the bioimaging of the cells treated with probe **16** (An et al., 2024). Jiang et al. synthesized an acridine-based probe capable of detection of Cr^{3+} and PO_4^{3-}. The potential of the probe (**17**) to detect Cr^{3+} and PO_4^{3-} was demonstrated in living MGC-803 cells (Figure 2.2). The cells were unable to display any fluorescence when initially incubated with 2.5 μM of **17**. Later the same cells were incubated with 50 μM of Cr^{3+} and a bright fluorescence was observed while the fluorescence was again quenched when the same cells were again incubated with 50 μM of PO_4^{3-}. The probe **17** and the **17**-Cr^{3+} complex exhibited excellent cell permeability and were effectively utilized in living MGC-803 cells to detect Cr^{3+} and PO_4^{3-}, respectively (Jiang et al., 2014). The structures of probes for biological applications are depicted in Figure 2.3.

2.3.3 Acridine-Based Fluorescent Probes for Nucleic Acid Binding

Nucleic acid-binding probes are molecules designed to specifically interact with DNA or RNA. They often contain fluorescent or other detectable labels that enable researchers to visualize and analyze nucleic acids in biological samples. These probes are essential tools in molecular biology, diagnostics, and drug development, providing insights into gene expression, genetic variations, and disease mechanisms. In 1961, Mayor and Hill introduced a method using the fluorescent dye AO for rapid identification of viral nucleic acids. This technique relies on color differences to distinguish between double-stranded DNA (which fluorescences

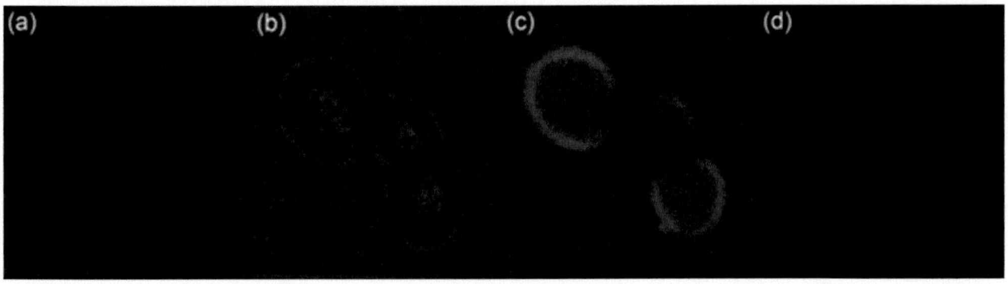

FIGURE 2.2 Images of MGC-803 cells in $CH_3OH–H_2O$ (8/2, v/v): (a) fluorescence image of MGC-803 cells incubated with probe 17 (2.5 μM) for 30 min; (b) bright field image of cells incubated with probe 17 (2.5 μM) and Cr^{3+} 50 μM; (c) fluorescence image of cells incubated with probe 17 (2.5 μM) and Cr^{3+} 50 μM; (d) fluorescence image of cells pretreated with probe 17 (2.5 μM) and Cr^{3+} 50 μM, further incubated with 5o μM PO_4^{3-}. Image reproduced from Jiang et. al., 2014 with permission Elsevier copyright 2014.

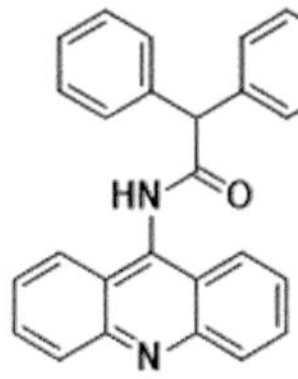
12

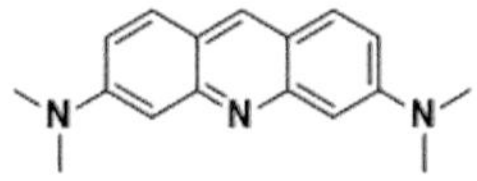
AO

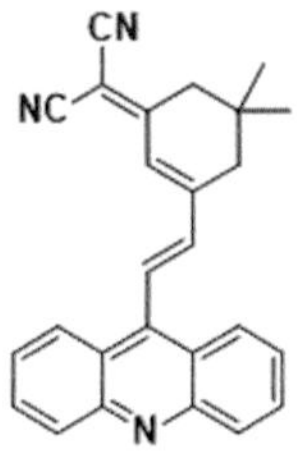
13a

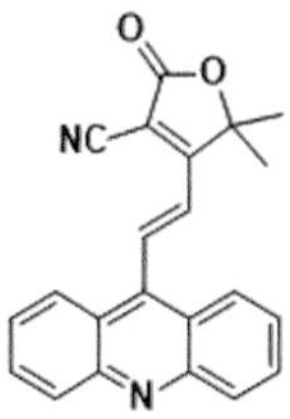
13b

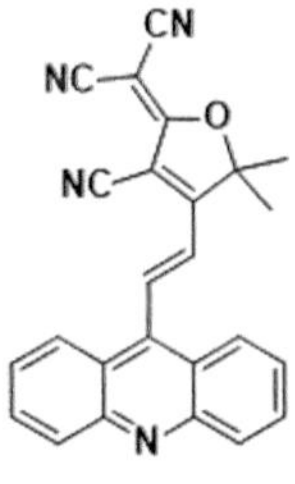
13c

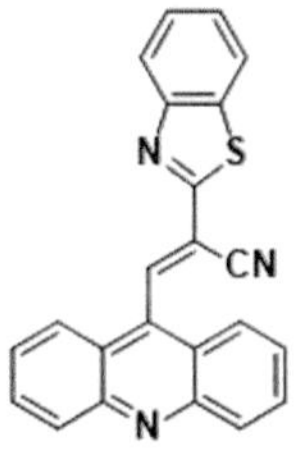
13d

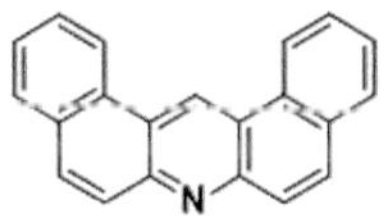
14

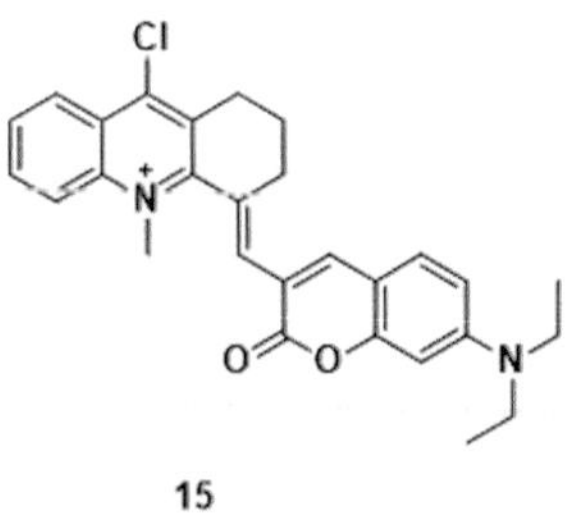
15

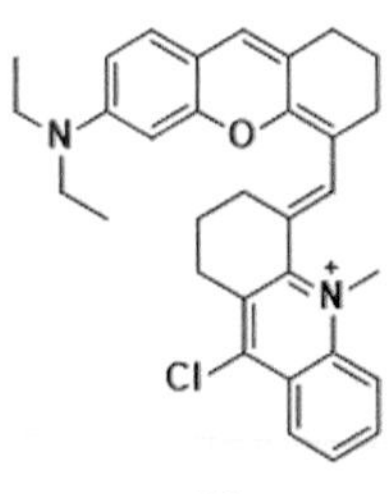
16

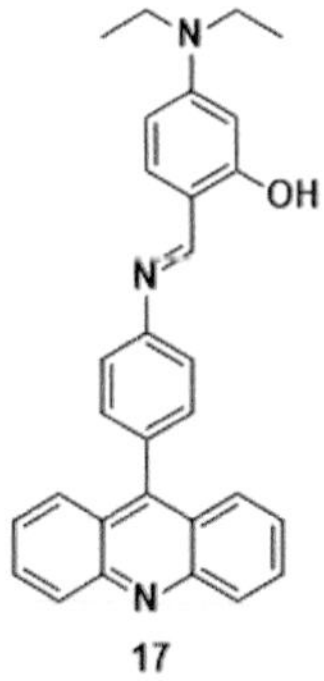
17

FIGURE 2.3 Structure of acridine-based fluorescent probes for biological applications.

green) and RNA/single-stranded DNA (which fluorescences red) (Mayor and Hill, 1961). Later D. E. Bradley in 1965 published a method of fluorescence staining bacteriophage nucleic acids using AO. In this communication, improvements to address some concerns, including substituting fluorescence microscopy with direct viewing under ultraviolet light, simplifying pre-staining treatment, and mitigating pH sensitivity issues, were addressed. Viruses containing double-stranded DNA are easily identified by their color change to a distinct shade after phosphate treatment. Both single-stranded DNA and RNA viruses exhibit a fiery red color following phosphate treatment, with RNA viruses appearing almost crimson and devoid of any green tint. Citric acid treatment aids in distinguishing between these two types of nucleic acids. The key distinction lies in the behavior over 3 minutes: RNA viruses maintain their color without fading, while single-stranded DNA viruses may fade noticeably and possibly shift toward a green color (Bradley, 1965). Lauretti et al. reported the use of AO in staining the RNA of rotavirus in polyacrylamide gel. In the study, a protocol was established to utilize AO for detecting rotavirus double-stranded RNA (ds

RNA) in polyacrylamide gels, comparing it with silver and ethidium bromide staining methods. It was found that the most effective condition involved staining the gels containing rotavirus ds RNA bands in green with 4.3 μM AO after electrophoresis, followed by destaining with distilled water at pH 6 and 37°C. This protocol enabled the calculation of rotavirus RNA concentration, with the minimum amounts detected by AO, ethidium bromide, and silver staining being 26.0 ± 4.29 ng, 15.6 ± 1.48 ng, and 1.06 ± 0.11 ng, respectively. Additionally, it was demonstrated that rotavirus ds RNA could be utilized for reverse transcription activity and subsequent PCR amplification after AO staining. In conclusion, despite being less sensitive than ethidium bromide and silver staining, the practicality, affordability, metachromatic properties, and compatibility with RT-PCR make AO a suitable stain for various virological and nucleic acid molecular biology applications (Lauretti et al., 2003). The structures of probes for biological applications are depicted in Figure 2.3.

2.4 ACRIDINE-BASED FLUORESCENT PROBES FOR CATION SENSING

Metals are ubiquitous in both organic and inorganic entities. Cations such as iron, nickel, and zinc are crucial for numerous biological functions and enzymatic processes within the human body. However, the presence of certain cations like cadmium, lead, and mercury in the environment poses health hazards. Detection of cations, essential for numerous biological and industrial processes, is critical for assessing environmental and health risks. While traditional spectrophotometric and electroanalytical methods are effective in detecting trace amounts of cations in relevant samples, they are time-consuming, require expensive setups, and can be challenging to grasp. Cations sensor that utilizes a fluorescence-based method to detect a trace amount of cations in the relevant sample displays an advantage over the aforementioned techniques (Chowdhury et al., 2018). Two fluorescent sensors based on acridine (**18a** and **18b**) have been developed by Wang et al. for the detection of Fe^{3+} and Ni^{2+} ions. Both **18a** and **18b** show a significant reduction in fluorescence intensity when exposed to Fe^{3+} and Ni^{2+}, with detection limits of 4.13 μM and 1.52 μM, respectively, in a mixture of DMSO and H_2O (1:1, v/v). These sensors are effective at neutral pH and demonstrate excellent selectivity for their respective ions (Wang et al., 2018). Carlos et al. reported an acridine-based chemosensor, 4,5-bis(phthalimidemethyl)acridine (**19**). The chemosensor was investigated as a chemosensor for metal ions in alcoholic solutions, demonstrating selectivity for Al^{3+}. It exhibited a linear increase in fluorescence intensity by 230% within the concentration range of 10–70 μmol L^{-1}. Analysis using Benesi-Hildebrand and Job's plot suggested the formation of a 1:1 complex between Al^{3+} and **19**, with a binding constant of 6.30 x 10^3 L mol^{-1}. Computational calculations using DFT/TDDFT (time-dependent density functional theory) provided insights into the energies of frontier orbitals, explaining the fluorescence enhancement upon complex formation due to the inhibition of PET processes. The limits of detection and quantification (validated with $R^2 = 0.998$ using the least squares method) were determined to be 1.130 and 3.768 μmol L^{-1}, respectively. Validation was conducted by varying analytical conditions, and practical application in spiked Brazilian sugarcane spirit demonstrated a recovery rate of 84 ± 0.42% with no significant interference from other ions (dos Santos Carlos et al., 2020). Wang et al. synthesized a novel acridine-based fluoroionophore, 4,5-bis(N,N-di(2-hydroxyethyl)iminomethyl)acridine (**20**), following the fluorophore-spacer-receptor design. Its ability to sense metal ions was investigated in buffered aqueous solutions. In the presence of Cd^{2+}, a 1:1 complex between the ligand and the metal ion forms at neutral pH, resulting in enhanced emission at 454 nm. The fluorescence intensity shows a linear relationship with Cd^{2+} concentration within the range of 1.0×10^{-6} to 3.0×10^{-5} M (with a correlation coefficient of 0.9967). Importantly, **20** exhibited minimal interference from other metal ions, indicating its selective response to Cd^{2+} ions. With a detection limit of 1.3×10^{-7} M, **20** proves to be effective for detecting Cd^{2+} in aqueous solutions. Furthermore, experimental findings suggest that **20** acts as a reversible chemosensor for Cd^{2+}, enhancing its potential for sensing applications (Wang et al., 2011). The probe **17** (Figure 2.3) developed by Jiang et al. showed a notable ability to discern and react to Cr^{3+} ions specifically when dissolved in methanol, indicating a strong selectivity and sensitivity. The titration study revealed that the fluorescence intensity of the probe increases with the increasing concentration of Cr^{3+} (Figure 2.4). Furthermore, when the probe-Cr^{3+} complex was formed in solution, it proved to have a remarkable preference for binding to phosphate ions (PO_4^{3-}) over other ions present, utilizing a displacement method. In addition to its effectiveness in methanol, probe **17** demonstrated utility in neutral methanol aqueous solutions, opening up possibilities for its application in various biological and environmental contexts. In essence, this novel fluorescent probe **17**, derived from acridine, showcases impressive selectivity, sensitivity, and versatility in solvent (Jiang et al., 2014).

Dai et al. designed and synthesized a probe N,N'-(acridine-3,6-diyl)dipicolinamide (**21**), which was then evaluated as a potential chemosensor for detecting copper ions (Cu^{2+}). This compound proved to be both efficient and selective in its detection capabilities. Specifically, when tested in an HEPES buffer aqueous solution, probe fluorescence emission decreased significantly in the presence of Cu^{2+} ions, while interference from other metal cations remained minimal. The detection limit for Cu^{2+} ions was determined to be 1.2 x 10^{-7} M. To understand how **21** interacts with Cu^{2+} ions, researchers employed various analytical techniques including ^{1}H NMR, HR-MS analysis, and density functional theory (DFT) calculations. These methods allowed them to probe the structural and electronic properties of **21** in the presence of Cu^{2+}, providing insights

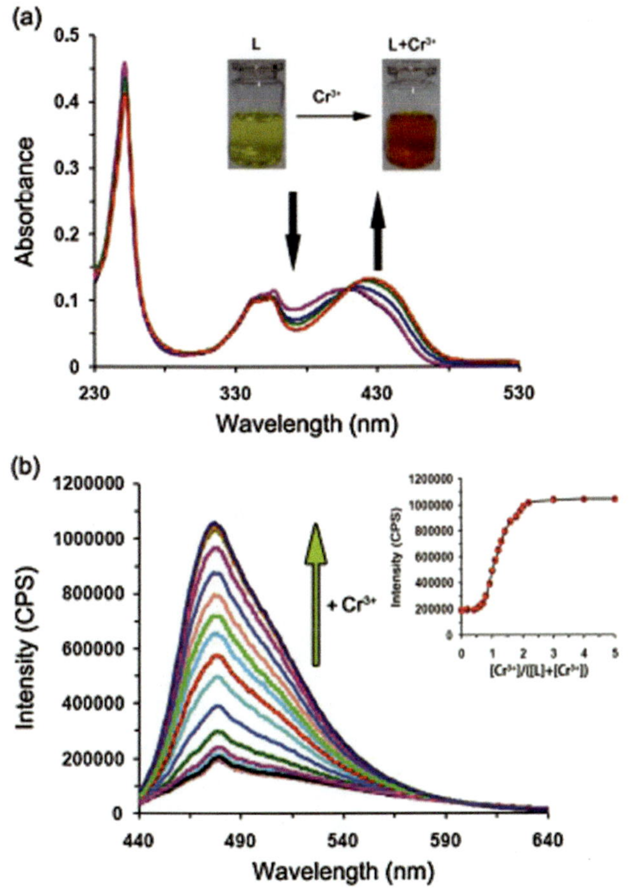

FIGURE 2.4 (a) UV–vis spectra and (b) fluorescence spectra of 17 (2.5 μM) on titration with increasing amounts of Cr^{3+} in methanol. Inset; (a) demonstrates the visual appearance of probe 17 before and after addition of Cr^{3+} and inset; (b) indicates a plot of concentration of Cr^{3+} vs fluorescence intensity. Image reproduced from Jiang et al., 2014 with permission Elsevier copyright 2014.

into the mechanism underlying its selective response to these ions (Dai et al., 2018). Furthermore, Zhou et al. reported the synthesis and development of a newly discovered compound, 4,5-bis(hydroxymethyl) acridine (**22**), to serve as a fluorescent probe for detecting Pd^{2+} ions. This compound exhibits a simple structure yet demonstrates highly selective recognition toward Pd^{2+} ions when compared to other metal ions in aqueous solutions. When tested under optimized conditions, the fluorescence intensity of sensor **22** exhibited a linear relationship with the concentration of Pd^{2+} ions in the range of 0–1 μM, with a detection limit as low as 0.021 μM. Further investigations involving the addition of EDTA and stoichiometry experiments revealed that sensor **22** functions as a reversible chemosensor for Pd^{2+} ions at neutral pH, forming a 2:1 ligand/metal complex. Moreover, sensor **22** was successfully employed in detecting Pd^{2+} ions in water samples as well as in palladium-containing catalysts. This practical application demonstrates its attractiveness for sensing purposes in various contexts (Zhou et al., 2015). Nunes et al. developed an acridine-based fluorescent probe (**23**) for the selective detection of Hg^{2+} ions in aqueous solutions. Computational methods, specifically DFT and TD-DFT calculations, revealed that fluorescence quenching occurred due to the formation of a coordination bond between Hg^{2+} ions and the thiolate groups present in the sensor molecule. This bond formation led to the creation of $[Hg\ probe\text{-}Cl_2]^{2-}$ as the most stable species. The sensitivity of this chemosensor was remarkable, with a limit of detection (LOD) of 4.40 μmol L^{-1} and a limit of quantification (LOQ) of 14.7 μmol L^{-1}, determined using the least squares method with an R^2 value of 0.9892. The linear range for concentration measurements extended from 14.7 to 100 μmol L^{-1}. Analysis using Benesi-Hildebrand and Job's plot supported the formation of a stable complex with a 1:1 ratio between the metal ion and the sensor, with a calculated binding constant of 5.14×10^3 L mol^{-1}. The robustness of the method was confirmed through variations in analytical conditions, with a maximum relative standard deviation of 4.6%. Furthermore, recovery experiments conducted with distilled water yielded results of 90.3 ± 4.6%, indicating minimal interference from other ions present in the solution. Overall, the analytical performance metrics demonstrate that this sensor offers a cost-effective and reliable alternative for the detection of Hg^{2+} ions in aqueous solutions, making it an attractive option for practical applications (Nunes et al., 2020). The structures of probes for cation sensing are depicted in Figure 2.5.

2.5 ACRIDINE-BASED FLUORESCENT PROBES FOR ANION SENSING

Anions are highly influential across chemistry, biology, medicine, and the environment. They participate in chemical reactions, vital biological processes, and medical functions like nerve signaling and pH regulation. Additionally, in environmental contexts, they impact pollutant behavior and soil health. Understanding anions is crucial for advancing these fields and addressing related challenges effectively (Pal et al., 2021). Hypochlorite (ClO^-) is a reactive oxygen species (ROS) generated through the oxidation of chloride ions (Cl^-) and hydrogen peroxide (H_2O_2), a reaction catalyzed by the heme protein myeloperoxidase. Lee et al. synthesized a new fluorescent chemosensor called **24**, comprised of acridine, for detecting ClO^-. Probe **24** was synthesized using a novel method involving the introduction of an aldehyde group using formic hydrazide. When ClO^- was present, **24** exhibited significant fluorescence quenching, demonstrating high selectivity compared to other substances. The sensor's detection limit was determined to be 7.65 μM, highlighting its sensitivity in detecting ClO^- ions. Probe **24** demonstrated turn-off fluorescence after interaction with ClO^- ions that can be visualized under UV lamp. The mechanism of sensing of ClO^- ions by probe **24** was proved by NMR. The sensing of ClO^- ions was not affected by the presence of other ions in the interference study (Lee et al., 2020). Further Isaac et al. synthesized three (**25a**, **25b**, **25c**) acridine-based fluorescent probes as

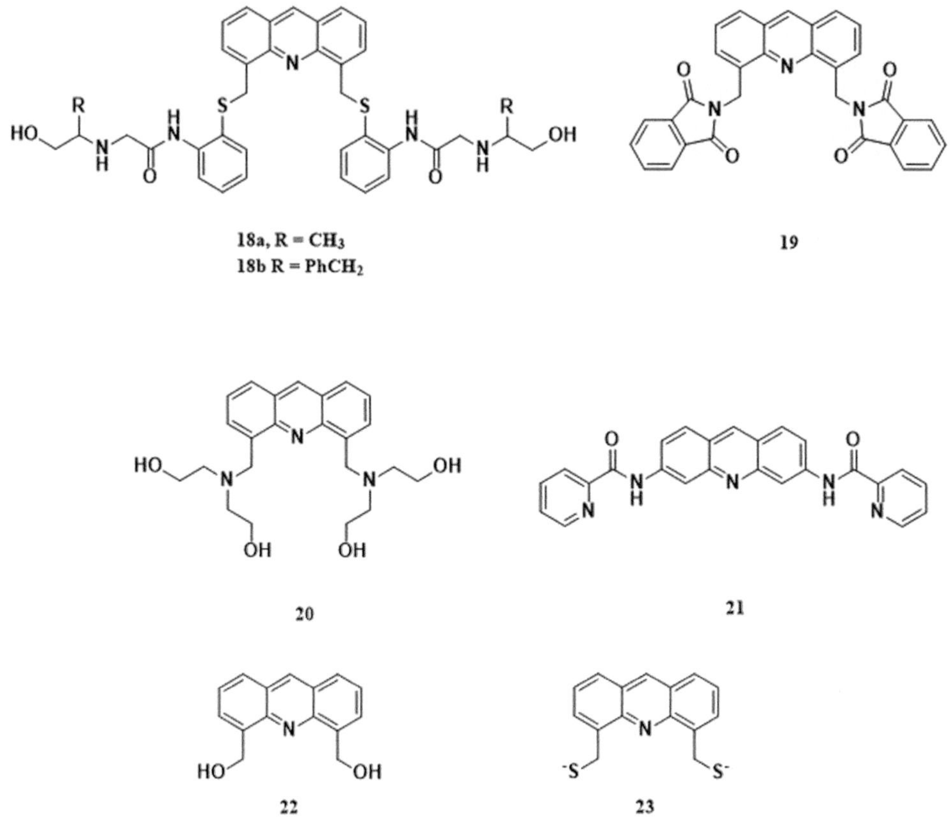

FIGURE 2.5 Structure of acridine-based fluorescent probes for cation sensing.

highly selective fluoride (F^-) ion sensors with no effect for sensing other competing anions. Additionally, when a fluoride anion (F^-) source (tetrabutylammonium fluoride) was gradually added, probes **25a–c** displayed a distinct color change in the solution. The sensing mechanism involving anions was thoroughly examined using UV–visible absorption, fluorescence, and ^{1}H-NMR experiments. Notably, nickel fluoride (NiF_2), an uncommon transition metal fluoride salt, was utilized as a catalyst during the synthesis of acridine probes **25a–c**. Theoretical investigations employing density functional theory were also conducted to further understand the sensing mechanism and the selectivity of these probes toward fluoride anions (F^-). The LOD for probe **25a** was found to be 6.879×10^{-5} M for the detection of F^- (Isaac et al., 2018).

The $H_2PO_4^-$ anions, among a range of anions, hold crucial significance in signal transmission, energy storage within metabolism, and forming the structural framework of DNA and RNA. Zang et al. synthesized a fluorescent sensor (**26**) using an acridine fluorophore. The sensor ability to bind to anions was investigated through various analytical techniques such as fluorescence, UV–visible spectroscopy, proton nuclear magnetic resonance (^{1}H NMR), and high-resolution mass spectrometry (HRMS). Notably, sensor **26** exhibited dual-responsive selectivity toward $H_2PO_4^-$ and HSO_4^- through distinctive fluorescence responses: bathochromic shift for $H_2PO_4^-$ and fluorescence quenching for HSO_4^- that can easily differentiate both the anions. The exceptional selectivity of sensor **26** toward $H_2PO_4^-$ and HSO_4^- was attributed to the combined binding effect of benzimidazolium and urea functional groups. Additionally, the formation of an additional hydrogen bond between the hydroxyl group in $H_2PO_4^-$/HSO_4^- and the nitrogen atom on the acridine moiety of sensor **26** contributed significantly to enhancing its affinity for these anions. The fluorescence response of **26** for both the anions was not interfered in the presence of other anions (Zhang et al., 2013). Two newly synthesized pseudopeptidic molecules comprising one macrocyclic (**27b**) and one open chain structure (**27a**), both incorporating an acridine unit, were developed by Centelles and his group. Their fluorescence behaviors in response to a range of acids were assessed in $CHCl_3$. These receptors demonstrated a preference for $H_2PO_4^-$ over HSO_4^-, exhibiting even higher selectivity when compared to other anions like Cl^-, Br^-, CH_3COO^-, and CF_3COO^-. The macrocyclic receptor exhibited superior selectivity for $H_2PO_4^-$ compared to its open-chain counterpart. Through in silico modeling and various experimental methods, the supramolecular interactions between the triprotonated receptors and different anions were investigated. Computational simulations provided optimized geometries consistent with experimental findings, especially in fluorescence experiments, suggesting that the selective supramolecular interactions involved coordination of the anions with the triprotonated form of the receptor (Marti-Centelles et al., 2012). A series of novel acridine-derived bis-benzimidazolium macrocyclic fluorescent sensors were developed and synthesized. The X-ray crystal structures illustrated the self-assembly

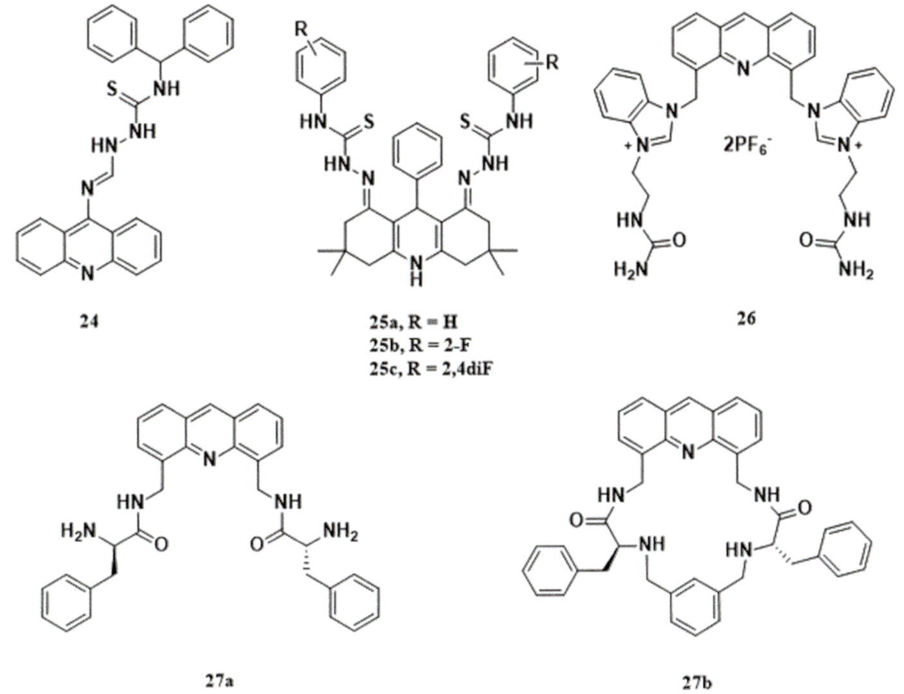

FIGURE 2.6 Structure of acridine-based anion sensors.

behavior of these cyclophanes in the solid state driven by hydrogen bonding and π-π interactions. Anion binding studies of these sensors indicated a significant impact of the macrocyclic size and rigidity on $H_2PO_4^-$ sensing, leading to an evident turn-on effect and bathochromic shift in fluorescence emission. Varied cavity sizes or rigidity of the sensors resulted in different degrees of bathochromic shift (ranging from 36 to 126 nm) in fluorescence emission induced by $H_2PO_4^-$, resulting in notable color changes from blue to orange-red, orange, green, and blue-green, respectively. The unique fluorescence response to $H_2PO_4^-$ could be attributed to the $H_2PO_4^-$-induced assembly of sensors, forming excimers between two acridine rings to varying extents (Zhang et al., 2013). The structures of probes for anion sensing are depicted in Figure 2.6.

2.6 ACRIDINE-BASED FLUORESCENT PROBES FOR MISCELLANEOUS APPLICATIONS

Accurate determination of intracellular pH is essential for investigating various physiological and pathological phenomena, including cellular apoptosis, cancer development, and cell growth regulation. Luminescent sensors are preferred for conducting precise quantitative pH measurements. These sensors offer a reliable means to assess pH levels within cells, enabling researchers to gain valuable insights into cellular processes with high precision and accuracy. Nagy et al. synthesized and developed a pH-sensitive acridine-based fluorescent probe, amino-isocyanoacridine (**28**), that emerged as a unique and versatile dye similar to AO with several advantageous properties. It exhibited an extended solvatochromic emission range in water and reduced basicity. Additionally, its optical property could be finely adjusted through complexation with Ag(I) ions. Utilizing both vibronic absorption and charge transfer bands, **28** demonstrated precise pH sensing capabilities with minimal error (2–3%) within the physiological pH range of 6–8, employing UV-vis and fluorescence detection methods. This dye was also capable of detecting pH changes in various microenvironments, such as the Stern layer, as evidenced by experiments with sodium lauryl sulfate micelles. The optical characteristics of **28** were elucidated through advanced quantum chemical calculations (Nagy et al., 2019).

Amino acids, which are the essential components of proteins, have vital functions in the signaling activities of living organisms. Therefore, improvements in the analysis of amino acids have notable implications for evaluating nutritional status. Different techniques have been utilized to analyze amino acids, such as spectroscopic assays and electrochemical methods. Nonetheless, fluorescence-based detection methods have become more prominent due to their enhanced sensitivity, convenience, and ability to provide visual results. As a result, considerable endeavors have been directed toward creating artificial fluorescence sensors that can identify amino acids. Dai et al. developed two acridine-based fluorescent probes that are capable of sensing amino acid anion after complexation with Cu^{2+} cations. The fluorescence of **29a** or **29b** was observed to be quenched upon the addition of Cu^{2+} cations. Confirmation of the ligand's bonding to the Cu^{2+} cation was achieved through UV-vis and fluorescence spectral studies. Energy-minimized structures of **29a** and its complex were simulated. The fluorescence quenching induced by chelation via a ligand-to-metal process was observed. These characteristics, along with the high selectivity displayed by **29a** and

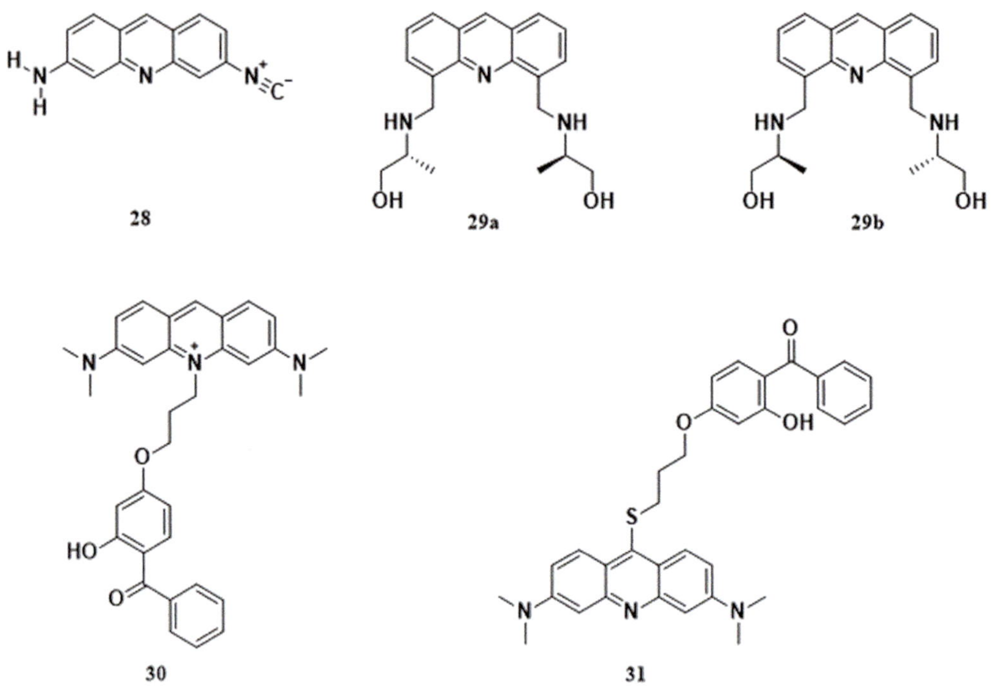

FIGURE 2.7 Structure of acridine-based fluorescent probes for miscellaneous applications.

29b, positioned them as promising sensor candidates for Cu^{2+} cations at physiological pH levels. Furthermore, the resulting **29a**–Cu^{2+} and **29b**–Cu^{2+} complexes were identified as excellent metal complex sensors for the potential detection of amino acid anions in aqueous solutions. The copper complex along with the molecules was capable to selectively sense the amino acid anion without the interference of any other cation that remains nonfluorescent after interaction with the complex (Dai et al., 2016).

Two novel AO-based dyes (**30** and **31**) were synthesized by Prabhakaran et al., by condensation of the commercial dye AO, to study their photophysical characteristics compared to AO. The dyes underwent characterization using techniques such as ^1H NMR, ^{13}C NMR, and HRMS, along with investigation of their photophysical properties, including solvatochromic effects in various solvents and pH effects ranging from 2 to 10. Exceptional results were observed in photophysical property studies conducted in alcoholic and acidic mediums. Absorption spectra were recorded for dyes AO, **30**, and **31**, indicating peak wavelengths around λ_{max} 488, 495, and 520 nm, respectively. Corresponding emission spectra showed peak wavelengths at λ_{em} 520, 520, and 560 nm, respectively. Molar extinction coefficients and Stokes shifts were calculated accordingly. Irradiation and lightfastness stability studies were performed and compared with a commercial dye, revealing that the synthesized dyes **30** and **31** exhibited superior stability and durability when exposed to UV-light irradiation compared to commercial dye. These findings were further supported by DFT computation calculations performed using the Gaussian 09 program. In this study, it was observed that while AO displayed a yellow color on modacrylic fabric, the synthesized dyes successfully achieved an orange color on the same fabric (Prabhakaran et al., 2021). The structures of probes for miscellaneous applications are depicted in Figure 2.7.

2.7 CONCLUSION

This chapter provides an overview of recent advancements in utilizing acridine for a wide range of fluorescence applications, spanning from biological to sensing applications. A diverse array of fluorescent probes incorporating the acridine core has been engineered for various bioimaging tasks, including DNA and RNA sensing. Additionally, some acridine-based fluorescent probes have been employed in detecting cations such as Cu^{2+}, Al^{3+}, Hg^{2+}, Fe^{3+}, Ni^{2+}, Cd^{2+}, Cr^{3+}, etc. Versatile fluorescent molecular chemosensors for sensing of anions, such as ClO^-, F^-, $H_2PO_4^-$, HSO_4^-, etc., have been reported. Furthermore, acridine-based fluorescent probes have been investigated for their utility as viscosity sensors and polarity sensors and their application in cell imaging. Various other applications leveraging the fluorescent properties of acridine are discussed in this chapter. However, while acridine has shown promise in fluorescence applications, its potential for disease diagnosis remains largely untapped, with only a few reported instances. This underscores the opportunity for future innovation and development of acridine-containing probes for disease diagnosis, particularly in Alzheimer's disease as the role of acridine has yet to be explored for disease diagnosis.

REFERENCES

Adan, A., Alizada, G., Kiraz, Y., Baran, Y., & Nalbant, A. (2017). Flow cytometry: Basic principles and applications. *Critical reviews in biotechnology*, *37*(2), 163–176.

An, S., Lin, Y., Wang, J., Ye, T., Mao, Y., Zhang, J., & Liu, H. (2023). Near-infrared mitochondria-targeted fluorescent probe with a large Stokes shift for rapid and sensitive detection of cysteine/homocysteine and its bioimaging application. *Sensors and Actuators B: Chemical*, *374*, 132799.

An, S., Lin, Y., Ye, T., Bai, T., He, D., Guo, L., & Wang, J. (2024). An extra-large Stokes shift near-infrared fluorescent probe for specific detection and imaging of cysteine. *Talanta*, *267*, 125247.

Bernthsen, A. (1884). Die acridine. *Justus Liebigs Annalen der Chemie*, *224*(1–2), 1–56.

Bradley, D. E. (1965). Staining of bacteriophage nucleic acids with acridine orange. *Nature*, *205*(4977), 1230–1230.

Bratulescu, G. (2016). Synthesis of acridines by an eco-friendly method. *Revista de Chimie*, *67*(3), 435–437.

Byvaltsev, V. A., Bardonova, L. A., Onaka, N. R., Polkin, R. A., Ochkal, S. V., Shepelev, V. V., Aliyev, M. A., & Potapov, A. A. (2019). Acridine orange: A review of novel applications for surgical cancer imaging and therapy. *Frontiers in Oncology*, *9*, 925.

Chowdhury, S., *et al.* (2018). Review on recent advances in metal ions sensing using different fluorescent probes. *Journal of Fluorescence*, *28*(4), 999–1021.

Dai, Q., *et al.* (2018). A one-step synthesized acridine-based fluorescent chemosensor for selective detection of copper(II) ions and living cell imaging. *New Journal of Chemistry*, *42*(1), 613–618.

Dai, Y., Xu, K., Li, Q., Wang, C., Liu, X., & Wang, P. (2016). Acridine-based complex as amino acid anion fluorescent sensor in aqueous solution. *Spectrochimica Acta Part A: Molecular and Biomolecular Spectroscopy*, *157*, 1–5.

Drummen, G. P. (2012). Fluorescent probes and fluorescence (microscopy) techniques—illuminating biological and bio medical research. *Molecules*, *17*(12), 14067–14090.

Enbanathan, S., & Sathiyanarayanan, K. I. (2024). The catalyst free synthesis of dibenzo [a, j] acridine and its applications in bioimaging of BF3 in HeLa cells. *Methods*, *221*, 65–72.

Gao, M., Yu, F., Lv, C., Choo, J., & Chen, L. (2017). Fluorescent chemical probes for accurate tumor diagnosis and targeting therapy. *Chemical Society Reviews*, *46*(8), 2237–2271.

Goni, L. K. M. O., *et al.* (2022). Acridine and its derivatives: Synthesis, biological, and anticorrosion properties. *Materials*, *15*(21), 1–15.

Herole, R. A., Jat, R. K., & Dighe, R. D. (2022). A review on acridine, xanthene and its derivatives: Synthesis, physical and pharmacological properties. *Tropical Journal of Pharmaceutical and Life Sciences*, *9*(6), 01–12.

Isaac, I. O., Munir, I., Al-Rashida, M., Ali, S. A., Shafiq, Z., Islam, M., Ludwig, R., Ayub, K., Khan, K. M., & Hameed, A. (2018). Novel acridine-based thiosemicarbazones as 'turn-on' chemosensors for selective recognition of fluoride anion: A spectroscopic and theoretical study. *Royal Society Open Science*, *5*(7), 180646.

Jahanmehr, S. A., Hyde, K., Geary, C. G., Cinkotai, K. I., & Maciver, J. E. (1987). Simple technique for fluorescence staining of blood cells with acridine orange. *Journal of Clinical Pathology*, *40*(8), 926.

Jiang, X. J., *et al.* (2014). A novel acridine-based fluorescent probe for the cascade recognition of Cr^{3+} and PO_4^{3-}. *Sensors and Actuators, B: Chemical*, *202*, 388–394.

Khandelwal, S., *et al.* (2020). Use of sustainable organic transformations in the construction of heterocyclic scaffolds. In *Green approaches in medicinal chemistry for sustainable drug design* (pp. 245–352), Elsevier Inc.

Kumar, R., Kaur, M., & Kumari, M. (2012). Acridine: A versatile heterocyclic nucleus. *Acta Poloniae Pharmaceutica - Drug Research*, *69*(1), 3–9.

Lauretti, F., de Melo, F. L., Benati, F. J., de Mello Volotao, E., Santos, N., Linhares, R. E. C., and Nozawa, C. (2003). Use of acridine orange staining for the detection of rotavirus RNA in polyacrylamide gels. *Journal of virological methods*, *114*(1), 29–35.

Lee, S. C., Park, S., So, H., Lee, G., Kim, K.T. and Kim, C., (2020). An acridine-based fluorescent sensor for monitoring ClO^- in water samples and zebrafish. *Sensors*, *20*(17), 4764.

Mahesh, K., Karpagam, S., & Pandian, K. (2019). How to design donor–acceptor based heterocyclic conjugated polymers for applications from organic electronics to sensors. *Topics in Current Chemistry*, *377*(3), 12.

Mayor, H. D., & Hill, N. O. (1961). Acridine orange staining of a single-stranded DNA bacteriophage. *Virology*, *14*(2), 264–266.

Martí-Centelles, V., Burguete, M. I., Galindo, F., Izquierdo, M. A., Kumar, D. K., White, A. J., Luis, S. V., & Vilar, R. (2012). Fluorescent acridine-based receptors for H2PO4. *The Journal of Organic Chemistry*, *77*(1), 490–500.

Nagy, M., Rácz, D., Nagy, Z. L., Fehér, P. P., Kovács, S. L., Bankó, C., Bacsó, Z., Kiss, A., Zsuga, M., & Kéki, S. (2019). Amino-isocyanoacridines: Novel, tunable solvatochromic fluorophores as physiological pH probes. *Scientific Reports*, *9*(1), 8250.

Nunes, M. C., *et al.* (2020). A facile preparation of a new water-soluble acridine derivative and application as a turn-off fluorescence chemosensor for selective detection of Hg^{2+}. *Journal of Fluorescence*, *30*(2), 235–247.

Pal, A., Karmakar, M., Bhatta, S. R., & Thakur, A. (2021). A detailed insight into anion sensing based on intramolecular charge transfer (ICT) mechanism: A comprehensive review of the years 2016 to 2021. *Coordination Chemistry Reviews*, *448*, 214167.

Peng, H. Y., Zhang, G., Sun, R., Xu, Y. J., & Ge, J. F., (2022). Fluorescent probes based on acridine derivatives and their application in dynamic monitoring of cell polarity variation. *Analyst*, *147*(22), 5231–5238.

Pirota, V., Stasi, M., Benassi, A., & Doria, F., (2020). An overview of quadruplex ligands: Their common features and chemotype diversity. *Annual Reports in Medicinal Chemistry*, *54*, 163–196.

Plemel, J. R., Caprariello, A. V., Keough, M. B., Henry, T. J., Tsutsui, S., Chu, T. H., Schenk, G. J., Klaver, R., Yong, V. W., & Stys, P. K. (2017). Unique spectral signatures of the nucleic acid dye acridine orange can distinguish cell death by apoptosis and necroptosis. *Journal of Cell Biology*, *216*(4), 1163–1181.

Pordel, M., Gheibi, H., & Sharif, A. (2024). Recent advances in the synthesis and optical applications of acridine-based hybrid fluorescent dyes. *Journal of Fluorescence*, 1–38.

Prabakaran, K., Manivannan, R., Oh, H., Parthiban, C., & Son, Y. A. (2021). Synthesis and characterisation of new acridine dye molecules combined UV absorber and exploring photophysical properties. *Dyes and Pigments*, *192*, 109391.

Prasher, P., & Sharma, M. (2018). Medicinal chemistry of acridine and its analogues. *MedChemComm*, *9*(10), 1589–1618.

Ravichandiran, P., Prabakaran, D. S., Maroli, N., Kim, A. R., Park, B. H., Han, M. K., Ramesh, T., Ponpandian, S., & Yoo, D. J. (2021). Mitochondria-targeted acridine-based dual-channel fluorescence chemosensor for detection of

Sn^{4+} and $Cr_2O_7^{2-}$ ions in water and its application in discriminative detection of cancer cells. *Journal of Hazardous Materials*, *419*, 126409.

Rizvi, S. B., Ghaderi, S., Keshtgar, M., & Seifalian, A. M. (2010). Semiconductor quantum dots as fluorescent probes for in vitro and in vivo bio-molecular and cellular imaging. *Nano Reviews*, *1*(1), 5161.

dos Santos Carlos, F., *et al.* (2020). A highly selective acridine-based fluorescent probe for detection of Al^{3+} in alcoholic beverage samples. *Spectrochimica Acta - Part A: Molecular and Biomolecular Spectroscopy*, *231*, 118119.

Spicer, J. A., *et al.* (1997). Structure-activity relationships for acridine-substituted analogues of the mixed topoisomerase I/II inhibitor N-[2-(dimethylamino)ethyl] acridine- 4-carboxamide. *Journal of Medicinal Chemistry*, *40*(12), 1919–1929.

Stuker, F., Ripoll, J., & Rudin, M. (2011). Fluorescence molecular tomography: principles and potential for pharmaceutical research. *Pharmaceutics*, *3*(2), 229–274.

Tsien, R. Y. (1989). Fluorescent probes of cell signaling. *Annual Review of Neuroscience*, *12*(1), 227–253.

Wang, C., *et al.* (2018). Acridine-based fluorescence chemosensors for selective sensing of Fe^{3+} and Ni^{2+} ions. *Spectrochimica Acta - Part A: Molecular and Biomolecular Spectroscopy*, *199*(2017), 403–411.

Wang, Y., *et al.* (2011). A new acridine derivative as a highly selective "off-on" fluorescence chemosensor for Cd^{2+} in aqueous media. *Sensors and Actuators, B: Chemical*, *156*(1), 126–131.

Wang, H. (2010). Acridine synthesis. In *Comprehensive organic name reactions and reagents* (pp. 2838–2841), Wiley.

Yao, J., Yang, M., & Duan, Y. (2014). Chemistry, biology, and medicine of fluorescent nanomaterials and related systems: new insights into biosensing, bioimaging, genomics, diagnostics, and therapy. *Chemical Reviews*, *114*(12), 6130–6178.

Zhang, D., Jiang, X., Dong, Z., Yang, H., Martinez, A., & Gao, G. (2013). A bifunctional acridine-based fluorescent sensor: ratiometric sensing of H_2PO^{4-} and obvious fluorescence quenching towards HSO^{4-} through a synergistic binding effect of benzimidazolium and urea moieties. *Tetrahedron*, *69*(48), 10457–10462.

Zhang, D., Jiang, X., Yang, H., Martinez, A., Feng, M., Dong, Z., & Gao, G. (2013). Acridine-based macrocyclic fluorescent sensors: Self-assembly behavior characterized by crystal structures and a tunable bathochromic-shift in emission induced by H_2PO^{4-} via adjusting the ring size and rigidity. *Organic & Biomolecular Chemistry*, *11*(20), 3375–3381.

Zhou, Y., *et al.* (2015). A simple-structured acridine derivative as a fluorescent enhancement chemosensor for the detection of Pd^{2+} in aqueous media. *Spectrochimica Acta - Part A: Molecular and Biomolecular Spectroscopy*, *137*, 33–38.

Zhu, H., Fan, J., Du, J., & Peng, X. (2016). Fluorescent probes for sensing and imaging within specific cellular organelles. *Accounts of Chemical Research*, *49*(10), 2115–2126.

3 Rhodamine-Based Molecules
Design and Applications

Esther Rani Motamarri and Abha Sharma
National Institute of Pharmaceutical Education and Research, Raebareli, Lucknow, UP, India

LIST OF ABBREVIATIONS

ADP	Adenosine diphosphate
AMP	Adenosine monophosphate
ATP	Adenosine triphosphate
BAPTA	1,2-bis(o-amino phenoxy)ethane-N, N, N′, N′-tetraacetic acid
CHEQ	Chelation-enhanced quenching
DFT	Density functional theory
EDTA	Ethylenediaminetetraacetic acid
EDX	Energy-dispersive X-ray
ESI-MS	Electrospray ionization mass spectrometry
FRET	Fluorescence resonance energy transfer
FT-IR	Fourier transform infrared spectroscopy
GECI	Genetically encoded calcium indicator
GFP-Halo	Green fluorescent protein – HaloTag
LOD	Limit of detection
MTT assay	3-(4,5-dimethylthiazol-2-yl)-2,5-diphenyltet razolium bromide
Na_2 EDTA	Sodium salt of ethylenediaminetetraacetic acid
NMR	Nuclear magnetic resonance
POI	Protein of interest
SDS-PAGE	Sodium dodecyl sulfate–polyacrylamide gel electrophoresis
SEM	Scanning electron microscopy
SPR	Surface plasmon resonance
ThT	Thioflavin T
UV-Vis	Ultraviolet-visible
WHO	World Health Organization
XRD	X-ray diffraction

3.1 INTRODUCTION

Fluorescent sensors based on organic molecules are popular for chemosensory applications due to their high sensitivity, selectivity, and easy construction. They have demonstrated their great use in various domains, particularly as biological probes, environmental sensors, and food safety (Nguyen et al., 2003; Maniyazagan et al., 2017). It is widely acknowledged that metal ions are essential for numerous vital biological processes. The proper physiological function of the human body is significantly impacted by the concentration and distribution of metal ions in our body. Metals are classified into two groups based on their impact on the biosystems, i.e., essential and non-essential metals. Our bodies need the right proportions of 10 metal ions, potassium (K), sodium (Na), calcium (Ca), magnesium (Mg), cobalt (Co), molybdenum (Mo), iron (Fe), copper (Cu), zinc (Zn), manganese (Mn), and sodium (Na), to survive, out of which, K, Na, Ca, and Mg constitute over 99% of the total metal ions present in our body (Shi et al., 2023).

Rhodamine dyes often exhibit high molar absorptivities in the visible spectrum and have good photostability, tremendous luminous intensity, and the capacity to control dye replacement. Diverse rhodamine dyes have been reported to date. The purest form of rhodamine has an excitation wavelength of 496 nm and a fluorescence emission spectrum of 517 nm (Jiao et al., 2015). Rhodamine dye, when in spiro ring-closed form, is colorless and non-fluorescent, but in the ring-opened form, pink-red color and intense fluorescence are observed, which has been extremely popular in the field of probe design during the last decade. Rhodamine dye is one of the optimal acceptor fluorophores in a system that uses the FRET mechanism. The characteristic emission of a rhodamine-based dye in the spiro ring-closed state is blocked, so the FRET dyad only reveals the donor's emission, which disallows the system's FRET process; nevertheless, post reaction/complexation with the necessary analyte, the spiro ring of the rhodamine dye opens, resulting in a strong emission and activation of the FRET process "ON." Based on this, several rhodamine-based FRET sensors have been developed for various analytes (Maniyazagan et al., 2017).

3.2 APPLICATIONS OF RHODAMINE DYES

3.2.1 Identification of Metal Ions

An important field of research is developing and producing effective chromogenic and fluorogenic sensors for desired metal ions as they affect living things or the environment. Fluorescent sensors have several advantages over other techniques that aid in identifying the metal ions showing particularly high sensitivity, specificity, simplicity, real-time monitoring, and quick response times. The rhodamine

DOI: 10.1201/9781032706245-3

B derivative changes from its colorless, nonfluorescent spirocyclic form to pink where a highly fluorescent ring is opened to amide form in the presence of metal ions. Rhodamine B scaffold is widely recognized for its long absorption and emission wavelengths, strong fluorescence quantum yield, significant absorption coefficient, and ring switching activity. Rhodamine B practical application in metal ion detection is thus interesting (Mondal et al., 2019).

3.2.1.1 Sensing $Al^{3+}$and Fe^{3+}

Due to their importance in biology and the environment, Al^{3+} and Fe^{3+} ion identification and recognition are of great interest. The Al^{3+} ion significantly affects human health. It influences the function of digestive enzymes and is linked to numerous brain illnesses including Parkinson's disease and Alzheimer's disease (AD). The harmful action of Al^{3+} also contributes to disorders associated with smoking, chronic renal failure, and bone weakening. Iron is the most prevalent vital element among transition metal ions. Iron is also crucial for human health due to its role in cellular metabolism and enzyme catalysis, as an oxygen carrier in haemoglobin, and as a cofactor in numerous enzymatic processes connected to the mitochondrial respiratory chain. Nevertheless, an increase in its level in the body is linked to some organ malfunction.

The development and synthesis of rhodamine dye 1 (Figure 3.1) with perimedine labeling were done by Mondal et al., and metal ion binding studies were carried out in a buffer of CH_3CN/H_2O (3:1, v/v, 10 mM tris-HCl, pH = 6.90). Under various experimental conditions like concentration fluctuation helped distinguish the complexation of dye 1 with Al^{3+} and Fe^{3+} based on fluorescence response because when these ions come in contact, the color of the solution turns from colorless to pink with dye 1. Only the Al^{3+} ions can demonstrate turn-on fluorescence with a strong color change at low concentrations ($c = 1 \times 10^{-4}$ M). Upon Fenton's reaction, they were able to detect Fe^{3+} ions by activating fluorescence and changing the color of the sample. The FTIR analysis gave a clear idea of the complexation occurring through spirolactam ring opening. Following the complexation of Al^{3+} and Fe^{3+}, the amide carbonyl stretching that appeared at 1,692 cm^{-1} of dye 1 was decreased to 1,646 and 1,647 cm^{-1}, respectively, and the C=N stretching, which was at 1615 cm^{-1}, slightly increased to 1,620 or 1,626 cm^{-1}. Different anions like F^-, Cl^-, I^-, OAc^-, and PO_3^{-4} were checked for their interference with the complex formed, that is, in a pink-colored solution. Of all these anions, F^-, OAc^-, and PO_3^{-4} were able to bring back the spirolactam ring resulting in a colorless solution, which was verified by observing the fluorescence absorption spectrum as it returned to the same as dye 1 that didn't complex with any cations of which only PO_3^{-4} showed immense predilection to restore the spirolactam ring.

To assess the effect of Cu^{2+} ions on the sensing of Al^{3+} and Fe^{3+} ions, a comparative experiment was conducted in which Al^{3+} and Fe^{3+} ions were added to a solution of rhodamine dye 1 along with other metal ions. The results of the study indicated that Cu^{2+} ions did interfere with the sensing of Al^{3+} and Fe^{3+} ions as Cu^{2+} forms a stable complex with perimidine ring (Mondal & Ghosh, 2019).

3.2.1.2 Sensing Al^{3+}, F^-, and CH_3COO^-

With a mass of about 8.3% aluminum, makes up the third most prevalent element in the crust of the Earth. Aluminum's principal targets in terms of toxicological effects are distinct from those of heavy metals. Aluminum is reported to impair the iron-sulfur respiratory chain protein's potential to function and to increase the possibility of Alzheimer's disease. People are frequently exposed to aluminum due to its extensive usage in water treatment, as a food additive, and in a variety of industrial processes, especially in the production of vehicles and computers. Therefore, the Al^{3+} ion's detection in the presence of numerous other cations has biological as well as environmental significance. Popular methods for detecting aluminum include inductively coupled plasma atomic emission spectroscopy and graphite furnace atomic absorption spectroscopy. Each of these methods is typically laborious and costly. In contrast, optical detection using fluorescence is not only exceptionally sensitive but also an operationally uncomplicated method.

Because anions are known to serve as an assortment of recognized biological activities, the detection of anions using emission readout has received a lot of attention. Fluoride, the smallest anion with the highest charge density, is especially vital due to its application in dental care and other fields. On the contrary, the acetate ion is a crucial element of many metabolic processes and plays vital functions in several enzymes and antibodies. Therefore, it is necessary to build sensors that are selective for fluoride and acetate ions. There aren't much fluorescence chemosensors or chemodosimeters for the ions of aluminum, fluoride, and acetate that are effective in just organic solutions, which

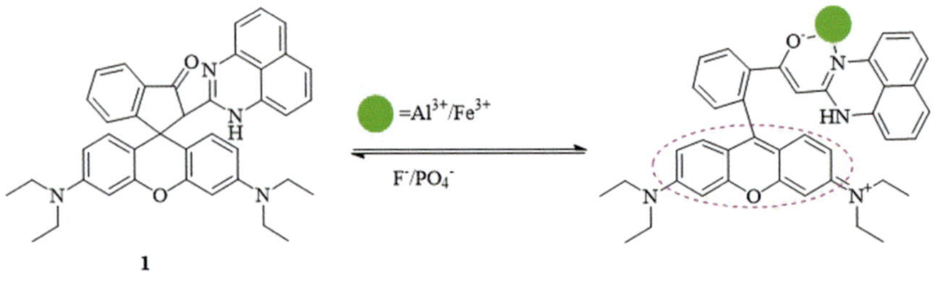

FIGURE 3.1 Rhodamine dye 1 sensing Al^{3+} and Fe^{3+}

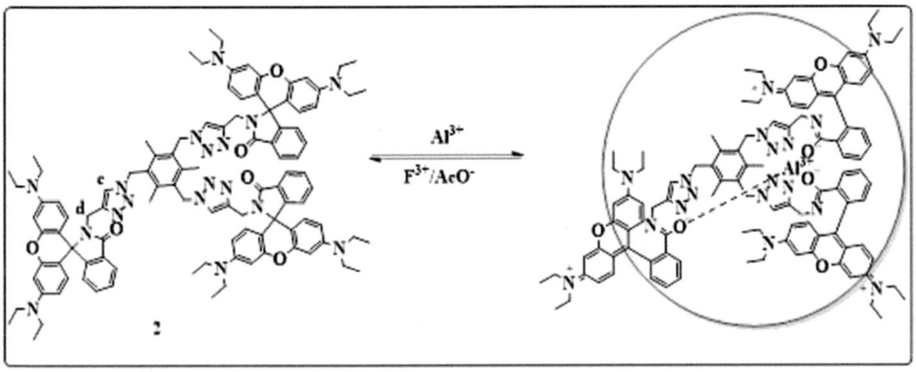

FIGURE 3.2 Rhodamine dye 2 sensing Al^{3+} and its reversibility using F^- and CH_3COO^-.

limits their use as sensors. A chemosensor's utilization range greatly expands when it is used in an aqueous media or a mixed aqueous and organic medium.

Maity and Bharadwaj synthesized and studied the photophysical properties of a rhodamine dye 2 with a 1,2,3-triazole ring (Figure 3.2) that is used to specifically detect the Al^{3+} ion in a methanolic aqueous media using both colorimetry and fluorimetry. They demonstrated that the Al^{3+} ion can distinguish between fluoride and acetate ions in the presence of a variety of different anions once it has been attached to the dye. The dye 2 was investigated using perchlorate salts of various metal ions, such as Na^+, K^+, Mg^{2+}, Ca^{2+}, Al^{3+}, Mn^{2+}, Fe^{2+}, Co^{2+}, Ni^{2+}, Cu^{2+}, Zn^{2+}, Cd^{2+}, Ag^+, Pb^{2+}, and Fe^{3+}, while the nitrate salts of the Ce^{3+} and Cr^{3+} ions were employed. None of the metal ions (other than Al^{3+}) listed above, including metal-free L, has exhibited any appreciable absorption band in the visible spectrum. Only Cr^{3+} and Fe^{3+} exhibited a minimal intensity enhancement among the transition metals utilized, because of their stronger Lewis acidic bonds (Figure 3.2). Both Job's plot and Benesi–Hildebrand plot experiments were conducted to examine the binding stoichiometry between L and the Al^{3+} ion, with the outcomes demonstrating a 1:1 complexation that was supported by ^1H NMR titration in a $CDCl_3CD_3OD$ (1:1, v/v) mixture in the presence of different equivalent concentrations of Al^{3+} ions and the presence of a large peak at 589.3210 nm that is attributed to [dye 2 + Al^{3+} + $H_2O]^{3+}$ in the mass spectrum. In the presence of the majority of competing metal ions, dye 2 is therefore proven to be a potential selective fluorescence sensor for Al^{3+}. The complexation mode of dye 2 toward the Al^{3+} ion, in addition to spectrum variations caused by the metal's binding to the dye, was observed where 1.0 equiv of Al^{3+} salt addition caused the signals of Hc and Hd to shift downfield by 0.62 and 0.11 ppm, correspondingly (Maity et al., 2013).

3.2.1.3 Sensing Al^{3+}

Aluminum-based products have been used extensively in manufacturing, medical equipment, aircraft, and other industries, and the widespread use of aluminum ions in daily life elevates the problems of aluminum poisoning, thereby causing calcium metabolism disorders, interfering with iron concentration in the blood, and leading to osteomalacia and microcellular hypochromic anemia. Al^{3+} can play a role in the generation of reactive oxygen species, which leads to neurodegenerative diseases like Alzheimer's disease, Wilson's disease, amyotrophic lateral scleroderma, and amyotrophic lateral sclerosis. According to the WHO, the maximum amount of Al^{3+} that humans can consume per day is roughly 3 mg, and the maximum concentration of Al^{3+} in drinking water is 7.41 mM (Rahman et al., 2017; Leng et al., 2019).

Fluorescent probe technology has been widely applied in the investigation of sensitive and speedy detection of Al^{3+}. This method can identify target ions even at extremely low ion concentrations without any special sample preparation. They offer great selectivity, sensitivity, and are easily miniaturized, making them adequate for bioimaging and medical diagnosis.

A Schiff base probe 3 based on Rhodamine B (RBGP, Figure 3.3) has been synthesized by Chu Kan et al. The probe showed a distinct fluorescence response to Al^{3+} that is both quick and persistent. The probe remained nonfluorescent from pH 6 to 9 but got fluorescent when Al^{3+} was added. Absorption spectra showed an intense peak at 560 nm, which was characteristic of the probe with Al^{3+}, and showed a strong emission peak at 582 nm, which wasn't observed in the other metals. This fluorescence was quenched by adding F^- where the probe's five-member spirolactam returns to the closed state demonstrating its reversibility. Al^{3+} had a detection limit of 0.146 mM and a binding constant of 1.01 × 10^{-4} M^{-1}. Job's plot analysis showed that the complex ratio of probe 3 to Al^{3+} is 1:1. The probe was checked in complex real-world water samples, which proved that this probe can also be used in intricate environments.

Confocal bioimaging analysis showed that the probe didn't enter the nucleus but remained in the cytoplasm of MCF-7 cells when incubated along with Al^{3+}, showing red-colored fluorescence. The probe was experimented on zebra fish where the probe was incubated for 30 minutes before performing confocal imaging, and the analysis showed that there wasn't any fluorescence but when incubated along with Al^{3+} (20 nM) for 30 minutes, there appeared red fluorescence suggesting that this probe can be used in vivo (Kan et al., 2021).

Mandal et al. developed four rhodamine 6G-based chemosensors, probes 4–7 (H3L1–H3L4) (Figure 3.4), for

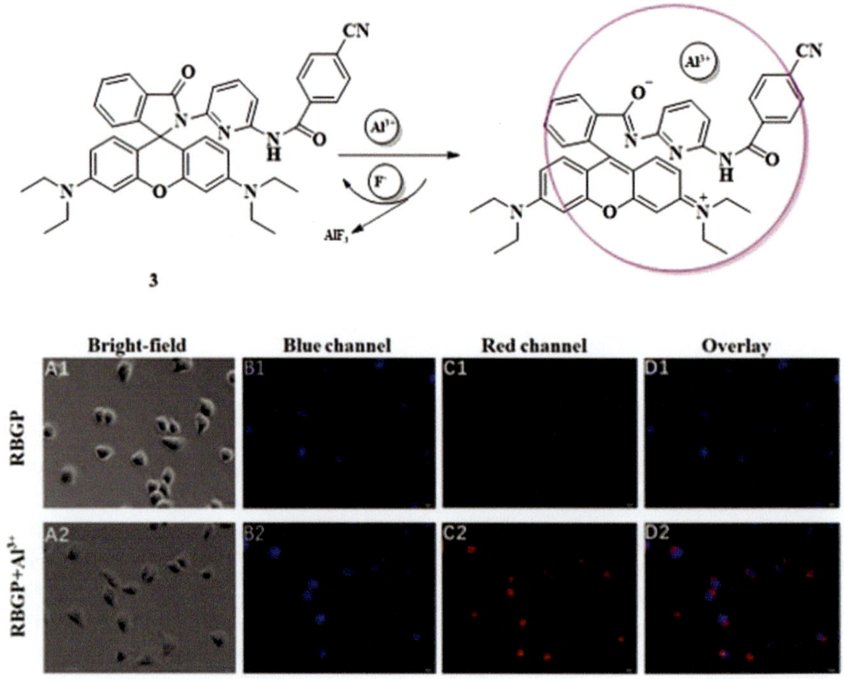

FIGURE 3.3 Rhodamine dye 3 sensing Al^{3+}, which was reversible in the presence of F^- and the confocal fluorescence images of Rhodamine dye 3 in living MCF-7 cells. Reprinted (adapted) with permission from Kan et al. (2021). Copyright: Elsevier, 2021.

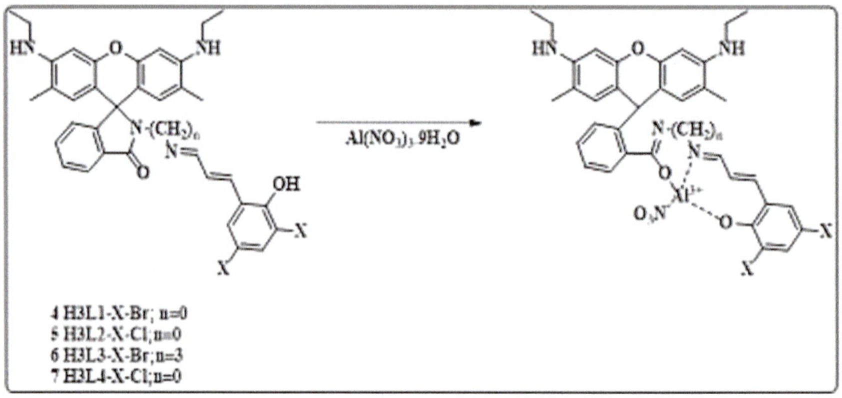

FIGURE 3.4 Rhodamine dyes 4–7 (H3L1–H3L4) sensing Al^{3+}.

identifying Al^{3+} ions preferentially, which was characterized by X-ray crystallography. The molecules, H3L1–H3L4, showed an intense band at 530 nm in their absorption spectra when Al^{3+} ions were present and without Al^{3+} ions showed absorbance around 290 nm and 345 nm for H3L1 and H3L2 while H3L3 and H3LR exhibited absorbance peaks around 420 nm. These characteristic peaks were reduced in intensity upon the addition of Al^{3+} ions and significant alterations in all chemosensor spectra resulting in color change from colorless to a brilliant pinkish yellow. The inclusion of Al^{3+} ions significantly increases the emission intensity at 555 nm. Job's plot analysis confirms the chemosensor's 1:1 binding stoichiometry to Al^{3+} ions whose observations are supported by the ESI-MS analysis. Adding Mn^{2+}, Fe^{3+}, Co^{2+}, Ni^{2+}, Na^+, K^+, Mg^{2+}, and Ca^{2+} did not noticeably alter the free chemosensor's absorption spectrum. Fluorescence properties were analyzed at ambient conditions for which the probes (10 μM) did not emit fluorescence when excited at 345 nm, but the addition of Al^{3+} ions (0–11 μM) to the probe resulted in a significant fluorescence increase at 555 nm. This indicates that these chemosensors (H3L1–H3L4) preferentially recognize Al^{3+} ions over other metal ions and common anions where Cr^{3+}, Mg^{2+}, Cu^{2+}, and Fe^{3+} ions significantly reduced fluorescence intensity. That real-time applications of chemosensors rely on reversibility and regeneration is proved by the addition of 1 equiv. of sodium salt of ethylenediaminetetraacetic acid (Na_2EDTA) solution resulting in the disappearance of the fluorescent pinkish yellow probe-Al^{3+} complex and a reduction in fluorescence intensity, suggesting the formation of the free chemosensor. The impact of pH on chemosensors (H3L1–H3L4) was analyzed in both free and Al^{3+} ion-containing conditions. The chemosensor's spirolactam ring opens under acidic conditions for which similar results were observed with Al^{3+}

ions present. At pH 2–4, both the free chemosensor and the chemosensor-Al^{3+} adduct showed high fluorescence intensity. At pH 5, the free chemosensor's fluorescence intensity decreases drastically. After pH 5–11, the fluorescence intensity decreases and the same is maintained supporting the rebuilding of the spirolactam ring in neutral and basic circumstances. Cells rapidly up take the ligands and Al^{3+} salts, resulting in a red fluorescent signal. The biocompatibility of the ligands suggests their potential for use in biomedical domains, as no substantial toxicity was identified at dosages up to 100 μM (Mandal et al., 2019).

3.2.1.4 Sensing Ca^{2+}

Calcium is an important second messenger in living cells. Changes in the endogenous concentration of Ca^{2+} have a major impact on signaling the transmission inside the cells (Clapham, 2007). Numerous methodologies for measuring calcium fluctuations in live cell studies have been reported that include genetically encoded calcium indicators (GECIs) that use calcium-sensitive fluorescent proteins or fusion proteins for which well-known examples are "cameleons" and "camgaroos." Although GECIs possess the unique feature of being genetically tailored to any protein of interest (POI), they do not respond quickly to changes in Ca^{2+} in living cells (Los et al., 2008).

Best et al. created a GFP-Halo protein-labeled calcium-sensitive probe RhoCa-Halo (Figure 3.5) that is site-specific and gets anchored to proteins in living cells using the bio-orthogonal HaloTag labeling technique. It is an iodo-derivatized BAPTA chelator with a tetramethyl rhodamine fluorophore that can be improved subsequently via Sonogashira cross-coupling reaction. This novel calcium sensor fluorescence is 200 times more when calcium is bound to it and can be activated in the green spectrum ($E_x max = 552$ nm; $E_m max = 580$ nm).

This newly constructed calcium sensor was checked for its protein labeling through in vitro experiment where it was incubated with green fluorescent protein (GFP-Halo) and its covalent bonding was analyzed using SDS- PAGE, which showed that this protein is covalently bonded with the probe, and live cell assays were done to check its utility where NIH/3T3 and HeLa cells were transiently transfected with H2B-GFP and Halo-H2B and then treated with the appropriate AM-Ester to give RhoCa-Halo AM (Figure 3.5). In both cell lines, epifluorescent microscopy revealed that the nucleus had been successfully labeled with the localizable calcium sensor. The calcium sensor's behavior in living cells was studied after ATP stimulation and induction of cytosolic influx of calcium ions from endogenous sources. A significant increase in fluorescence was observed in the nucleus of transfected and labeled cells when compared to nontransfected cells, which showed no detectable increase in fluorescence in the nucleus, demonstrating that the fluorescent sensor RhoCa-Halo can be used as a localizable calcium sensor to monitor spatiotemporal changes in free calcium ions in live cell experiments (Best et al., 2016).

3.2.1.5 For Sensing Cd^{2+}

Cadmium (Cd^{2+}) is widely employed in agriculture and industry and also recognized for its toxicity as it can cause medical issues such as renal failure, calcium metabolism issues, prostate cancer, and so on. The EPA (United States

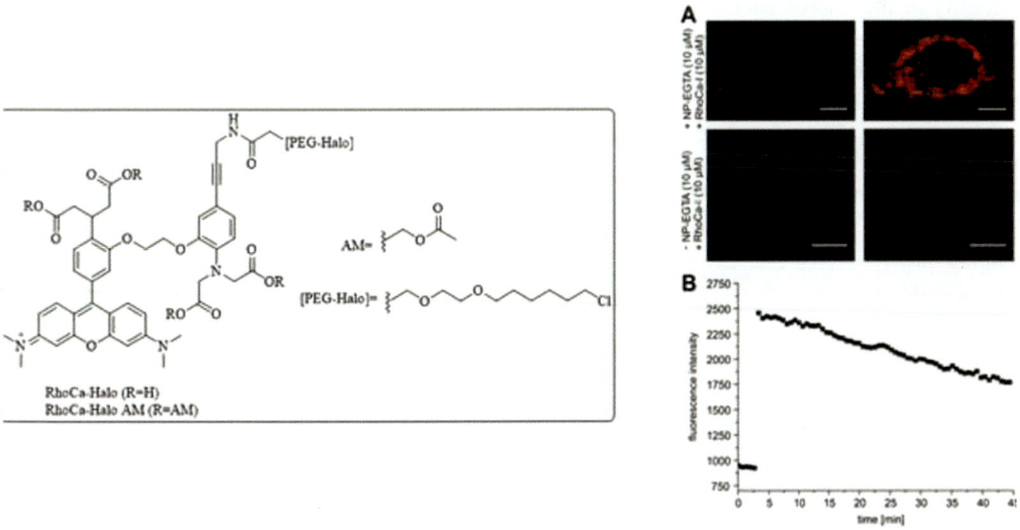

FIGURE 3.5 Rhodamine dyes RhoCa-Halo and RhoCa-Halo AM for sensing Ca^{2+}. (A) Live cell imaging of UV-mediated release of Ca^{2+}, λex = 559 nm, detection at 572–672 nm. Upper panel: HeLa cells were incubated with 10 μM NP-EGTA AM and endogenous release of calcium ions from cellular stores was stimulated by ATP. After incubation with 10 μM RhoCa-I AM (upper panel, left), NP-EGTA-chelated Ca^{2+} was liberated by UV radiation (405 nm), leading to a 2.5-fold increase of fluorescence (upper panel, right). Lower panel: Cells were incubated with RhoCa-I AM, and the sensor shows clear subcellular localization (lower panel, left). NP-EGTA AM is omitted in the experiment, and no fluorescence increase is monitored upon UV radiation (lower panel, right). (B) Corresponding fluorescence intensity trace. UV-mediated uncaging was conducted after 2 min with a short laser pulse. Reprinted (adapted) with permission from Best et al. (2016). Copyright: Royal Society of Chemistry, 2016.

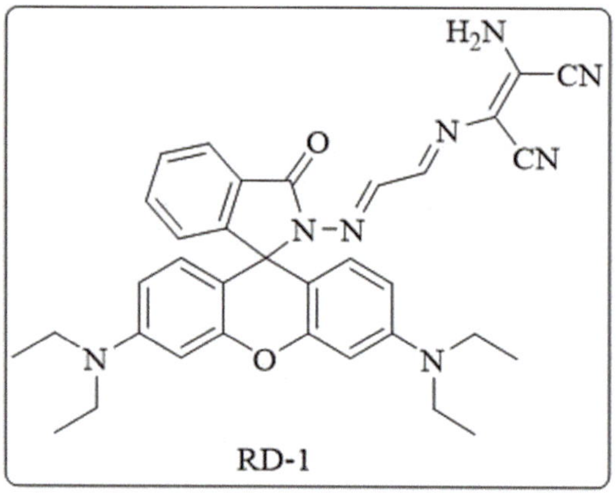

FIGURE 3.6 Rhodamine RD-1 for sensing Cd^{2+}.

Environmental Protection Agency) sets an enforceable Cd^{2+} drinking water standard of 5 ppb to prevent kidney damage and other related disorders; however, the World Health Organization (WHO) sets a more stringent Cd^{2+} drinking water guideline value of 3 ppb (McElroy et al., 2006; Verougstraete et al., 2003).

Sakthivel et al. developed a rhodamine-based probe RD-1 (Figure 3.6), which is coupled to the diaminomaleonitrile moiety. This probe can be utilized to detect Cd^{2+} with excellent sensitivity and selectivity as well as good linearity and a considerable colorimetric and fluorometric response. Fluorescence titration was done to analyze RD-1's Cd^{2+} binding capacities. Because the probe RD-1 spirocyclic structure did not open after excitation at 530 nm, its fluorescence was comparatively less intense. The progressive introduction of Cd^{2+} produced a very strong emission band at 553 nm for which the intensity of the fluorescence was nearly elevated by 200 fold, which is most likely caused by the Cd^{2+}-induced rhodamine spiro-ring opening process. The molar fraction was 0.5, and Job's plots showed the same maximum absorbance, confirming a 1:1 stoichiometry. The Benesi–Hildebrand equation was used to compute the association constant (Ka) of the probe RD-1 with Cd^{2+}, which was found to be 2.33×10^5 M^{-1} from UV-vis titration data. MTT assay and cell imaging proved RD-1 low cytotoxicity and cell membrane permeability subsequently (Sakthivel et al., 2017).

3.2.1.6 Sensing Cd^{2+} and S^{2-}

Cadmium is one of the superfluous elements, widely utilized in batteries and fertilizers causing significant air, water, and soil pollution; thus, it is counted as an environmental hazard that can pose serious health risks to all living beings (Mendes et al., 2006). The sulfide anion is commonly found in water due to industrial operations and microbial reduction of sulfate by anaerobic bacteria or sulfur-containing amino acids in animal proteins. Sulfide can cause mucous membrane irritation and possibly unconsciousness and respiratory paralysis. It becomes even more poisonous when it is protonated to form HS or H_2S. H_2S, at high concentrations, can cause loss of consciousness, lasting brain damage, or even death by asphyxiation and at low concentrations, personal distress. As a result, it is vital to create a fast and precise approach for monitoring sulfide levels in aqueous media and biological systems (Gosselin et al., 1984).

Maniyazagan et al. reported a new fluorescence sensor based on rhodamine B and 2-pyridinecarboxaldehyde (RBD4, Figure 3.7). The RBD4 incorporates Cd^{2+} to produce complex RBD4-Cd^{2+}, which exhibits excellent sensitivity and selectivity for S^{2-} over other competitive anions by generating CdS. XRD, SEM, EDX, FT-IR, and Job's plot characterized the 1:1 complex of RBD4-Cd^{2+}. In ACN/HEPES buffer (2:8, v/v, pH 7.2, 10 mM), RBD4 showed an absorption maximum at 308 nm and an emission maximum near 480 nm, indicating that the spirolactam ring of the rhodamine B unit is closed. Its selectivity was evaluated using multiple biologically important metal cations such as Na^+, K^+, Mg^{2+}, Hg^{2+}, Cu^{2+}, Pb^{2+}, Fe^{2+}, Fe^{3+}, Ca^{2+}, Ni^{2+}, Cd^{2+}, and Al^{3+} in ACN/HEPES buffer, which proved that there was a prominent change in UV-vis spectral pattern only when Cd^{2+} is present that caused the red shift in the spectra from 480 nm to 590 nm indicating spirolactam ring opening, which leads to color changes from colorless to deep magenta. Association constant of RBD4 with Cd^{2+} is 4.2524×10^4 M^{-1} at 25°C. Free fluorescent probe RBD4 suppresses the FRET pathway but when the probe RBD4 binds to Cd^{2+}, it activates the FRET process resulting in intense rhodamine-based red emission. This energy transfer from pyridine to

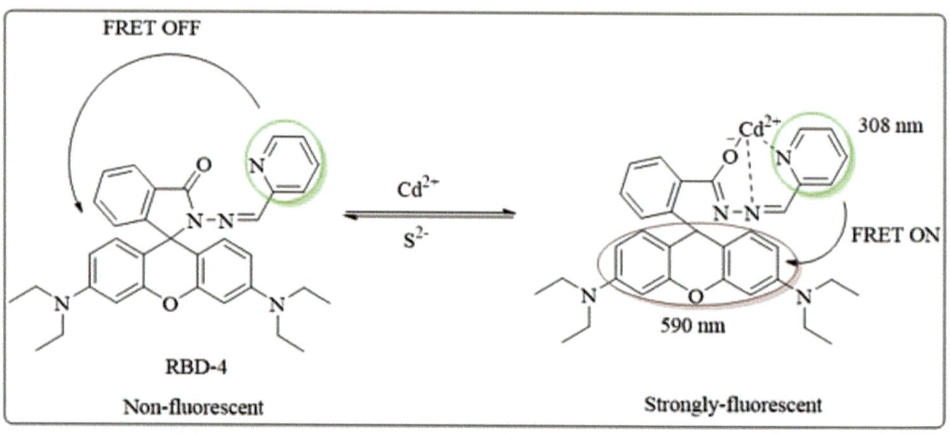

FIGURE 3.7 Rhodamine dye RBD-4 sensing Cd^{2+} and its reversibility in the presence of S^{2-}

xanthene occurs through ring opening, which increases the overlap integral between the pyridine and xanthene moiety. Investigating the effect of the presence of metal ions, such as Na^+, K^+, Mg^{2+}, Hg^{2+}, Cu^{2+}, Pb^{2+}, Fe^{2+}, Fe^{3+}, Ca^{2+}, Ni^{2+}, Cd^{2+}, and Al^{3+}, showed weaker spectrum response compared to the Cd^{2+} ion at the same concentration. Upon successive titrations (0–100 mM Cd^{2+}), the emission band peaked at 590 nm, indicating delocalization in the xanthene moiety of the rhodamine, resulting in an increase in emission intensity due to the metal complexation, which induced an electronic rearrangement within the dye, opening the spirolactam and yielding a fully conjugated rhodamine dye. It exhibited bright orange fluorescence, with a 20-fold increase at 590 nm. This suggests that RBD4 could function as a "off-on" fluorescent chemosensor for Cd^{2+}. DFT simulations reveal that the binding mode of Cd^{2+} ion is in NON coordination with bonds formed through one nitrogen of the imine group, one nitrogen of the pyridine moiety, and one oxygen atom of the carbonyl group. This results in a roughly planar shape around Cd^{2+}. The optimized structures of Cd^{2+} with RBD4 probe indicate individual low-energy complexation between Cd^{2+} and NON atoms in free RBD4. No color or fluorescence properties change was detected for free RBD4 in pH range of 3–9 but RBD4s' fluorescence was intense in high acidic pH < 3 in the presence of Cd^{2+} ions but exhibited no fluorescence in the basic pH range due to the production of $Cd(OH)_2$, making this probe useful in the detection of Cd^{2+} in physiological pH. EDX measurements showed the existence of Cd^{2+} in the RBD4-Cd^{2+} complex. The spectra displayed stretching vibrations of rhodamine spirolactam amide carbonyl (C=O) at 1,689 cm^{-1} as well as xanthene-NH at 3,440 and 2,960 cm^{-1}. The addition of 1 eq. of Cd^{2+} altered the stretching vibrations of the functional groups, indicating a 1:1 stoichiometry for the RBD4-Cd^{2+} complex. In the free RBD4, the band at 1,650 cm^{-1} corresponds to (C=N) frequency. After complexation, this band shifts to lower frequencies, 534–577 cm^{-1}, corresponding to O→M and 425–468 cm^{-1}, respectively, indicating the involvement of N and O atoms in the complexation with metal ions under study. Cytotoxicity of the probe RBD4 in the cells was tested and 80% cell viability was seen even after 12 hours, indicating its biocompatibility at all dosages and potential for fluorescence bioimaging. Fluorescence microscopic analysis suggests that RBD4 can effectively cross the membrane barrier, traverse HeLa cells, and detect intracellular Cd^{2+} (Maniyazagan et al., 2017).

3.2.1.7 Sensing Cr^{3+}

Chromium can be introduced to soil by anthropogenic and industrial activities, in addition to its natural existence. While Cr^{3+} is less hazardous to humans, chromium in higher oxidation states (Cr^{3+} and Cr^{6+}) poses a significant health risk. Cr^{3+}, a trivalent chromium, plays a role in various metabolic activities at the cellular level. Cr^{3+}, a vital trace element in human nutrition, serves a crucial role in the metabolism of carbohydrates, lipids, proteins, and nucleic acids. It activates enzymes, stabilizes proteins and nucleic acids, and helps create haemoglobin in red cells. It can also assist in stimulating insulin thereby lowering blood sugar. High quantities of Cr^{3+} can bind to DNA and harm cellular structures and components. However, the toxicity of Cr^{3+} is lower than that of Cr^{4+} in vivo (Li et al., 2011).

UV-vis spectroscopy and fluorescence spectroscopy were used to study the interaction of metal ions with the RDC-1 probe in a CH_3CN-H_2O (2:8 v/v) buffered solution containing HEPES (Figure 3.8). The absence of any change in visible absorption indicates that the probe RDC-1 remains in its lactam form in the solution. The probe RDC-1 is very sensitive to Cr^{3+} ions, as evidenced by an absorption peak at 526 nm and a color change from colorless to pink in the visible region. This was not observed with solutions containing Ag^+, Al^{3+}, Ba^{2+}, Ca^{2+}, Cd^{2+}, Co^{2+}, Cu^{2+}, Fe^{3+}, Hg^{2+}, K^+, Mg^{2+}, Mn^{2+}, Na^+, Ni^{2+}, Pb^{2+}, and Zn^{2+} salts. In the absence of metal ions, the fluorescence spectra of RDC-1 (10 μM) in a CH_3CN-H_2O solution (2:8, v/v, HEPES buffer, pH = 7.54) showed no fluorescence at 554 nm (E_x = 500 nm). When Cr^{3+} was added to the RDC-1 solution, there was a new emission band at 554 nm and fluorescence increased by almost 100 times. The color also progressively altered from colorless to pink. The probe RDC-1 was studied for its selectivity over other metal ions at increasing concentrations for which there was no cause of fluorescent variations. The Job's plot analysis shows a maximum at a mole fraction of 0.5, implying a 1:1 stoichiometry for the RDC-1-Cr^{3+} complex was also verified by the ESI-MS peak at 630.1578 and FTIR measurements on RDC-1 with and without Cr^{3+} ions confirming that the addition of Cr^{3+} ions causes a structural change in the probe. Plotting fluorescence intensity versus Cr^{3+} ion concentration yielded a linear pattern, suggesting that the RDC-1 probe can accurately measure Cr^{3+} ion concentration. The association constant (Ka) for Cr^{3+} indicated an excellent interaction between the chemosensor RDC-1 and Cr^{3+} at a 1:1 stoichiometry. The RDC-1 probe

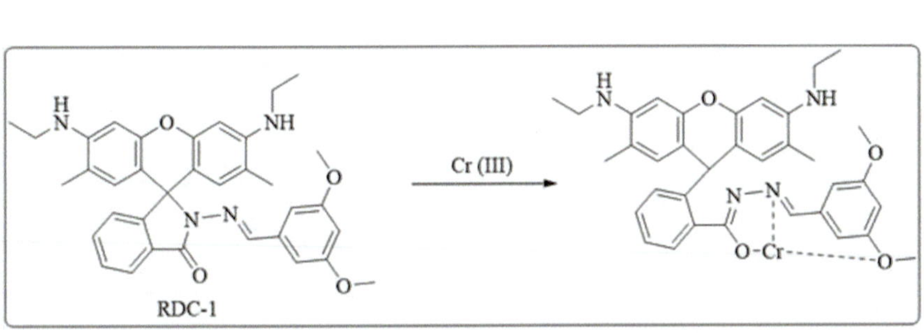

FIGURE 3.8 Complex formation of RDC-1 dye with Cr^{3+}.

detects Cr^{3+} ions in living cells and biological systems with an acceptable limit of detection (LOD). The fluorescence amplification was found to be caused by coordination of Cr^{3+} with RDC-1, resulting in the development of a highly fluorescent ring-opened RDC-1-Cr^{3+} complex, rather than a change in pH when checked for the effect of pH on the Cr^{3+} ions in the solutions. To observe the reversibility of the RDC-1 probe, EDTA titrations were performed. The addition of EDTA reduces the fluorescence of the RDC-1-Cr^{3+} complex and alters the color from pink to colorless. The results demonstrate that the probe for Cr^{3+} sensing is reversible. DFT studies were done using B3LYP/6-311G (d,p) and LANL2DZ basis sets to optimize the geometries of RDC-1 and its Cr^{3+} complex. The results show that the Cr^{3+} interaction enhances the structural stability of the complex, implying its stability. The cytotoxicity of RDC-1 to HeLa cells was assessed using the conventional MTT (3-(4,5-di methylthiazol-2-yl)-2,5-diphenyltetrazolium bromide) test concluding that RDC-1 was not harmful to HeLa cells. Incubating HeLa cells with RDC-1 (10 μM, 2% DMSO) resulted in minimal fluorescence. However, when $CrCl_3$ (10 μM) was added to the cells, intracellular fluorescence increased significantly suggesting that the probe is permeable to the cell membrane. The sensor was tested for its capacity to quantify Cr^{3+} in blood serum and water samples. Adding blood and water samples to the probe RDC-1 (10 μM) did not increase emission intensity, while the samples that were treated with Cr^{3+} salts showed elevated fluorescence (Sunnapu et al., 2017a).

3.2.1.8 Sensing Cu^{2+}

Copper being the third most prevalent heavy metal in the human body plays an important role in numerous physiological processes, namely the activation of dioxygen that is essential for the survival of all living species and also helping in iron absorption, hemopoiesis, enzyme activity, redox reactions etc. Copper ion inconsistencies may lead to ailments such as nausea, fatigue, elevated blood pressure, hemolysis, neurotoxicity, and neurodegenerative disorders. So, Cu^{2+} detection through implementing fluorescent and colorimetric sensors has lately attained scientific importance (Tian et al., 2019).

Abebe et al. synthesized L1 and L2 (Figure 3.9) utilizing microwave irradiation to recognize Cu^{2+} in aqueous solutions precisely. The L1 and L2 solutions were colorless because of ring-closed spirolactam, exhibiting no absorption above 450 nm, while the addition of Cu^{2+} ions resulted in a new absorption band at 560 nm with a shoulder peak at 520 nm that boosted as the concentration of Cu^{2+} increased as the ring-opened amide forms of L1 and L2 are formed through ion interaction, thereby causing the solutions to turn pink instantly, indicating that they can be used for detecting Cu^{2+} ions as naked eye indicators. Cr^{3+} and Ni^{2+} showed very little impact on the probe compared to the other metal ions such as Na^+, K^+, Mg^{2+}, Ca^{2+}, Ni^{2+}, Zn^{2+}, Co^{2+}, Hg^{2+}, Pb^{2+}, Fe^{2+}, Fe^{3+}, Cr^{2+}, and Cu^{2+} in interference study. The conjugated xanthene tautomer in the rhodamine moiety of these compounds caused the emission wavelengths for L1 and L2 sensors to be 580 nm and 578 nm, respectively. L1 was a highly effective Cu^{2+} turn-on sensor given that it displayed a sixfold increase in fluorescence emission intensity in the presence of Cu^{2+} due to the electron withdrawing group on the chromone ring with a LOD of 2.11 μM. Job's plot proved L1 and L2 binding to Cu^{2+} ions in a 1:1 stoichiometry. The fluorescence intensity of L1 and L2 increases, indicating that the sensors' spirolactum ring opens due to intense protonation at lower pH. These analysis studies imply that L1 and L2 can be utilized to collect environmental and biological materials. The L1-Cu^{2+} complex was tested for its stability with anions such as Cl^-, I^-, F^-, ClO_4^-, CH_3COO^-, HSO_4^-, $H_2SO_4^-$, PO_4^{3-}, SCN^-, S^{2-}, HS^-, and OH^- found response for CN^{-1} with color change from pink to colurless (Abebe et al., 2020).

Goswami et al. developed a new rhodamine-based colorimetric and chelation-enhanced (CHEF) "off-on" fluorescent probe 12 for Cu^{2+} (Figure 3.10). Absorbance and fluorescence spectra of this compound with the addition of metal ions like Na^+, Mg^{2+}, Mn^{2+}, Ni^{2+}, Co^{2+}, Zn^{2+}, Cu^{2+}, Cd^{2+}, Hg^{2+}, Pb^{2+}, and Fe^{3+}, caused the probes' absorption bands at 351 and 528 nm to surge substantially with a modest drop at 235 nm in case of its incubation with Cu^{2+} ions. Five isosbestic points were found at around 244, 255, 269, 301, and 320 nm, indicating the formation of probe 12-Cu^{2+} complex in 1:1 ratio as confirmed by ESI-MS. The probe 12 has an emission band at 582 nm demonstrating a significant fluorescence increase upon the progressive addition of Cu^{2+} ions after which significant CHEF impact is seen that is caused by the opening of the spirolactam ring followed by a change into a pink coloration. The competition experiment in the presence of the other metal ions like Cd^{2+}, Co^{2+},

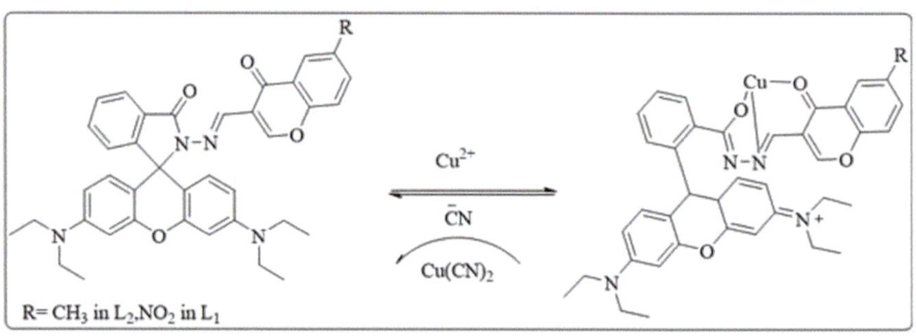

FIGURE 3.9 Rhodamine dye 10 sensing Cu^{2+} that showed reversibility in the presence of CN^-.

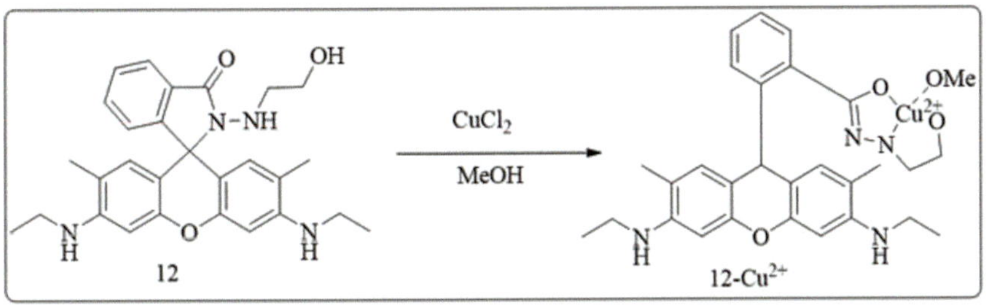

FIGURE 3.10 Rhodamine dye 12 for sensing Cu^{2+}.

Zn^{2+}, Fe^{3+}, Hg^{2+}, Mn^{2+}, Ni^{2+}, Pb^{2+}, Na^{+}, and Mg^{2+} proved the selectivity of the probe towards Cu^{2+}-induced fluorescence enhancement doesn't change in the presence of other metal ions apart from Fe^{3+} proving that the sensor could be employed for the detection of Cu^{2+} ions due to its excellent accuracy (Goswami et al., 2012).

3.2.1.9 Sensing Zn^{2+}

Zn^{2+} impacts biological processes, notably gene transcription, immunological and brain functions, and mammalian reproduction. The metal ions in enzymes and cofactors actively take part in complex cellular functions and biological activities. So, dysregulation of zinc ion concentrations causes cerebral ischemia, ALS, Parkinson's disease, epilepsy, diabetes, and Alzheimer's disease. Therefore, methods for detecting Zn^{2+} influx, aggregation, and outflow have attracted a lot of interest lately (Wechakorn et al., 2016).

Chellappa et al. tested their probe ZRP-1 (Figure 3.11) for its activity against metal ions such as Na^{+}, K^{+}, Ca^{2+}, Cu^{2+}, Zn^{2+}, Cd^{2+}, Hg^{2+}, Pb^{2+}, Fe^{3+}, Cr^{3+}, Ni^{2+}, Ag^{+}, Co^{2+}, and Zn^{2+} and found response only for Zn^{2+} ions, turns the solution red from colorless. The probe becomes highly fluorescent while Fe^{3+} also turns the solution red but doesn't make the probe fluorescent because of the CHEQ effect as Fe^{3+} forms a Fe^{3+}-ZRP-1 complex. Job's plot illustrates that the Zn^{2+}–ZRP-1 complex has a 1:1 binding stoichiometry, which was further proved by the position of ZRP-1 + Zn^{2+} + H^{+} at m/z 565.42 in the ESI mass spectrum. To investigate the reversibility of the probe, the interaction of the Zn^{2+}–ZRP-1 complex with EDTA was analyzed during which fluorescence quenching was observed along with pink color disappearance. Live cell imaging was done to check the cell permeability of the probe in E. coli in which they were incubated with the probe and $ZnCl_2$ showed red fluorescence in the intracellular region (Sivaraman et al., 2012).

Rhodamine-based probe ZRL1 (Figure 3.11) was developed by Lippard et al. It was colorless and nonfluorescent in aqueous media or organic solvents and exhibited no significant absorption or emission band above 500 nm. However, a visible absorption band was formed once zinc ions were added to ZRL1, with a maximum of 595 nm contributing to the formation of pink color and red fluorescence, suggesting that ZRL1 could identify zinc ions by generating the spirolactam ring. ZRL1's fluorescence turn-on is specific toward Zn^{2+} when compared to different monovalent metal ions such as Na^{+}, K^{+}, Ca^{2+}, and Mg^{2+}, the most biologically essential and possible competing metal ions. The fluorescence of the zinc-free and zinc-ZRL1 complex is barely perturbed by divalent transition metal ions apart from Ni^{2+}; Co^{2+} does not induce significant fluorescence turn-on. ZRL1's capacity to monitor zinc ions in living cells was tested using HeLa cells, and it was found that incubation with just the probe had very little fluorescence (Figure 3.11). However, when Zn^{2+} ions were added, these cells showed immense fluorescence upon forming a complex with the probe in vivo, which was quenched upon adding TPEN (Du et al., 2010).

3.2.1.10 Sensing Fe^{3+}

Iron is one of the prevalent transition metals in nature and a necessary component of the human body being utilized in numerous biological activities, including photosynthesis, electron transport, DNA synthesis and repair, oxygen transport, and nitrogen fixation. Ferric ions (Fe^{3+}) are extensively stored in cells in protein-bound or "free" form useful in structural processes or as components of catalytic sites.

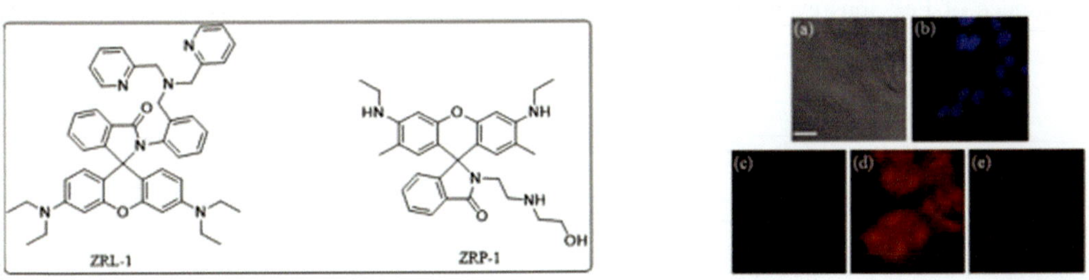

FIGURE 3.11 Rhodamine dyes ZRL-1 and ZRP-1 for sensing Zn^{2+}. Live-cell imaging of HeLa cells after incubation with ZRL1 (10 μM) and Hoechst 33258 nuclear stain (blue). (a) Bright-field transmission image. (b) Nuclear image stained by Hoechst 33258. (c) Fluorescence image without the addition of Zn^{2+}. (d) Fluorescence image of cells incubated with ZRL1 for 15 min, washed twice, and then treated with 50 μM Zn^{2+} pyrithione for 20 min. (e) ZRL1-loaded, 50 μM Zn^{2+}-supplemented cells treated with 80 μM TPEN after 10 min. Scale bar = 25 μm. Reprinted (adapted) with permission from Du et al. (2010). Copyright: ACS, 2010.

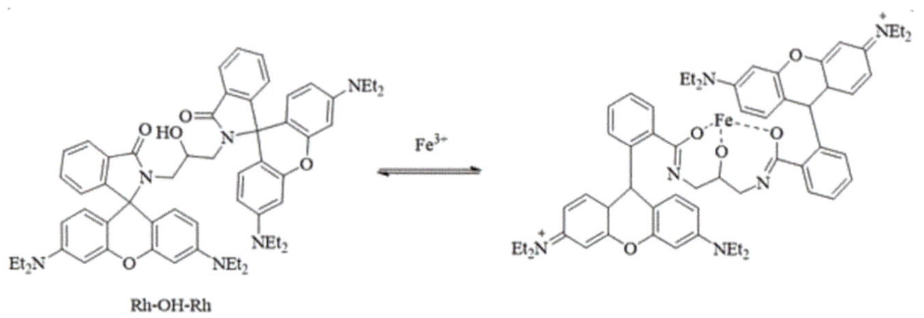

FIGURE 3.12 Rhodamine dye Rh-OH-Rh sensing Fe^{3+}.

Even at concentrations lower than 10^{-18} M, free iron (Fe^{2+} or Fe^{3+}) is perilous because it causes oxidative wreckage within cells. Thus, developing precise and targeted detection techniques for detecting iron is extremely important. (Yang et al., 2015; Dong et al., 2010)

Keeping this in mind Aydin et al. synthesized a turn-on fluorescent sensor Rh-OH-Rh (Figure 3.12) that involves metal-coordination-induced fluorescent activation (CIFA) of rhodamine derivatives. This probe exhibited very faint absorbance beyond 500 nm (in ACN/DMF) portraying that it is spirolactum form but in the presence of Fe^{3+}, an intense absorption band was observed at 556 nm indicating the opening of the spirolactum ring in rhodamine molecules. Titration studies of the probe Rh-OH-Rh with Fe^{3+} indicated 1:1 Fe^{3+}-Rh-OH-Rh complex formation, which was confirmed by Jobs plot, proved by ESI-MS, and its binding constant was found to be 5.2×10^6 M^{-1} while its binding mode is supported by molecular modeling studies that indicated that Fe^{3+} is present in the center of the ring-shaped conformation of the probe. This complex can be unbound in the presence of salicylaldehyde isonicotinoyl hydrazone (SIH). Fluorescence analysis showed that this molecule, in the absence of metal ions, exhibited less intense fluorescence at 592 nm, whereas in the presence of Fe^{3+} ions it showed increased fluorescence intensity at 585 nm up to more than 60-fold, which was not seen in other metals apart from Cu^{2+} whose enhancement was comparatively very inferior (Aydin et al., 2010).

Kan et al. designed and synthesized a reversible "turn-off-on" fluorescence probe N-TC (Figure 3.13) for detecting Fe^{3+} using a rhodamine B analogue. The addition of Fe^{3+} ions disrupted the spirolactum ring, resulting in the emission intensity at 582 nm, indicating that N-TC is responsive to the presence of Fe^{3+} ions even in the presence of other competing ions, making this probe specific for Fe^{3+} detection. A ^{1}H NMR titration experiment proved that they observed proton field shifts because of N-TC-Fe^{3+} complex formation. MTT assay proved that the probe was not cytotoxic at lower concentrations and had well-established cell membrane permeability. N-TC showed a 1:1 binding stoichiometry to Fe^{3+} and had a greater association constant and a lower detection limit. N-TC was effective for detecting trace Fe^{3+} in lake water, with high recovery efficiency. Biological experiments have revealed their potential applicability in living cells (MCF-7 cells), animals (zebra fish), and plant tissues (soyabean) (Aydin et al., 2012).

3.2.1.11 Sensing Hg^{2+}

Mercury is one of the most significant heavy metal ions produced due to different natural processes, industrial spills, and human endeavors including coal mining, gold mining, and volcanic eruptions. Microorganisms can absorb it and convert it into methylmercury (Harada, 1995), which can then enter the food chain and be linked to multiple fatal illnesses (Renzoni et al., 1998; Liu et al., 2017). Jiang et al. synthesized a fluorescent probe ThioRh-1 (Figure 3.14) based on rhodamine for sensing mercury in buffer solution as well as in living cells. The probe showed comparatively faint absorption when compared to in the presence of Hg^{2+} ions where it showed an intense peak with a visible color change into red. The above change is because of spirolactum ring disruption caused by the complexation of the probe with Hg^{2+} ions, which wasn't observed in the case of other metal ions like Cu^{2+}, Pb^{2+}, Cd^{2+}, Zn^{2+}, Ag^{+}, Mn^{2+}, Co^{2+}, Ni^{2+}, Fe^{3+}, Cr^{3+}, Ca^{2+}, Mg^{2+}, and K^{+}. The emission study showed that there was an intense peak at 590 nm upon the addition of 3 equivalents of Hg^{2+}. The complexation was

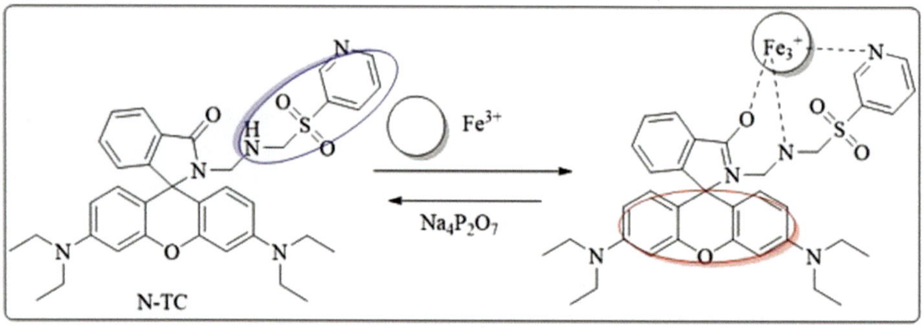

FIGURE 3.13 Rhodamine dye N-TC sensing Fe^{3+}.

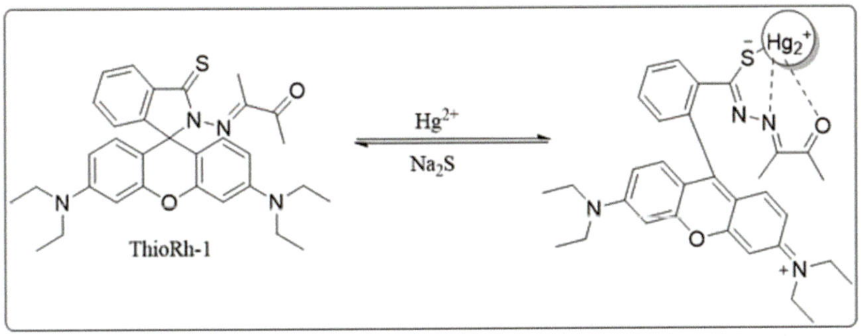

FIGURE 3.14 Rhodamine dye ThioRh-1 sensing Hg^{2+}.

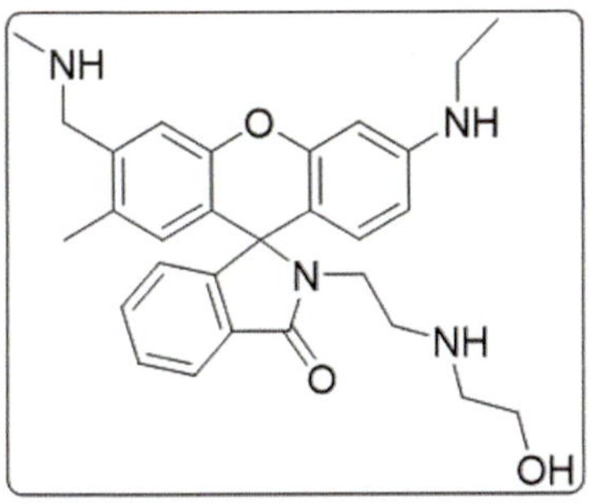

FIGURE 3.15 Rhodamine dye ARP1 for sensing abnormalities in ATP functioning.

accomplished in 3 minutes, and the addition of Na_2S caused the decomplexation of the probe from Hg^{2+} making it facile as a reversible probe. To check the probe usage in living cells, the probe was incubated in A549 cells that didn't show any fluorescence but when Hg^{2+} was added, the cells showed strong fluorescence. This proves that ThioRh-1 is cell permeable and can be used for biological applications (Dong et al., 2010).

3.2.2 Adenosine 5′-Triphosphate Malfunction

Mitochondria synthesize adenosine 5'-triphosphate (ATP) as an energy source in eukaryotic cells (Mitchell, et al., 1961), where ATP synthase takes advantage of a membrane potential within the inner membrane of the mitochondria developed amid oxidative phosphorylation. Mitochondrial disorders (MDs) are irregularities in the membrane potential of eukaryotic cells, resulting in malfunction (de la Fuente-Herreruela et al., 2017). A "turn-on" fluorescent probe ARP-1 (Figure 3.15) was designed and synthesized by Sivaraman et al., which serves as a colorimetric and fluorescent chemosensor for ATP via hydrogen bond interactions. Concerning other nucleoside polyphosphates like ADP and AMP, this probe demonstrates a high degree of selectivity toward ATP. H-bonding is responsible for the advantageous connection between the N atoms of the probe ARP-1 and the triphosphate unit of ATP. The addition of ATP resulted in a new absorption peak at 527 nm coupled with a noticeable color change and increased emission at 557 nm while different anions like Cl^-, Br^-, I^-, NO_3^-, NO_2^-, OAc^-, PF_6^-, ClO_4^-, HPO_4^{2-} ,$P_2O_7^{4-}$, PO_4^{3-}, and F^- ascribed to spirolactam ring opening, although the probe itself didn't show any absorption in the visible area. Job's plot showed that ATP and ARP-1 demonstrated a 1:1 stoichiometry and its LOD was calculated to be 45 nM. HeLa cells were used in the MTT assay to evaluate the probe's cytotoxicity. The results showed that even at 50 micromolar concentration, 90% of the cells were still viable, indicating that the probe is not lethal and is potent enough to be used in biological applications.

Colocalization studies for ARP-1 selectivity in mitochondria in HeLa cells were done where intensity scatter plots indicated that there is a localized association between ARP-1 followed by a strong correlation primarily on mitochondria (Sunnapu et al., 2017b).

3.2.3 Amyloid Beta Detection

According to the World Health Organization data, there are approximately 50 million dementia patients globally, and by 2050, it is predicted to reach 152 million, with 60%–70% of those cases being Alzheimer's disease (AD) (Los et al., 2008). AD patients' brain experiences a comprehensive decline in brain function, which involves memory loss, aphasia, apraxia, agnosia, impaired visuospatial function, decreased executive function, and behavioral and personality changes. The disease has a significant impact on a person's day-to-day functioning (Wood, 2015; Wilson et al., 2012).

Xiang et al. synthesized rhodamine-based fluorescent probe Rho1-4 (Figure 3.16) to detect amyloid beta ($A\beta_{42}$), which is counted as one of the main biomarkers of Alzheimer's disease.

Titration using Cu^{2+} salt caused reduced emission intensity due to the PET effect with a visible color change into purple, while in absorbance spectra, a red shift was observed as Rho-1-Rho-4-Cu metal complexes were formed. As these complexes were added to $A\beta_{42}$, there was reversibility observed due to PET inhibition between rhodamine moiety and copper ions. Molecular docking studies showed that among all four, Rho4-Cu had the lowest binding free energy comparatively. Displacement assay with ThT showed that Rho4-Cu is a much better probe. SPR assay was done to check the affinity of the probe to $A\beta_{42}$. It has been known from this that these molecules are dose dependent. Job's plot showed that Rho4 and Cu^{2+} formed a 1:1 complex. In vitro and in vivo studies proved that the probe Rho4-Cu is capable of effectively differentiating

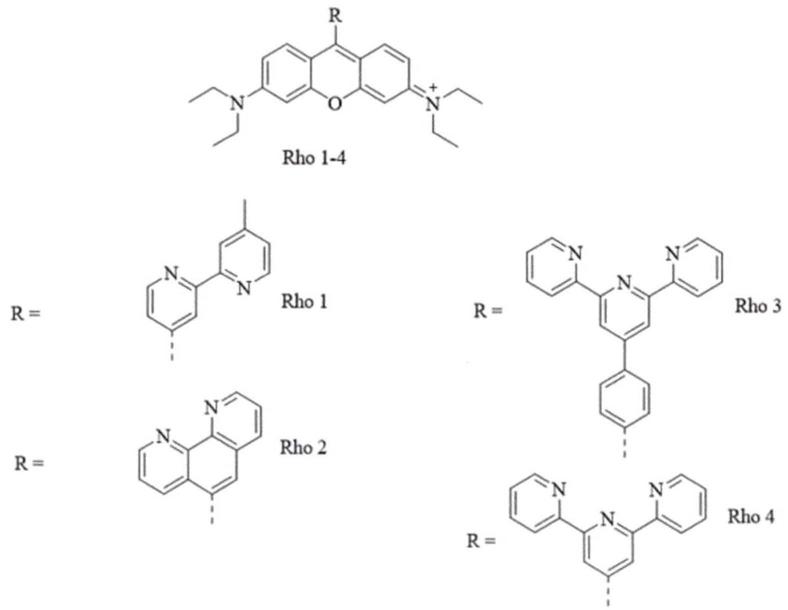

FIGURE 3.16 Rhodamine dyes Rho 1-4 for detection of $A\beta_{42}$.

between AD and normal mice exhibiting great potential for AD diagnosis in the future (Xiang et al., 2022).

3.3 CONCLUSION

Rhodamine dyes are a class of fluorescent dyes known for their widespread applications in biological and chemical research. They are bioconjugated with biomolecules such as proteins and antibodies, allowing for targeted imaging of specific cells or tissues. They are mostly useful in fluorescent microscopy, staining of live and fixed cells, labeling of DNA, RNA, and proteins, tracking cellular processes, clinical diagnostics, environmental monitoring for tracking pollutants in water, industrial use such as textile dyes and marker inks, chemical analysis where they are used in fluorescence microscopy and sensor development, and in nanotechnology as drug delivery systems.

REFERENCES

Abebe, F., Perkins, P., Shaw, R., & Tadesse, S. (2020). A rhodamine-based fluorescent sensor for selective detection of Cu2+ in aqueous media: Synthesis and spectroscopic properties. *Journal of Molecular Structure*, *1205*, 127594.

Aydin, Z., Wei, Y., & Guo, M. (2012). A highly selective rhodamine based turn-on optical sensor for Fe3+. *Inorganic Chemistry Communications*, *20*, 93–96.

Best, M., Porth, I., Hauke, S., Braun, F., Herten, D. P., & Wombacher, R. (2016). Protein-specific localization of a rhodamine-based calcium-sensor in living cells. *Organic & Biomolecular Chemistry*, *14*(24), 5606–5611.

Clapham, D. E. (2007). Calcium signaling. *Cell*, *131*(6), 1047–1058.

de la Fuente-Herreruela, D., Gónzalez-Charro, V., Almendro-Vedia, V. G., Morán, M., Martín, M. Á., Lillo, M. P., ... López-Montero, I. (2017). Rhodamine-based sensor for real-time imaging of mitochondrial ATP in living fibroblasts. *Biochimica et Biophysica Acta (BBA)-Bioenergetics*, *1858*(12), 999–1006.

Dong, L., Zeng, X., Mu, L., Xue, S. F., Tao, Z., & Zhang, J. X. (2010). The synthesis of a rhodamine B schiff-base chemosensor and recognition properties for Fe3+ in neutral ethanol aqueous solution. *Sensors and Actuators B: Chemical*, *145*(1), 433–437.

Du, P., & Lippard, S. J. (2010). A highly selective turn-on colorimetric, red fluorescent sensor for detecting mobile zinc in living cells. *Inorganic Chemistry*, *49*(23), 10753–10755.

Gosselin, R. E. (1984). Hydrogen sulfide. *Clinical Toxicology of Commercial Products*, 198–202.

Goswami, S., Manna, A., Aich, K., & Paul, S. (2012). A simple rhodamine-based naked-eye and fluorescence "off–on" sensor for Cu (II) in aqueous solution. *Chemistry Letters*, *41*(12), 1600–1602.

Harada, M. (1995). Minamata disease: methylmercury poisoning in Japan caused by environmental pollution. *Critical Reviews in Toxicology*, *25*(1), 1–24.

Jiao, Y., Zhu, B., Chen, J., & Duan, X. (2015). Fluorescent sensing of fluoride in cellular system. *Theranostics*, *5*(2), 173.

Kan, C., Wu, L., Wang, X., Shao, X., Qiu, S., & Zhu, J. (2021). Rhodamine B-based chemiluminescence sensor for aluminum ion monitoring and bioimaging applications. *Tetrahedron, 85*, 132054.

Leng, X., Xu, W., Qiao, C., Jia, X., Long, Y., & Yang, B. (2019). New rhodamine B-based chromo-fluorogenic probes for highly selective detection of aluminum (III) ions and their application in living cell imaging. *RSC Advances*, *9*(11), 6027–6034.

Li, Z., Zhao, W., Zhang, Y., Zhang, L., Yu, M., Liu, J., & Zhang, H. (2011). An 'off–on'fluorescent chemosensor of selectivity to Cr3+ and its application to MCF-7 cells. *Tetrahedron*, *67*(37), 7096–7100.

Liu, J., & Qian, Y. (2017). A novel naphthalimide-rhodamine dye: Intramolecular fluorescence resonance energy transfer and ratiometric chemodosimeter for Hg2+ and Fe3+. *Dyes and Pigments*, *136*, 782–790.

Los, G. V., Encell, L. P., McDougall, M. G., Hartzell, D. D., Karassina, N., & Zimprich, C., Zhu, J., Darzins, A., Klaubert, D. H., Bulleit, R. F., & Wood, K. V. (2008). HaloTag: A novel protein labeling technology for cell imaging and protein analysis. *ACS Chemical Biology*, *3*(6), 373–382.

Maity, S. B., & Bharadwaj, P. K. (2013, February). A chemosensor built with rhodamine derivatives appended to an aromatic platform via 1,2,3-triazoles: Dual detection of aluminum(III) and fluoride/acetate ions. *Inorganic Chemistry*, *52*(3), 1161–1163. doi:10.1021/ic301915s. PMID: 23347119.

Mandal, J., Ghorai, P., Pal, K., Bhaumik, T., Karmakar, P., & Saha, A. (2019). Development of rhodamine 6G-based fluorescent chemosensors for Al3+-ion detection: Effect of ring strain and substituent in enhancing its sensing performance. *ACS Omega*, *5*(1), 145–157.

Maniyazagan, M., Mariadasse, R., Jeyakanthan, J., Lokanath, N. K., Naveen, S., Premkumar, K., ... Stalin, T. (2017). Rhodamine based "turn–on" molecular switch FRET–sensor for cadmium and sulfide ions and live cell imaging study. *Sensors and Actuators B: Chemical*, *238*, 565–577.

McElroy, J. A., Shafer, M. M., Trentham-Dietz, A., Hampton, J. M., & Newcomb, P. A. (2006). Cadmium exposure and breast cancer risk. *Journal of the National Cancer Institute*, *98*(12), 869–873.

Mendes, A. M. S., Duda, G. P., Nascimento, C. W. A. D., & Silva, M. O. (2006). Bioavailability of cadmium and lead in a soil amended with phosphorus fertilizers. *Scientia Agricola*, *63*, 328–332.

Mitchell, P. (1961). Coupling of phosphorylation to electron and hydrogen transfer by a chemi-osmotic type of mechanism. *Nature*, *191*(4784), 144–148.

Mondal, S., & Ghosh, K. (2019). Perimedine-linked rhodamine dye in visual sensing of Al3+, Fe3+ and Fe2+ ions in aqueous organic medium under different experimental conditions. *Supramolecular Chemistry*, *31*(10), 645–652.

Nguyen, T., & Francis, M. B. (2003). Practical synthetic route to functionalized rhodamine dyes. *Organic Letters*, *5*(18), 3245–3248.

Rahman, F. U., Ali, A., Khalil, S. K., Guo, R., Zhang, P., Wang, H., ... Zhang, D. W. (2017). Tuning sensitivity of a simple hydrazone for selective fluorescent "turn on" chemo-sensing of Al3+ and its application in living cells imaging. *Talanta*, *164*, 307–313.

Renzoni, A., Zino, F., & Franchi, E. (1998). Mercury levels along the food chain and risk for exposed populations. *Environmental Research*, *77*(2), 68–72.

Sakthivel, P., Sekar, K., Sivaraman, G., & Singaravadivel, S. (2017). Rhodamine diaminomaleonitrile conjugate as a novel colorimetric fluorescent sensor for recognition of Cd 2+ ion. *Journal of Fluorescence*, *27*, 1109–1115.

Shi, Y., Zhang, W., Xue, Y., & Zhang, J. (2023). Fluorescent sensors for detecting and imaging metal ions in biological systems: Recent advances and future perspectives. *Chemosensors*, *11*(4), 226.

Sivaraman, G., Anand, T., & Chellappa, D. (2012). Turn-on fluorescent chemosensor for Zn (II) via ring opening of rhodamine spirolactam and their live cell imaging. *Analyst*, *137*(24), 5881 5884.

Sunnapu, O., Kotla, N. G., Maddiboyina, B., Asthana, G. S., Shanmugapriya, J., Sekar, K., ... & Sivaraman, G. (2017a). Rhodamine based effective chemosensor for Chromium (III) and their application in live cell imaging. *Sensors and Actuators B: Chemical*, *246*, 761–768.

Sunnapu, O., Kotla, N. G., Maddiboyina, B., Marepally, S., Shanmugapriya, J., Sekar, K., ... Sivaraman, G. (2017b). Rhodamine-based fluorescent turn-on probe for facile sensing and imaging of ATP in mitochondria. *ChemistrySelect*, *2*(25), 7654–7658.

Tian, M., He, H., Wang, B. B., Wang, X., Liu, Y., & Jiang, F. L. (2019). A reaction-based turn-on fluorescent sensor for the detection of Cu (II) with excellent sensitivity and selectivity: Synthesis, DFT calculations, kinetics and application in real water samples. *Dyes and Pigments*, *165*, 383–390.

Verougstraete, V., Lison, D., & Hotz, P. (2003). Cadmium, lung and prostate cancer: A systematic review of recent epidemiological data. *Journal of Toxicology and Environmental Health, Part B*, *6*(3), 227–256.

Wechakorn, K., Suksen, K., Piyachaturawat, P., & Kongsaeree, P. (2016). Rhodamine-based fluorescent and colorimetric sensor for zinc and its application in bioimaging. *Sensors and Actuators B: Chemical*, *228*, 270–277.

Wood, H. (2015). Evidence for trans-synaptic and exo-synaptic tau propagation in Alzheimer disease. *Nature Reviews Neurology*, *11*(12), 665–666.

Wilson, R. S., Segawa, E., Boyle, P. A., Anagnos, S. E., Hizel, L. P., & Bennett, D. A. (2012). The natural history of cognitive decline in Alzheimer's disease. *Psychology and Aging*, *27*(4), 1008.

Xiang, J., Xiang, C., Zhou, L., Sun, M., Feng, L., Liu, C., ... Gong, P. (2022). Rational design, synthesis of fluorescence probes for quantitative detection of amyloid-β in Alzheimer's disease based on rhodamine-metal complex. *Analytical Chemistry*, *94*(34), 11791–11797.

Yang, X. H., Li, S., Tang, Z. S., Yu, X. D., Huang, T., & Gao, Y. (2015). A simple, water-soluble, Fe3+-selective fluorescent probe and its application in bioimaging. *Chinese Chemical Letters*, *26*(1), 129–132.

4 BODIPY-Based Fluorescent Probes and their Applications

Mareechika Gaddam and Abha Sharma
National Institute of Pharmaceutical Education and Research, Raebareli, Lucknow, UP, India

LIST OF ABBREVIATIONS

AD	Alzheimer's disease
PET	Photoinduced electron transfer
EDTA	Ethylenediaminetetraacetic acid
BODIPY	Boron-dipyrromethene
HOMO	Highest occupied molecular orbital
LUMO	Lowest unoccupied molecular orbital
EET	Excitation energy transfer
NIR	Near-infrared
ICT	Intermolecular charge transfer
OLEDs	Organic light-emitting Diodes
OPVs	Organic photovoltaics
ACN	Acetonitrile
PD	Parkinson's disease
DPEN	1,2-diphenyl-1,2-ethylenediamine
BPA	Bisphenol-A
Cys	Cysteine
Hcy	Homocysteine
MTG	MitoTracker Green
MTO	MitoTracker Orange
MTR	MitoTracker Red
MTDR	MitoTracker Deep Red
STED	Stimulated emission depletion microscopy
PEG	Polyethylene glycol
A*β*	Amyloid *β*
ThS	Thioflavin S
BBB	Blood brain barrier

4.1 INTRODUCTION

Treibs and Kreuzer were the first to describe BODIPY fluorescent probes as 4,4-difluoro-4-borata-3a,4a-diaza-*s*-indacene in 1968, which has gained success in the last few decades (Treibs & Kreuzer, 1968). BODIPYs belong to the organo difluoroboron complexes class of fluorescent probes, broadly used in bioimaging fields, due to their unique properties such as high molar extinction coefficient, intense absorption, high fluorescence quantum yield, high solubility, photostability, and good biocompatibility (Yang et al., 2019; Debnath et al., 2016). Fluorescent probes that show absorption and emission at far-red (FR) and near-infrared (NIR) (650–900 nm) are preferred for bioimaging in living systems. Usage of these FR/NIR probes is advantageous as there is minimal photo damage to tissues, minimal autoabsorption and autofluorescence from biomolecules in presence of these probes (Bandi et al., 2014).

A common strategy for designing NIR-based probes is intermolecular charge transfer (ICT) where a conjugated structure is used to connect the strong electron-withdrawing and electron-donating groups to form a push-pull structure. The ICT excitation of an electron cloud from the highest occupied molecular orbital (HOMO) to the lowest unoccupied molecular orbital (LUMO) results in a larger Stokes shift, beneficial for bioimaging. As we look into the structure of BODIPY the boron atom is electron deficient, which readily accepts electrons, along with the electron-negative atom fluorine contributes to the electron-withdrawing backbone unit of BODIPY to construct ICT-based probes (Ma et al., 2023).

The main reason of BODIPY's success and adaptability is its chromophoric core, which may be easily accessed by a number of different routes for organic synthesis. The BODIPY core can be specifically functionalized using electrophilic and nucleophilic substitutions, Pd-mediated C-C coupling, Knoevenagel, Suzuki, Sonogashira, and Liebeskind reactions (Benstead et al., 2011). Consequently, a great deal of functional groups can be precisely regioselectively connected to the indacene core. Such substitution patterns also modify the photophysical characteristics of the generated derivative in parallel. Consequently, it is possible to develop: (1) BODIPY, which has spectral bands that are tunable throughout the visible spectrum and even in the near-infrared (Kowada et al., 2015; Lu et al., 2014); (2) novel photophysical processes that can be induced for sensing activity, such as ICT or photoinduced electronic transfer (PET) (Boens *et al.,* 2012), which are sensitive to particular environmental properties or the presence of a particular analyte in the surrounding environment; (3) molecular assemblies that can be tailored to have well-separated absorption and emission regions through excitation energy transfer (EET) (Fan et al., 2013); (4) oligomeric structures can be produced for electrogenerated chemiluminescence (Nepomnyashchii *et al.,* 2011); (v) a multitude of connecting units can be added to enable the grafting of BODIPY to certain targets, such as biomolecules, polymers, nanoparticles, and graphene; (6) due to the "quasi-aromatic" π-system

DOI: 10.1201/9781032706245-4

of BODIPY, which results in a low triplet-state population (Pavlopoulos, 2002), the heavy atom effect can significantly increase the intersystem crossing probability and encourage the formation of singlet oxygen (Kamkaew et al., 2013). In another way, after a thorough and well-thought-out design, customized BODIPYs with the right qualities can be built for the needed application. Theoretical chemists have also become interested in BODIPYs due to their fascinating spectroscopic characteristics.

Derivatives of BODIPY have been applied in a variety of fields outside of biology (Debnath et al., 2015). Optoelectronic devices, such as organic light-emitting diodes (OLEDs) (Chapran *et al.,* 2017) and organic photovoltaics (OPVs) (Ho et al., 2019), have been employed as fluorescent dyes because of their great fluorescence quantum yields and remarkable thermal and chemical stability. Furthermore, BODIPY-based sensors have been used for pollution detection, food quality management, and environmental monitoring (Wang et al., 2022). Biological imaging frequently makes use of BODIPY derivatives (Debela et al., 2021). They are ideal for viewing cellular structures and processes because of their fluorescent properties. They can be used to track the dynamics and localization of specific macromolecules within cells, such as proteins or nucleic acids, and to identify them (Li et al., 2022). Aside from being employed as sensors for various analytes like ions, pH, reactive oxygen species, and enzymatic activities, BODIPY derivatives can also be used to monitor biochemical processes in living systems in real time. Figure 4.1 presents the applications of BODIPY in different fields.

4.2 METAL ION DETECTION

4.2.1 Detection of Hg^{2+} and Cu^{2+} Ions

Detection of transition and heavy metal ions using fluorescent probes has gained attention in recent years, of which Hg^{2+} and Cu^{2+} ions have been explored as two common metal pollutants. Most fluorescent probes are used for the detection of either Hg^{2+} (Zhu et al., 2012; Sirilaksanapong et al., 2012) or Cu^{2+} ion (Chen et al., 2012; Xuan et al., 2012).

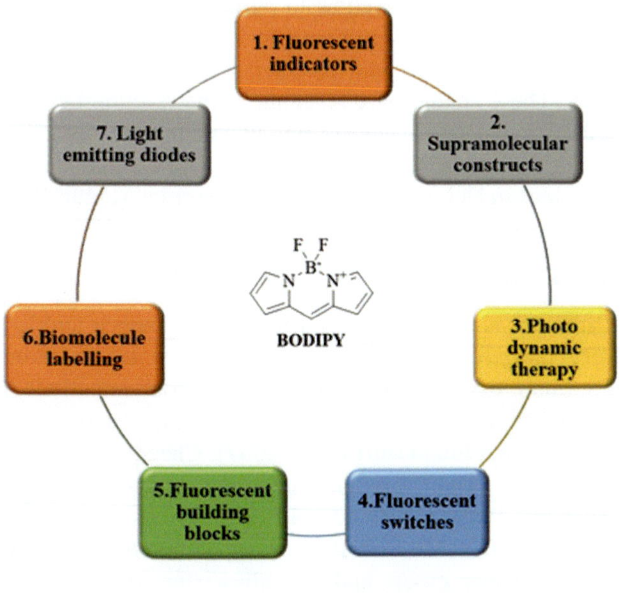

FIGURE 4.1 Applications of BODIPY containing small molecules.

Huang et al. (2019) presented a click-derived near-infrared BODIPY-based probe for the highly selective, ratiometric, and fluorescent detection of Hg^{2+} and Cu^{2+} ions in aqueous solution and live cells. It contains a single bis-(1,2,3-triazole) amino group and can bind to two metal ions (Figure 4.2, Probe 1). This sensor could detect Hg^{2+} and Cu^{2+} ions with varying degrees of spectroscopic change and be visible to the naked eye, as well as act as a ratiometric and discriminative fluorescent probe for dual sensing of Hg^{2+} and Cu^{2+} ions in living cells (Figure 4.2). This probe's exceptional qualities make it a versatile and promising technique for detecting Hg^{2+} and Cu^{2+} ions in biological and environmental applications. He et al. (2014) developed a BODIPY-based dual-mode sensor (Figure 4.3) for the detection of Hg^{2+} and Cu^{2+} ions. This chemosensor consists of bis-[2-(phenylseleno)ethyl] amine as binding site and shows high selectivity toward mercury and copper ions over alkali and transition metal ions. Initially Se-1 shows the absorption at two different wavelengths, maximum at 525 nm and a weaker band at 350 nm in acetonitrile (ACN). The visual appearance of Se-1 in ACN is pink color initially, which changes to purple in the presence of Hg^{2+} ions. The UV spectral changes in the presence of Hg^{2+} include the appearance of a new band at 408 nm and 571 nm. As the concentration of Hg^{2+} increases, the fluorescence intensity is increased with a blue shift of 2 nm. In case of Cu^{2+} addition, great spectral changes have been observed. The two new absorption bands appeared at 404 nm and 589 nm. As the concentration of Cu^{2+} ions increased, the fluorescence intensity enhancement was observed by 13 fold with a visual color change from pink to blue. The detection limits of Se-1 for Hg^{2+} and Cu^{2+} ions are 0.08 μM and 0.53 μM, respectively.

Huang et al. (2013) synthesized a BODIPY-based colorimetric and fluorometric turn-on chemosensor for the detection of Hg^{2+} and Cu^{2+} ions in the presence of various metal ions. The RS-1 shows absorption maxima at 522 nm in ACN solvent, which might correspond to the S1←S0 transition of BODIPY chromophore. RS-1 does not show any changes to metal ions except Hg^{2+} and Cu^{2+}. Strong absorption peaks were induced by Hg^{2+} at 581 nm and 544 nm whose intensity gradually increases with increasing concentration of Hg^{2+} till 20 μM while in case of Cu^{2+} new absorption bands were observed at 564 and 600 nm, the intensity of which increases as the Cu^{2+} concentration increases till 6 μM after which saturation occurs. The pink color solution of RS-1 in ACN turned to purple and blue in the presence of Hg^{2+} and Cu^{2+}, respectively (Figure 4.3). Initially when excited at 522 nm, RS-1 in ACN displays a weak fluorescence peak at 576 nm due to ICT from the nitrogen atom of the picolyl amine moiety to BODIPY core. Upon addition of Hg^{2+} and Cu^{2+} the red shift in emission wavelength was observed with increased intensity by 13 nm and 34 nm, respectively.

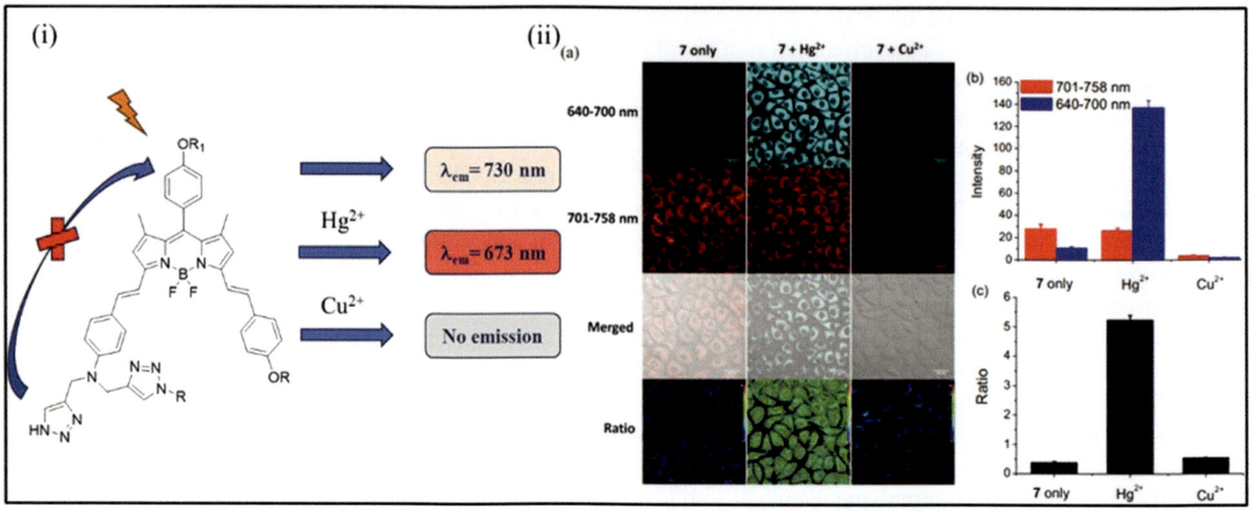

FIGURE 4.2 (i) Structure of Probe 1. (ii) Confocal fluorescence images of living HeLa cells. (Left) Incubated with Probe 1 (6 μM) for 0.5 hour and then further incubated with (middle) Hg^{2+} (120 μM) or (right) Cu^{2+} (120 μM) for 30 minutes. Channel 1 (λ_{em} = 640–700 nm) and Channel 2 (λ_{em} = 701–758 nm) upon excitations at 630 nm. Ratio image was generated from Channel 1 and Channel 2 (I640–700/I701–758). Comparisons of (b) fluorescence intensity of two channels and (c) ratio values ((I640–700/I701–758)) of 7 without and with the treatment of Hg^{2+} or Cu^{2+} ion. Data are expressed as the mean ± standard deviation based on three independent experiments. Reprinted (adapted) with permission from Huang et al. (2019). Copyright 2019, Elsevier.

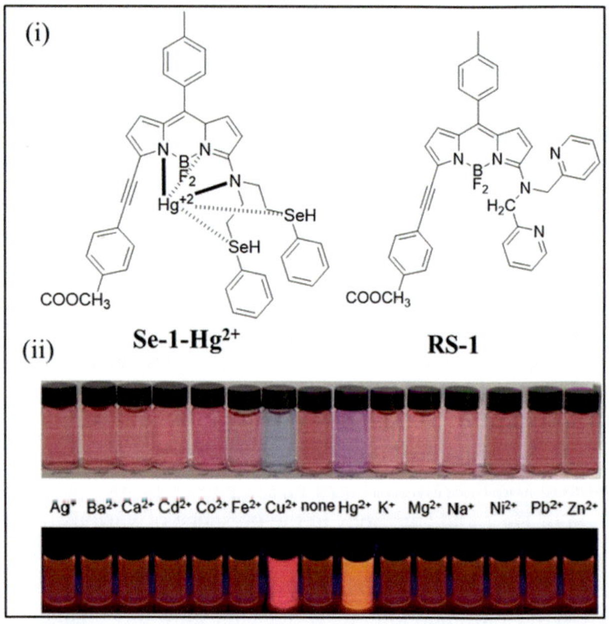

FIGURE 4.3 (i) Structures of Se-1-Hg and RS-1. (ii) Photographs of solutions of RS-1 in ACN upon the addition of various metal ions under UV lamp. Reprinted (adapted) with permission from Huang et al. (2013). Copyright 2013, Elsevier.

4.2.2 Detection of Zn^{2+} Ions

As they are cofactors for metalloproteins and regulators of enzyme performance in brain signal transmission and gene expression, zinc ions (Zn^{2+}) are important for basic physiological functions (Berg et al., 1996; Que et al., 2008; Falchuk, 1998). Numerous neurological conditions and growth retardations, including cerebral ischemia, Alzheimer's, Parkinson's, hypoxia-ischemia, and epilepsy, have been intimately linked to dysregulations in zinc homeostasis (Cuajungco et al., 1997; Choi et al., 1998; Weiss et al., 2000). Though free zinc is also present in various organs, including the brain, pancreas, and the intestine, it is generally accepted that Zn^{2+}, enzyme, and protein activity are related (Xu et al., 2010).

Xia et al. (2017) developed a novel BODIPY-based probe BNDP for selective detection of Zn^{2+} ions (Figure 4.4). BNDP exhibits a significant absorption at 525 nm in CH_3CN/0.02 M HEPES buffer (1:1 v/v, pH 7.0). With increasing concentration of Zn^{2+}, the intensity of the main absorption peak at 525 nm gradually reduced, followed by the creation of a new band centered at 552 nm, with isosbestic point at 535 nm. The solution's color changed noticeably from pale yellow to purple in tandem with the absorption change in response to Zn^{2+} ions. The fluorescence color change was observed from green to salmon pink. These changes are resultant of the reaction between DPEN and methyl acetate of BNDP with the Zn^{2+} ions. The detection limit was found to be 87 nM. Probe BNDP detects Zn^{2+} ions in HeLa cells at 5 μM concentration (Figure 4.6).

4.2.3 Detection of Cd^{2+} Ions

Cadmium is categorized as a human carcinogen and is a very dangerous industrial and environmental pollutant. Cadmium exposure may result in cancerous mutations in a number of organs, including the kidney, endometrium, prostate, and lung (Huff et al., 2007). Cheng et al. (2017) developed BODIPY-based fluorescent probes for Cd^{2+} detection. Probe C-1 (Figure 4.5) is based on tetramethylated BODIPY (Me_4BODIPY) incorporated with BPA, an electron donor group, and also allows for strong binding with Cd^{2+}. There are two prominent peaks in the absorption

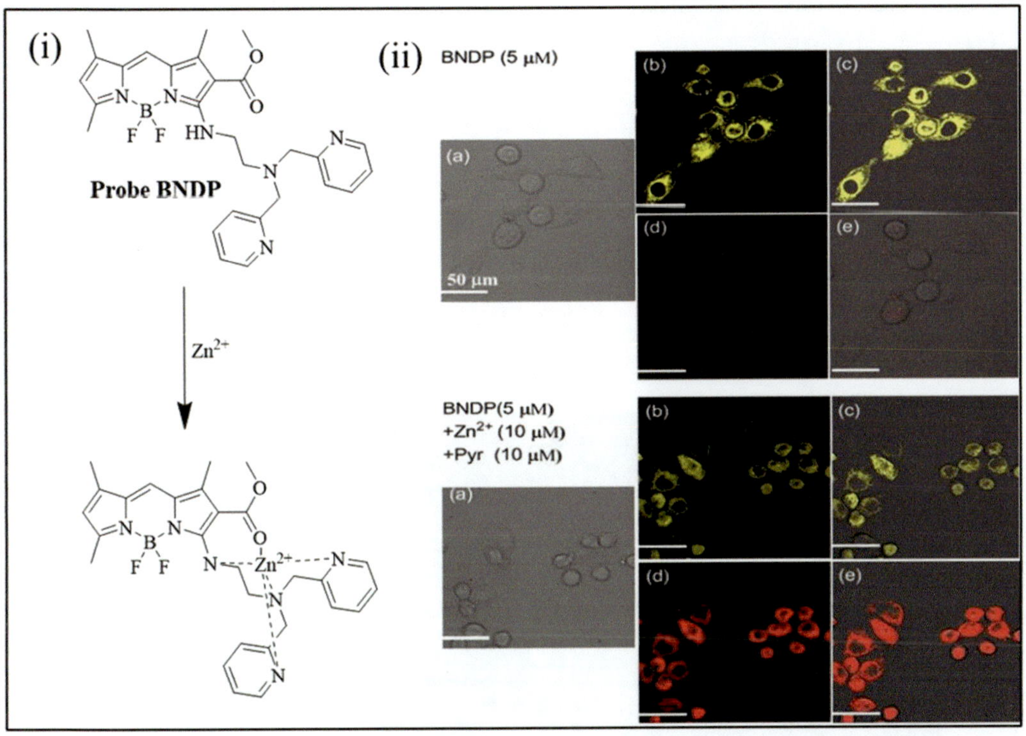

FIGURE 4.4 (i) Probe BNDP mechanism for Zn^{2+} detection. (ii) Fluorescence imaging of HeLa cells with BNDP (5 μM). Cells were supplemented with 10 μM each of zinc (II) chloride and sodium pyrithione (Pyr) for 30 minutes before acquiring images. (a) Bright field; (b) fluorescence recorded at 530–570 nm emission, 515 nm excitation (mKusabira-Orange); (c) overlay images for (a) and (b); (d) fluorescence recorded at 580–620 nm emission, 559 nm excitation (Alexa fluor 564); (e) overlay images for (a) and (d). Reprinted (adapted) with permission from Xia et al. (2018). Copyright 2018, Elsevier.

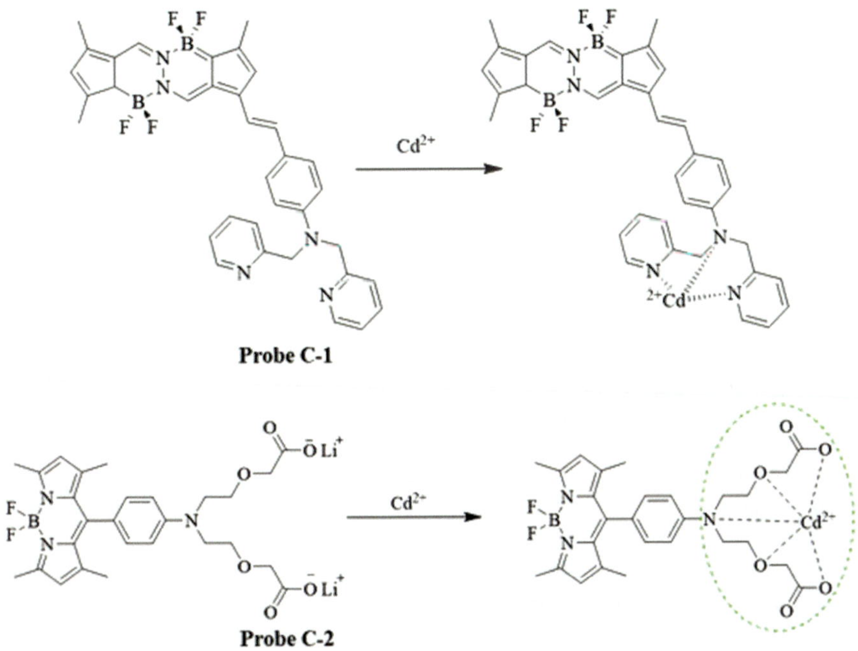

FIGURE 4.5 BODIPY-based probes for cadmium ions detection.

spectra of Probe C-1 at approximately 505 and 550 nm. When compared to the absorption peaks of Me_4BODIPY, which have equivalent peaks at 444 and 467 nm, these two peaks are noticeably red-shifted. As it has been previously noted in other electron donor-acceptor compounds, this spectral red shift is caused by the ICT electronic transition. A rising, blue-shifted absorption peak with a center at 475 nm accompanied the slow decrease of absorption at 550 nm

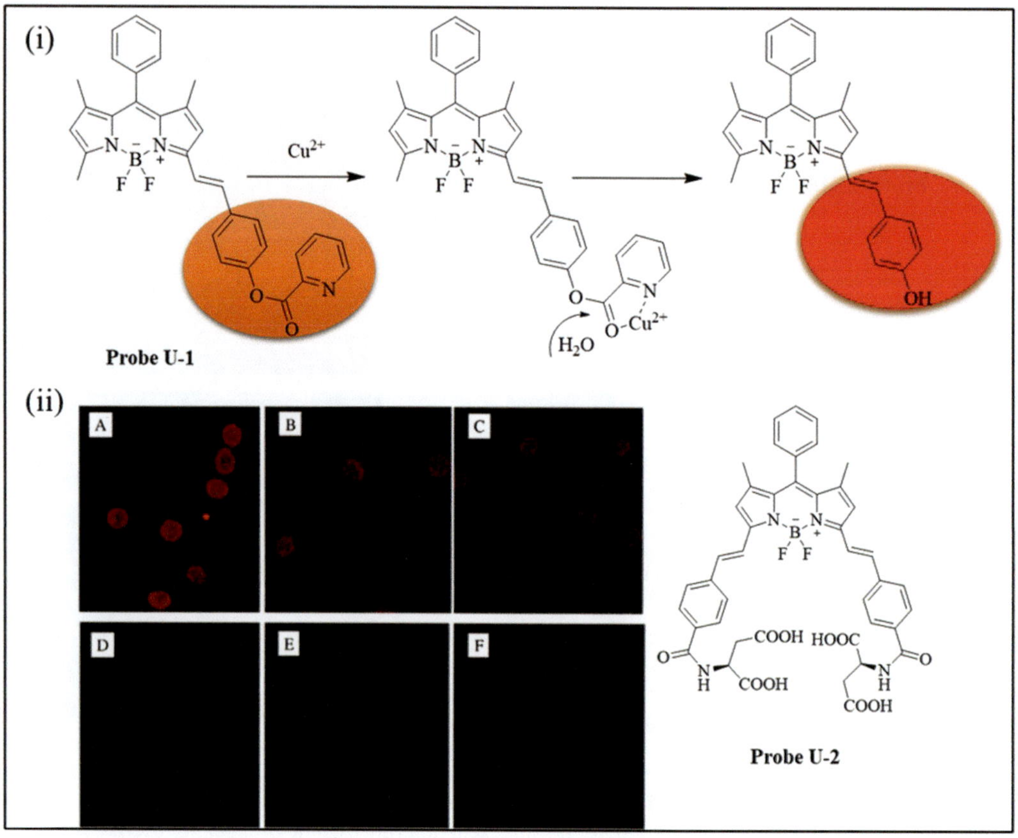

FIGURE 4.6 (i) BODIPY-based probes for copper ions detection. (ii) Fluorescence microscopy images of erythrocytes incubated with 10 (A), 5 (B), 1 (C), 0.5 (D), 0.1 (E), and 0.05 (F) μM (final concentrations) of Probe U-2. Reprinted (adapted) with permission from Papalia et al. (2017). Copyright 2017, Elsevier.

during titration with Cd^{2+} ion. At 520 nm, an isosbestic point was visible, signifying the molecule's stoichiometric conversion. The solution's color changed from red to brilliant yellow as the Cd^{2+} concentration increased, in line with the absorption spectrum shift. The aniline moiety's ability to donate electrons is diminished as a result of Cd^{2+} binding at the BPA chelator, which causes the spectrum shift that has been observed. As a result, the ICT transition of Probe C-1 is decreased. The detection limit was found to be 0.77 ppb.

Maity et al. (2019) synthesized water-soluble OFF-ON BODIPY-based fluorescent probe for Cd^{2+} detection. Probe C-2 (Figure 4.5) is dilithium salt BODIPY derivative, appended with dicarboxylate pseudo crown ether. The normal S_0–S_1 transition for BODIPY dyes was shown by Probe C-2 to have absorption centered at 494 nm. The stimulation of the S_0–S_1 absorption band results in the very faint fluorescence band of the free chromophore appearing at 512 nm. The free probe's faint fluorescence can be attributed to the PET, which functions by transferring electrons from the amine that donates electrons to the BODIPY skeleton. The response of Probe C-2 to a wide range of metal ions, such as alkali (K+ and Na+), alkaline earth metal ions (Ca^{2+}, Mg^{2+}, Sr^{2+}, and Ba^{2+}), and heavy/transition metal ions (Hg^{2+}, Ni^{2+}, Zn^{2+}, Cd^{2+}, Co^{2+}, Cu^{2+}, Fe^{3+}, Cr^{3+}, and Ag^{+}), was measured in aqueous solution. No significant changes were observed in the intensity or wavelength of the absorption band of Probe C-2 upon the addition of these different metal ions. However, the addition of Cd^{2+} resulted in a dramatic increase in fluorescence intensity by about four folds with a negligible red shift (2 nm) of the emission wavelength. For other metal ions, except Zn^{2+}, which showed very slight enhancement in intensity under UV irradiation, a discernible color shift to green was also seen at 365 nm. Following the addition of Cd^{2+}, the probe's green fluorescence is activated by the suppression of the PET process. The detection limit was found to be 35 ppb.

4.2.4 Detection of Cu^{2+} Ions

Cu^{2+} plays a significant role in chemical, physiological, and environmental fields, mainly in enzyme-catalyzed reactions and hemopoiesis. However, major problems or pathological conditions were reported by enhanced or decreased levels of copper; deficiency causes Menke's disease and enhanced levels lead to Wilson's disease. In addition, copper also plays a role in the pathogenesis of Alzheimer's disease and Parkinson's disease (Stern, 2010; Gaggelli et al., 2006; Georgopoulos et al., 2001).

Shu et al. (2018) developed a novel fluorescent probe based on BODIPY bearing 2-picolinic ester for the detection of Cu^{2+} ions in aqueous media (Figure 4.6: Probe U-1). Probe U-1 capacity to detect Cu^{2+} was determined using absorbance and fluorescence titration in a buffer solution (0.02 M HEPES, pH 7.2, 50% DMSO). The absorption band at 561 nm steadily diminished as Cu^{2+} aqueous solution was added to Probe U-1 concurrently; a new absorption band centered at 572 nm developed and increased. This new absorption band is hypothesized to be a part of Probe U-1's hydrolysis product. The fact that the spectra remained unchanged upon adding an excessive amount of EDTA to the solution suggested that the interaction between Cu^{2+} and Probe U-1 was irreversible. The emission band at 574 nm shrank as Cu^{2+} was added slowly, while a new band at 593 nm progressively emerged. The fluorescence quantum yield dropped from 0.93 to 0.52 in the emission band, exhibiting a red shift from 574 to 593 nm. Visual fluorescence color changes were observed under UV lamp from orange to red in case of Cu^{2+} ions, whereas other Probe U-1 fluorescence remains unchanged in case of other metal ions. The detection limit was found to be 0.08 μM. Papalia et al. (2016) synthesized BODIPY derivative for selective detection of Cu^{2+} ions. Probe U-2 emits in red region; upon addition of the Cu^{2+} ions, the fluorescence quenching has been observed. Its capacity to complex Cu^{2+} and internalization within human erythrocytes were both studied (Figure 4.6). When human erythrocytes are exposed to Cu^{2+} solution, Probe U-2 functions as a Cu^{2+} turn-off chemosensor without compromising its cellular viability.

4.2.5 Detection of Fe^{3+} Ions

Iron plays an essential role in many biological processes, which include oxygen transport, energy metabolism, nucleotide synthesis, cell growth, and synthesis of vitamins and hormones (Wilson et al., 2008; Grillo et al., 2017; Sornjai et al., 2020). Imbalance of iron concentrations results in pathological conditions such as Alzheimer's disease and Parkinson's disease in humans. The mechanism of Fe^{3+} in various biological processes is not clear; therefore, it is of great significance to regulate its concentration in living cells (Dornelles et al., 2010).

Leng et al. (2022) developed a BODIPY-based fluorescent probe (BODIPY-Cl) (Figure 4.7) for selective detection

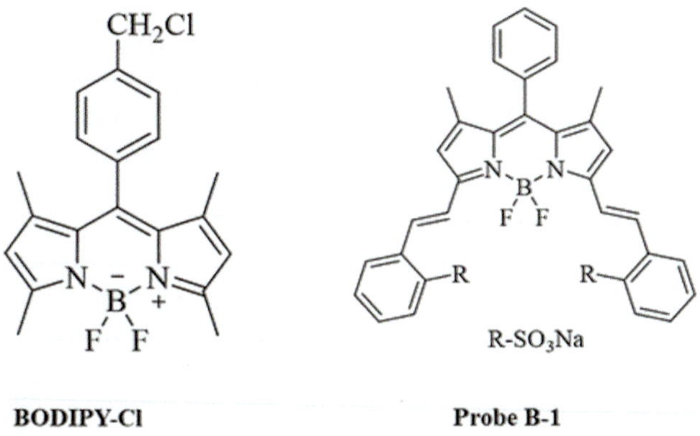

FIGURE 4.7 BODIPY-based fluorescent probes for ferric ions detection.

of Fe^{3+} ions. BODIPY-Cl shows maximum absorption at 498 nm and emission at 516 nm. The BODIPY-Cl has shown a selective response toward Fe^{3+} ions. Quenching of fluorescence was observed at 516 nm in case of Fe^{3+} ions compared to other metal ions. Colorimetric changes were observed in case of both Fe^{3+} and Pb^{2+} ions, i.e., from yellow green to light yellow. Whereas the BODIPY-Cl has shown selectivity toward Fe^{3+} ions by quenching of fluorescence when viewed under UV lamp (green-OFF). The para-chloromethyl benzene unit can form a complex with Fe^{3+} ions and cause quenching in fluorescence. The stoichiometric binding ratio of BODIPY-Cl and Fe^{3+} ions was determined to be 2:1. The detection limit was found to be 2.9 μM. BODIPY-Cl was successfully used to image Fe^{3+} ions in cells.

Ji et al. (2018) developed a highly sensitive BODIPY-based fluorescent probe (Probe B-1) (Figure 4.7) for the detection of Fe^{3+} ions in aqueous media. Probe B-1 consists of a styryl sulfonate side chain responsible for the probe's water solubility. It exhibited absorption maxima at 615 nm and emission at 628 nm. On addition of various metal ions to Probe B-1, slight decrease in fluorescence intensity was observed for other metal ions except Al^{3+} and Fe^{3+}. In case of Fe^{3+}, quenching was observed at the concentration of 40 μM. Fluorescence changes were observed under UV lamp: red fluorescence was quenched in case of Fe^{3+}, whereas slight red fluorescence can be seen in case of Al^{3+}. The detection limit was reported to be 14.2 nM. The binding of Probe B-1 with Fe^{3+} ions is reversible upon the addition of strong chelating agent EDTA.

4.3 OTHER APPLICATIONS

4.3.1 Detection of Cysteine and Homocysteine

The abnormal levels of cysteine (Cys) result in risk factors such as liver damage, skin lesions, etc., whereas homocysteine (Hcy) is a risk factor for diseases such as Behet's, cardiovascular diseases, and AD (Ozkan et al., 2007). Thus, the research is ongoing for the development of sensitive and selective methods for their detection. Recently different methods have been reported using dyes containing aldehydes as sensors.

Yue et al. (2011) have developed a simple method for the detection of Cys and Hcy based on the reaction of aldehyde-containing BODIPY dye with Cys/Hcy. Two fluorescent probes were designed with varied aldehyde positions (Figure 4.8: Probes 1 and 2). The emission wavelength of the probes was found to be 513 nm in 0.1 M HEPES buffer of pH 7.4/ACN (2:3 v/v). On addition of the series of amino acids and GSH to Probe 1, no distinct changes were observed in emission spectra except in the case of Cys and Hcy. Enhanced intensity was observed due to the reaction of aldehyde with Cys/Hcy to form a stable ring of thiazinane or thiazolidine. In case of Probe 2 no discriminate changes were observed in the presence of Cys/Hcy, indicating the fact that the position of aldehyde affects the reactivity of BODIPY.

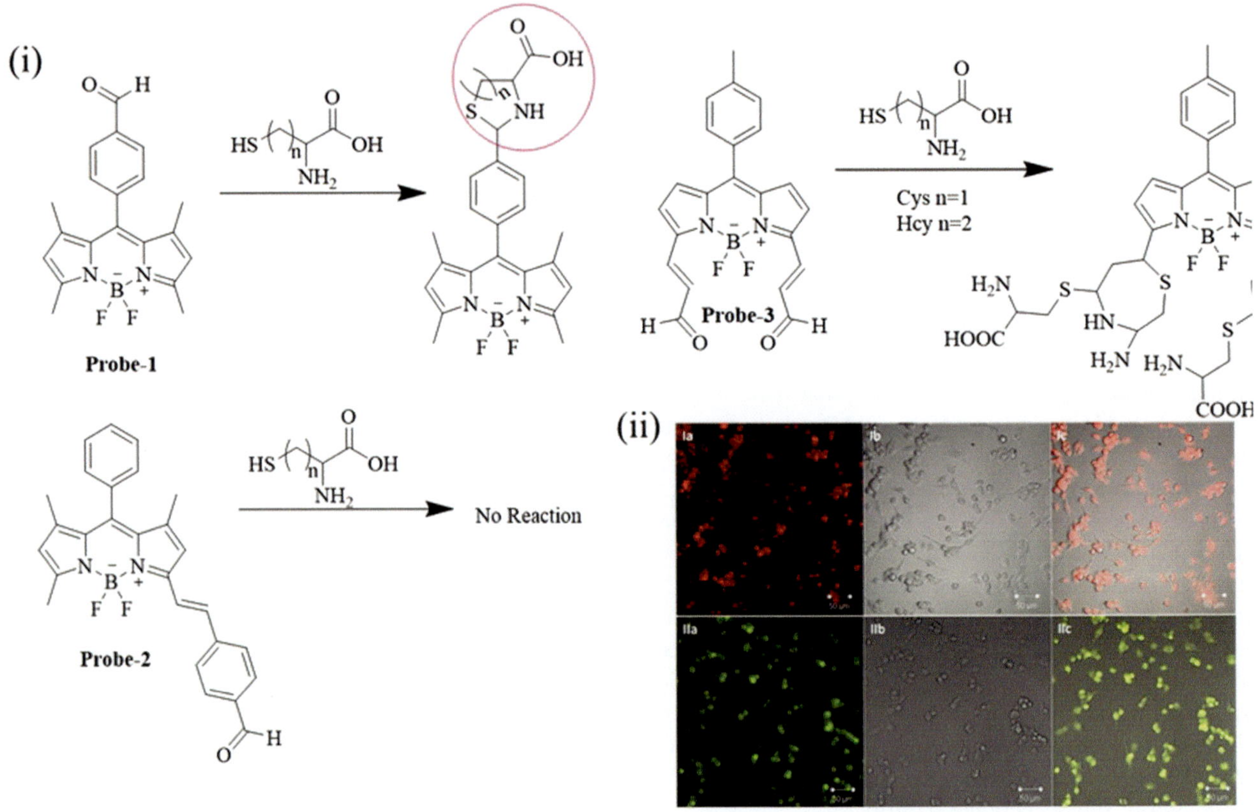

FIGURE 4.8 (i) BODIPY-based probes for cysteine and homocysteine detection. (ii) Confocal laser scanning fluorescence images of living HepG2 cells before (I) and after (II) incubation with Probe 3 (10 μM) with Cys. The left panel represents the fluorescence, the center panel the bright-field transmission images, and the right panel respective bright-field and overlay images; incubation was performed at 37°C for 30 minutes. Scale 50 μm. Reprinted (adapted) with permission from Madhu et al. (2013). Copyright 2013, American Chemical Society.

Madhu et al. (2013) developed red fluorescent probes consisting of 3,5-bis(acrylaldehyde) for the specific detection of Cys and Hcy in living cells. Probe 3 (Figure 4.8) has shown higher emission wavelength and high selectivity compared to others. Probe 3 showed a sharp absorption band at 602 nm and shifted hypsochromically as the polarity of dispersed solvent increased. The absorption maxima was found to be 596 nm in pH 7.4 PBS buffer/ACN (9:1, v/v). Upon addition of Cys to the probe, a decreased intensity was observed at 596 nm absorption band and a new band appeared at 552 nm with a clear isosbestic point at 568 nm. In fluorescence spectra, a hypsochromic shift was observed where the initial emission wavelength of 612 nm is shifted to 567 nm with an isoemissive point at 598 nm. A similar change was observed in the case of Hcy. The general mechanism behind the detection is based on the known fact that the aldehydes react with the thiol groups to form thiazinane or thiazolidine derivatives. The appearance of a new band in absorption and emission spectra indicates the formation of hexahydro-1,4-thiazepine derivative. Ratiometric fluorescence changes were observed in the presence of Cys and Hcy, that is, red color fluorescence turned to green under a UV lamp. This probe had practical applications for visualizing Cys in living cells. The living HepG2 cells treated with the probe showed intense intracellular red fluorescence. In contrast, cells pretreated with Cys and incubated with the probe showed intense yellow-green intracellular fluorescence.

4.3.2 Mitochondria Labeling

Mitochondrial imaging is crucial for understanding illness causation and diagnostics, as abnormal morphology can lead to various diseases (Youle et al., 2012; Chan, 2006). Advancements in microscopy and fluorescent labeling have led to the development of commercial MitoTrackers, including MitoTracker Green (MTG), MitoTracker Orange (MTO), MitoTracker Red (MTR), and MitoTracker Deep Red (MTDR) (Poot et al., 1996; Shim et al., 2012). Positive charges attract mitochondria by electrostatic contact, while benzyl chloride groups establish strong covalent connections with mitochondrial proteins for long-lasting staining and labeling. However, the presence of benzyl chloride groups in these compounds inhibits respiration, leading to cytotoxicity. These dyes have low photostability, making them unsuitable for STED imaging of mitochondrial ultrastructures.

STED, a common super-resolution imaging technique, requires a high depletion laser intensity for optimal resolution (Tam et al., 2015). Developing mitochondrial tags with high imaging capabilities and biocompatibility is important.

Gayatri et al. (2018) developed neutral BODIPY probes for the imaging of mitochondria selectively. Probe M-1 (Figure 4.9)

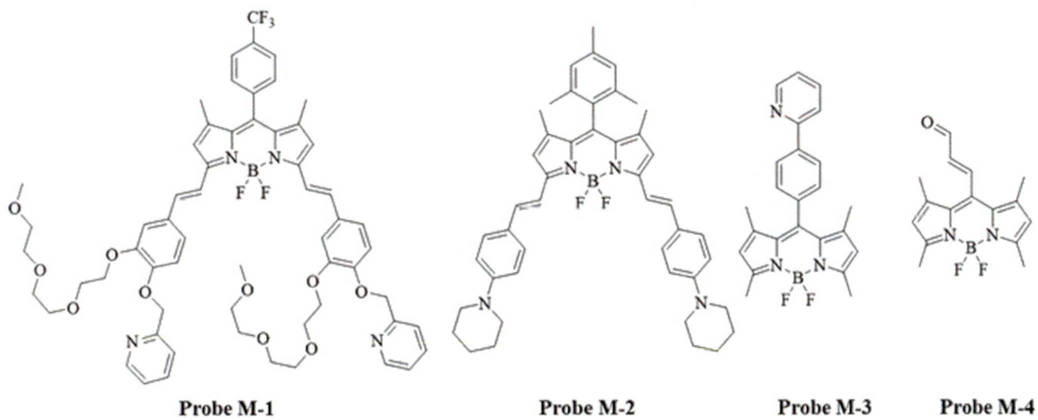

FIGURE 4.9 BODIPY-based probes for mitochondria labeling.

uses short PEG chains to promote cell permeability and biocompatibility. The probe's strong electron-drawing group CF_3 at the meso position results in a +δ character over a specific region, endorsing its enrichment in the mitochondria. Compared to the commercially available mitochondria tracker Red CMXRos, Probe M-1 has shown more specificity. Yang et al. (2021) developed BODIPY-based NIR Probe M-2 (Figure 4.9) consisting of piperidinyl groups. Fluorescence intensity was enhanced gradually, and it tends to be stable with the increased incubation time. Probe M-2 selectively accumulated in mitochondria in living HepG2 cells. Zhu et al. (2021) developed pyridine-BODIPY Probe M-3 (Figure 4.9) for STED imaging of mitochondria. The probe labels mitochondria selectively by protonating the pyridine ring and forming persistent N-S interactions with membrane proteins. In comparison to commercial dyes, Probe M-3 exhibits cristae specificity and excellent photostability, allowing it to observe the dynamic nanostructure of mitochondria at unprecedented resolution and successfully witness the evolution of mitochondria with increased width and reduced cristae under oxidative phosphorylation conditions.

Lincoln et al. (2017) developed a lipophilic BODIPY-acrolein probe to image alkylation events in mitochondria at high resolution. When the probe's acrolein warhead produces a protein adduct in the cell, fluorescence intensity of the Probe M-4 (Figure 4.9) increased by 350 fold. Finally, with the help of Probe M-4, the excretion of mitochondria electrophile adducts and cellular trafficking pathways were studied.

4.3.3 Monitoring ROO· (Peroxy Radicals)

Disrupting redox homeostasis is a key pathogenic factor in many diseases. Research indicates that lipid peroxidation plays a significant role in the development of various diseases, including neurodegenerative disorders and brain tumors (Peña-Bautista et al., 2019; Jaganjac et al., 2021). It promotes the development of oxidized lipids and free radical reactions, leading to damage to cell membranes and other biomolecules. Developing sensors that can specifically monitor lipid peroxyl radical fluctuations in mitochondria is important.

Prime et al. (2012) developed a lipophilic Probe R-1 (Figure 4.10) from C11-BODIPY581/591 to detect lipid peroxidation in mitochondria. The probe's TPP+ moiety specifically binds to the matrix-facing leaflet of the inner

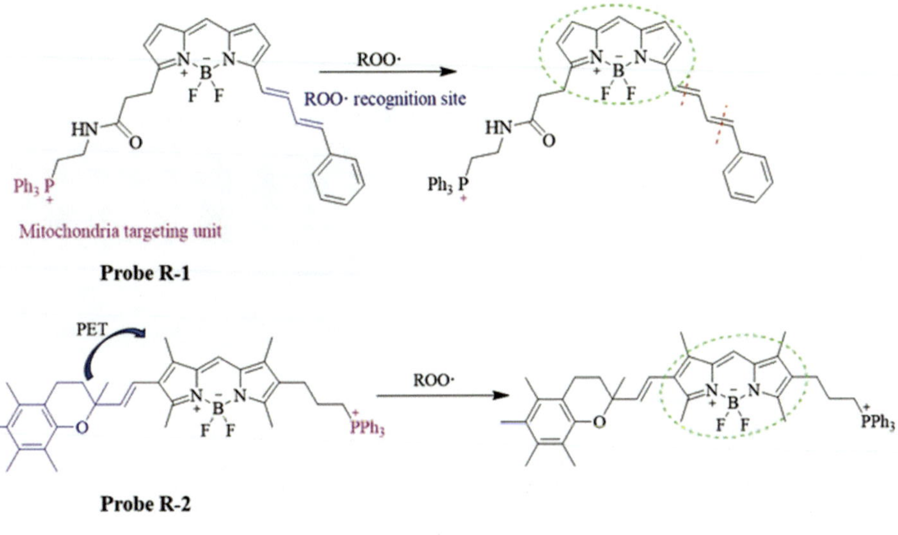

FIGURE 4.10 BODIPY probes for detection of peroxy radicals.

mitochondrial membrane. Exposure to ROO· causes cleavage of the probe's dienyl link, resulting in enhanced emission at 520 nm and a small rise at 590 nm. The emission ratio at 520/590 nm responds to changes in mitochondrial lipid peroxidation. Krumova et al. (2013) used a pre-functionalization method to create a mitochondria-targeted Probe R-2 (Figure 4.10) for lipid peroxyl radicals. The PET process from chromanol to BODIPY initially reduced the probe's fluorescence. However, in the presence of ROO·, the sensing group was oxidized, interrupting the PET mechanism and restoring fluorescence.

The probe enabled real-time monitoring of mitochondrial lipid peroxyl radicals in NIH-3T3 mouse embryonic cells.

4.3.4 Detection of Aβ

Alzheimer's disease is characterized by the accumulation of β-amyloid (Aβ) and tau tangles.

Although numerous Aβ-targeted treatment candidates failed clinical trials, it is considered that Aβ buildup is the first biomarker in the malignant neurodegenerative processes of AD (Hawkes, 2016; Staderini et al., 2015). Early detection is crucial for developing treatments of AD. Currently, the FDA and EMA have approved three amyloid-based radiolabeled tracers (florbetapir, florbetaben, and flutemetamol) (Camus et al., 2012; Richards et al., 2014; Heurling et al., 2016) for measuring Aβ levels in the brain during PET imaging.

Ma et al. (2023) have developed BODIPY-based three different probes consisting of cyclic amines of different sizes, which include TPyrBDP, TPipBDP, and THAIBDP comprising pyrrolidine, piperidine, and azepane, respectively. All three derivatives show maximum emission wavelengths of 752, 751, and 758 nm in DMSO (dimethyl sulfoxide) solvent when excited at 675, 653, and 676 nm, respectively. The ability of these probes to bind Aβ_{1-42} aggregates was determined in comparison to Thioflavin-T. ThT is the standard probe, generally used for the staining of Aβ_{1-42} aggregates in vitro. Enhanced fluorescence intensity is observed after binding to the Aβ_{1-42} aggregates due to the restriction of the molecular rotor. The fluorescence response of TPyrBDP, TPipBDP, and THAIBDP against Aβ_{1-42} was determined in PBS (phosphate buffer saline). The 42.5-fold fluorescence enhancement with a slight red shift of 13 nm was observed in case of TPyrBDP, and 75.5-fold and 27.2-fold enhancement with a blue shift of emission wavelength (27 and 22 nm) was observed in case of TPipBDP and THAIBDP, respectively. All three probes show selectivity toward Aβ_{1-42} aggregates in the presence of BSA, metal ions, and amino acids. TPipBDP crosses the BBB and image Aβplaques in vivo. The PC12 cell viability was found to be above 80% when incubated with the three probes (Figure 4.11). The hemolysis rate of the probes was found to be less than 5 % at the concentration of 100 μM. The K_d (binding affinity) value for the probes TPyrBDP, TPipBDP, and THAIBDP was found to be 54.16 ± 5.82 nm, 28.30 ± 5.94, and 28.16 ± 3.07 nm, respectively. As lower the K_d value better the affinity, the TPipBDP shows best response against Aβ_{1-42} aggregates, and it is chosen for further cell and animal experiments. The results presented in Figure 4.11 indicate that TPipBDP could image Aβ_{1-42} aggregates.

Ren et al. (2018) developed BODIPY-based near-infrared OFF-ON fluorescent probe QAD-1 for staining of amyloid-β plaques. This probe consists of BODIPY as fluorophore and quenching group tetrahydroquinoxaline. QAD-1 shows maximum absorption and emission of 645 and 755 nm in DMSO, respectively. The QAD-1 (5 μM) shows faint fluorescence in PBS (pH 7.4), whereas the fluorescence intensity of QAD-1 has increased in the presence of Aβ_{1-42} fibrils (2 μM, 4 μM, 6 μM, and 8 μM) of varied concentrations. Significant changes were observed in the emission spectra of QAD-1 for different forms of Aβ in case of fibrils and oligomers blue shift of 30 nm in emission wavelength was observed and 70 nm blue shift in case of monomers (Figure 4.12). The reason behind the increased intensity might be an elongated distance between BODIPY and tetrahydroquinoxaline (quencher) when QAD-1 binds to Aβ. Aβ plaques staining ability of QAD-1 was determined in brain tissues of 15-month-old transgenic mouse in comparison with ThS (Thioflavin S). QAD-1 stains broader

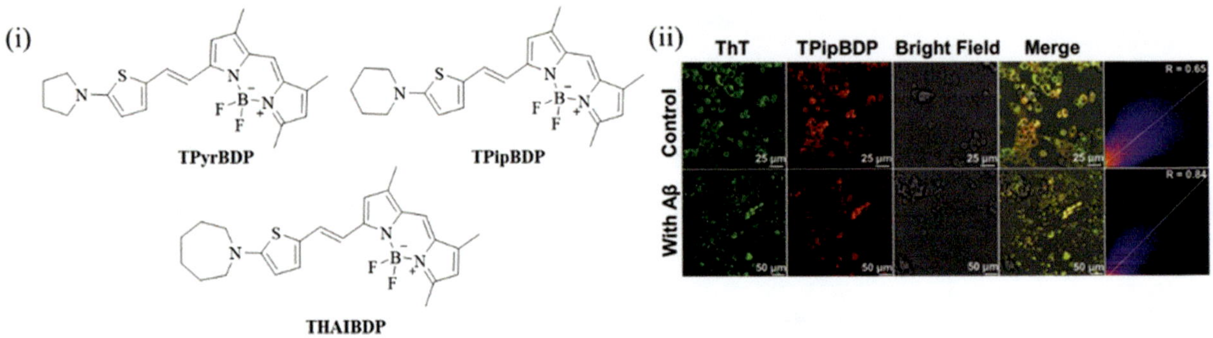

FIGURE 4.11 (i) BODIPY-based probe for amyloid β_{1-42} aggregates detection. (ii) TPipBDP staining Aβ_{1-42} aggregates. Reprinted (adapted) with permission from Ma et al. (2023). Copyright 2023, Wiley-VCH GmbH, Weinheim.

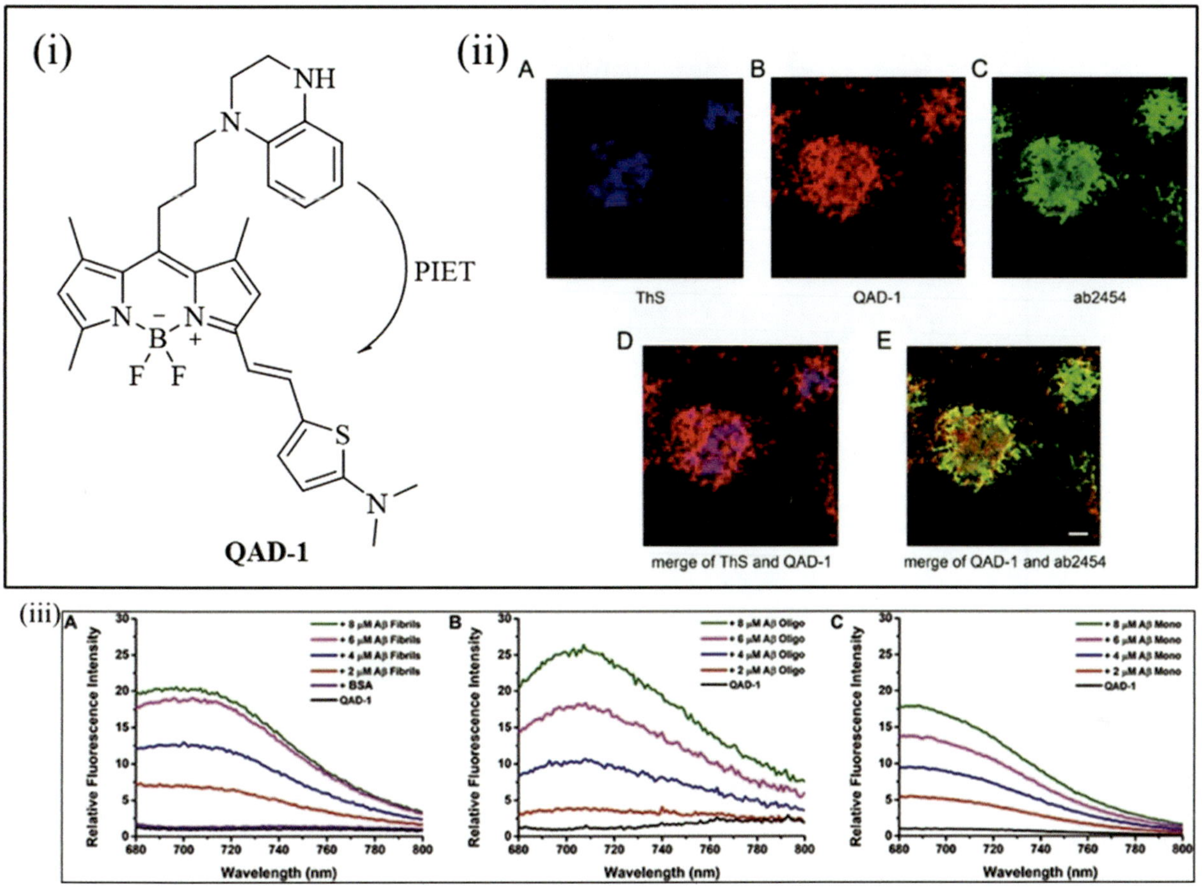

FIGURE 4.12 (i) Structure of QAD-1. (ii) Fluorescent staining of β -amyloid deposits with ThS, QAD-1, and Aβ antibody in the brain slice of 12-month-old APPSWE/PSEN 1dE9 mice. A: ThS staining. B: QAD-1 staining. C: Aβ antibody (ab2454) staining. D: The merge of ThS and QAD-1. E: the merge of QAD-1 and ab2454. Excitation channel = 488 nm, emission channel = 550 nm for ThS; excitation channel = 635 nm, emission channel = 700 nm for QAD-1; excitation channel = 546 nm, emission channel = 573 nm for ab2454; scale bar = 10 μm. (iii) Fluorescent spectra of QAD-1 (5 μM) upon interaction with Aβ (1–42) fibrils (A), oligomers (B), and monomers (C) in PBS (10 mM, pH 7.4) with 10% EtOH. Excitation at 645 nm. Reprinted (adapted) with permission from Ren et al. (2018). Copyright 2018, American Chemical Society.

regions compared to ThS; in order to determine the staining patterns of the probe, triple staining assay was performed using ThS, QAD-1, and Aβ antibody ab2454 using the same brain tissues. The ThS is known to stain Aβ plaques primarily, whereas ab2454 stains both the species of Aβ including monomers and oligomers (Figure 4.12). Comparable staining pattern was observed in case of QAD-1 indicating the interaction between QAD-1 and Aβ monomers and oligomers along with aggregates. Rapid BBB penetration of QAD-1 was observed, which was determined by biodistribution experiments.

A series of BODIPY-based probes were synthesized based on the following strategies: (1) position of methyl group on BODIPY scaffold, (b) conjugated system, and (c) functional group (Figure 4.13). The fluorescent properties of the BAOP series were determined where the emission maxima were found to be more than 650 nm and the quantum yield of all the probes ranges from (8.96% to 55% in $CHCl_3$. The fluorescence changes were observed in the presence of Aβ oligomers or fibrils, where all the probes showed increased intensity by 3.25–140-fold after binding to Aβ oligomers. BAOP-1 and BAOP-14 selectively bind to Aβ oligomers over fibrils with a selectivity index of 1.37–3.06. Whereas , the probes BAP-1, BAOP-1, and BAOP-2 have selective indexes of 2.78, 0.800, and 1.75, respectively, toward Aβ oligomers, indicating that the steric hindrance of methyl position of the BODIPY scaffold affects the selectivity toward Aβ oligomers. The extension of conjugation may not be a necessity for the Aβ oligomer selective probes development as BAOP-3 selectivity was found to be less in comparison to BAP-1. As the excitation and emission wavelengths in chloroform ($CHCl_3$) were found to be higher in case of BAOP-15, BAOP-16, and BAOP-17, these were used for in vivo imaging in comparison to other derivatives. BAOP-16 has shown the highest binding affinity toward Aβ oligomers in comparison to the all other derivatives and reported probe PTO-41. BAOP-16 exhibits no cytotoxicity and good BBB permeability and has been found to selectively detect Aβ oligomers in the brains of Aβ inoculated model mice (Akasaka et al., 2023).

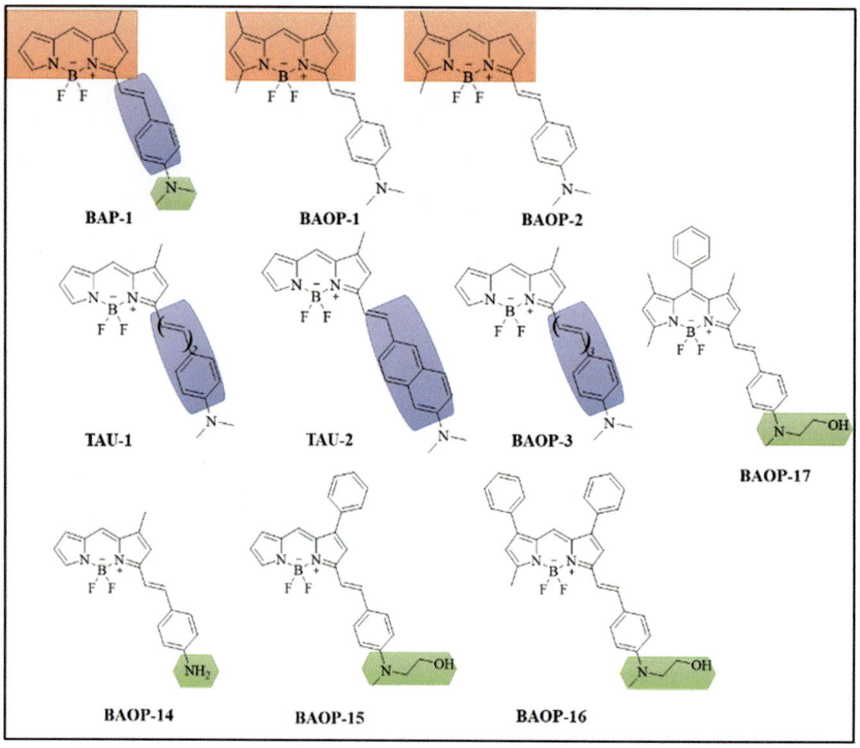

FIGURE 4.13 BODIPY-based probes with modification of BAP-1.

REFERENCES

Akasaka, T., Watanabe, H., & Ono, M. (2023). In vivo near-infrared fluorescence imaging selective for soluble amyloid β aggregates using y-shaped BODIPY derivative. *Journal of Medicinal Chemistry*, *66*(20), 14029–14046.

Bandi, V., Das, S. K., Awuah, S. G., You, Y., & D'Souza, F. (2014). Thieno-pyrrole-fused 4, 4-Difluoro-4-bora-3a, 4a-diaza-s-indacene–Fullerene dyads: Utilization of near-infrared sensitizers for ultrafast charge separation in donor–acceptor systems. *Journal of the American Chemical Society*, *136*(21), 7571–7574.

Benstead, M., Mehl, G. H., & Boyle, R. W. (2011). 4, 4′-Difluoro-4-bora-3a, 4a-diaza-s-indacenes (BODIPYs) as components of novel light active materials. *Tetrahedron*, *67*(20), 3573–3601.

Berg, J. M., & Shi, Y. (1996). The galvanization of biology: A growing appreciation for the roles of zinc. *Science*, *271*(5252), 1081–1085.

Boens, N., Leen, V., & Dehaen, W. (2012). Fluorescent indicators based on BODIPY. *Chemical Society Reviews*, *41*(3), 1130–1172.

Camus, V., Payoux, P., Barré, L., Desgranges, B., Voisin, T., Tauber, C., ... Guilloteau, D. (2012). Using PET with 18 F-AV-45 (florbetapir) to quantify brain amyloid load in a clinical environment. *European Journal of Nuclear Medicine and Molecular Imaging*, *39*, 621–631.

Chan, D. C. (2006). Mitochondria: Dynamic organelles in disease, aging, and development. *Cell*, *125*(7), 1241–1252.

Chapran, M., Angioni, E., Findlay, N. J., Breig, B., Cherpak, V., Stakhira, P., ... Skabara, P. J. (2017). An ambipolar BODIPY derivative for a white exciplex OLED and cholesteric liquid crystal laser toward multifunctional devices. *ACS Applied Materials & Interfaces*, *9*(5), 4750–4757.

Chen, Y., Zhu, C., Yang, Z., Li, J., Jiao, Y., He, W., ... & Guo, Z. (2012). A new "turn-on" chemodosimeter for Hg $^{2+}$: ICT fluorophore formation via Hg $^{2+}$-induced carbaldehyde recovery from 1, 3-dithiane. *Chemical Communications*, *48*(42), 5094–5096.

Cheng, D., Liu, X., Xie, Y., Lv, H., Wang, Z., Yang, H., ... Zang, L. (2017). A ratiometric fluorescent sensor for Cd^{2+} based on internal charge transfer. *Sensors*, *17*(11), 2517.

Choi, D. W., & Koh, J. Y. (1998). Zinc and brain injury. *Annual Review of Neuroscience*, *21*(1), 347–375.

Cuajungco, M. P., & Lees, G. J. (1997). Zinc metabolism in the brain: Relevance to human neurodegenerative disorders. *Neurobiology of Disease*, *4*(3–4), 137–169.

Debela, D. T., Muzazu, S. G., Heraro, K. D., Ndalama, M. T., Mesele, B. W., Haile, D. C., ... Manyazewal, T. (2021). New approaches and procedures for cancer treatment: Current perspectives. *SAGE Open Medicine*, *9*, 20503121211034366.

Debnath, S., Singh, S., Bedi, A., Krishnamoorthy, K., & Zade, S. S. (2015). Synthesis, optoelectronic, and transistor properties of BODIPY-and cyclopenta [c] thiophene-containing π-conjugated copolymers. *The Journal of Physical Chemistry C*, *119*(28), 15859–15867.

Debnath, S., Singh, S., Bedi, A., Krishnamoorthy, K., & Zade, S. S. (2016). Site-selective synthesis and characterization of BODIPY–acetylene copolymers and their transistor properties. *Journal of Polymer Science Part A: Polymer Chemistry*, *54*(13), 1978–1986.

Dornelles, A. S., Garcia, V. A., de Lima, M. N., Vedana, G., Alcalde, L. A., Bogo, M. R., & Schröder, N. (2010). mRNA expression of proteins involved in iron homeostasis in brain regions is altered by age and by iron overloading in the neonatal period. *Neurochemical Research*, *35*, 564–571.

Falchuk, K. H. (1998). The molecular basis for the role of zinc in developmental biology. In *Molecular and cellular effects of nutrition on disease processes* (pp. 41–48), Springer.

Fan, J., Hu, M., Zhan, P., & Peng, X. (2013). Energy transfer cassettes based on organic fluorophores: Construction and applications in ratiometric sensing. *Chemical Society Reviews*, *42*(1), 29–43.

G. Georgopoulos, A., Roy, M. J., Yonone-Lioy, R. E., Opiekun, P. J., & Lioy, P. (2001). Environmental copper: Its dynamics and human exposure issues. *Journal of Toxicology and Environmental Health Part B: Critical Reviews*, *4*(4), 341–394.

Gaggelli, E., Kozlowski, H., Valensin, D., & Valensin, G. (2006). Copper homeostasis and neurodegenerative disorders (Alzheimer's, prion, and Parkinson's diseases and amyotrophic lateral sclerosis). *Chemical Reviews*, *106*(6), 1995–2044.

Gayathri, T., Karnewar, S., Kotamraju, S., & Singh, S. P. (2018). High affinity neutral bodipy fluorophores for mitochondrial tracking. *ACS Medicinal Chemistry Letters*, *9*(7), 618–622.

Grillo, A. S., SantaMaria, A. M., Kafina, M. D., Cioffi, A. G., Huston, N. C., Han, M., ... Burke, M. D. (2017). Restored iron transport by a small molecule promotes absorption and hemoglobinization in animals. *Science*, *356*(6338), 608–616.

Hawkes, N. (2016). Amyloid plaques are still main target for Alzheimer's drugs. *British Medical Journal*, *352*, i214.

He, X., Zhang, J., Liu, X., Dong, L., Li, D., Qiu, H., & Yin, S. (2014). A novel BODIPY-based colorimetric and fluorometric dual-mode chemosensor for Hg^{2+} and Cu^{2+}. *Sensors and Actuators B: Chemical*, *192*, 29–35.

Heurling, K., Leuzy, A., Zimmer, E. R., Lubberink, M., & Nordberg, A. (2016). Imaging β -amyloid using [18 F] flutemetamol positron emission tomography: From dosimetry to clinical diagnosis. *European Journal of Nuclear Medicine and Molecular Imaging*, *43*, 362–373.

Ho, D., Özdemir, R., Kim, H., Earmme, T., Usta, H., & Kim, C. (2019). BODIPY-based semiconducting materials for organic bulk heterojunction photovoltaics and thin-film transistors. *ChemPlusChem*, *84*(1), 18–37.

Huang, J., Ma, X., Liu, B., Cai, L., Li, Q., Zhang, Y., ... Yin, S. (2013). A colorimetric and ratiometric turn-on BODIPY-based fluorescent probe for double-channel detection of Cu^{2+} and Hg^{2+}. *Journal of Luminescence*, *141*, 130–136.

Huang, Y., Li, C. F., Shi, W. J., Tan, H. Y., He, Z. Z., Zheng, L., ... Yan, J. W. (2019). A near–infrared BODIPY-based fluorescent probe for ratiometric and discriminative detection of Hg^{2+} and Cu^{2+} ions in living cells. *Talanta*, *198*, 390–397.

Huff, J., Lunn, R. M., Waalkes, M. P., Tomatis, L., & Infante, P. F. (2007). Cadmium-induced cancers in animals and in humans. *International Journal of Occupational and Environmental Health*, *13*(2), 202–212.

Jaganjac, M., Cindrić, M., Jakovčević, A., Žarković, K., & Žarković, N. (2021). Lipid peroxidation in brain tumors. *Neurochemistry International*, *149*, 105118.

Ji, J., Chereddy, S. S., Ren, Y., Chen, X., Su, D., Zhong, Z., ... Yang, C. (2018). A BODIPY-based near infrared fluorescent probe for Fe^{3+} in water. *Journal of Photochemistry and Photobiology A: Chemistry*, *355*, 78–83.

Kamkaew, A., Lim, S. H., Lee, H. B., Kiew, L. V., Chung, L. Y., & Burgess, K. (2013). BODIPY dyes in photodynamic therapy. *Chemical Society Reviews*, *42*(1), 77–88.

Kowada, T., Maeda, H., & Kikuchi, K. (2015). BODIPY-based probes for the fluorescence imaging of biomolecules in living cells. *Chemical Society Reviews*, *44*(14), 4953–4972.

Krumova, K., Greene, L. E., & Cosa, G. (2013). Fluorogenic α-tocopherol analogue for monitoring the antioxidant status within the inner mitochondrial membrane of live cells. *Journal of the American Chemical Society*, *135*(45), 17135–17143.

Leng, J., Lan, X., Liu, S., Jia, W., Cheng, W., Cheng, J., & Liu, Z. (2022). Synthesis and bioimaging of a BODIPY-based fluorescence quenching probe for Fe $^{3+}$. *RSC Advances*, *12*(33), 21332–21339.

Li, F. Z., Wu, Z., Lin, C., Wang, Q., & Kuang, G. C. (2022). Photophysical properties regulation and applications of BODIPY-based derivatives with electron donor-acceptor system. *Results in Chemistry*, *4*, 100384.

Lincoln, R., Greene, L. E., Zhang, W., Louisia, S., & Cosa, G. (2017). Mitochondria alkylation and cellular trafficking mapped with a lipophilic BODIPY–acrolein fluorogenic probe. *Journal of the American Chemical Society*, *139*(45), 16273–16281.

Lu, H., Mack, J., Yang, Y., & Shen, Z. (2014). Structural modification strategies for the rational design of red/NIR region BODIPYs. *Chemical Society Reviews*, *43*(13), 4778–4823.

Ma, L., Geng, Y., Zhang, G., Hu, Z., James, T. D., Wang, X., & Wang, Z. (2023). Near-infrared bodipy-based molecular rotors for β -Amyloid imaging in Vivo. *Advanced Healthcare Materials*, *12*(25), 2300733.

Madhu, S., Gonnade, R., & Ravikanth, M. (2013). Synthesis of 3, 5-bis (acrylaldehyde) boron-dipyrromethene and application in detection of cysteine and homocysteine in living cells. *The Journal of Organic Chemistry*, *78*(10), 5056–5060.

Maity, A., Ghosh, U., Giri, D., Mukherjee, D., Maiti, T. K., & Patra, S. K. (2019). A water-soluble BODIPY based 'OFF/ON' fluorescent probe for the detection of Cd $^{2+}$ ions with high selectivity and sensitivity. *Dalton Transactions*, *48*(6), 2108–2117.

Nepomnyashchii, A. B., Bröring, M., Ahrens, J., & Bard, A. J. (2011). Synthesis, photophysical, electrochemical, and electrogenerated chemiluminescence studies. Multiple sequential electron transfers in BODIPY monomers, dimers, trimers, and polymer. *Journal of the American Chemical Society*, *133*(22), 8633–8645.

Özkan, Y., Yardım-Akaydın, S., Sepici, A., Engin, B., Sepici, V., & Şimşek, B. (2007). Assessment of homocysteine, neopterin and nitric oxide levels in Behcet's disease. *Clinical Chemistry and Laboratory Medicine*, *45*(1), 73–77.

Papalia, T., Barattucci, A., Barreca, D., Bellocco, E., Bonaccorsi, P., Minuti, L., ... Foti, C. (2017). Sequestering ability to Cu^{2+} of a new bodipy-based dye and its behavior as in vitro fluorescent sensor. *Journal of Inorganic Biochemistry*, *167*, 116–123.

Pavlopoulos, T. G. (2002). Scaling of dye lasers with improved laser dyes. *Progress in Quantum Electronics*, *26*(4–5), 193–224.

Peña-Bautista, C., Vento, M., Baquero, M., & Cháfer-Pericás, C. (2019). Lipid peroxidation in neurodegeneration. *Clinica Chimica Acta*, *497*, 178–188.

Poot, M., Zhang, Y. Z., Krämer, J. A., Wells, K. S., Jones, L. J., Hanzel, D. K., ... Haugland, R. P. (1996). Analysis of mitochondrial morphology and function with novel fixable fluorescent stains. *Journal of Histochemistry & Cytochemistry*, *44*(12), 1363–1372.

Prime, T. A., Forkink, M., Logan, A., Finichiu, P. G., McLachlan, J., Pun, P. B. L., ... Murphy, M. P. (2012). A ratiometric fluorescent probe for assessing mitochondrial phospholipid peroxidation within living cells. *Free Radical Biology and Medicine*, *53*(3), 544–553.

Que, E. L., Domaille, D. W., & Chang, C. J. (2008). Metals in neurobiology: probing their chemistry and biology with molecular imaging. *Chemical Reviews*, *108*(5), 1517–1549.

Ren, W., Zhang, J., Peng, C., Xiang, H., Chen, J., Peng, C., ... Hu, Y. (2018). Fluorescent imaging of β -amyloid using BODIPY based near-infrared off–on fluorescent probe. *Bioconjugate Chemistry*, *29*(10), 3459–3466.

Richards, D., & Sabbagh, M. N. (2014). Florbetaben for PET imaging of beta-amyloid plaques in the brain. *Neurology and Therapy*, *3*, 79–88.

Shim, S. H., Xia, C., Zhong, G., Babcock, H. P., Vaughan, J. C., Huang, B., ... Zhuang, X. (2012). Super-resolution fluorescence imaging of organelles in live cells with photoswitchable membrane probes. *Proceedings of the National Academy of Sciences*, *109*(35), 13978–13983.

Shu, T., Yang, Z., Cen, Z., Deng, X., Deng, Y., Dong, C., & Yu, Y. (2018). A novel ratiometric fluorescent probe based on a BODIPY derivative for Cu $^{2+}$ detection in aqueous solution. *Analytical Methods*, *10*(48), 5755–5762.

Sirilaksanapong, S., Sukwattanasinitt, M., & Rashatasakhon, P. (2012). 1, 3, 5-Triphenylbenzene fluorophore as a selective Cu $^{2+}$ sensor in aqueous media. *Chemical Communications*, *48*(2), 293–295.

Sornjai, W., Van Long, F. N., Pion, N., Pasquer, A., Saurin, J. C., Marcel, V., ... Smith, D. R. (2020). Iron and hepcidin mediate human colorectal cancer cell growth. *Chemico-Biological Interactions*, *319*, 109021.

Staderini, M., Martín, M. A., Bolognesi, M. L., & Menéndez, J. C. (2015). Imaging of β -amyloid plaques by near infrared fluorescent tracers: a new frontier for chemical neuroscience. *Chemical Society Reviews*, *44*(7), 1807–1819.

Stern, B. R. (2010). Essentiality and toxicity in copper health risk assessment: overview, update and regulatory considerations. *Journal of Toxicology and Environmental Health, Part A*, *73*(2–3), 114–127.

Tam, J., & Merino, D. (2015). Stochastic optical reconstruction microscopy (STORM) in comparison with stimulated emission depletion (STED) and other imaging methods. *Journal of Neurochemistry*, *135*(4), 643–658.

Treibs, A., & Kreuzer, F. H. (1968). Difluoroboryl complexes of di- and tripyrrylmethenes. *Justus Liebigs Annalen der Chemie*, *718* (1), 208–223.

Wang, F., Zhang, X., Huangfu, C., Zhu, M., Li, C., & Feng, L. (2022). The BODIPY-based chemosensor for the fluorometric determination of organochlorine pesticide dicofol. *Food Chemistry*, *370*, 131033.

Weiss, J. H., Sensi, S. L., & Koh, J. Y. (2000). Zn^{2+}: A novel ionic mediator of neural injury in brain disease. *Trends in Pharmacological Sciences*, *21*(10), 395–401.

Wilson, M. T., & Reeder, B. J. (2008). Oxygen-binding haem proteins. *Experimental Physiology*, *93*(1), 128–132.

Xia, S., Shen, J., Wang, J., Wang, H., Fang, M., Zhou, H., & Tanasova, M. (2018). Ratiometric fluorescent and colorimetric BODIPY-based sensor for zinc ions in solution and living cells. *Sensors and Actuators B: Chemical*, *258*, 1279–1286.

Xu, Z., Yoon, J., & Spring, D. R. (2010). Fluorescent chemosensors for Zn^{2+}. *Chemical Society Reviews*, *39*(6), 1996–2006.

Xuan, W., Chen, C., Cao, Y., He, W., Jiang, W., Liu, K., & Wang, W. (2012). Rational design of a ratiometric fluorescent probe with a large emission shift for the facile detection of Hg $^{2+}$. *Chemical Communications*, *48*(58), 7292–7294.

Yang, J., Fan, Y., Cai, F., Xu, X., Fu, B., Wang, S., ... Xu, H. (2019). BODIPY derivatives bearing borneol moieties: Enhancing cell membrane permeability for living cell imaging. *Dyes and Pigments*, *164*, 105–111.

Yang, J., Zhang, R., Zhao, Y., Tian, J., Wang, S., Gros, C. P., & Xu, H. (2021). Red/NIR neutral BODIPY-based fluorescent probes for lighting up mitochondria. *Spectrochimica Acta Part A: Molecular and Biomolecular Spectroscopy*, *248*, 119199.

Youle, R. J., & Van Der Bliek, A. M. (2012). Mitochondrial fission, fusion, and stress. *Science*, *337*(6098), 1062–1065.

Yue, Y., Guo, Y., Xu, J., & Shao, S. (2011). A Bodipy-based derivative for selective fluorescence sensing of homocysteine and cysteine. *New Journal of Chemistry*, *35*(1), 61–64.

Zhu, T., Yang, G., Liu, X., Xiang, P., Yang, Z., Zhang, S., ... Tian, X. (2021). Live cell mitochondrial 3-dimensional dynamic ultrastructures under oxidative phosphorylation revealed by a Pyridine-BODIPY probe. *Biosensors and Bioelectronics*, *178*, 113036.

Zhu, W., Huang, X., Guo, Z., Wu, X., Yu, H., & Tian, H. (2012). A novel NIR fluorescent turn-on sensor for the detection of pyrophosphate anion in complete water system. *Chemical Communications*, *48*(12), 1784–1786.

5 Julolidine-Based Molecules
Design, Synthesis, and Applications

Krishna Kumar, Dheeraj Pandey and Abha Sharma
National Institute of Pharmaceutical Education and Research,
Raebareli, Lucknow, UP, India

LIST OF ABBREVIATIONS

AD	Alzheimer's disease
ICT	Intramolecular charge transfer
CHEF	Chelation enhanced fluorescence
ESIPT	Excited-state intramolecular proton transfer
PET	Photoinduced electron transfer
PJI	Phenyl julolidineimine
LOD	Limit of detection
RFAP	Ratiometric formaldehyde probe
WHO	World Health Organization
TICT	Twisted internal charge transfer
NIR	Near-infrared
MTT	(3-[4,5-dimethylthiazole-2-yl]-2,5 diphenyl tetrazolium bromide
PBS	Phosphate-buffered saline
GSH	Glutathione
NEM	N-Ethylmaleimide
UMRs	Ultrafast molecular rotor
UBER	Universal base excision reporter
BER	Base excision repair
EGCG	Epigallocatechin gallate
DCVJ	9-dicyanovinyljulolidine
DFT	Density functional theory
RFAP	Ratiometric formaldehyde probe
FA	Formaldehyde
TTR	Transthyretin
ADH	Adenosine diphosphate
ATP	Adenosine triphosphate

5.1 INTRODUCTION

Julolidine is a nitrogen-containing heterocyclic aromatic organic compound, and its first synthesis was reported by G. Pinkus in 1892. The chemical formula of julolidine is $C_{12}H_{15}N$, and the IUPAC nomenclature describes julolidine as 2,3,6,7-tetrahydro-1H,5H-pyrido[3,2,1-ij]quinolone. Julolidine acts as a fluorescent probe for monitor solid-state polymerization (Jbilou et al., 2016), sensing volatile organic compounds (VOCs) (Martini et al., 2015), and detecting different ions, for example, Cu^{2+}, Zn^{2+}, Mn^{2+}, Al^{3+}, Hg^{2+}, CH_3COO^-, and F^-, in organic and aqueous solutions (Lee et al., 2016). Julolidine is widely used as a fluorescent probe to assess the pharmaceutical formulations for stability of protein aggregation within them. Also, many other structurally modified julolidines have displayed promising biomedical applications. The utility of julolidine-based fluorescent probes for measurement of viscosity of blood, photoinduced release of neurotransmitter amino acids (Piloto et al., 2013), and as a sensor for bioimaging has been explored (Brewer et al., 2017). They were was also used to measure cell membrane viscosity (Haidekker et al., 2001); the diagnosis of certain complex diseases such as diabetes and Alzheimer's disease (AD) can also be performed using julolidine-based probes. In addition to this, certain studies demonstrate that some julolidine derivatives display biological activities such as inhibition of self-assembly of amyloid-β protein that involves a prime therapeutic strategy for AD (Davis et al., 2009) and possess antifungal activities (Raju et al., 2020).

Julolidine is indeed a compound that has gained attention for its fluorescence properties, and it has been studied for its potential as a fluorescent probe for various applications. Fluorescence is a phenomenon where a substance absorbs light at one wavelength and emits it at another, typically longer wavelength. This property has made julolidine valuable in several fields. Julolidine-based probes help to detect and quantify specific molecules in a complex sample, biological imaging, labeling, and visualization of specific cellular components or biomolecules. It is employed in the development of sensors for various analytes and as an environmental monitoring probe to detect and quantify environmental pollutants. The julolidine is a popular fluorophore moiety while julolidine-based probes commonly display fine applicability in aqueous solutions. The fluorescence-based analysis has driven major curiosity due to its edge of simple operation, real-time analysis, economic affordability, and higher sensitivity. However, some of the reported probes even display certain disadvantages including lower Stokes shift, shorter emission wavelength, and are affected easily through environmental background. As a result, near-infrared (NIR) probes (emission wavelength 650–900 nm) have fascinated more attention because of their unique distinctive advantages such as higher emission wavelength, depth of tissue penetration, lower fluorescence interference, and comparatively less damage to biological samples (Chen et al., 2017; Yang et al., 2020).

DOI: 10.1201/9781032706245-5

5.2 SYNTHESIS OF JULOLIDINE

The traditional synthetic path for julolidine (**2**) involves the alkylation of a tetrahydroquinoline (**1**) with 1-bromo-3-chloropropane (Figure 5.1: Scheme 1) (Glass et al., 1946). The *p*-substituted benzotriazole (**3**) was involved in synthesizing 1,7,9-trisubstituted julolidines (**4**) (Figure 5.1: Scheme 2) (Katritzky et al., 1999). In Scheme 3, author reported the synthesis of unsubstituted and monosubstituted julolidine (**6**) using different substituted anilines (**5**), 1-bromo-3-chloropropane, Na_2CO_3 with heating (Katayama et al., 1982). Further, in Scheme 4, synthesis of 8-hydroxyjulolidine (**8**) was reported by Van Gompel and Schuster in 1987. In both of their reactions, the reaction of 3-methoxy aniline (**7**) with 1-bromo-3-chloropropane was performed under the influence of Na_2CO_3 in order to prevent the breakdown of the ether group (Gompel & Schuster, 1987). In 1996 Katritzky and his team reported mono-substituted, di-substituted, and tri-substituted julolidine synthesis by using benzotriazoles (Katritzky et al., 1996). Katritzky et al. (1995) synthesized 2,3-disubstituted (**10**) and 1,2,3-trisubstituted julolidines (**11**) from 1,2,3-tetrahydroquinoline (**9**) (Figure 5.1: Scheme 5) (Katritzky et al., 1995). The 1,7-disubstituted julolidines (**13** and **15**) were synthesized by reaction of N,N-bis-[(benzotriazol-1-yl)methyl]aniline (**12**) with **14** in the presence of ethyl vinyl ether, THF, and 1-vinylpyrrolidin-2-one (Figure 5.1: Scheme 6).

5.3 PROBES BASED ON JULOLIDINE FOR ION SENSING

5.3.1 Cation Sensing Probe

Metal ions play crucial functions, particularly in terms of their biological, environmental, and chemical importance. There is much to be explored in the development of fluorescent and colorimetric molecular probes for the selective detection of transition metal ions. The selectivity of julolidine-based molecular systems toward metal ions has distinguished them from other fluorescent and colorimetric molecular probes. Jo et al. in 2015 synthesized and characterized a 2,3-diaminomaleonitrile and julolidine hybrid

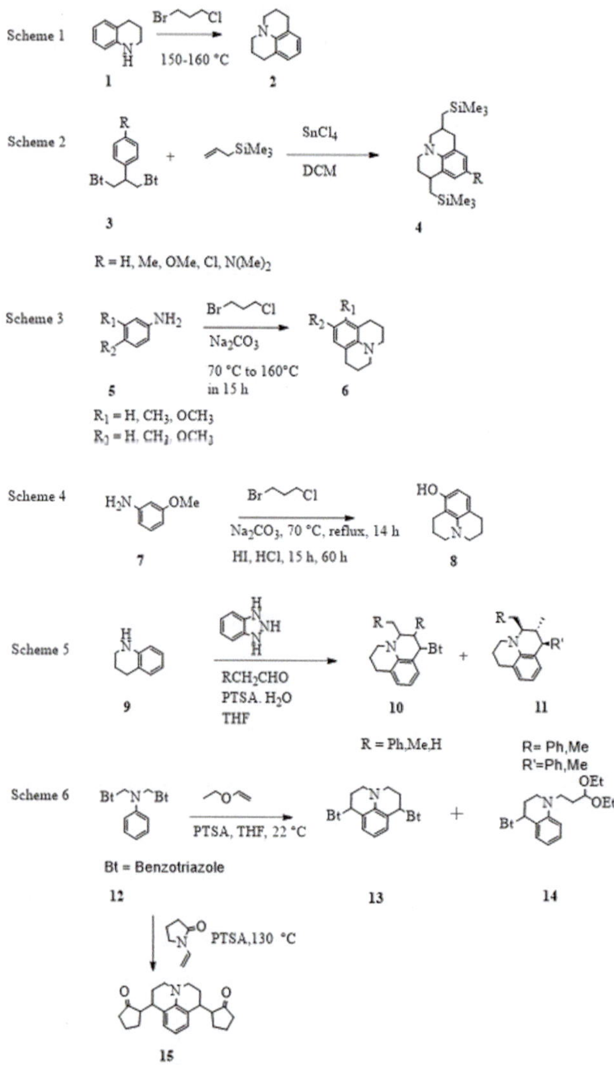

FIGURE 5.1 Different reaction scheme for the synthesis of julolidine.

moiety (**16**) with selective colorimetric sensing ability for Cu^{+2} by displaying a color change from yellow to colorless, and it can be utilized for the sensing of Cu^{+2} at a pH range of 4–12. Probe **16** displayed a limit of detection (LOD) of 2.1 mM for Cu^{+2}, which is comparatively lesser as compared to the maximum permissible limit recommended by the WHO of 30 mM (Jo et al., 2015). Ganesan et. al. in 2019 made a julolidine-based bifunctional sensor probe (**17**) for the detection of Zn^{2+} and Cu^{+2}. The unbound probe **17** displayed a fluorescence emission at 510 nm on excitation at 380 nm. When the probe was utilized for sensing biologically relevant ions, significant changes in the emission intensity were observed only in the case of Zn^{2+} and Cu^{2+} ions. The interaction of probe **17** with Zn^{2+} ions developed an enhancement in the fluorescence signals with a turn-on ratio of over 9.1 folds, while a quenching in the emission spectra with a turn-off ratio of approximately 10 folds was observed followed by the interaction of probe **17** with Cu^{2+} ions. The LOD of probe **17** for sensing Zn^{2+} cations was determined to be 3.5×10^{-8} M while the quantification limit was found to be 1.15×10^{-7} M. The probe **17** displayed a binding constant (Ka) of 3.93×10^{4} M^{-1} toward Zn^{2+} ions. The interference of various metal ions for change in the fluorescence intensity of probe **17** along with approximately one equivalent of Zn^{2+} was performed. The results displayed no interference of various metal ions on fluorescence response of probe **17** brought about by Zn^{2+} ions except for Cu^{2+} ions that entirely diminishe the emission band due to their nonradiative decay of the excited state. The quenching of emission intensity and a slight hypsochromic shift in spectra were observed with gradual increase in the concentration of Cu^{2+} ions from 1 µM to 30 µM. The LOD of probe **17** for detection of Cu^{+2} was calculated to be 1.46×10^{-6} M while the quenching constant was observed to be 3.77×10^{5} M^{-1} (Ganesan et al., 2019).

Singhal et al. in 2016 developed a julolidine-containing fluorescent probe (**18**) for selective sensing of Al^{+3} and Mn^{2+} cations. Probe **18**, on interaction with Al^{3+} ions, displayed a noticeable fluorescence turn-on. Probe **18** displayed a dual sensing behavior with a fluorescence change in the case of Al^{3+} metal ions and a colorimetric response for Mn^{2+} and Al^{3+} metal ions. The sensing of Al^{3+} is very difficult with chelating ligands due to its low coordination. Also, the interfering ions affect the detection of Al^{+3}. Probe **18** displayed an enhancement in the fluorescence signals with Al^{3+} ions with 50 nm hypsochromic shift. Probe **18** alone displayed an absorption band at 475 nm while a new absorption band appeared at 527 nm on addition of Mn^{2+} ions and the absorption band at 475 nm disappeared after gradual addition of Mn^{2+} ions. Probe **18** on interaction with Mn^{2+} ions displayed a distinct isosbestic point at 501 nm, which is an indication for production of new chemical species. A formation of dark yellow color on interaction of probe **18** with Mn^{2+} ions and Al^{3+} metal ions confirmed probe **18** to be a colorimetric naked eye detector of both the aforesaid ions. The binding constants (Ka) for Mn^{+2} and Al^{+3} ions were found to be 1.0×10^{4} and 6.1×10^{3}, respectively, and the LODs were calculated as 8.1 µM and 1.5 µM for Mn^{+2} and Al^{+3}, respectively. The selectivity study was performed by using 10 µM of probe **18** and 100 µM of other interfering metal ions, and no interference was observed due to the presence of metal ions in sensing Mn^{+2} and Al^{+3} ions. Probe **18** was observed to be capable of selectively sensing Al^{3+} and Mn^{2+} ions via a tridentate interaction, without displaying interference in the presence of other most common interfering cations (Singhal et al., 2016). In 2016 Choi et al. developed a julolidine- and tryptophan-based hybrid molecule (**19**) for selective and sensitive detection of Al^{3+} ions in an aqueous medium through "OFF-ON" fluorescence activity. The sensing mechanism was proposed to be the initiation of the chelation-enhanced fluorescence (CHEF) mechanism after interaction and hindrance in the C=N isomerization. The mechanism for detection of Al^{3+} ions using probe **19** was also proved by theoretical calculations. The LOD of probe **19** for sensing Al^{3+} ions was calculated to be 6.4 µM, which is lower compared to the levels recommended by the WHO (7.41 µM) in drinking water (Vasjari et al., 2021). Importantly, probe **19** was also observed to be capable of detecting Al^{3+} in living cells (Choi et al., 2016a).

In 2023 Kumar et al. designed and synthesized a julolidine-hydrazone-based Zn^{2+}-sensitive chemosensor (**20**). Probe **20** was observed to be initially nonemissive; however, it displayed fluorescence after interaction with Zn^{2+} ions. The **Commission Internationale de l'Eclairage** (Judd, 1933) chromaticity diagram was used to determine the chromaticity coordinates derived from the emission colors of probe **20**-Zn^{2+} complex in solution, hybrid polymer thin film, and powder phase. It was discovered that, in comparison to the emission in the solution phase, the powder's emission was red-shifted and thin film's was blue-shifted. In addition, Zn^{2+} sensing showed a reversible change in the presence of pyrophosphate (PPi), enabling PPi measurement. It's interesting to note that the procedure was selective and distinguished PPi from adenosine diphosphate (ADP) and adenosine triphosphate (ATP) in addition to detecting PPi utilizing the in situ generated **20**-Zn^{2+} complex. Also, it has been shown that the probe **20**-coated sensor paper may be utilized for the on-site detection of Zn^{2+} and the subsequent **20**-Zn^{2+} complex for PPi ions. Whenever possible, there was a strong correlation between the experimental data and the theoretical calculations (Kumar et al., 2023).

In 2024, Deng et al. and his team developed a novel julolidine-based fluorescent sensor (**21**) for sensing Al^{+3}. Probe **21** showed 132 nm Stokes shift and excellent selectivity toward Al^{3+}. The LOD for Al^{3+} was calculated to be 1.48×10^{-7} M. Furthermore, probe **21** was successfully used to identify Al^{3+} in a variety of food samples (such as edible alum, baking powder, fritters, and jellyfish), herbal teas, and environmentally significant water samples. Crucially, **21** showed minimal toxicity and permeability through cell membranes to find Al^{3+} in living cells (Deng et al., 2024). In 2022, Li et al. developed a julolidine-chalcone derivative (**22**), which can identify Al^{3+} without any interference with high selectivity. It can detect Al^{+3} with turn-on fluorescence in a wide pH range of 4.0–9.0. The mechanism of sensing involves the formation of complex between Al^{3+} ions and hydroxyl oxygen and carbonyl oxygen of probe **22**. Probe **22** can also detect Al^{3+} in actual water samples and applied for sensing exogenous Al^{3+} in A549 cells (Li et al., 2022). In 2023, Megha and colleagues developed a novel red-emitting molecule based on julolidine that exhibits both excited state intramolecular proton transfer (ESIPT) and aggregation-induced emission (AIE) characteristics, showing luminescence beyond traditional excimer emission and displaying an ON-OFF emission response specific to Cu^{2+} ions. They synthesized a julolidine-fluorene-based Schiff base, active for ESIPT-induced emission, and thoroughly analyzed its photophysical behavior in both solution and solid states. For comparison, they also created a methylated non-ESIPT variant of the probe, designated as probe **23**, which exhibited excimer emission only in the solid state. In solution, probe **23** displayed both normal and keto emissions, and upon the addition of water, the resulting aggregation led to a fluorescence emission with an excimer band in the lower energy region. Notably, in its solid form, probe **23** demonstrated emission that extended beyond typical excimer emission, with wavelength dependency. This extended emission is thought to result from the formation of various molecular clusters, enabling increased delocalization beyond the excimer effect and resulting in clustering-triggered emission, which gives probe **23** a bright red color in the solid state. Furthermore, probe **23** selectively detected Cu^{2+} ions, evidenced by quenching of its emission as Cu^{2+} interacted with its oxygen and nitrogen atoms, thereby preventing excimer formation (Megha et al., 2023). Paisuwan et al. in 2020 synthesized a series of "turn-on" fluorescent sensors hybrid of julolidine and amino benzyl dipicolylamine by simple Schiff base reaction. They synthesized three compounds, (**24**), (**25**), and (**26**), out of which one compound (**24**) displayed a strong green fluorescence increment upon the addition of Al^{3+} or Mg^{2+} in various media. The responses of fluorescence were selective for Al^{3+} or Mg^{2+} ions over other metal ions with LOD of 0.17 and 1.32 μM, respectively. The binding stoichiometry of $[Al{\cdot}L]^{2+}$ and $[Mg{\cdot}HL]^{2+}$ was proved by Job's plots in 1:1 ratio, and their association constants were calculated to be 2.25×10^4 and 4.00×10^4 M^{-1}, respectively. The fluorescence enhancement observed in the complexes $[Al{\cdot}L]^{2+}$ in its phenolate form and $[Mg{\cdot}HL]^{2+}$ in its phenolic form was attributed to the inhibition of ESIPT and C=N bond isomerization, driven by the CHEF effect. The application of probe **24**, which was based on array-type dual-sensing system, was developed and used for sensing Al^{3+} and Zn^{2+} as an on-site kit (Paisuwan et al., 2020a). In 2013, Na et al. developed a probe (**27**), which is a hybrid of julolidine and furan as a Zn^{2+} selective fluorescent chemosensor. Probe **27** was synthesized by condensation of furfurylamine with 8-hydroxyjulolidine-9-carboxaldehyde. The probe displayed an absorbance maxima at 380 nm, and after interaction with Zn^{+2} ions a new band was formed at 357 nm. Probe **27** alone displayed a fluorescence emission band at 452 nm ($\Phi = 5.11 \times 10^{-2}$), but the interaction with Zn^{2+} resulted in a four-fold increase of the emission ($\Phi = 1.9 \times 10^{-1}$). At 375 nm isobestic point was observed, which implies there was only one product formed after interaction with Zn^{+2} ions. The complexation of probe **27** with Zn^{2+} displayed an increase in the fluorescence emission in methanol, while other divalent metal ions had no effect. The LOD was calculated by using fluorescence titration study and was observed to be 7.5 μM for Zn^{+2} ions. The binding stoichiometry of the **27**-Zn^{2+} complex was calculated to be 1:1 ratio by Job's plot (Na et al., 2013).

Gallium compounds have potential applications as diagnostic and anticancer agents. The chemical properties of aluminum and gallium are quite similar, as both predominantly exhibit a +3 oxidation state. While fluorescent chemosensors for aluminum ions are relatively common, those for gallium ions remain less developed. Other techniques for trace gallium detection, such as ion exchange, adsorption, extraction, and atomic absorption spectrometry, have been established. Although a few fluorescent chemosensors for Ga^{3+} detection exist, they generally show low selectivity for gallium over aluminum. Additionally, water-soluble sensors suitable for cell imaging and biological analysis are rare. In 2013, Noh and collaborators reported a selective fluorescence-based method for detecting aluminum and gallium ions using probe **28** in both aqueous and nonaqueous solvents. Probe **28**, a π-conjugated Schiff base receptor, displayed significant fluorescence enhancement with high selectivity upon binding to Al^{3+} in water and DMF, whereas its selective response to Ga^{3+} was observed in methanol. The fluorescence intensity ratio (I_{Al}/I_{Ga}) of probe **28** in the presence of Al^{3+} and Ga^{3+} was 32, the highest among similar chemosensors reported at that time. Probe **28** demonstrated notable selectivity for Al^{3+} over both Ga^{3+} and In^{3+}. The Al^{3+} binding induced a "turn-on" fluorescence response with emission bands at 482 and 512 nm, where the fluorescence intensity of probe **28** increased by a factor of 100. Fluorescence titration experiments were used to determine the binding affinity of probe **28** for Al^{3+}, yielding an association constant (Ka) of 1×10^7 M^{-1}. Upon the addition of Al^{3+}, the absorption spectrum of probe **28** changed, with a decrease in the band at 430 nm and a weak UV band appearing around 300 nm. Furthermore, probe **28** was shown to bind intracellular Al^{3+}, emitting fluorescence that indicates its utility in assessing cellular aluminum exposure. The fluorescence titration of probe **28** with Ga^{3+} produced an association constant of 2.5×10^6 M^{-1} and a detection limit of 0.10 mM. UV-visible absorption spectra showed a bathochromic shift from 392 nm to 440 nm as Ga^{3+} concentration increased, with the formation of isosbestic point at 410 nm (Noh et al., 2013).

Liu et al. in 2022 developed a julolidine-based Schiff base (**29**), which can determine Co^{+2}, Ni^{+2}, Zn^{+2}, and Cu^{+2} in water samples. The Schiff base includes 2-hydrazinylpyridine structure that is capable of recognizing metal ions by its hydrazine group, and their recognition was confirmed by colorimetric and fluorescence spectroscopy. The association constants (Ka) for **29** were calculated as 2.10×10^7, $1.16 \times$

10^5, 1.02×10^5, and $1.35 \times 10^5\ M^{-1}$ for Zn^{+2}, Cu^{+2}, Co^{+2}, and Ni^{+2}, respectively. The detection limits for probe **29** were calculated to be 1.09×10^{-7}, 6.23×10^{-7}, 6.00×10^{-8}, and 3.20×10^{-8} for Zn^{+2}, Cu^{+2}, Ni^{+2}, and Co^{+2}, respectively. This probe **29** could be used to detect trace amounts of Cu^{+2}, Ni^{+2}, Co^{+2}, and Zn^{+2} in water bodies such as river and tap water (Liu et al., 2022). Liu et al., in 2018, developed a novel multifunctional chemosensor (**30**), which has been synthesized with julolidine as a core moiety and a Schiff base unit. Using a single receptor to detect multiple targets is both cost-effective and efficient compared to traditional one-to-one detection methods. The julolidine group is recognized as a highly effective fluorophore, and Schiff base compounds with hydrazide units provide N and O donor sites ideal for binding with Al^{3+} ions. In addition to hydrazide units, julolidine includes a phenolic hydroxyl group, allowing for potential hydrogen bonding interactions with F^- and CN^- ions through its NH proton and hydroxyl group, enabling their detection. As a fluorescent sensor, probe **30** can sensitively and selectively detect Al^{3+}, F^-, and CN^- ions in acetonitrile with a low detection limit. Moreover, the relatively low cytotoxicity of probe **30** makes it suitable for cell imaging applications to identify Al^{3+} ions (Liu et al., 2018). Akarasareenon et al., in 2023, reported julolidine-hydrazone-based fluorescent probe for selective detection of Cu^{+2}. They developed three julolidine hydrazone-based probe derivatives, among which the picolinohydrazide-derived compound (**31**) functioned as a turn-on fluorescent sensor for Cu^{2+} in aqueous DMSO media. The detection mechanism relied on the hydrolysis of the picolinohydrazide moiety, producing a highly fluorescent compound, julolidine-9-carboxaldehyde, with an emission wavelength at 420 nm and a detection limit of 0.1 ppm (Akarasareenon et al., 2023).

Hebbar et al. developed a new donor-π-acceptor (D-π-A) probe (**32**), incorporating julolidine and hydrazide derivatives. The resulting probe **32** showed strong potential for fluorescence-based detection of Al^{3+} ions. The probe **32** exhibited excellent selectivity for Al^{3+}, with detection limits of 0.0485 ppm. The turn-on fluorescence response of probe to Al^{3+} ions is due to the CHEF effect. This probe was successfully utilized for environmental detection of Al^{3+} ions in various environmental samples such as natural water, pharmaceutical tablets, soil, and cell sample including E. coli cells. Additionally, a portable test strip was created for rapid Al^{3+} detection. Furthermore, live cell imaging experiments verified that the probe could reliably detect Al^{3+} ions in E. coli cells using a cell imaging reader (Hebbar et al., 2024). Ahamed et al. developed a Schiff base fluorescent probe (**33**), synthesized by a straightforward condensation reaction between julolidine aldehyde and dansyl hydrazine, which demonstrated high selectivity and a fast response time (170 seconds) for Zn^{2+} ions in DMSO/H_2O solutions (8:2, v/v, buffered to pH 7.2). Fluorometric titration revealed that the probe-Zn^{2+} complex forms in a 1:1 binding ratio, with an association constant (K_a) of $1.03 \times 10^5\ M^{-1}$. The detection limit of probe for Zn^{2+} was calculated to be 15 nM, which is significantly below the World Health Organization's standard of 76 nM. Probe **33** effectively detected Zn^{2+} in real water samples with strong detection performance. Furthermore, probe-coated test strips enabled rapid, visual, and qualitative monitoring of Zn^{2+} ions (Ahamed et al., 2024). Man et al. developed a novel fluorescent sensor probe (**34**), based on a half-salamo and julolidine structure, and tested for its responsiveness to various metal ions. Spectroscopic studies revealed that probe **34** showed a significant increase in fluorescence when Zn^{2+} or Al^{3+} ions were added to the sensor solution, with distinct emission colors that could be distinguished by the naked eye, 451 nm for Zn^{2+}, showing a blue color, and 426 nm for Al^{3+}, showing a lighter blue color. Using a combination of experimental data and theoretical calculations, a sensing mechanism was proposed. The probe selectively identifies Zn^{2+} and Al^{3+} through coordination binding, which blocks PET and activates a chelation-enhanced fluorescence effect. The detection limits for Zn^{2+} and Al^{3+} were calculated to be 0.21 μM and 0.097 μM, respectively. The probe also showed effective Zn^{2+} and Al^{3+} detection in real water samples, indicating its practical application potential (Man et al., 2024). The structures of julolidine-based cation sensors are depicted in Figure 5.2.

5.3.2 Anion Sensing Probe

Due to the biological importance, some of the anions such as F^- and CN^- are the most concerning of the physiologically significant anions. In addition to sensing of Cu^{2+}, the probe (**16**), developed by Jo et al., was also capable of sensing F^- ions in solution colorimetrically through changes in color from colorless to orange, yellow to orange, and orange to blue. The chemosensor **30** developed by Liu et al., in 2018, apart from Al^{3+} detection, was also used for detection of F^- and CN^-. The selectivity study of probe **30** demonstrated the appearance of two new absorption bands at 467 nm and 439 nm after interaction with F^-/CN^- ions without showing any significant change with other anions. Along with the spectral changes, the color of probe **30** also changed from colorless to visible light yellow while the color of the solution of probe **30** with other anions did not change (Liu et al., 2018).

In 2017, Jeong et al. developed a turn-on fluorescence chemosensor (**35**) based on furan and julolidine, specifically for detecting fluoride ions (F^-) in aqueous solution. According to the WHO guidelines, the maximum allowable fluoride concentration in drinking water is 79 μM. In this chemosensor, furan units serve as fluorescence signaling agents, while julolidine acts as a hydrophilic chromophore and electron donor. Schiff bases containing phenolic groups function effectively as anion sensors, as the phenolic group can interact with anions via hydrogen bonding. The sensing mechanism of probe **35** toward F^- is attributed to an intramolecular charge transfer (ICT) process enabled by hydrogen bonding. Upon increasing the F^- concentration in a solution containing probe **35**, a significant 400-fold increase in fluorescence intensity was observed at 483

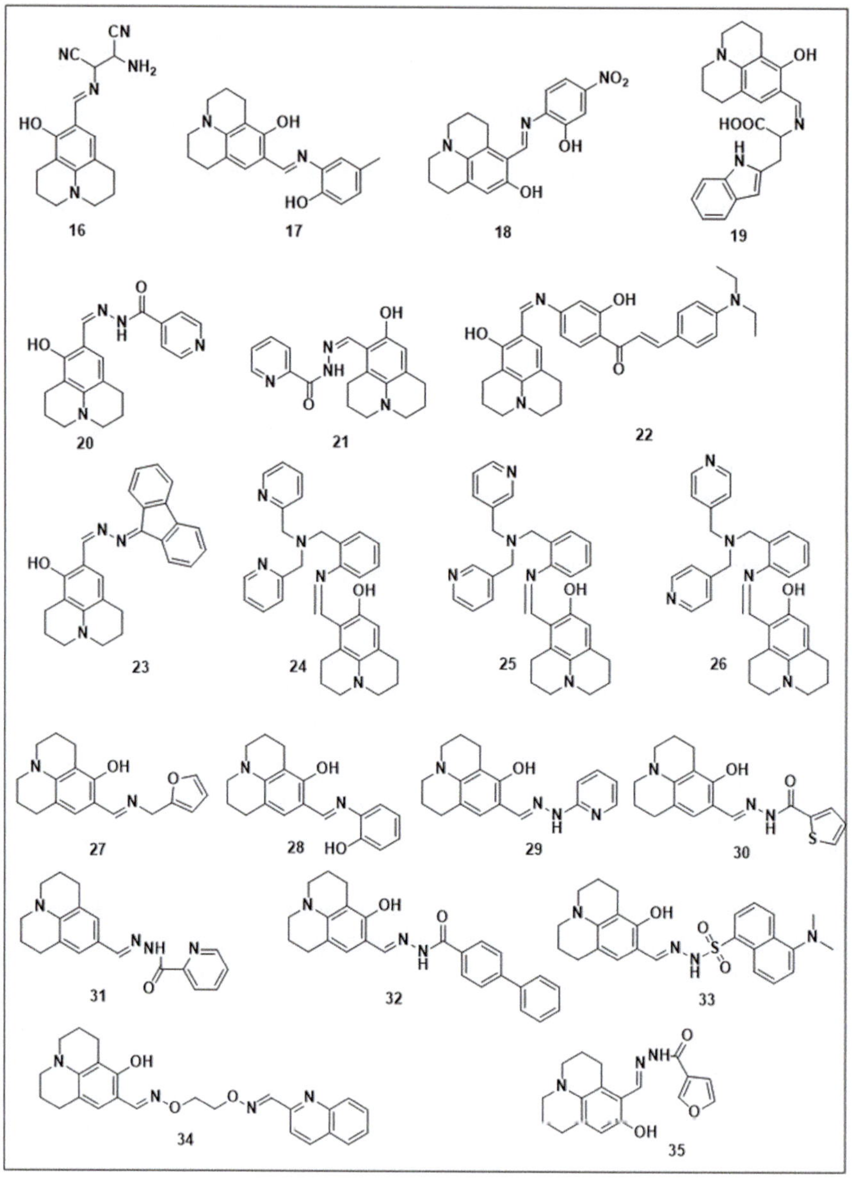

FIGURE 5.2 Structures of described cation and anion sensors.

nm (excitation at 384 nm). UV-visible titration experiments demonstrated that, in the presence of F^-, the absorption bands at 250 nm and 375 nm decreased, while new bands appeared at 325 nm and 450 nm, showing a slight increase. The reaction between probe **35** and F^- was confirmed using ^1H NMR titration, which also helped to determine an association constant of $7.5 \times 10^2\ M^{-1}$ and a detection limit of 10.02 μM. Based on Job's plot, ^1H NMR titration, and theoretical calculations, the mechanism of F^- sensing by probe **35** was proposed (Jeong et al., 2017). The structure of julolidine-based anion sensors is depicted in Figure 5.2.

5.4 FLUORESCENT LABEL MOLECULAR ROTORS

Kalel and colleagues investigated the interaction of DNA with julolidine-based ultrafast molecular rotors (UMRs). UMRs are a unique class of molecules characterized by an ultrafast twisting motion in their excited states, resulting in nonemissive twisted states and exceptionally low emission yields. However, in viscous solvents or confined spaces, this twisting motion slows down, allowing a significant increase in emission from the locally excited state. This property enables UMRs to detect subtle changes in DNA's secondary structure, which are often undetectable by other techniques like circular dichroism spectroscopy. Probe **36**, a 9-dicyano-vinyl julolidine UMR, exhibits a very low emission yield (~10^{-4}) in low-viscosity solvents. In its ground state, the molecule remains largely planar, with a dihedral angle of only 8° between the vinyl and julolidine groups. Upon excitation, however, the molecule undergoes rapid twisting around the vinyl bond, leading to an orthogonal configuration between the vinyl and julolidine units. This twisting process occurs without an energetic barrier, making it highly efficient and fast. The twisted state is nonemissive due to the molecule's rapid return to the ground state through internal conversion. In viscous environments or confined spaces, this twisting motion is hindered, leading to a considerable rise in emission yield (as observed by Allen et al., 2005). Due to these unique properties, probe **36** has numerous biological applications, such as assessing microviscosity in cell membranes, blood plasma, protein aggregation, and supramolecular assemblies to measure bilayer viscosity and monitor tubulin polymerization. When bound to free tubulin, probe **36** showed an increase in quantum yield, signifying partial immobilization. At 4°C, it bound to a single class of high-affinity hydrophobic sites with a dissociation constant (Kd) of 1.12 ± 0.26 pM, displaying a protein concentration-dependent stoichiometry. In an amyloid system, probe **36** was utilized to trace the aggregation pathway of transthyretin (TTR) under acidic conditions. While it exhibited minimal steady-state emission with the native tetrameric TTR, probe **36** showed an enhanced signal when bound to isolated misfolded aggregates formed from a specific molten globule state of the protein (Kalel et al., 2016)..

Goh et al., in 2014, developed molecular rotors as fluorescent labels for rapid detection of biomolecular interactions. Molecular rotors are known for their ability to undergo twisted intramolecular charge transfer (TICT). Structurally, they consist of three key components: an electron-donating unit, an electron-accepting unit, and a π-conjugated linker, which facilitates electron transfer in the molecule's planar state. In one study, researchers linked a molecular rotor, 9-(2-carboxy-2-cyanovinyl) julolidine (**37**), to biotin and observed fluorescence changes when it bound to streptavidin. The addition of streptavidin led to a two-fold increase in fluorescence, highlighting the effectiveness of this biotin-rotor probe. They further enhanced the application by attaching the rotor **37** to peptides JP1 and JP2, which differ by a single amino acid but exhibit a 10-fold difference in binding affinities. To enable conjugation, an extra lysine was added to the C-terminal of each peptide, which was then reacted with the N-hydroxysuccinimidyl-activated ester of **37**, creating rotor-peptide probes (**38**). The effectiveness of these probes was tested with recombinant Mdm2 protein. Binding of the JP1 rotor to Mdm2 resulted in a concentration-dependent increase in fluorescence. This ability of molecular rotors to change fluorescence upon target binding demonstrates their versatility as tools for detecting specific biomolecular interactions (Goh et al., 2014). The structures of these fluorescent molecular rotors are depicted in Figure 5.3.

5.5 JULOLIDINE-BASED PROBE FOR BIOIMAGING APPLICATION

Bioimaging is a multidisciplinary field that encompasses various techniques and methods for visualizing and analyzing biological structures and processes at different scales. It plays a crucial role in advancing our understanding of biology, medicine, and other life sciences. Bioimaging studies utilize a wide range of imaging modalities and technologies to observe and study biological samples, from individual cells to entire organisms. Formaldehyde (FA) is a type of reactive carbonyl species (RCS) with various impacts on human health and disease. Additionally, it is a widespread environmental pollutant that poses substantial health risks. To mitigate these risks, the Occupational Safety and Health Administration (OSHA) has established a safe exposure limit for FA at 0.75 ppm. In 2017 Brewer et al. reported the design, synthesis, and biological evaluation of a ratiometric formaldehyde probe (**39**) indicator. Anomalous increases in FA levels contribute to a variety of disorders, including diabetes, cancer, heart, liver, and neurological illnesses. Identifying the relationship between FA physiology and pathology drives the creation of chemical instruments that provide the precise identification of this response component in biological settings while maintaining temporal and spatial accuracy. In the human body, FA production is estimated to be up to 1.4 mM/min. Utilizing flow cytometry or live cell imaging, a probe (**39**) can be used to track disturbances in FA levels in biological samples. Probes **40** and **41** used FA-dependent aza-Cope reactivity for the conversion of alkylamine-containing coumarin into its aldehyde congener with approximately 50 nm shift in the excitation wavelength. With a fluorescence ratio change of up to six times, the probes displayed observable excitation and emission spectrum along with strong selectivity for FA. Probes **40** and **41** possess several appealing photophysical properties along with visible excitation and emission profiles, large optical brightness of both dyes, very strong insensitivity to pH variations within the physiological range of basic to acidic circumstances, and high sensitivity to FA detection (Brewer et al., 2017).

In 2023 Liu et al. developed lysosome-targeted julolidine- and coumarin-fused probe (**42**) for the detection of SO_2. The SO_2 gas is a toxic environmental pollutant, and

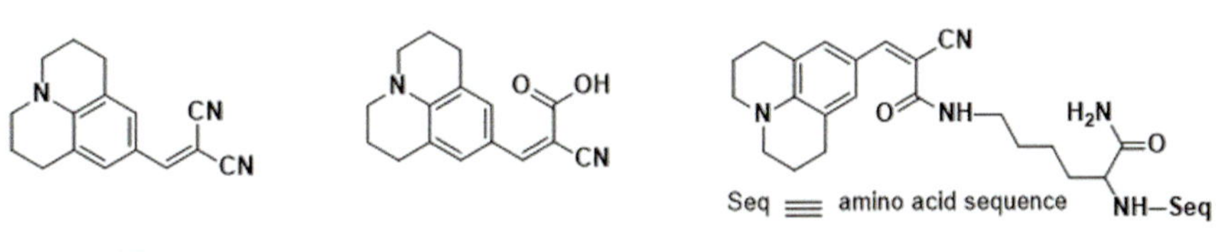

36 37 38

FIGURE 5.3 Structures of julolidine based fluorescent molecular rotors.

high levels of sulfite intake can cause diarrhea, low blood pressure, as well as allergic and asthmatic reactions in sensitive individuals. To address this, the researchers developed a ratiometric NIR fluorescent probe **42**, specifically designed to detect SO_2 derivatives. In this molecular setup, coumarin serves as the donor due to its excellent photostability, high fluorescence quantum yield, large Stokes shift, and adjustable optical properties. The acceptor of probe consists of a julolidine electron-donating group and a quinolinium salt electron-accepting group, forming a hemicyanine-like structure that enables an ICT system. Upon interaction with HSO_3^-/SO_3^{2-} ions, probe **42** exhibits a reduction in red fluorescence at 662 nm along with an increase in blue fluorescence at 489 nm, covering a detection range of 2.8–80 μM. Additionally, probe **42** demonstrates high energy transfer efficiency, a substantial Stokes shift, and a low detection limit, allowing it to detect subtle changes in SO_2 levels within cells and effectively target lysosomes. Furthermore, probe **42** was applied not only to image both exogenous and endogenous HSO_3^-/SO_3^{2-} in live cells but also to differentiate between normal and cancer cells and target lysosomal regions (Liu et al., 2023). The structures of julolidine-based fluorescent probes for biological applications are depicted in Figure 5.4.

5.5.1 Bioimaging of Amino Acid

Cysteine (Cys) is a sulfur-containing amino acid that plays a crucial role in numerous physiological functions, including maintaining oxidative balance, supporting metabolism, and aiding protein synthesis. Typically, intracellular levels of Cys range from 30 to 200 μM, and deviations from this range are linked to various health disorders. A deficiency in Cys can lead to growth issues, skin damage, edema, fatigue, and liver dysfunction, while an excess can contribute to cardiovascular neurotoxicity and is associated with

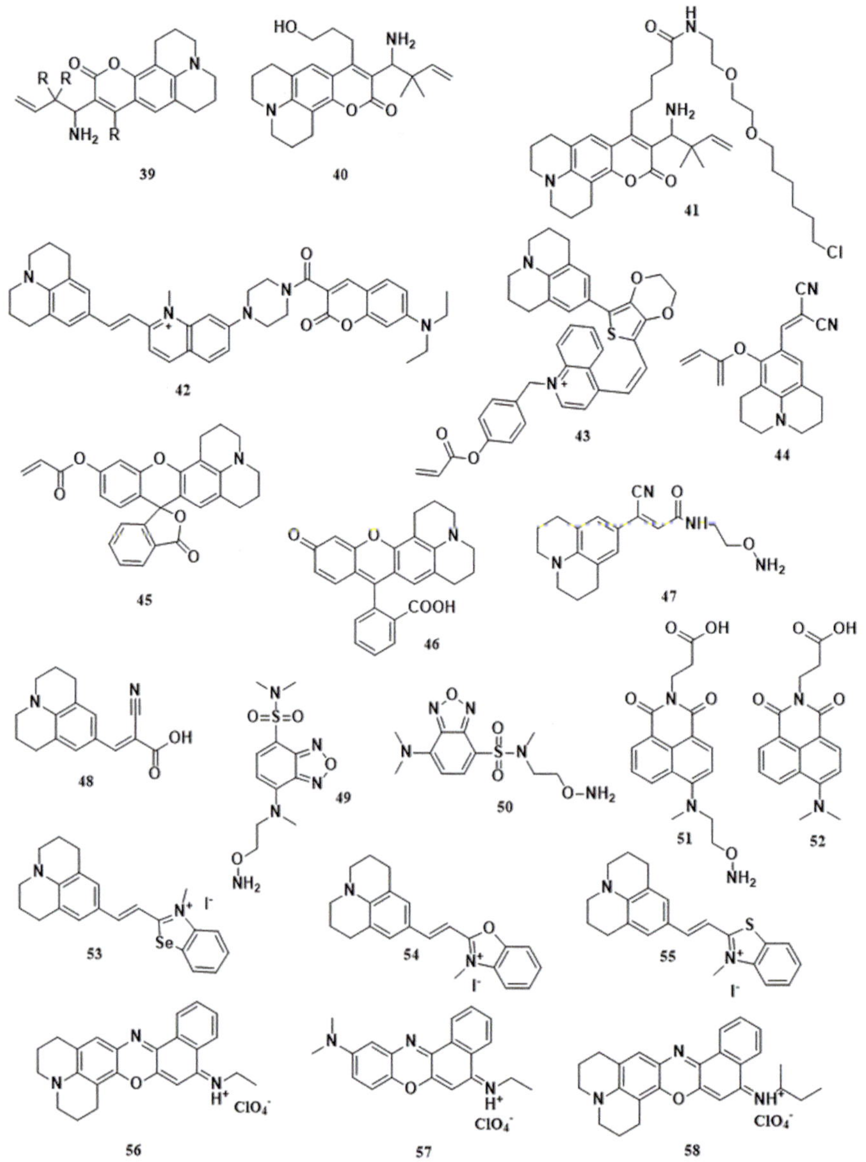

FIGURE 5.4 Structures of julolidine based fluorescent probes for biological applications, bioimaging of amino acids, DNA and RNA.

conditions like Parkinson's disease. Thus, accurately and sensitively measuring Cys levels under physiological conditions is essential in clinical and biochemical research. However, distinguishing Cys from other biothiols, specifically homocysteine (Hcy) and glutathione (GSH), is challenging due to their similar structures and reaction characteristics. Donor-π-acceptor dyes offer an effective approach to designing near-infrared fluorescent probes, as they allow for a large Stokes shift and provide an adjustable emission wavelength, making them ideal for selective Cys detection amidst similar molecules. Fang et al. developed a julolidine-based NIR fluorescent probe (**43**) for sensing Cys. In this probe, julolidine acts as the electron donor, while quinoline serves as an electron acceptor, with 3,4-ethylenedioxythiophene forming a bridge to connect them. This bridge increases π-conjugation and also increases its flexibility through vibrational and rotational movements. These structural enhancements enable probe **43** to emit NIR light and achieve a large Stokes shift. Probe **43** produced a strong "turn-on" fluorescence signal at 741 nm, with a significant Stokes shift of 268 nm after detection of Cys. This probe was capable of detecting very low levels of Cys, with a detection limit of 24 nM. Probe **43** was observed to be highly stable under light (photostable), less toxic to cells, and produced a clear signal with minimal background interference (high signal-to-noise ratio); probe **43** had been successfully used in imaging studies to track Cys levels in cells and zebra fish (Fang et al., 2024). Meng et al. developed a julolidine-based new probe (**44**), chemically 9-(2,2-dicyanovinyl)-2,3,6,7-tetrahydro-1H,5H-pyrido[3,2,1-ij]quinolin-8-yl acrylate. The probe undergoes a series of reactions (addition, cleavage, and cyclization rearrangement) when it encounters the amino acid cysteine (Cys), resulting in the formation of a bright green iminocoumarin containing fluorescent compound. The fluorescence efficiency of this product, measured by the quantum yield (Φ_f), raises significantly from 0.007 to 0.793. With this reaction mechanism, probe **44** delivers a highly sensitive "turn-on" fluorescence response, showing a 172-fold increase in signal when Cys is present. It can detect very low levels of Cys (down to 4.1 nM) and has a broad detection range from 5 to 6,000 nM. Inspired from the impressive sensitivity of probe **44**, it has been used to visualize natural Cys levels in live HeLa cells, yielding clear imaging results (Meng et al., 2024). Ji et al., in 2019, developed a highly sensitive fluorescent probe (**45**), which is a hybrid based on julolidine and fluorescein for sensing and bioimaging of Cys in living cells. The Cys plays a crucial part in intracellular redox homeostasis. Probe **45** reacted with Cys more readily as compared with GSH and Hcy. The discrimination of Cys from GSH and Hcy is because of the intramolecular cyclization reaction of their thioethers. The sensing mechanism is based on Michael's addition of Cys to acrylates to form thioethers after cyclization to seven-membered S,N heterocycles, which forms free fluorophores (**46**). This probe is based on a turn-on mechanism, and it allows a turn-on red fluorescence response (567 nm) in phosphate buffer solution (PBS). It has a very low LOD of 39.2 nM, and it also displayed the selective detection of Cys. Before the study of selective imaging, the cytotoxicity of **45** was identified by 3-[4,5-dimethylthiazol-2-yl]-2,5 diphenyl tetrazolium bromide (MTT) assay in HepG2 cells. The MTT assay displayed low cytotoxicity in the concentration range of 0–30 μM. On incubation of HepG2 cells with **45** (5 μM) for 30 minutes, a deep red fluorescence appeared. However, when it was pretreated with N-ethylmaleimide (NEM), inhibition of the red fluorescence was observed, which suggested that the signal produced was due to the presence of intracellular biothiols including Cys. After the pretreatment of cells with NEM for 30 minutes, followed by treatment with biothiols such as Cys, GSH, or Hcy for another 30 minutes, and lastly incubation with **45** for 30 minutes, a bright red fluorescence was observed in the Cys-treated group while in the GSH- and Hcy-treated groups the red fluorescence was absent. These results confirm that probe **45** renders sufficient cell permeability and is sufficiently selective for detection of Cys in living cells (Figure 5.6ii and iii)) (Ji et al., 2019). The structures of julolidine-based fluorescent probes for bioimaging of amino acids are depicted in Figure 5.4.

5.5.2 Bioimaging DNA

Jun et. al., in 2022, found an advanced DNA labeling method based on fluorescence via base excision trapping (BETr). The use of fluorescence labeling of DNAs is very useful. Covalent labeling of DNA has been extensively studied and is broadly divided into strategies for direct fluorophore incorporation into DNA during chemical or enzymatic DNA synthesis. Incorporating fluorophores into oligodeoxynucleotides (ODNs) via direct chemical synthesis using fluorophore-conjugated phosphonamidite reagents allows for precise positioning and high yields. However, these reagents are both costly and unstable, and they are generally suitable only for synthesizing shorter DNA sequences, typically under 200 nucleotides. For longer DNA sequences, fluorophores can be introduced directly during synthesis by using fluorescent nucleoside triphosphates. Yet, this method is limited by the need for structures that are compatible with the active site of enzyme, restricting the possible dye positions. Post-synthetic fluorophore labeling, on the other hand, bypasses these constraints related to enzyme-substrate compatibility. Jun and their coworkers reported a DNA labeling strategy that was developed for both oligonucleotides and double-stranded DNA (dsDNA) using aldehyde-reactive rotor dyes, also known as universal base excision reporters (UBERs), to target apurinic/apyrimidinic (AP) sites created from deaminated base excision. This method involved three key steps. First, DNA strands containing one or more deaminated bases were chemically or enzymatically synthesized. Second, base excision repair (BER) was used to generate reactive AP sites. Finally, oxime bond formation occurred between the AP site and the light-up UBER dyes. Among the probes tested, probe **47** demonstrated the best performance. It contained a cyanovinyl julolidine dye attached to a three-atom ligand and an aminooxy group. The reaction of probe **47** with AP sites in DNA showed two unique properties: (1) an exceptionally rapid reaction rate, with rate constants reaching approximately 400 M/s at neutral pH, and (2) a high degree of lightup of the dye. A substantial fluorescence enhancement of about 500-fold was observed upon covalent bonding of probe **47** to AP sites in comparision with the dye alone in solution. This enhancement made it suitable for imaging DNA repair processes within cells and tissues. The dye in DNA was found to be highly photostable, observable with standard fluorescein filter sets, and compatible with two-photon excitation imaging. These favorable properties led researchers to envision using probe **47**, along with other UBER probes, as general fluorescent labels for oligodeoxynucleotides and dsDNA. They explored a single-reaction method to substitute deaminated DNA bases with these dyes using affordable bacterial repair enzymes for base excision. For DNA synthesis in the BETr approach, they proposed incorporating deaminated bases at specific sites, either through chemical synthesis with phosphonamidites or enzymatic synthesis using DNA polymerase and deaminated nucleoside triphosphates. Chemical synthesis offered precise control over base positioning and could be commercially sourced at a reasonable cost.

Polymerase-driven synthesis enables the creation of extended DNA strands with multiple labeling sites and can be carried out in the biological mixtures using primers that are specific to the target. A DNA polymerase reaction with an enzymatic-based excision and aminooxy labeling could be carried out in a single step, simplifying the labeling process. To test these labeling possibilities, the author performed a series of experiments. One of the advantages of this approach is that all the three steps needed for the labeling (DNA lesions insertion, AP site trapping, and DNA repair) are parallel and orthogonal and can be performed at the same time, allowing for DNA synthesis as well as fluorescence labeling in a single tube reaction by simply combining all the steps.The fluorescence increase from the light-up properties of UBER dyes upon reacting with DNA was used to monitor reaction progress in real time. By combining two different colors of UBER dyes with two distinct damage/glycosylase pairs, researchers achieved site-specific two-color labeling within a single DNA strand by sequentially applying the BETr method with dU and dI. Finally, the double-strand specificity of a repair enzyme was utilized to sensitively detect a target DNA or RNA in a mixture, with isothermal turnover and single-nucleotide precision. This approach provided a convenient and versatile method for general fluorescent labeling of DNA (Jun et al., 2022).

Wilson and Kool in 2019 developed ultrafast oxime formation that allowed for efficient fluorescence light-up DNA base excision measurement. DNA glycosylases are an important group of DNA repair enzymes that are involved in the initiation of DNA base excision repair (BER). Measuring the activity of these enzymes can be useful in understanding the mechanism of DNA damage and repair, as well as in practical applications in the fields of cancer diagnosis and drug screening, where previous fluorescence assays are often complicated and restricted to a specific enzyme type. For measuring DNA glycosylase activity, they designed a UBER probe based on light-up mechanism. After the glycosylase removed the damaged DNA base of interest, the resulting hemiacetal form of the DNA AP site reacted with the aminooxy linker of the UBER probe, forming a highly fluorescent probe-DNA oxime conjugate. Prior to oxime formation with the AP site of DNA, the UBER probe remained mostly nonfluorescent due to free bond rotation at the linker attachment site. The bases adjacent to the DNA duplex restricted this bond rotation, producing a fluorescence response that was highly selective for the AP site of DNA, over small-molecule aldehydes and ketones.

The design of the UBER probe was motivated by a number of recent TICT and molecular rotor probes. In TICT as well as rotor probes, the fluorophore undergoes nonradiographic relaxation through bond rotation, which is then conjugated with a reactive function that targets the probe to the desired biomolecule. The design of the probe is such that when the probe binds to the target molecule, the bond rotation inside the probe constraints, which results in a particular fluorescence enhancement. They selected a well-characterized environmentally sensitive 1,8-naphthalimide fluorophore, the molecular rotor 9-(2-carboxy-2-cyanovinyl)-julolidine **(48)**, and some benzoxadiazole dye, yielding probes **51**, **47**, and **49**, respectively. Out of this **47** afforded a notable 256-fold rise in fluorescence when it binded to the AP site. Excitation was at 470 nm for **47** and 450 nm for **51/52** and **49/50**. Other than this, probe **44** is accessed readily via four-step synthesis and displayed a well spectral overlap with fluorescein. This property renders the probe available and suitable for researchers to broadly study the probe on majority of fluorescence-based instruments (Wilson & Kool, 2019). The structures of julolidine-based fluorescent probes for bioimaging of DNA are depicted in Figure 5.4.

5.5.3 Bioimaging Study of RNA

Mondal et al., in 2023, developed a julolidine-based fluorescent probe for RNA imaging in living cells. They designed and synthesized three cationic julolidine-azolium conjugates, Benzselenazole-Julolidine (**53**), Benzoxazole-Julolidine (**54**), and Benzthiazole-Julolidine (**55**). These all are turn-on fluorescent probes that displayed large quantum yields and high brightness when they interact with RNA. The free rotation between acceptor and donor gets restricted when it binds with the RNA resulting in high fluorescence emission and higher wavelength on excitation. The author reported promising photophysical properties and optical responses of **55** and **53** toward RNA, and the probes were selected for biological studies for cell imaging application. The probes enabled imaging of intracellular RNA, as they stained the nucleoli in live cells recorded under a wide range of physiological conditions. Probe **53** proved to be a cationic molecular probe permeable to cell membrane that can be applied for staining of intracellular RNA with a dark red color displaying high selectivity (Mondal et al., 2023).

Yao et al. in 2018 developed an NIR fluorescent probe for differentiating RNA from DNA, based on the "door-bolt" mechanism biomaterial. So far, while many high-quality DNA fluorescence probes (Han et al., 2015; Hu et al., 2017) have been developed, RNA probes are very rare (Liu et al., 2014; Shirinfar et al., 2013; Wang et al., 2016). The commercial dye for RNA staining is a green dye called "SYTO RNA-Select." The lack of RNA probes is caused by DNA interference. An RNA molecule is made up of many pockets such as hairpins, internal loops, and bulges. The authors reported NIR probe (**56**) with a bulky julolidine as a stereo group linked to the Nile blue core. They compared **57** (a molecule without julolidine core) with **58** having two bulky groups at every terminal of the molecule. With a molecular thickness of up to 5 Å, the proposed probe (**56**) can't penetrate the double-strain gap or DNA minor groove due to Van der Waals resistance. For RNA molecules, the probe fit into a large pocket within the RNA structure, blocking internal conversion and generating intense fluorescence. This mechanism resembled the function of a traditional Chinese wooden door and bolt, where the bolt must fit the socket precisely to slide in and secure the door. In this "door-bolt mechanism," probe **56** successfully differentiated RNA from DNA in both solutions and cellular environments. Probe **56** self-associated in aqueous solution (with absorption peaks at 618 and 667 nm) and displayed a very low fluorescent quantum yield ($\Phi = 0.04$, emission at 680 nm), likely due to quenching from H-aggregation. Other probes with varied stereo substituents, specifically **57** (lacking the julolidine stereo group) and **58** (with two bulky stereo groups at each end of the molecule), did not show any fluorescence distinction between yeast RNA and ct-DNA. This indicated that the distinct stereo structure of probe **56,** with a large julolidine group at one end and a smaller group (such as ethyl) at the other, enabled it to selectively differentiate RNA from DNA. In addition to its high RNA selectivity, probe **56** exhibited rapid cell membrane permeability, high photostability, and pH stability. Its fluorescence signal became visible within seconds. Following the addition of probe **56**, absorbance remained at 97% after 3 hours of exposure to 500 W light and at 93% after 6 hours. These properties made probe **56** highly suitable for RNA staining in live cells and tissues. Surprisingly, probe **56** produced more fluorescence on cancer cells than on normal cells, which can help in cancer cell differentiation (Jögi et al., 2012; Yao et al., 2018). The structures of julolidine-based fluorescent probes for bioimaging of RNA are depicted in Figure 5.4.

5.4.4 Bioimaging of Amyloid-β Peptide

Nagarajan and Lapidus in 2017 developed a fluorescent probe (**36**) with high sensitivity for the identification of amyloid-β peptide early in the lag phase. Monomers and low oligomers, which are mainly present in the lag phase, are majorly dynamic and heterogeneous and make it very challenging to observe in real time. As a result, most drug development assays show that a compound is active in inhibiting the formation of inert fibrils rather than in inhibiting toxic oligomer formation. Therefore, there is an urgent requirement to develop a prodrug sensitive to the early stage of aggregation. The authors tested a number of fluorescence probes that have previously been reported to be sensitive to a change in microenvironment and aggregation during the Aβ lag phase. Among all the tested molecules, only **36** displayed a significant spectral change and large sensitivity for oligomer formation. Probe **36** belongs to a category of molecular rotors that display various applications such as cellular imaging, food processing, and polymer chemistry. Due to its responsiveness towards viscosity change in the local environment, the quantum yield of **36** can be increased up to 30 times upon transfer from propane-1-ol to glycerol. An intramolecular quencher dicyano group connected to the julolidine moiety of **36** through a sigma bond. Free rotation around the single bond reduces the quantum yield of **36** due to intramolecular quenching in a low-viscosity solution but because of high viscosity or entrapment in a cavity or a cage, the rotation is restricted and as a result the quantum yield enhances. This displayed that probe **36** fluorescence can be used to sense the aggregation of actin and tubulin (Tuszynski et al., 2003). Probe **36** was also applied for the sensing of oligomer formation in transthyretin and α-synuclein. Two isoforms of amyloid-β peptide, Aβ_{42} and Aβ_{40}, were selected for the kinetics of the lag phase. For Aβ_{40}, the increase in probe **36** intensity in the lag phase indicated that it is highly sensitive to the early oligomers or some other nonthioflavinophilic species.

The recording of probe **36** spectra at different concentrations in the presence of Aβ_{40} demonstrated the formation of another peak at 570 nM above 6 mM concentration. This peak was denoted as the excimer peak. The presence of 570 nm peak displayed the formation of oligomers by Aβ_{40} that can accept two or more molecules of probe **36**, which together produce the excimer. During monitoring the entire length of its lag phase, there is no effect of probe **36** upon the addition of Aβ_{42}; **36**/oligomer binding rate was slow when compared with the Aβ_{42} lag time. However, the addition of 1 mM of compound **36** to preformed Aβ_{40} oligomers caused an immediate increase in fluorescence signals at 510 and 570 nm, indicating that probe **36** bound to Aβ_{40} very rapidly. Based on these findings, the researchers concluded that the predominant oligomers formed by Aβ_{42} had distinct molecular structures compared to Aβ_{40}, preventing probe **36** from binding during the lag phase. This difference was attributed to variations in both the kinetics and the structural formations of Aβ_{42}, as seen in the divergent fluorescence spectra of probe **36** for Aβ_{40} and Aβ_{42} during the lag phase. At 16 minutes, no change in fluorescence of probe **36** was observed with Aβ_{42}, suggesting that probe **36** did not interact with any of these particles, including protofibrils, although it did bind to fibrils. The data demonstrated that in the early lag phase, wild-type Aβ_{40} formed oligomers with a partially open structure that permitted probe **36** excimer formation. During the late lag phase or pre-transition growth phase, these oligomers underwent substantial structural shifts, transforming into β-sheet-rich aggregates. However, these oligomers did not form fibrils or undergo structural changes detectable by probe **36** binding. In contrast, late in the lag phase or during the pre-transition growth phase, Aβ_{42} formed spherical, curved, and rod-shaped particles that quickly transformed into β-sheet-rich aggregates (Nagarajan & Lapidus, 2017). The structure of julolidine-based fluorescent probe for bioimaging of amino acids is depicted in Figure 5.3.

5.5 MISCELLANEOUS

In 2022, Ye et al. developed a fluorescent probe to assess meat freshness, a critical factor in food safety and public health that has recently gained heightened attention. The authors created a set of fluorescence probes, labeled **59**, **60**, and **61**, featuring unique electron-withdrawing substituents based on a julolidine-fused coumarin framework. These probes reacted with cadaverine through an aza-Michael addition followed by an elimination process, producing a colorimetric and ratiometric fluorescence response. The sensing capabilities of these probes were enhanced by strengthening the electron-withdrawing properties of the substituent groups. Notably, probe **59**, which included a dicyanovinyl group as the reactive site, displayed exceptional sensing characteristics, including a fast response time (approximately 60 seconds), high selectivity, and a low detection threshold of 14 nM. Probe **59** was incorporated into test kits capable of detecting cadaverine vapor, indicated by a striking fluorescence shift from red to green. Through this color change, on-site, noncontact, and noninvasive monitoring of meat freshness was effectively achieved. The strong performance of the **59** test kits suggests a promising tool for rapid, real-time, and precise assessment of meat freshness through fluorescence screening (Figure 5.6v) (Ye et al., 2022).

In 2018, Xia et al. reported a julolidine-fused anthracene derivative, referred to as **63**, which displayed notable photophysical characteristics absent in its dimethyl-amino-substituted counterpart, **62**. Compound **63** exhibited significant red shifts in both absorption and emission spectra, with shifts of 74 nm and 131 nm, respectively, compared to the unmodified anthracene. In toluene, the fluorescence quantum yield of **63** was measured at 0.55, whereas **62** showed no fluorescence as its emission was entirely quenched. The integration of julolidine greatly influenced anthracene's photophysical properties, as seen in the peak absorption wavelength of **63** at 450 nm, about 70 nm red-shifted relative to the original anthracene. While anthracene emits blue light (with max emission at 380 nm), compound **63** emitted green light with a peak at 518 nm and a quantum yield of 0.55 (Xia et al., 2018). The structures of julolidine-based fluorescent probes for miscellaneous applications are depicted in Figure 5.5.

Probe **17** was utilized to visualize the Zn^{2+} ions in living cells. The cell viability of the probe **17** and **17**-Zn^{2+} was studied by MTT assay at different concentrations. The cells were incubated with **17** (25 µM) for 1 hour at 37°C, then after treated with Zn^{2+} (25 µM) and incubated for another 0.5 hour. To clean any residual probe, the cells were washed with PBS buffer. The images of probe **17** before and after the addition of Zn^{2+} are depicted in Figure 5.6i. The results confirmed that probe **17** can penetrate the cell, so it can be used to track Zn^{2+} in living cells. Probe **30** was also utilized to visualize the presence of Al^{3+} in living cells. The cell cytotoxicity of probe **30** was studied by MTT assay on HeLa cells. Cell viability gradually decreased as the concentration of probe 30 increased. Probe 30 was utilized for fluorescence imaging of intracellular Al^{3+} in living cells. HeLa cells were incubated with probe 30 (20 µM) for 30 minutes at 37°C, during which almost no fluorescence was detected. However, when Al^{3+} (50 µM) was added to the HeLa cells, a significant enhancement in fluorescence occurred, resulting in green fluorescence (Figure 5.6iv). The results demonstrated that probe **30** could play a role in intracellular fluorescence imaging of Al^{3+}.

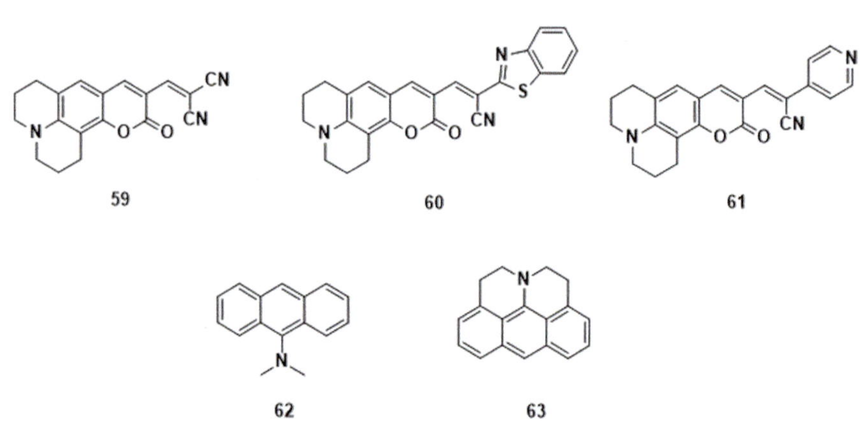

FIGURE 5.5 Structures of julolidine-based fluorescent probes for miscellaneous applications.

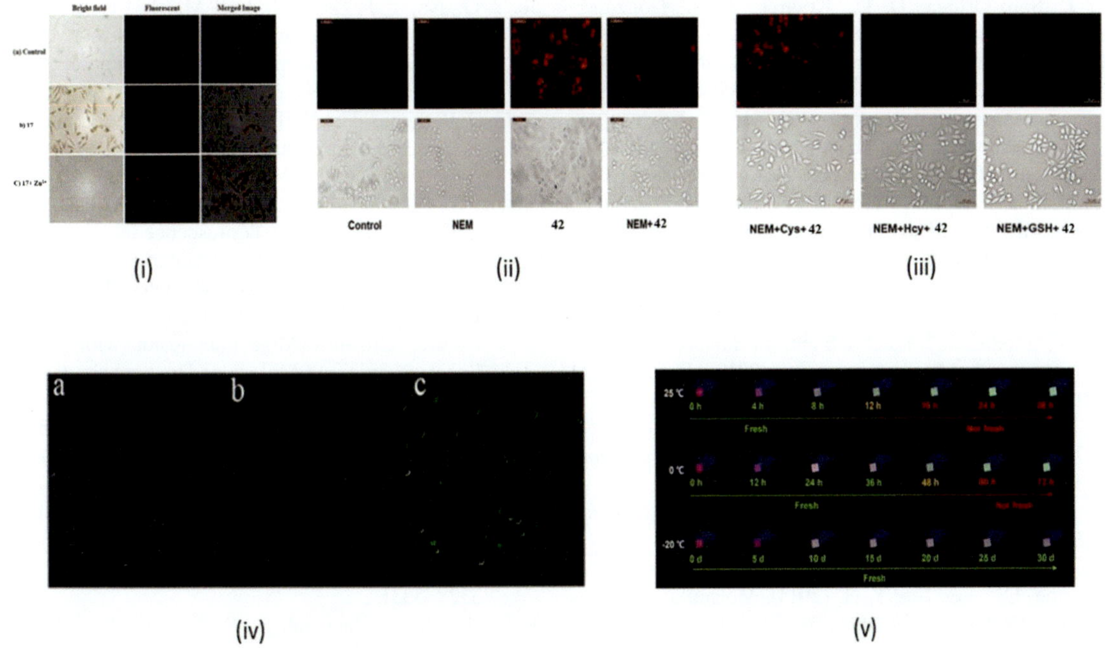

FIGURE 5.6 Fluorescent image of Hela cells. (i)(a) Cells in the absence of probe 17 (b) cell with probe 17 (25 μM) for 1 h (c) probe 17 + Zn^{2+} (25 μM) for 30 min. (ii) Fluorescence image and bright field images of HepG2 cells control, incubated with probe 45 (5 μM), incubated with N-ethylmaleimide (NEM) and incubated with probe 45 and NEM. (iii) Fluorescence image and bright field images and of HepG2 cells incubated with NEM and probe 45 for sensing Cys, GSH or Hcy. (iv) Fluorescence images of HeLa cells. (a) treated with probe 30 (20 μM); (b) treated with probe 30 (20 μM) in the presence of Al^{3+} (50 μM) (c) merge of (a) and (b). (v) Fluorescence pictures of probe 59 test strips stored with beef at different temperatures and for different times 365 nm UV light. The images are reproduced from Ganesan et al. (2019) and Ji et al. (2019) with permission. Elsevier copyright 2019, Liu et al. (2018) with permission from Royal Society of Chemistry (2018) and Ye et al. (2022) with permission from American Chemical Society (2022).

5.6 CONCLUSION AND FUTURE PERSPECTIVES

This chapter summarizes the recent development of julolidine for its fluorescence applications ranging from biological to sensing applications. A variety of fluorescent probes based on julolidine core have been designed and developed for various bioimaging applications including DNA and RNA sensing, amino acid detection, etc. Some of the julolidine-containing fluorescent probes are also reported for their application in sensing cations and anions including Cu^{2+}, Al^{3+}, Zn^{2+}, F^-, CN^-, and HSO_4^-. Julolidine-based fluorescent molecular rotors are also explored for their role in viscosity sensors. These probes display changes in the fluorescence intensity with changing viscosity that can be utilized for the sensing of viscosity of blood and cell microenvironment. Various other applications related to the fluorescent properties of julolidine are included in this chapter. Apart from the conventional applications of julolidine in fluorescence, very few applications related to the diagnosis of diseases have been reported. This raises the scope for the development and innovation of julolidine-containing probes for the diagnosis of diseases.

REFERENCES

Ahamed, A. A., Alharbi, S. A., & Venkatesan, G. (2024). A julolidine aldehyde dansyl hydrazine schiff base as fluorescence chemosensor for Zn^{2+} ions recognition and its application. *Journal of Fluorescence*, 1–11.

Akarasareenon, W., Chanmungkalakul, S., Xiaogang, L., & Rashatasakhon, P. (2023). Selective fluorescent sensors for copper (II) ion from julolidine hydrazone derivatives. *Journal of Photochemistry and Photobiology A: Chemistry*, *437*, 114422.

Allen, B. D., Benniston, A. C., Harriman, A., Rostron, S. A., & Yu, C. (2005). The photophysical properties of a julolidene-based molecular rotor. *Physical Chemistry Chemical Physics*, *7*(16), 3035–3040.

Brewer, T. F., Burgos-Barragan, G., Wit, N., Patel, K. J., & Chang, C. J. (2017). A 2-aza-Cope reactivity-based platform for ratiometric fluorescence imaging of formaldehyde in living cells. *Chemical Science*, *8*(5), 4073–4081.

Chen, W., Xu, S., Day, J. J., Wang, D., & Xian, M. (2017). A general strategy for development of near-infrared fluorescent probes for bioimaging. *Angewandte Chemie*, *129*(52), 16838–16842.

Choi, Y. W., Lee, J. J., Nam, E., Lim, M. H., & Kim, C. (2016a). A fluorescent chemosensor for Al^{3+} based on julolidine and tryptophan moieties. *Tetrahedron*, *72*(16), 1998–2005.

Davis, T. J., Soto-Ortega, D. D., Kotarek, J. A., Gonzalez-Velasquez, F. J., Sivakumar, K., Wu, L., & Moss, M. A. (2009). Comparative study of inhibition at multiple stages of amyloid-β self-assembly provides mechanistic insight. *Molecular Pharmacology*, *76*(2), 405–413.

Deng, L., Yao, Y., Xue, L., Hu, H., & Wang, H. (2024). A novel "off-on-off" sensor for consecutive detection of Al^{3+} and H_2PO^{4-} as well as application in food and living cells. *Journal of Photochemistry and Photobiology A: Chemistry*, *446*, 115180.

Fang, W. L., Liang, Z. Y., Guo, X. F., & Wang, H. (2024). A D-π-A-based near-infrared fluorescent probe with large Stoke's shift for the detection of Cys in vivo. *Talanta*, *268*, 125354.

Ganesan, J. S., Gandhi, S., Radhakrishnan, K., Balasubramaniem, A., Sepperumal, M., & Ayyanar, S. (2019). Execution of julolidine based derivative as bifunctional chemosensor for Zn^{2+} and Cu^{2+} ions: Applications in bio-imaging and molecular logic gate. *Spectrochimica Acta Part A: Molecular and Biomolecular Spectroscopy*, *219*, 33–43.

Glass, D. B., & Weissberger, A. (1946). Julolidine. *Organic Syntheses*, *26*, 40.

Goh, W. L., Lee, M. Y., Joseph, T. L., Quah, S. T., Brown, C. J., Verma, C., ... & Teo, Y. N. (2014). Molecular rotors as conditionally fluorescent labels for rapid detection of biomolecular interactions. *Journal of the American Chemical Society*, *136*(17), 6159–6162.

Gompel, J. Van, & Schuster, G. B. (1987). Chemiluminescence of organic peroxides: Intramolecular electron-exchange luminescence from a secondary perester. *Journal of Organic Chemistry*, *52*(8), 1465–1468.

Haidekker, M. A., Ling, T., Anglo, M., Stevens, H. Y., Frangos, J. A., & Theodorakis, E. A. (2001). New fluorescent probes for the measurement of cell membrane viscosity. *Chemistry and Biology*, *8*(2), 123–131.

Han, G., Kim, D., Park, Y., Bouffard, J., & Kim, Y. (2015). Excimers beyond pyrene: A far-red optical proximity reporter and its application to the label-free detection of DNA. *Angewandte Chemie*, *127*(13), 3984–3988.

Hebbar, S. D., Sparsha, K. A., Raval, K., & Trivedi, D. R. (2024). Julolidine-hydrazide-based D-π-A fluorescent chemoprobes for detection of Al^{3+} and differentiation of arsenic species: Applications in portable test strips, ecological assessment, and DFT studies. *Microchemical Journal*, *207*, 111776.

Hu, P., Wu, T., Fan, W., Chen, L., Liu, Y., Ni, D., Bu, W., & Shi, J. (2017). Near infrared-assisted fenton reaction for tumor-specific and mitochondrial DNA-targeted photochemotherapy. *Biomaterials*, *141*, 86–95.

Jbilou, F., Georgousopoulou, I. N., Marinkovic, S., Vouyiouka, S., Papaspyrides, C. D., Estrine, B., Dole, P., Cottaz, A., & Joly, C. (2016). Intelligent monitoring of solid state polymerization via molecular rotors: The case of poly(butylene succinate). *European Polymer Journal*, *78*, 61–71.

Jeong, H. Y., Lee, S. Y., & Kim, C. (2017). Furan and julolidine-based "turn-on" fluorescence chemosensor for detection of F^- in a near-perfect aqueous solution. *Journal of Fluorescence*, *27*(4), 1457–1466.

Ji, Y., Dai, F., & Zhou, B. (2019, January). Developing a julolidine-fluorescein-based hybrid as a highly sensitive fluorescent probe for sensing and bioimaging Cys in living cells. *Talanta*, *197*, 631–637.

Jögi, A., Vaapil, M., Johansson, M., & Påhlman, S. (2012). Cancer cell differentiation heterogeneity and aggressive behavior in solid tumors. *Upsala Journal of Medical Sciences*, *117*(2), 217–224.

Jo, T. G., Na, Y. J., Lee, J. J., Lee, M. M., Lee, S. Y., & Kim, C. (2015). A diaminomaleonitrile based selective colorimetric chemosensor for copper (II) and fluoride ions. *New Journal of Chemistry*, *39*(4), 2580–2587.

Judd, D. B. (1933). The 1931 ICI standard observer and coordinate system for colorimetry. *Journal of the Optical Society of America*, *23*(10), 359–374.

Jun, Y. W., Harcourt, E. M., Xiao, L., Wilson, D. L., & Kool, E. T. (2022). Efficient DNA fluorescence labeling via base excision trapping. *Nature Communications*, *13*(1), 5043.

Kalel, R., Mora, A. K., Ghosh, R., Dhavale, D. D., Palit, D. K., & Nath, S. (2016). Interaction of a Julolidine-Based Neutral Ultrafast Molecular Rotor with Natural DNA: Spectroscopic and Molecular Docking Studies. *Journal of Physical Chemistry B*, *120*(37), 9843–9853.

Katayama, H., Abe, E., & Kaneko, K. (1982). Synthesis of julolidines from anilines. *Journal of Heterocyclic Chemistry*, *19*(4), 925–926.

Katritzky, A. R., Luo, Z., & Cui, X. L. (1999). Synthesis of substituted piperidines from N, N-Bis [(benzotriazol-1-yl) methyl] amines. *The Journal of Organic Chemistry*, *64*(9), 3328–3331.

Katritzky, A. R., Rachwal, B., & Rachwal, S. (1995). Reactions of N-Alkyl-N-phenyl-1H-benzotriazole-1-methanamines with. alpha.,. beta.-Unsaturated Ethers. A Novel Route to 1, 4-and 1, 3-Disubstituted 1, 2, 3, 4-Tetrahydroquinolines. *The Journal of Organic Chemistry*, *60*(8), 2588–2596.

Katritzky, A. R., Rachwal, B., Rachwal, S., & Abboud, K. A. (1996). Convenient synthesis of julolidines using benzotriazole methodology. *Journal of Organic Chemistry*, *61*(9), 3117–3126.

Kumar, V., Kaur, P., & Singh, K. (2023). Julolidine-hydrazone based chemosensor for detection of Zn^{2+}: Fluorescent in-situ formed Zn^{2+} complex discriminates PPi from ADP and ATP. *Analytica Chimica Acta*, *1240*, 340758.

Lee, S. Y., Bok, K. H., Kim, J. A., Kim, S. Y., & Kim, C. (2016). Simultaneous detection of Cu^{2+} and Cr^{3+} by a simple Schiff-base colorimetric chemosensor bearing NBD (7-nitrobenzo-2-oxa-1,3-diazolyl) and julolidine moieties. *Tetrahedron*, *72*(35), 5563–5570.

Li, P., Li, R., Wang, K., Liu, Q., Ren, B., Ding, Y., Guan, R., & Cao, D. (2022). A julolidine-chalcone-based fluorescent probe for detection of Al^{3+} in real water sample and cell imaging. *Spectrochimica Acta Part A: Molecular and Biomolecular Spectroscopy*, *276*, 121213.

Liu, F. T., Fan, C., Tu, Y., & Pu, S. (2018). A new fluorescent and colorimetric chemosensor for Al^{3+} and F^-/CN^- based on a julolidine unit and its bioimaging in living cells. *RSC Advances*, *8*(54), 31113–31120.

Liu, F.-T., Han, W.-W., Ren, H., Yang, W.-J., Wang, R.-N., Miao, J.-Y., Zhao, B.-X., & Lin, Z.-M. (2023). A ratiometric lysosome-targeted fluorescent probe for imaging SO_2 based on the coumarin-quinoline-julolidine molecular system. *Dyes and Pigments*, *210*, 111017.

Liu, H., Ding, S., Lu, Q., Jian, Y., Wei, G., & Yuan, Z. (2022). A versatile Schiff Base Chemosensor for the determination of trace Co^{2+}, Ni^{2+}, Cu^{2+}, and Zn^{2+} in the water and its bioimaging applications. *ACS Omega*, *7*(9), 7585–7594.

Liu, Y., Zhang, W., Sun, Y., Song, G., Miao, F., Guo, F., Tian, M., Yu, X., & Sun, J. Z. (2014). Two-photon fluorescence imaging of RNA in nucleoli and cytoplasm in living cells based on low molecular weight probes. *Dyes and Pigments*, *103*, 191–201.

Man, L. L., Wang, T. L., Liu, K., Hou, X., Wang, Z. X., Tong, L., & Dong, W. K. (2024). A novel dual-functional half salamo-based fluorescent turn-on sensor for selectively recognition and differentiation of Zn^{2+} and Al^{3+} ions. *Journal of Molecular Structure*, *1303*, 137586.

Martini, G., Martinelli, E., Ruggeri, G., Galli, G., & Pucci, A. (2015). Julolidine fluorescent molecular rotors as vapour sensing probes in polystyrene films. *Dyes and Pigments*, *113*, 47–54.

Megha, K. V., Kaur, P., & Singh, K. (2023). Julolidine based red emitting ESIPT/AIE active material showing luminescence beyond excimer emission: An "on-off" emission response to Cu^{2+}. *Spectrochimica Acta Part A: Molecular and Biomolecular Spectroscopy*, *290*, 122239.

Meng, L., Xu, Z. Y., Chen, J. R., Luo, H. Q., & Li, N. B. (2024). Efficient turn-on fluorescent probe cooperated by cascade response for disclosing the fluctuation of cysteine in cells. *Analytica Chimica Acta*, *1308*, 342660.

Mondal, I. C., Rawat, P., Galkin, M., Deka, S., Karmakar, A., Mondal, P., & Ghosh, S. (2023). Julolidine-based small molecular probes for fluorescence imaging of RNA in live cells. *Organic & Biomolecular Chemistry*, *21*(38), 7831–7840.

Na, Y. J., Hwang, I. H., Jo, H. Y., Lee, S. A., Park, G. J., & Kim, C. (2013). Fluorescent chemosensor based-on the combination of julolidine and furan for selective detection of zinc ion. *Inorganic Chemistry Communications*, *35*, 342–345.

Nagarajan, S., & Lapidus, L. J. (2017). Fluorescent probe DCVJ displayed high sensitivity for characterization of amyloid-β peptide early in the lag phase. *ChemBioChem*, *18*(22), 2205–2211.

Noh, J. Y., Kim, S., Hwang, I. H., Lee, G. Y., Kang, J., Kim, S. H., Min, J., Park, S., Kim, C., & Kim, J. (2013). Solvent-dependent selective fluorescence assay of aluminum and gallium ions using julolidine-based probe. *Dyes and Pigments*, *99*(3), 1016–1021.

Paisuwan, W., Lertpiriyasakulkit, T., Ruangpornvisuti, V., Sukwattanasinitt, M., & Ajavakom, A. (2020a). 8-Hydroxyjulolidine aldimine as a fluorescent sensor for the dual detection of Al^{3+} and Mg^{2+}. *Sensing and Bio-Sensing Research*, *29*, 100358.

Piloto, A. M., Hungerford, G., Costa, S. P. G., & Gonçalves, M. S. T. (2013). Photoinduced release of neurotransmitter amino acids from coumarin-fused julolidine ester cages. *European Journal of Organic Chemistry*, *2013*(34), 7715–7723.

Raju, B. R., Leitao, M. I. P., Sousa, M. J., Coutinho, P. J., & Gonçalves, M. S. T. (2020). New NIR dyes based on quinolizino [1, 9-hi] phenoxazin-6-iminium chlorides: synthesis, photophysics and antifungal activity. *Dyes and Pigments*, *173*, 107870.

Shirinfar, B., Ahmed, N., Park, Y. S., Cho, G.-S., Youn, I. S., Han, J.-K., Nam, H. G., & Kim, K. S. (2013). Selective fluorescent detection of RNA in living cells by using Imidazolium-based cyclophane. *Journal of the American Chemical Society*, *135*(1), 90–93.

Singhal, D., Gupta, N., & Singh, A. K. (2016). Fluorescent sensor for Al^{3+} ion in partially aqueous media using julolidine based probe. *New Journal of Chemistry*, *40*(9), 7536–7541.

Tuszynski, J. A., Brown, J. A., & Sept, D. (2003). Models of the collective behavior of proteins in cells: Tubulin, actin and motor proteins. *Journal of Biological Physics*, *29*(4), 401–428.

Vasjari, M., Duka, S., Shehu, A., Vallja, L., & Broli, N. (2021). Evaluation of well water used for drinking water in relation to metal content, dietary exposure and impact on the recommended daily amount. *Journal of Hygienic Engineering & Design*, *37*.

Wang, H., Tian, X., Du, W., Zhang, Q., Guan, L., Wang, A., Zhang, Y., Wang, C., Zhou, H., Wu, J., & Tian, Y. (2016). A two-photon fluorescent RNA probe screened from a series of oxime-functionalized 2,2′:6′,2′′-terpyridine ZnX_2 (X = Cl, Br, I) complexes. *Journal of Materials Chemistry B*, *4*(28), 4818–4825.

Wilson, D. L., & Kool, E. T. (2019). Ultrafast oxime formation enables efficient fluorescence light-up measurement of DNA base excision, *Journal of the American Chemical Society*, *141*(49), 19379–19388.

Xia, Z., Guo, X., Zhu, Y., Wang, Y., & Wang, J. (2018). A julolidine-fused anthracene derivative: Synthesis, photophysical properties, and oxidative dimerization. *RSC Advances*, *8*(24), 13588–13591.

Yang, F., Zhang, Q., Huang, S., & Ma, D. (2020). Recent advances of near-infrared inorganic fluorescent probes for biomedical applications. *Journal of Materials Chemistry B*, *8*(35), 7856–7879.

Yao, Q., Li, H., Xian, L., Xu, F., Xia, J., Fan, J., Du, J., Wang, J., & Peng, X. (2018). Differentiating RNA from DNA by a molecular fluorescent probe based on the "door-bolt" mechanism biomaterials. *Biomaterials*, *177*, 78–87.

Ye, H., Koo, S., Beitong Zhu, Ke, Y., Sheng, R., Duan, T., Zeng, L., & Kim, J. S. (2022). Real-time fluorescence screening platform for meat freshness. *Analytical Chemistry*, *94*(44), 15423–15432.

6 Perylene-Based Fluorescent and Colorimetric Probes for Multimode Sensing Applications

Madhu Bala, Mareechika Gaddam and Abha Sharma
National Institute of Pharmaceutical Education and Research,
Raebareli, Lucknow, UP, India

LIST OF ABBREVIATIONS

PDI	Perylene tetracarboxylic diimide
PMI	Perylene monoimide
PBI	Perylene bisimide
UV	Ultraviolet
LDL	Lowest detection limit
ROS	Reactive oxygen species
OFET	Organic field-effect transistor
OSC	Organic solar cell
PTE	Perylene tetraester
OLED	Organic light-emitting diode
PEI	Perylene diester monoamide
PVA	Polyvinyl alcohol
PET	Photoinduced electron transfer
EDTA	Ethylenediaminetetraacetic acid
EDPA	Ethyl dipicolylamine aniline
PTCDA	3,4,9,10-perylene tetracarboxylic dianhydride
ATP	Adenosine triphosphate
AuNP	Gold nanoparticle

6.1 INTRODUCTION

Perylene or perilene ($C_{20}H_{12}$) is a fused polycyclic hydrocarbon compound found in natural sources including river sediment, crude oils, fossils, coals, as well as lake environments. It is the backbone of perylene quinones and their related derivatives, which exist in certain plants, fungi, and insects (Nagao et al., 1997). Perylene and its quinone-related compounds have a significant UV-visible absorption as well as strong fluorescence emission both in solutions and in solid phase. It exhibited strong photosensitizing properties, which are capable of generating the reactive oxygen species (ROS). Due to their fluorescence emission properties, perylene and its derivatives are utilized as fluorophores in the development of chemical sensors for the trace-level detection of environmentally hazardous materials such as organic pollutants, heavy metal ions, etc. These developing techniques have great response in comparison with conventional analytical instrumental techniques like mass spectrometry, chromatography, infrared spectrometry, and electrochemistry. Chemosensors attract much more research attention due to their unique properties like small size, high selectivity and sensitivity, low cost, and quick response for on-site real-time detection. Based on colorimetric or fluorescence signals, optical sensors (Chen et al., 2023) exhibit superiority regarding nondestructive detection with high selectivity, sensitivity and facile operation. The major challenging aspects in designing this molecule include stability against photobleaching, visible light absorption, photoluminescence quantum yield, and flexibility of structure with side functional groups that can be able to bind with analytes specifically (Chen et al., 2019; Wang et al., 2016; Chen et al., 2015; Avlasevich et al., 2010). In most cases, the specific binding is driven by weak host–guest interactions, metal coordination, hydrogen bonding, electrostatic forces, Van der Waals forces, or other noncovalent interactions.

In the past two decades, perylene tetracarboxylic diimide (PDI) and its derivatives (PDIs) have formed a category of high-grade dyes, stable semiconductor materials, and attractive research attention for developing either colorimetric or fluorescence chemical sensors (Kohl et al., 2004). PDI molecules have good tailorable structures, desirable excited-state lifetime, and versatile electronic and optical features such as strong optical absorption, remarkable fluorescence with 100% quantum yield in solvents, etc. (Renner et al., 2022). PDIs have also unique photochemical and thermal stability, i.e., resistance to heat and photobleaching (Rosso et al., 2021), which both are very critical conditions for practical use as optical sensors (Zhang et al., 2018). The two imide positions of PDI are node in π-orbital, and substitution at these positions does not produce any change to the electronic property of PDI, i.e., UV-vis absorption. This provides many options to change side groups for optimizing the binding with different analytes (related to both selectivity and sensitivity), but without any alteration in absorption wavelength, allowing for quantitative comparison of fluorescence change because of analyte's binding, i.e., a fluorescence sensor, works. In addition, PDI molecules are more prone to form supramolecular H-aggregates via π-π stacking with hypsochromic shift of the absorption band due to strong π-π electronic interaction (Kaufmann et

DOI: 10.1201/9781032706245-6

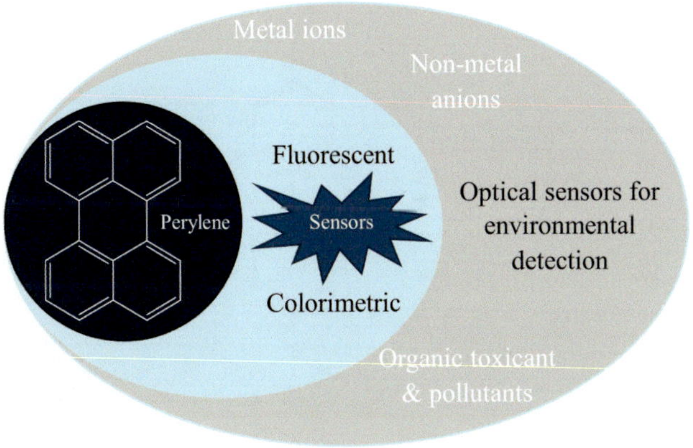

FIGURE 6.1 Schematic illustration of PDI-based optical sensors for environmental detection.

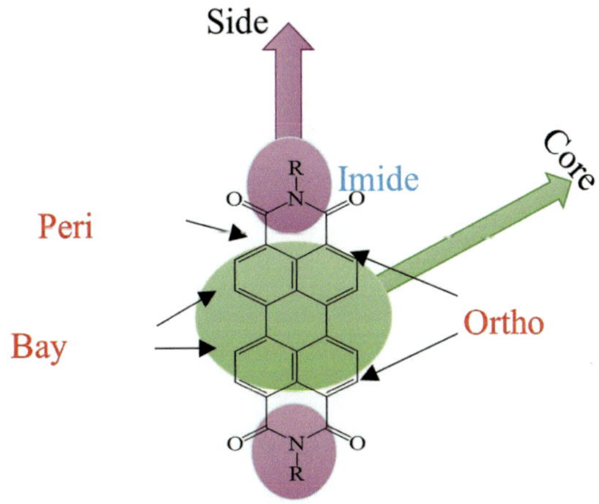

FIGURE 6.2 Detailed structure of perylene-based molecule, i.e., perylene tetracarboxylic diimide.

al., 2018; Tang et al., 2023; Wang et al., 2021). PDIs are especially appropriate colorimetric as well as fluorescence sensors along with high sensitivity and selectivity (Schubert et al., 2013). PDI-based optical sensors can work either as liquid-phase media or solid-phase thin films. So far, various reviews have been published on the solid-phase PDI-based sensors, which include self-assembled nanofibers and thin-film fluorescence chemosensors for the detection of volatile organic amines (Goerl et al., 2012). The PDI-based supramolecular sensors have also been published for the detection of ions. In this review, we provide an overview of current progress in research and development of PDI-based colorimetric and fluorescent sensors for environmental detection of metal or nonmetal ions, organic pollutants, and amines (Krupka et al., 2023; Soh et al., 2011; Sriramulu and Valiyaveettil, 2016) (Figure 6.1). Considering the fact that analyte's binding produces alteration in both fluorescence and absorption spectra of PDIs (Singh et al., 2021; Chen et al., 2020). The PDI-based optical sensors are also utilized for chemical monitoring in various fields like food safety, public security, healthcare, biomedicine, etc. Thus, the modulable structure of PDIs plays an important factor in the development of recent research generation of optical sensors.

There are various perylene-based molecules in which bay, ortho (Li et al., 2017), and peri-substitution have occurred or two imide groups are attached at the peri-position of perylene structure in PDI-based molecules (Marcon et al., 2009; Lin et al.,2012; Li et al., 2019; Wurthner et al., 2006) (Figure 6.2).

6.2 PERYLENE-BASED MOLECULES

Perylene provides a flexible structure to prepare different fluorescent or colorimetric probes or dyes such as perylene bisimide (PBI) dyes (Wurthner et al., 2006) and perylene tetracarboxylic diester monoimides (PEIs). PBI-based molecules exhibit interesting features, and they have found multiple applications as highly photoluminescent materials, photostable dyes, photoreceptors of laser printers and photocopiers, active components in organic field-effect transistors (OFETs), and electron acceptors in organic solar cells (OSCs). Most of the functionalized PBIs have been utilized for the functionalization/passivation of gold nanorods that show photophysical behavior. Two types of substitution of PBIs are available, i.e., ortho and bay. The bay substitution of the PBIs provided to shift the emission band toward the red region along with an increase in the Stokes shift as well as fluorescence lifetime. Tetra-esters of perylene tetracarboxylic acids (PTEs), related to PBIs, were also used in the area of liquid crystals, electron acceptors in the organic light-emitting diodes (OLEDs), and fluorescent dyes. Due to the presence of four carboxylate groups, PTEs have significant electron-accepting properties similar to PBIs, and they also can form aggregates with strong π-π overlap in both solid and solution states.

Recently, perylene tetracarboxylic diester monoimides (PEIs) have been attracting interest due to their intermediate electron-deficient feature between PTEs and PBIs. A few reports regarding the synthesis of PEIs have been obtained. Yang et al. (2008) reported the perylene imido-diesters with three *n*-alkyl substituents by reaction of unsubstituted perylene tetracarboxylic monoimido-monoanhydride with an excess amount of alkyl bromide (Figure 6.3). Xue et al. prepared perylene tetracarboxylic monoanhydride diester from corresponding PTE with the treatment of n-alkyl amine and yielded the PEI (Gupta et al., 2018).

A high-yielding microwave-assisted approach was followed to obtain unsymmetrical perylene diester monoamide (PEI) by treating perylene tetra-ester (PTE) and amine. Compared to perylene bisimides (PBIs), PEIs have been less studied in self-assembly due to synthetic difficulty and low-yield production. PEIs have an unsymmetric structure and are less electron deficient as compared to PBIs. Furthermore, PEIs have better solubility than PBIs. This method to prepare PEIs and PBIs is very easy and green. This can be able to produce tuning of photophysical and electronic properties and new n-type materials for self-assembly (Figure 6.4).

The microwave-assisted approach has attracted synthetic chemists, as it provides production in higher yields under

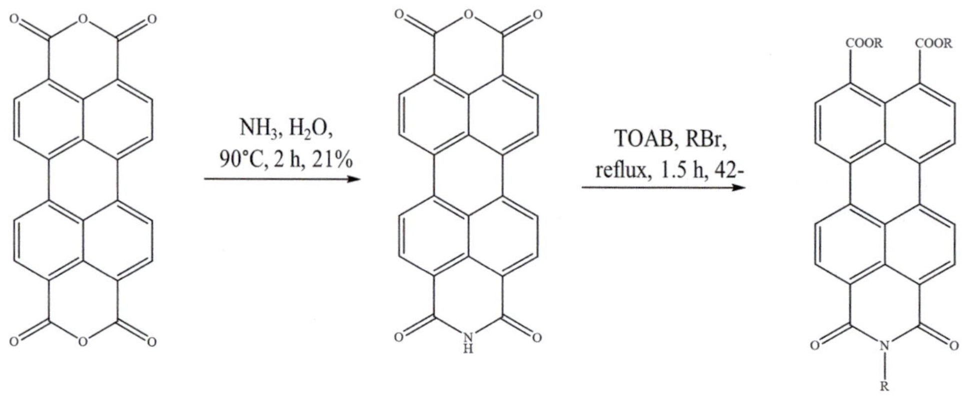

FIGURE 6.3 Scheme illustration for preparing perylene imido-diesters.

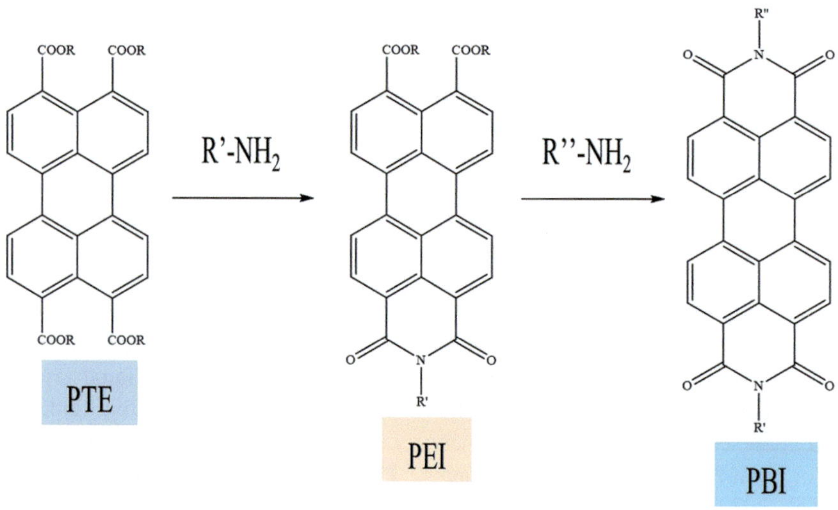

FIGURE 6.4 Green synthesis of perylene-based molecules.

milder reaction conditions. Compared to other conventional heating methods, microwaves greatly accelerate the rate of reactions to obtain unsymmetrical derivatives of perylene molecules. This method is completely straightforward and provides a good yield of 50–55% (Avlasevich et al., 2004; Seguy et al., 2001). Different types of perylene ester imides and unsymmetrical perylene bisimides were synthesized by using a microwave-assisted approach with PEI as the starting material (Figure 6.5) (Nowak-Krol et al., 2019). Thus, a microwave-assisted method for the synthesis of unsymmetrical PEIs by treating PTEs with amines produced a good yield. This method provides easy access to synthesize unsymmetrical perylene bisimides (PBIs), which are very important materials of supramolecular self-assembly.

Recently, nanoparticles have become the field of interest as they exhibit unique chemical as well as optical properties that differ from the isolated bulk solid-state molecules (Latterini et al., 2006).

Perylene was also utilized as active material to synthesize the nanoparticles via the precipitation method in order to test the tunability related to the particle properties, i.e., particle size (Fu et al., 2001). These particles were prepared with perylene as active material and polyvinyl alcohol (PVA) as stabilizing agent. Reprecipitation

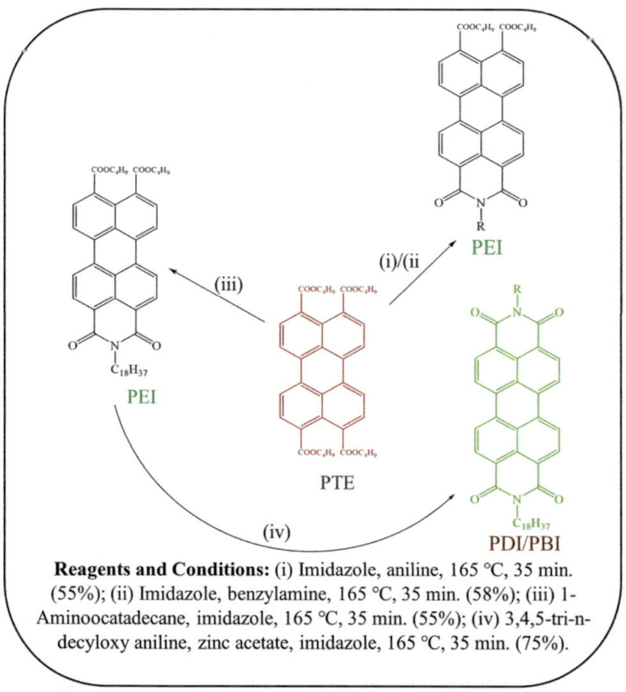

FIGURE 6.5 Synthesis of unsymmetrical perylene bisimides (PBIs/PDIs).

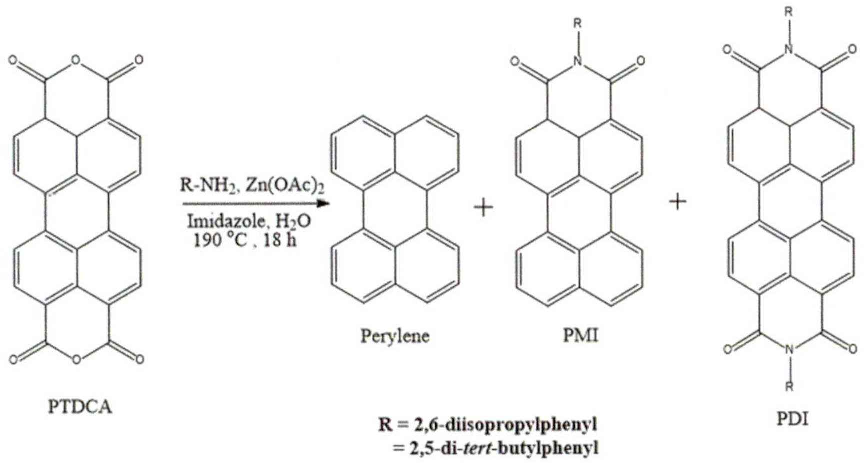

FIGURE 6.6 Synthetic pathway of PMIs developed by Langhals.

method was carried out to prepare these particles, i.e., the solution of perylene in acetone was injected in an aqueous solution containing PVA. The concentrations of both PVA and perylene should be optimized to obtain small particles with reproducible features along with higher fluorescence efficiency. By using the optimized concentrations, the nanoparticles were synthesized at different temperature levels with the addition of perylene acetone solution to the aqueous solution of PVA.

In the small family of perylene, PMI has a few unique properties over its different analogues of PDI, although they showed a plenty of common properties. Firstly, in comparison with PDI, PMI has an inherent higher ground-state dipole moment (6.4 D) along with a long conjugated axis. The self-assembled nano-framework structures produced by different PMI derivatives experience a dipolar nature in a bulkier state, which affects their transformation of the energy process. Secondly, the functional aspects of the asymmetrical position of imide moiety in the parent PMI scaffold furnish ICT characteristics on being conjugated with any other moieties. In 2007, Mullen and coworkers alleged this concept by synthesizing donor-acceptor PMI derivatives where the perspective of ICT characteristics through different donors influenced the photoelectrochemical properties along with the efficiency of solar cell systems. PMI shows fluorescence mainly in the red-shifted region compared with parent PDI's scaffold emission region; the Stokes shift of PMI scaffolds is comparatively larger than PDIs.

Marks et al. described the nature of Stokes shift of PMI moieties, which can be extremely large via the strong involvement of effective donors in peri-position. They synthesized a series of PMI-based donor (D)-acceptor (A) analogues (donors: carbazole, indolizine, dibenzoazepine, diphenylamine, and dialkylamine) (Zagranyarski et al., 2012; Vasilev et al., 2012) having Stokes shift greater than 100 nm, which can be utilized as suitable candidates in luminescent solar cell system. They are comparatively more photostable as compared to PDI moieties. As per electrochemical property, 0.4 V more negative one-electron reduction potential makes them more advantageous in energy transfer processes along with the ability to achieve greater open-circuit voltage in organic solar cell devices. Synthetically, PMI is recognized as an important intermediate to prepare higher perylene imides and their analogues for desired applications by functionalization of a peri-position (Roy et al., 2021).

Generally, the synthetic chemistry of dyes is based on perylene monoimides (PMIs), which has been developed very quickly over the past few years. The synthesis of PMIs was first initiated by Langhals, who developed a method/approach to synthesize the different derivatives of PMIs. According to this, when the commercially available 3,4,9,10-perylene tetracarboxylic dianhydride (PTCDA) was reacted with a primary amine in the presence of zinc acetate and molten imidazole at a higher temperature and pressure, the PMI was obtained. Then the side products of this reaction were found as PDI and perylene (as shown in Figure 6.6). Their low intrinsic solubility is the major challenge facing with PMIs. A widespread approach to improve their solubility has been developed by introducing sterically demanding residues at the imide positions, which minimizes the strong aggregation potential. Especially, ortho-substituted anilines (2,5-di-tert-butyl aniline, 2,6-diisopropyl aniline, etc.) and branched amines (2-ethylhexylamine, etc.) are excellent aggregation blockers, and they produce PMIs with excellent solubility in organic solvents.

6.3 STRUCTURAL-BASED SEVERAL APPLICATIONS OF PERYLENE MOLECULE

PDI-Based Fluorescent Sensors

PDI-based fluorescent sensors can produce a great response to external stimuli, which includes toxic organic pollutants or inorganic ions in the environment. The selectivity and sensitivity are measured from the change in fluorescence intensity. Turn ON the fluorescence of PDI by turning OFF the intramolecular PET process upon the coordination or binding with an analyte can be one of the mechanisms of sensing (Chen et al., 2012; Wang et al., 2014). These strong fluorescence turn-ON systems of sensors can be utilized for various analytes like metal ion sensing, nonmetal anion sensing, and organic pollutants.

6.3.1 Metal Ion Sensing

Toxic metal ion detection is important for testing water pollution, diagnosis of disease, and food safety. There are major challenges to studying the performances of sensing in pure aqueous solutions (He et al., 2005; Yan et al., 2009; Erdemir et al., 2015; Lin et al., 2014; Kumar et al., 2018). Noticeably, the fluorescence alterations of PDIs are pH-dependent, and they should be operated under a well-controlled pH working solution system, where the addition of buffer reagent, such as 2-[4-(2-hydroxyethyl)-1-pipera zinyl]-ethane sulfonic acid (HEPES), is necessary (Liu et al., 2013; Elmas Karuk et al., 2018; Xiong et al., 2019). In case of PDI, both fluorescence turn-OFF and turn-ON mechanism systems have been applied, and their sensing mechanism highly depends on the molecular structure of PDIs, solution conditions, types of analytes, i.e., concentration, pH, composition, etc. The interference study with other metal ions, especially those having similar groups in a periodic table, should be considered.

6.3.1.1 Detection of Copper Ions

Copper ion is one of the major contaminants responsible for environmental pollution that leads to biological toxicity. PDI-based fluorescence chemosensors sense Cu^{2+} via dual mode, i.e., fluorescent as well as colorimetric modulation. The binding area involves amine groups and is attached at imide positions of PDIs. In 2014, Singh's group reported the quenched fluorescence of the PDI 2-Cu^{2+} complex, which can be utilized as a novel platform for the ON-OFF sensing mechanism process. At the bay area of PDI, hydroxyquinoline group is attached as a Cu^{2+} recognition receptor, which senses Cu^{2+} ions via colorimetric change (from coral red to light pink color) and fluorescence turn-OFF. The LDLs of 0.5 μM and 1.0 μM were obtained for the sensing via colorimetric and fluorescent modulation, respectively, in $CHCl_3$ solutions (Figure 6.7). The obtained PDI 2-Cu^{2+} complex with 1:2 stoichiometry can be utilized as a ratiometric chemosensor to detect the cyanide ions based on color change and fluorescence turn-ON along with LDLs of 10 μM and 8.0 μM, respectively (Singh et al., 2014). Due to the strong copper cyanide binding affinity, they can form a stable complex, $[Cu\ (CN)_x]$, which is responsible for recovering the fluorescence of PDI 2. Zhu et al. also reported Au nanoparticle (AuNP)-mediated PDI-based fluorescent chemosensor for the detection of Cu^{2+} ions in organic media. The complexation of PDI 1 with AuNPs through the weak N ... Au interactions which leads to colloid formation between PDI-1 and AuNPs resulting in fluorescence quenching (He et al., 2005). The fluorescence was recovered upon the addition of Cu^{2+} ions, because the Cu^{2+} ions form a stronger coordination interaction with the pyridyl moiety of PDI 1, and 100-fold fluorescence intensity of PDI 1 was boost up in a solution of ACN/$CHCl_3$ along with LDL of 1.0 μM.

6.3.1.2 Detection of Barium Ions

Wurthner et al. reported a fluorescence turn-OFF sensor based on a modification of PDI with 15-crown-5 ether that was incorporated for selective detection of Ba^{2+} ions (Weißenstein et al., 2015). The sensing mechanism involves metal ion-induced self-assembly of PDI molecules, and they cause quenching in fluorescence (turn-OFF). The coordination interaction between Ba^{2+} and 15-crown-5 ether causes the formation of H-type aggregates of PDIs, and modification of the side group at imide positions helps to enhance the solubility of PDI 3.

6.3.1.3 Detection of Aluminum and Ferric Ions

Aluminim is the most prevalent element in the earth's crust and also causes environmental contamination. Anthony et al. reported PDIs substituted with pyridine isomers at the region of bay area, and these PDIs allowed for selective fluorescent and colorimetric sensing of Al^{3+} and Fe^{3+} ions in dimethylformamide (DMF). It was found that PDIs are not involved in the abovementioned metal coordination interactions but the Lewis acidic character of Fe^{3+} and Al^{3+} ions, i.e., responsible for sensing mechanism (Kundu et al., 2017). In the presence of two acidic ions, the protonation of base pyridine nitrogen atoms on PDI 4 occurred which hinders the intramolecular charge transfer from nitrogen lone pair electrons to PDI 4 core and then causes alteration in fluorescence quenching and electronic structure (Figure 6.7).

6.3.1.4 Detection of Cadmium and Zinc Ions

The most abundant transition metal ion in human body is the zinc ion, and, on the other side, the cadmium ion causes typically heavy metal toxicity in environment (Liu et al., 2013). Shangguan reported a fluorescent chemosensor (PDI 5) that can selectively detect the Zn^{2+} and Cd^{2+} ions, and their discrimination has been realized via adjustment of pH, i.e., fluorescence turn-ON at pH 9.0 for Cd^{2+} ions and at pH 6.0–7.0 for Zn^{2+} ions. A rapid response for both Zn^{2+} and Cd^{2+} ions was shown along with LDLs of 0.032 μM for Zn^{2+} ions and 0.048 μM for Cd^{2+} ions. Dipicolylamine (DPA) moiety has been widely utilized as a complexing agent to Zn^{2+} ions with a 2:1 binding stoichiometry. This strong complexation of DPA with Cd^{2+} or Zn^{2+} ions can lower the energy level of a lone pair of electrons on the amine group and then block the intramolecular PET process, which turns on the fluorescence (Figure 6.7). Furthermore, the two DPA groups are substituted at imide position of PDI 5, and protonation or deprotonation of DPA caused by alteration at pH level does not change the electronic structure of PDI 5.

6.3.1.5 Detection of Mercury Ions

Mercury ions have also much more intention toward serious toxicity rather than other heavy metal ions and cause contamination in soil and water systems. Hg^{2+} ions can be accumulated via food chains. Based on PDIs, fluorescence turn-ON and turn-OFF chemosensors have been developed for the detection of Hg^{2+} ions (Erdemir et al., 2015). In 2008, an ultraselective detector for Hg^{2+} ions based on a quenching fluorescence mechanism was reported. The PDI

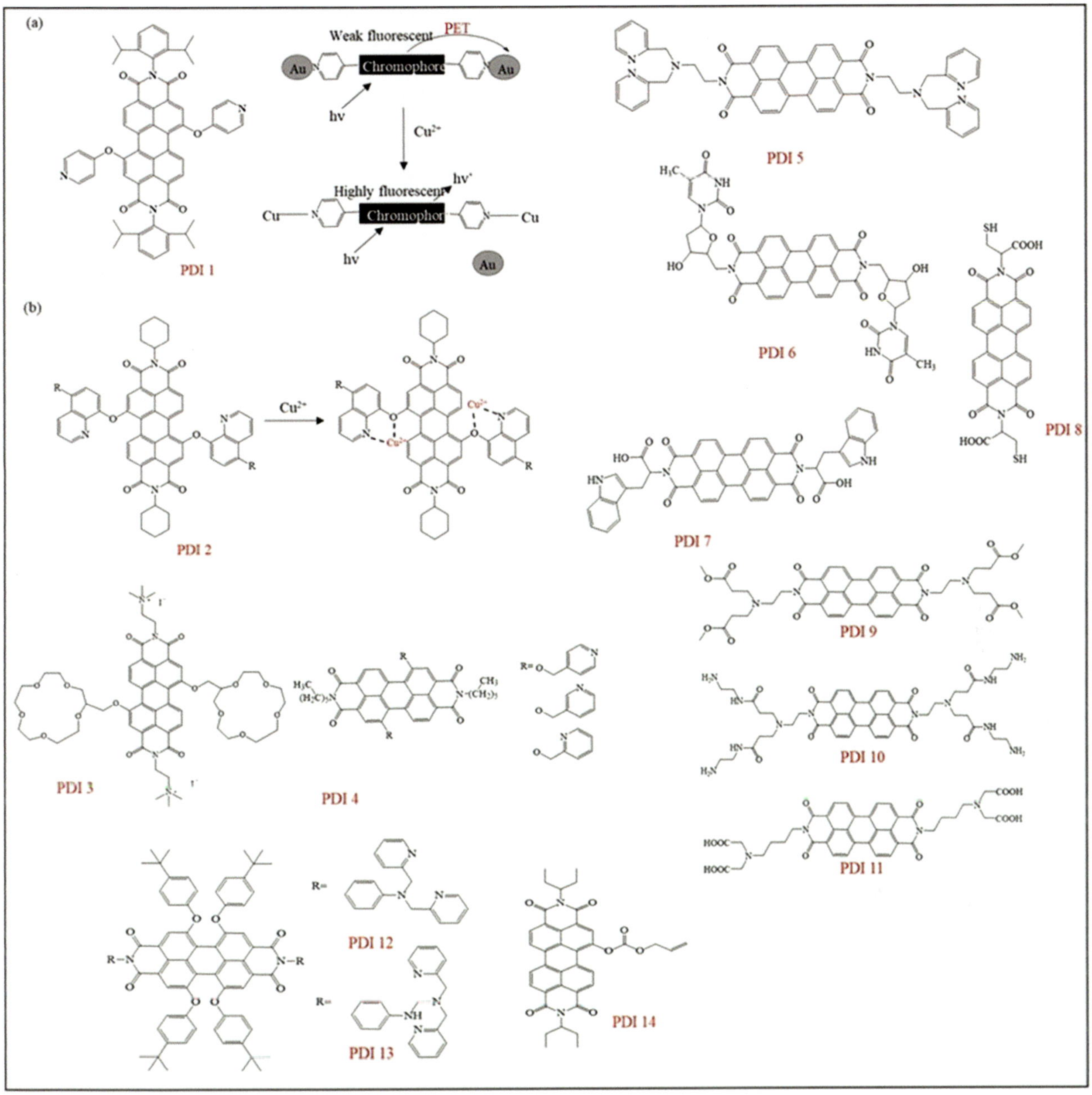

FIGURE 6.7 Metal ion sensors.

molecule is substituted with two thymine moieties that form a selective complexation with Hg^{2+} ions. This complexation causes aggregation/polymerization of PDIs, which leads to significant quenching in fluorescence. The LDL of 0.005 μM was obtained in the DMF/H_2O solution system. This aggregation-induced fluorescence quenching is in sharp contrast to abovementioned quenching of fluorescence due to electron transfer process. The fluorescence of PDI 6 has been recovered upon the addition of acid in the PDI 6-Hg^{2+} complex, because the thymine-Hg^{2+}-thymine coordination interactions break up by competitive protonation of the thymine moiety, i.e., lowering the complexing ability of thymine and making the chemosensor system as recyclable as well as reversible in nature (Che et al., 2008). One more example has also been reported by Karuk (Elmas et al., 2018), i.e., PDI 7 containing tryptophan as side chains. The obtained Hg^{2+}-PDI 7 complex (aggregate) is not fluorescent in nature, but upon the addition of thiol-containing compounds, the fluorescence turn-ON occur due to dissociation of the PDI aggregate via competitive binding with Hg^{2+} ions. Grisci et al. (2016) developed PDI 8 substituted with two cysteine groups as side chains at the imide positions. The PDI 8 tends to aggregate in an aqueous solution and remains a weak fluorescent molecule. Upon the addition of Hg^{2+} ions, the molecular aggregates of PDI 8 can be disrupted/dissociated, which causes yellow color fluorescence. LDL of 0.1 μM or 20 ppb has been obtained for the detection of Hg^{2+} ions. Noticeably, PDI 8 chemosensor can be operated in a water system without any utilization of organic solvents like THF or DMF (Figure 6.7).

6.3.1.6 Detection of Ferric Ions

Fe^{3+} ion is also a typical essential metal ion for supporting human health. This metal ion plays an important role in physiological and biological growth. Georgiev et al. (2011) developed two PDI 9 and PDI 10 fluorescence turn-ON chemosensors for the detection of Fe^{3+} ions in an aqueous solution. PDI 9 and PDI 10 with tetra-ester and poly amido amine moieties at the imide position can act as strong binding receptors to protons and metal ions. Turning-ON the fluorescence of PDI core occurred due to the blocking of intramolecular PET process upon the addition of metal ions because the electron-donating ability of side groups has been decreased. The turn-ON efficacy of PDI 9 is about 30 times greater than PDI 10, mainly due to the scaffold and conformational difference between the two chemosensors. Similarly, the enhancement of fluorescence sensing for the detection of Fe^{3+} ions by PDI 11 was reported by Marcia et al. (2014). PDI 11 contains two side groups, i.e., ethylenediaminetetraacetic acid (EDTA), for the complexation with metal ions. As a common metal ion, Ni^{2+} leads to sensing interference with Fe^{3+} ions, especially for optical chemosensors based on complexation or chemical binding.

To improve the discrimination between Ni^{2+} and Fe^{3+}, Wang et al. (2010) developed PDI 12 and PDI 13 as fluorescence turn-ON sensors for selective detection of two metal ions. Based on PDI backbone with four substitutions at the region of the bay area, two binding side groups, N-ethyldipicolylamine aniline (EDPA) and N-di picolylamine aniline (DPA), are attached at the two imide positions. PDI 12 showed higher selectivity as well as sensitivity toward Ni^{2+} ions, while PDI 13 was selective toward Fe^{3+} ions. With an extra diamino ethylene group attached between the DPA and phenyl group, PDI 13 is preferable for binding with Fe^{3+} as compared to Ni^{2+} ions. Later on, one of the fluorescent sensors PDI 14 was reported and optimized through modification of the side reactive group to allyl carbonate (Zhang et al., 2009) (Figure 6.7). This new chemosensor responded to Pd^0 in ACN/H_2O and DMSO/H_2O systems via a Pd^0-induced deallylation process.

6.3.2 Nonmetal Anion Sensing

For the detection of nonmetal anions, the pH-dependent PET process has been applied for the sensing mechanism of PDI-based fluorescent chemosensor (Dey et al., 2019; Du et al., 2013). In these cases, the hydrophilicity factor has been focused for designing as well as synthesis of PDI molecules.

6.3.2.1 Perchlorate Ions

Perchlorates are widely present in soil and water and cause serious health problems at higher concentrations. These ions can impair the function of thyroid glands. Singh et al. (2014) reported a fluorescence turn-OFF chemosensor (PDI 15, Figure 6.8) that leads to quenching of fluorescence with ClO_4^- ions only in HEPES buffer. The LDL of 0.06 μM was obtained under an optimal condition. The same chemosensor can also be utilized for solid-phase detection of ClO_4^- ions by spotting PDI 15 on TLC silica plate. For applications of PDI 15, chemosensor was tested for firework samples and drinking water.

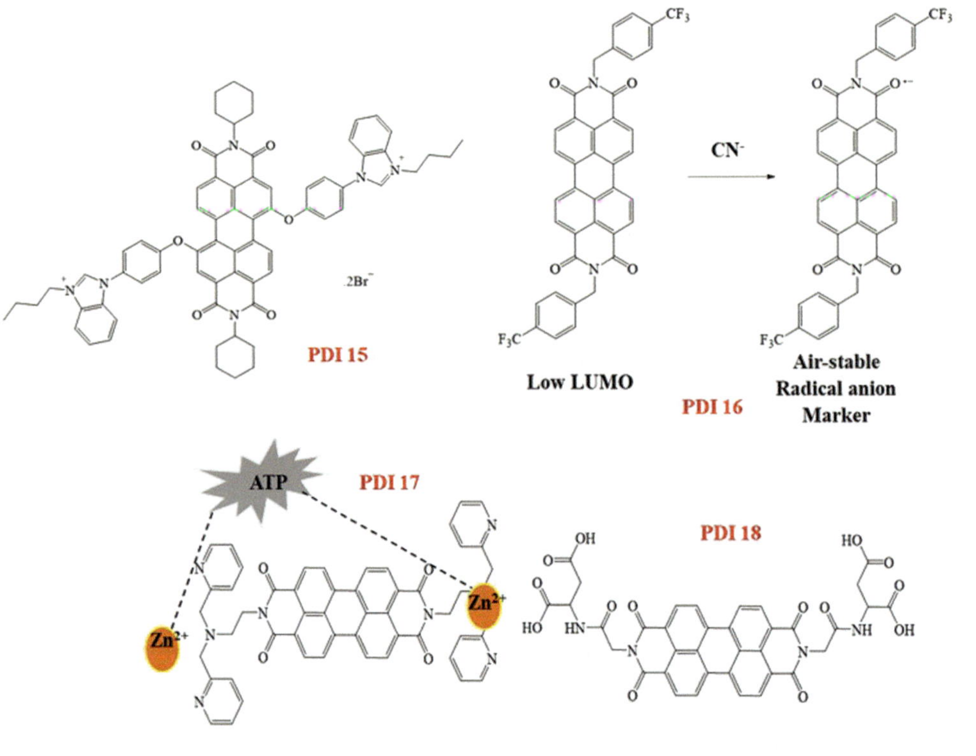

FIGURE 6.8 Structures of nonmetal anion sensors.

6.3.2.2 Cyanide Ions

Cyanide anion is the most toxic ion to living organisms. Ajaykumar et al. (2010) reported a multimodal chemosensor molecule (PDI 16) for the detection of CN^- in THF/H_2O solvent system. In this case, trifluoromethylbenzene groups were attached at imide position of PDI 16 to lower the energy level of LUMO state and therefore enhance the electron affinity of the PDI core. Upon the reaction with CN^-, PDI 16 can be produced by a single-electron transfer process. A combination of the charge, spin, and singly occupied molecular orbital (SOMO)-LUMO electronic transition produced a multimodal sensing mechanism for the detection of CN^- with higher sensitivity as well as selectivity. This sensing signal was realized by color change, i.e., colorimetric sensing mechanism process (Figure 6.8).

6.3.2.3 Phosphate Ions

Phosphate compounds like adenosine triphosphate (ATP), while crucial for life at the cellular level, can contribute to environmental and health issues when present in excessive amounts. One of the examples was reported by Yan et al. (2012) for sensing the ATP molecule by PDI 17, which has been modified with Zn^{2+}-dipicolylethylenediamine (Zn^{2+}-DPEN) moiety at two imide sites. The sensing mechanism involves complexation between ATP and two Zn^{2+} sites. The fluorescence of PDI 17 is low, but it increases upon the addition of ATP.

Similarly, Cu^{2+} ions can also be studied to mediate the sensing of phosphate. Disruption of aggregate of PDI 17-Cu^{2+} complex occurred by competitive binding between Cu^{2+} and PO_4^-, which leads to turn-ON the fluorescence. Feng et al. (2012) developed an aggregation caused quenching PDI 18 probe for detection of Cu^{2+} ions in aqueous solution. Due to the competitive binding of inorganic pyrophosphate (PPi) with Cu^{2+}, the aggregate of PDI 18/Cu^{2+} was disassembled upon the addition of PPi into the solution, which led to the recovery of PDI fluorescence (Figure 6.8). Turn-ON chemosensor showed higher selectivity and sensitivity toward PPi along with LDL of 0.2 μM.

6.3.3 Organic Pollutant Sensing

PDIs-based chemosensors have also been developed for the detection of toxic organic compounds like nitroaromatics, hydrazine, and other amines. These compounds cause toxicity as well as pollution-related issues to human health and water environment (Chen et al., 2020). So, there is a huge demand to develop low-cost chemosensors, portable types that are suited for on-site detection of organic compounds with high selectivity as well as sensitivity.

6.3.3.1 Nitroaromatics

It involves nitroaniline (NA), picric acid (PA), nitrophenol (NP), etc., causing severe contamination in soil and water. In 2016, Hariharan et al. (2016) reported PDI 19 chemosensor that has been substituted with two ethylene trimethyl ammonium iodide groups (Figure 6.9). PDI 19 is highly soluble in water and acts as a fluorescence quenching

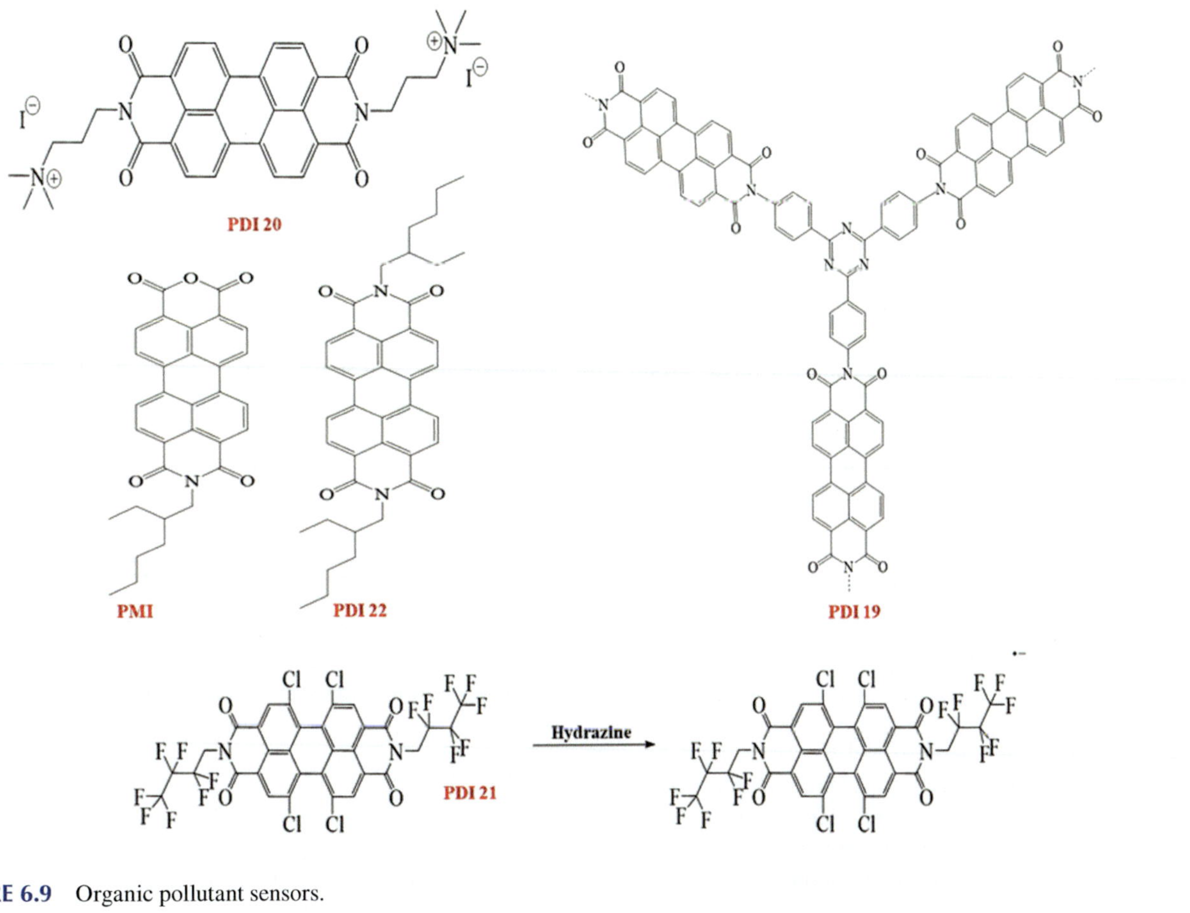

FIGURE 6.9 Organic pollutant sensors.

chemosensor for the detection of 4-nitroaniline (4-NA) and PA with higher selectivity in DMF solvent. In comparison to other PDI molecules, the salt of PDI 19 showed extremely stable fluorescence emission in water and DMF solvent along with LDL of 1 μM. Similarly, the fluorescence quenching detection of nitrophenols has been studied in organic solvents by using a PDI trimer structure, namely PDI 20, as developed by Geng et al. (2019) (Figure 6.9). The fluorescence quenching has occurred upon the addition of *o*-nitrophenol in chloroform solvent, which provides *LDL* of 0.017 nM. The sensing mechanism involves fluorescence quenching due to the aggregation of PDIs by strong complexation of *o*-NP with PDI 20 via hydrogen bonding.

6.3.3.2 Hydrazine

Hydrazine is a common industrial reagent that adversely affects human organs including respiratory system, kidney, central nervous system, and liver. Considered to be carcinogenic in nature, many PDI chemosensors have been designed to be fluorescence turn-ON-type sensors, which is triggered by a reaction with hydrazine. For example, Yu et al. (2018) has reported the multimodal sensor based on PDI 21 (Figure 6.9) for the trace detection of hydrazine. The oxidation power of PDI core was improved by bay substitution with four chlorines of PDI 21, which is very conductive for sensing of hydrazine. A dual modulation system was developed in a DMF solvent involving both fluorescence quenching and color alteration from yellow to green.

6.3.3.3 Amines

Organic amines are industrial chemicals, well utilized in drugs/medicines and in other few explosives. Extensive use of amines causes common pollution in the environment, i.e., from water to air and other matrixes. Organic amines include primary, secondary, tertiary, or aromatic amines with the substitution at nitrogen can create some difficulty for discrimination sensing. Sriramulu et al. (2016) have reported that PDI 22 and an alkyl-substituted PMI showed a fluorescence quenching upon the addition of different amines such as primary, secondary, and tertiary amines. PMI showed a better sensitivity as compared to PDI 22 due to the strong binding affinity of the anhydride group with amines (Figure 6.9), but PDI 22 exhibited higher selectivity for tertiary/aromatic amines because of their hydrophobic properties. Thus, PDI 22 and PMI provide great response for detecting amines like phenylethylamine, diethylenetriamine, and spermine in THF solvent.

6.4 BIOIMAGING

Fluorescence imaging is a versatile practical technique for biological analysis, drug delivery, and clinical diagnosis (Qu et al., 2004; Wang et al., 2015). This technique can be utilized for in vitro and in vivo studies by using light excitation at different wavelength regions like visible region (400–700 nm), the first NIR window (700–1,000 nm), and the second NIR window (1,000–1,700 nm). The most essential factor for fluorophores is that excitation wavelength with red light and an emission in NIR window of biological aspects, interestingly allowing to penetrate tissues, reduce autofluorescence and imaging depth, i.e., allow to visualize the biological molecules in deeper tissues (Roy et al., 2021).

Perylene monoimide (PMI) dye is also well known for its aggregation phenomenon (Kozma et al., 2016; Bettini et al., 2019), which can lead to changes in excitation and emission properties. There are some reported studies on the optical properties of aggregated PMI for bioimaging. Aggregation-induced excimer NIR emission (~700 nm) from a properly substituted PMI derivative in an aqueous media was reported by Jana et al. (2016). Organic NPs (PMI-NPs) are created when PMI derivatives aggregate. They made use of PMI-NPs, which were created using a reprecipitation procedure, and a self-assembling method was used to adjust their size and shape. A nonionic surfactant called pluronic F127 (EM-PMI@Plu127) has been used to stop PMI-NPs from self-assembling to form large NPs (colloidosomes). Remarkably, by enabling it to form the excimer in the aggregated state, the green emissive PMI (emission maxima ~550 nm) can be converted into an NIR-emitting fluorophore (emission maxima ~700 nm) in an aqueous solution. A green-emitting particle with a size distribution of 5–15 nm was also produced via a controlled reprecipitation of PMI with F127 (MM-PMI@Plu127) (Figure 6.10).

An amphiphilic copolymer polystyrene-PEG was utilized to induce the aggregation of PMI derivative, which causes a good emission of NIR region. Furthermore, the polymer was functionalized with folic acid to specifically focus on folate receptors overexpressing cancer cells. In order to demonstrate the cooling-induced emission (CIE) enhancement, Zhao et al. (2019) developed a biperylene monoimide (BPMI) derivative with a conjugated lactose-based glycodendrimer (BPMI-18Lac). When the temperature was changed from 80 to −40°C, BPMI-18Lac showed the CIE phenomenon with a seven-fold intensity enhancement (Figure 6.11). This enhancement was linked to the inhibition of the intramolecular electron interactions between the PMI moieties connected by the C–C single bond. BPMI-18Lac demonstrated selective fluorescence imaging for Hep G2 cells through the asiaglycoprotein receptor (ASGPR) on the cell surface. Liu et al. (2017) designed and synthesized PEGylated perylenemonoimide-dithienylethene (PEG-PMI-DTE), an amphiphilic photo switchable fluorophore with noticeable fatigue resistance, fluorescence switching, and bistable photochromism. The amphiphilic photo-switchable fluorophore, which has an optical resolution of 30 nm, can be used as a staining agent to directly study the fine nanostructures of liposomes under superresolution fluorescence microscopy (Figure 6.12).

6.5 CONCLUSION

PDI-based colorimetric and fluorescence sensors are described for detecting various types of analytes with higher sensitivity as well as selectivity along with the

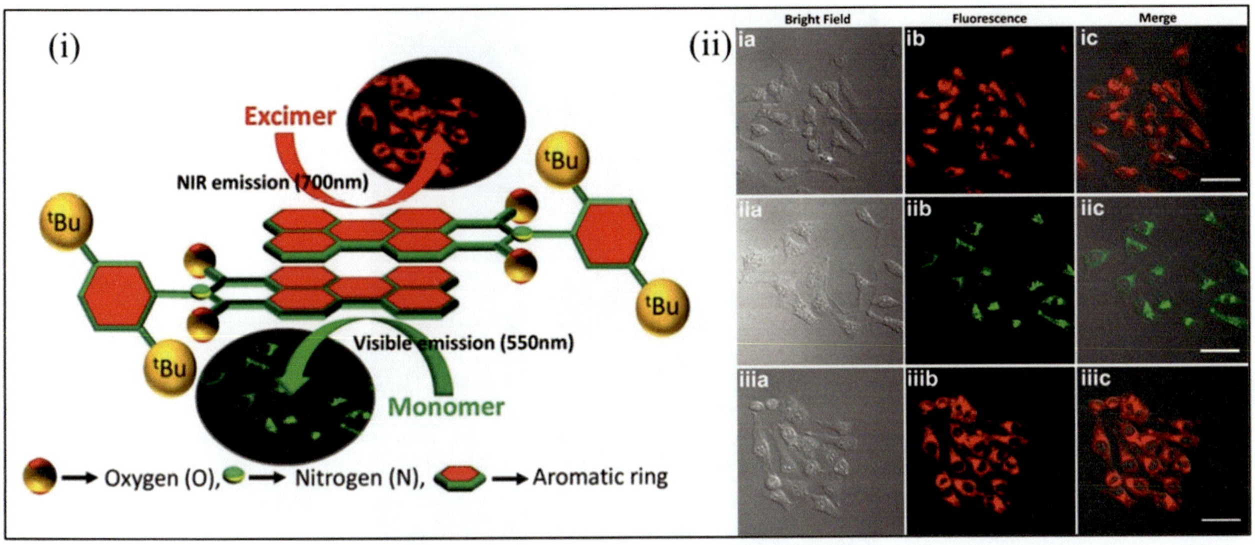

FIGURE 6.10 (i) Schematic representation of aggregation-induced excimeric NIR emission and preservation of monomeric emission from PMI. (ii) Live cell CLSM imaging with HeLa cells: (a) bright field, (b) fluorescence, and (c) merged images incubated with (i) PMI-NPs in water, (ii) green-emitting MM-PMI@Plu127, and (iii) NIR-emitting EM-PMI@Plu127 (scale bar = 50 μm). Emission was recorded using a 550-nm channel for the green emission and a 700-nm channel for the NIR emission. Reprinted with permission from Jana et al. (2016). Copyright 2016 American Chemical Society.

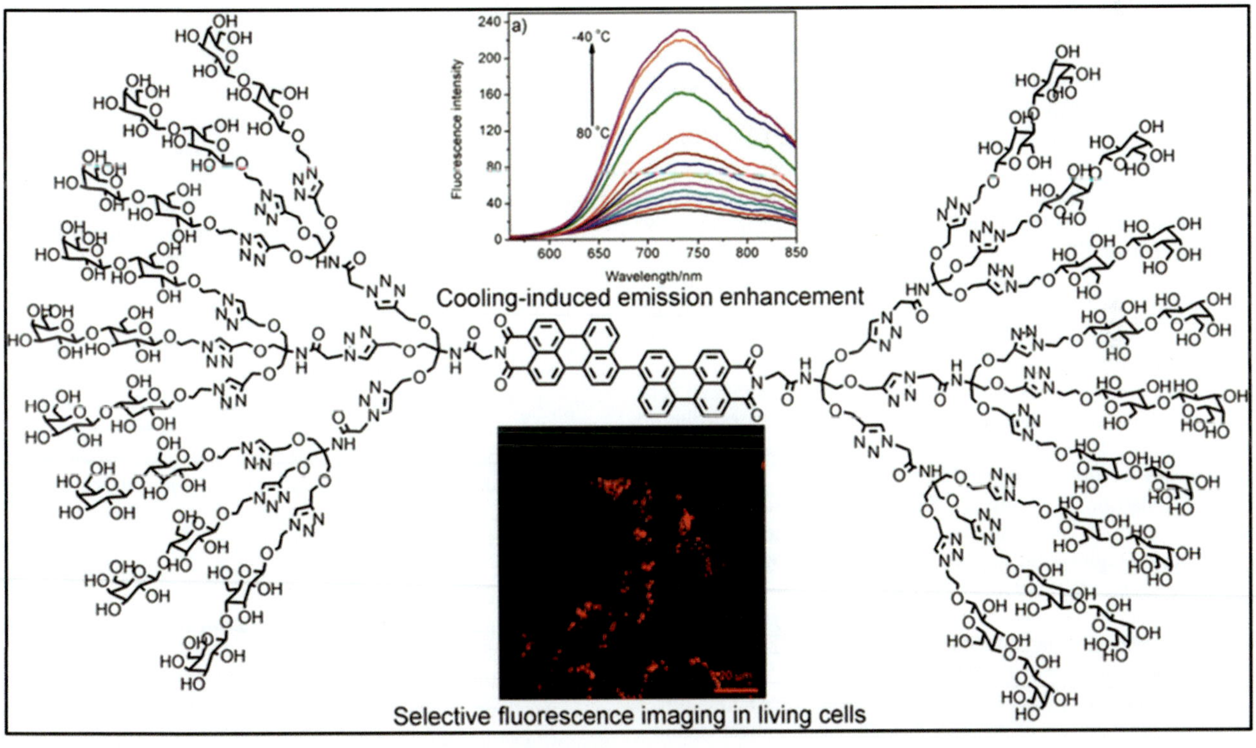

FIGURE 6.11 Structure of BPMI-18Lac applied for cooling-induced emission (CIE) enhancement and live cell fluorescence imaging. Reprinted with permission from Zhao et al. (2019). Copyright 2019 American Chemical Society.

measurement of LDL in this chapter. PDI-based materials possess several unique features like strong chemical, photochemical, and thermal stability, excellent biocompatibility, and low cytotoxicity. Due to their superior properties in comparison to other chemosensors, they have been extensively used in food and physiological detections, including cellular imaging, other relevant biological probing, biomedical imaging, and photodynamic treatment. PDIs along with a wide range of different electronic and binding structures would allow for enhancing the detection with higher selectivity. Thus, these types of chemosensors have been utilized for the detection of several environmental toxicants.

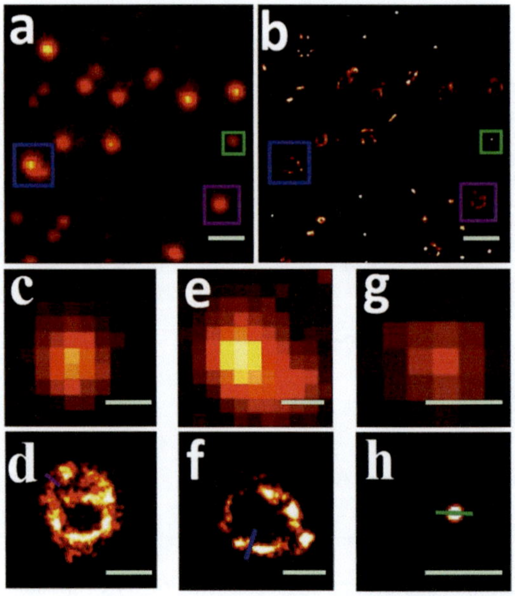

FIGURE 6.12 Optical imaging of liposomes. (a) Fluorescence image. (b) Reconstructed superresolution images. Scale bar: 1 μm. (c, d) Expansion of the blue box in a and b, respectively. (e, f) Expansion of the green box in a and b, respectively. (g, h) are the expansion of the purple box in a and b, respectively. Reprinted with permission from Liu et al. (2017). Copyright 2017 American Chemical Society.

REFERENCES

Avlasevich, Y.; Li, C.; Müllen, K., Synthesis and applications of core-enlarged perylene dyes. *Journal of Materials Chemistry* **2010,** *20* (19), 3814–3826.

Bettini, S.; Syrgiannis, Z.; Pagano, R.; D-ord-ević, L.; Salvatore, L.; Prato, M.; Giancane, G.; Valli, L., Perylene bisimide aggregates as probes for subnanomolar discrimination of aromatic biogenic amines. *ACS Applied Materials & Interfaces* **2019,** *11* (18), 17079–17089.

Che, Y.; Yang, X.; Zang, L., Ultraselective fluorescent sensing of Hg 2+ through metal coordination-induced molecular aggregation. *Chemical Communications* **2008,** *12*, 1413–1415.

Chen, S.; Gao, N.; Bunes, B. R.; Zang, L., Tunable nanofibril heterojunctions for controlling interfacial charge transfer in chemiresistive gas sensors. *Journal of Materials Chemistry C* **2019,** *7* (44), 13709–13735.

Chen, S.; Slattum, P.; Wang, C.; Zang, L., Self-assembly of perylene imide molecules into 1D nanostructures: methods, morphologies, and applications. *Chemical reviews* **2015,** *115* (21), 11967–11998.

Chen, S.; Xue, Z.; Gao, N.; Yang, X.; Zang, L., Perylene diimide-based fluorescent and colorimetric sensors for environmental detection. *Sensors* **2020,** *20* (3), 917.

Chen, S.; Zhou, M.; Zhu, L.; Yang, X.; Zang, L., Architectures and mechanisms of perylene diimide-based optical chemosensors for pH probing. *Chemosensors* **2023,** *11* (5), 293.

Chen, Z. J.; Wang, L. M.; Zou, G.; Zhang, L.; Zhang, G. J.; Cai, X. F.; Teng, M. S., Colorimetric and ratiometric fluorescent chemosensor for fluoride ion based on perylene diimide derivatives. *Dyes and Pigments* **2012,** *94* (3), 410–415.

Dey, S.; Sukul, P. K., Selective detection of pyrophosphate anions in aqueous medium using aggregation of perylene diimide as a fluorescent probe. *ACS Omega* **2019,** *4* (14), 16191–16200.

Du, F.; Bao, Y.; Liu, B.; Tian, J.; Li, Q.; Bai, R., POSS-containing red fluorescent nanoparticles for rapid detection of aqueous fluoride ions. *Chemical Communications* **2013,** *49* (41), 4631–4633.

Elmas Karuk, Ş. N.; Yılmaz, İ., A turn off-on fluorescent chemosensor for sequential determination of mercury and biothiols. *Journal of Fluorescence* **2018,** *28* (6), 1451–1458.

Erdemir, S.; Kocyigit, O.; Karakurt, S., A new perylene bisimide-armed calix [4]-aza-crown as "turn on" fluorescent sensor for Hg2+ ion and its application to living cells. *Sensors and Actuators B: Chemical* **2015,** *220*, 381–388.

Feng, X.; An, Y.; Yao, Z.; Li, C.; Shi, G., A turn-on fluorescent sensor for pyrophosphate based on the disassembly of Cu2+-mediated perylene diimide aggregates. *ACS Applied Materials & Interfaces* **2012,** *4* (2), 614–618.

Fu, H.-B.; Yao, J.-N., Size effects on the optical properties of organic nanoparticles. *Journal of the American Chemical Society* **2001,** *123* (7), 1434–1439.

Geng, T.; Chen, G.; Zhang, C.; Ma, L.; Zhang, W.; Xia, H., A superacid-catalyzed synthesis of fluorescent covalent triazine based framework containing perylene tetraanhydride bisimide for sensing to o-nitrophenol with ultrahigh sensitivity. *Journal of Macromolecular Science, Part A* **2019,** *56* (11), 1004–1011.

Georgiev, N. I.; Sakr, A. R.; Bojinov, V. B., Design and synthesis of novel fluorescence sensing perylene diimides based on photoinduced electron transfer. *Dyes and Pigments* **2011,** *91* (3), 332–339.

Goerl, D.; Zhang, X.; Wuerthner, F., Molecular assemblies of perylene bisimide dyes in water. *Angewandte Chemie International Edition* **2012,** *51* (26), 6328–6348.

Grisci, G.; Mróz, W.; Catellani, M.; Kozma, E.; Galeotti, F., Off-on fluorescence response of a cysteine–based perylene diimide for mercury detection in water. *ChemistrySelect* **2016,** *1* (12), 3033–3037.

Gupta, R. K.; Achalkumar, A. S., Microwave-assisted method for the synthesis of perylene ester imides as a gateway toward unsymmetrical perylene bisimides. *The Journal of Organic Chemistry* **2018,** *83* (12), 6290–6300.

Hariharan, P.; Pitchaimani, J.; Madhu, V.; Anthony, S. P., Perylene diimide based fluorescent dyes for selective sensing of nitroaromatic compounds: selective sensing in aqueous medium across wide pH range. *Journal of Fluorescence* **2016,** *26*, 395–401.

He, X.; Liu, H.; Li, Y.; Wang, S.; Li, Y.; Wang, N.; Xiao, J.; Xu, X.; Zhu, D., Gold nanoparticle-based fluorometric and colorimetric sensing of copper (II) ions. *Advanced Materials* **2005,** *17* (23), 2811–2815.

Jana, A., Bai, L., Li, X., Ågren, H., & Zhao, Y. (2016). Morphology tuning of self-assembled perylene monoimide from nanoparticles to colloidosomes with enhanced excimeric NIR emission for bioimaging. *ACS Applied Materials & Interfaces*, *8*(3), 2336–2347.

Kaufmann, C.; Bialas, D.; Stolte, M.; Wu̇rthner, F., Discrete π-stacks of perylene bisimide dyes within folda-dimers: insight into long-and short-range exciton coupling. *Journal of the American Chemical Society* **2018,** *140* (31), 9986–9995.

Kohl, C.; Weil, T.; Qu, J.; Müllen, K., Towards highly fluorescent and water-soluble perylene dyes. *Chemistry–A European Journal* **2004,** *10* (21), 5297–5310.

Kozma, E.; Grisci, G.; Mróz, W.; Catellani, M.; Eckstein-Andicsovà, A.; Pagano, K.; Galeotti, F., Water-soluble aminoacid functionalized perylene diimides: The effect of aggregation on the optical properties in organic and aqueous media. *Dyes and Pigments* **2016,** *125*, 201–209.

Krupka, O.; Hudhomme, P., Recent advances in applications of fluorescent perylenediimide and perylenemonoimide dyes in bioimaging, photothermal and photodynamic therapy. *International Journal of Molecular Sciences* **2023,** *24* (7), 6308.

Kumar, K.; Bhargava, G.; Kumar, S.; Singh, P., Controllable supramolecular self-assemblies (rods–wires–spheres) and ICT/PET based perylene probes for palladium detection in solution and the solid state. *New Journal of Chemistry* **2018,** *42* (2), 1010–1020.

Kundu, A.; Pitchaimani, J.; Madhu, V.; Sakthivel, P.; Ganesamoorthy, R.; Anthony, S. P., Bay functionalized perylenediimide with pyridine positional isomers: NIR absorption and selective colorimetric/fluorescent sensing of Fe 3+ and Al 3+ Ions. *Journal of Fluorescence* **2017,** *27*, 491–500.

Latterini, L.; Roscini, C.; Carlotti, B.; Aloisi, G.; Elisei, F., Synthesis and characterization of perylene nanoparticles. *Physica Status Solidi (a)* **2006,** *203* (6), 1470–1475.

Li, L.; Zhu, J.-Y.; Hong, Y.-J.; Li, M.-H.; Xiao, W.-C.; Lin, M.-J., 7, 8-Dichlorobenzo [ghi] perylenetriimide: a versatile synthon for bay-substituted π-extended perylene dyes. *Dyes and Pigments* **2019,** *167*, 83–88.

Li, Y.; Hong, Y.; Guo, J.; Huang, X.; Wei, H.; Zhou, J.; Qiu, T.; Wu, J.; Zeng, Z., Bay-and ortho-octasubstituted perylenes. *Organic Letters* **2017,** *19* (19), 5094–5097.

Lin, J.; Zhu, C.; Liu, X.; Chen, B.; Zhang, Y.; Xue, J.; Liu, J., A highly selective and turn-on fluorescent probe for Fe3+ Ion based on perylene tetracarboxylic diimide. *Chinese Journal of Chemistry* **2014,** *32* (11), 1116–1120.

Lin, M.-J.; Jiménez, Á. J.; Burschka, C.; Würthner, F., Bay-substituted perylene bisimide dye with an undistorted planar scaffold and outstanding solid state fluorescence properties. *Chemical Communications* **2012,** *48* (99), 12050–12052.

Liu, J. X.; Xin, B.; Li, C.; Xie, N. H.; Gong, W. L.; Huang, Z. L.; Zhu, M. Q., PEGylated perylenemonoimide-dithienylethene for super-resolution imaging of liposomes. *ACS Applied Materials & Interfaces* 2017, *9* (12), 10338–10343.

Liu, J. X.; Zhang, N.; Zhou, J.; Chang, T.; Fang, C.; Shangguan, D., A turn-on fluorescent sensor for zinc and cadmium ions based on perylene tetracarboxylic diimide. *Analyst* **2013,** *138* (3), 901–906.

Marcia, M.; Singh, P.; Hauke, F.; Maggini, M.; Hirsch, A., Novel EDTA-ligands containing an integral perylene bisimide (PBI) core as an optical reporter unit. *Organic & Biomolecular Chemistry* **2014,** *12* (36), 7045–7058.

Marcon, V.; Breiby, D. W.; Pisula, W.; Dahl, J.; Kirkpatrick, J.; Patwardhan, S.; Grozema, F.; Andrienko, D., Understanding structure– mobility relations for perylene tetracarboxydiimide derivatives. *Journal of the American Chemical Society* **2009,** *131* (32), 11426–11432.

Nagao, Y., Synthesis and properties of perylene pigments. *Progress in Organic Coatings* **1997,** *31* (1–2), 43–49.

Nowak-Król, A.; Würthner, F., Progress in the synthesis of perylene bisimide dyes. *Organic Chemistry Frontiers* **2019,** *6* (8), 1272–1318.

Qu, J.; Kohl, C.; Pottek, M.; Müllen, K., Ionic perylenetetracarboxdiimides: highly fluorescent and water-soluble dyes for biolabeling. *Angewandte Chemie International Edition* **2004,** *43* (12), 1528–1531.

Renner, R.; Stolte, M.; Heitmüller, J.; Brixner, T.; Lambert, C.; Würthner, F., Substituent-dependent absorption and fluorescence properties of perylene bisimide radical anions and dianions. *Materials Horizons* **2022,** *9* (1), 350–359.

Rosso, C.; Filippini, G.; Prato, M., Use of perylene diimides in synthetic photochemistry. *European Journal of Organic Chemistry* **2021,** *2021* (8), 1193–1200.

Roy, R.; Khan, A.; Chatterjee, O.; Bhunia, S.; Koner, A. L., Perylene monoimide as a versatile fluoroprobe: The past, present, and future. *Organic Materials* **2021,** *3* (03), 417–454.

Schubert, A.; Settels, V.; Liu, W.; Wu̇rthner, F.; Meier, C.; Fink, R. F.; Schindlbeck, S.; Lochbrunner, S.; Engels, B.; Engel, V., Ultrafast exciton self-trapping upon geometry deformation in perylene based molecular aggregates. *The Journal of Physical Chemistry Letters* **2013,** *4* (5), 792–796.

Seguy, I.; Jolinat, P.; Destruel, P.; Mamy, R.; Allouchi, H.; Courseille, C.; Cotrait, M.; Bock, H., Crystal and electronic structure of a fluorescent columnar liquid crystalline electron transport material. *ChemPhysChem* **2001,** *2* (7), 448–452.

Singh, P.; Hirsch, A.; Kumar, S., Perylene diimide-based chemosensors emerging in recent years: From design to sensing. *TrAC Trends in Analytical Chemistry* **2021,** *138*, 116237.

Singh, P.; Mittal, L. S.; Kumar, S.; Bhargava, G.; Kumar, S., Perylene diimide appended with 8-Hydroxyquinoline for ratiometric detection of Cu 2+ Ions and metal displacement driven "turn on" cyanide sensing. *Journal of Fluorescence* **2014,** *24*, 909–915.

Singh, P.; Mittal, L. S.; Vanita, V.; Kumar, R.; Bhargava, G.; Walia, A.; Kumar, S., Bay functionalized perylenediimide as a deaggregation based intracellular fluorescent probe for perchlorate. *Chemical Communications* **2014,** *50* (90), 13994–13997.

Soh, N.; Ueda, T., Perylene bisimide as a versatile fluorescent tool for environmental and biological analysis: A review. *Talanta* **2011,** *85* (3), 1233–1237.

Sriramulu, D.; Valiyaveettil, S., Perylene derivatives as a fluorescent probe for sensing of amines in solution. *Dyes and Pigments* **2016,** *134*, 306–314.

Tang, N.; Zhou, J.; Wang, L.; Stolte, M.; Xie, G.; Wen, X.; Liu, L.; Würthner, F.; Gierschner, J.; Xie, Z., Anomalous deep-red luminescence of perylene black analogues with strong π-π interactions. *Nature Communications* **2023,** *14* (1), 1922.

Vasilev, A. A.; De Mey, K.; Asselberghs, I.; Clays, K.; Champagne, B.; Angelova, S. E.; Spassova, M. I.; Li, C.; Mu'llen, K., Enhanced intramolecular charge transfer in new type donor–acceptor substituted perylenes. *The Journal of Physical Chemistry C* **2012,** *116* (43), 22711–22719.

Wang, H.; Wang, D.; Wang, Q.; Li, X.; Schalley, C. A., Nickel (II) and iron (III) selective off-on-type fluorescence probes based on perylene tetracarboxylic diimide. *Organic & Biomolecular Chemistry* **2010,** *8* (5), 1017–1026.

Wang, J.; Zhong, S.; Duan, W.; Gao, B., Fluorinated perylene diimides: synthesis, electrochemical–photophysical properties, and cellular imaging. *Tetrahedron letters* **2015,** *56* (6), 824–827.

Wang, Q.; Li, Z.; Tao, D.-D.; Zhang, Q.; Zhang, P.; Guo, D.-P.; Jiang, Y.-B., Supramolecular aggregates as sensory ensembles. *Chemical Communications* **2016,** *52* (88), 12929–12939.

Wang, Y.; Zhang, L.; Zhang, G.; Wu, Y.; Wu, S.; Yu, J.; Wang, L., A new colorimetric and fluorescent bifunctional probe for Cu2+ and F− ions based on perylene bisimide derivatives. *Tetrahedron Letters* **2014,** *55* (21), 3218–3222.

Wang, Y.; Zhang, Q.; Gong, J.; Zhang, X., Modulating optical and electrochemical properties of perylene dyes by twisting aromatic π-system structures. *Dyes and Pigments* **2021,** *189*, 109261.

Weißenstein, A.; Würthner, F., Metal ion templated self-assembly of crown ether functionalized perylene bisimide dyes. *Chemical Communications* **2015,** *51* (16), 3415–3418.

Würthner, F., Bay-substituted perylene bisimides: Twisted fluorophores for supramolecular chemistry. *Pure and Applied Chemistry* **2006,** *78* (12), 2341–2349.

Würthner, F.; Osswald, P.; Schmidt, R.; Kaiser, T. E.; Mansikkamäki, H.; Könemann, M., Synthesis and optical and electrochemical properties of core-fluorinated perylene bisimides. *Organic Letters* **2006,** *8* (17), 3765–3768.

Xiong, J.-K.; Wang, K.-R.; Wang, K.-X.; Han, T.-L.; Zhu, H.-Y.; Rong, R.-X.; Cao, Z.-R.; Li, X.-L., Fluorescent enhancement sensing of cadmium (II) ion based on a perylene bisimide derivative. *Sensors and Actuators B: Chemical* **2019,** *297*, 126802.

Yan, L.; Yang, L.; Lan, J.; You, J., A new perylene diimide-based colorimetric and fluorescent sensor for selective detection of Cu2+ cation. *Science in China Series B: Chemistry* **2009,** *52* (4), 518–522.

Yan, L.; Ye, Z.; Peng, C.; Zhang, S., A new perylene diimide-based fluorescent chemosensor for selective detection of ATP in aqueous solution. *Tetrahedron* **2012,** *68* (12), 2725–2727.

Yang, L.; Shi, M.; Wang, M.; Chen, H., Synthesis, electrochemical, and spectroscopic properties of soluble perylene monoimide diesters. *Tetrahedron* **2008,** *64* (22), 5404–5409.

Yu, L.; Liu, H.; Liu, X.; Wang, J.; Xu, J.; Wang, H.; Hou, W.; Zhang, H., Multiple detection for hydrazine based on reduction of the 1, 6, 7, 12-tetrachloroperylene diimide derivative. *Chemical Papers* **2018,** *72*, 1927–1933.

Zagranyarski, Y.; Chen, L.; Zhao, Y.; Wonneberger, H.; Li, C.; Mu'llen, K., Facile transformation of perylene tetracarboxylic acid dianhydride into strong donor–acceptor chromophores. *Organic Letters* **2012,** *14* (21), 5444–5447.

Zhang, F.; Ma, Y.; Chi, Y.; Yu, H.; Li, Y.; Jiang, T.; Wei, X.; Shi, J., Self-assembly, optical and electrical properties of perylene diimide dyes bearing unsymmetrical substituents at bay position. *Scientific Reports* **2018,** *8* (1), 8208.

Zhang, X.; Rehm, S.; Safont-Sempere, M. M.; Würthner, F., Vesicular perylene dye nanocapsules as supramolecular fluorescent pH sensor systems. *Nature Chemistry* **2009,** *1* (8), 623–629.

Zhao, C. M.; Wang, K. R.; Wang, C.; He, X.; Li, X. L. Cooling-induced NIR emission enhancement and targeting fluorescence imaging of biperylene monoimide and glycodendrimer conjugates. *ACS Macro Letters* **2019,** *8* (4), 381–386.

7 Cyanine-Based Fluorescent Molecules
Design and Applications

Puja Vinayak Bochare, Esther Rani Motamarri and Abha Sharma
National Institute of Pharmaceutical Education and Research,
Raebareli, Lucknow, UP, India

LIST OF ABBREVIATIONS

CZE	Capillary zone electrophoresis
DAF-2	Diaminofluorescein-2
DPA	Diphenylamine
EDTA	Ethylenediaminetetraacetic acid
FRET	Fluorescence resonance energy transfer
GSH	Glutathione
HEPES	4-(2-hydroxyethyl)-1-piperazineethanesul fonic acid
HPLC	High-performance liquid chromatography
ICT	Intramolecular charge transfer
LCFM	Laser confocal fluorescence microscopy
MS	Mass spectrum
MTT assay	3-(4,5-dimethylthiazol-2-yl)-2,5-dipheny ltetrazolium bromide
NBD	7-nitro-1,2,3-benzoxadiazole
NIR	Near-infrared
NMM	N-methylmaleimide
NMR	Nuclear magnetic resonance
PET	Positron emission tomography
ROS	Reactive oxygen species
TPEN	N,N,N′,N′-tetrakis(2-pyridylmethyl) ethylenediamine
UV	Ultraviolet
WHO	World Health Organization

7.1 INTRODUCTION

Fluorescent probes change their fluorescence emission due to binding event, chemical reaction, or when change happens in their immediate environment having an inherent capacity to release fluorescent photons elucidating their application for the development of new drugs, cellular imaging, environmental research, and several medicinal applications that are of extensive use today (You et al., 2015; Wu et al., 2017). They are used as imaging and fluorescent-based biosensors after conjugation to a biomolecule via a reactive site on the fluorophore. Fluorescent chemicals can have very selective emission properties after being excited by light; identification of proteins with them is viewed as an attractive option for these sorts of uses (Wu et al., 2017). Fluorescence emission can be detected with high sensitivity, and fluorescence microscopy enables extraordinary spatial and temporal resolution in cell imaging, making many fluorescent molecules available (Fu et al., 2018). Fluorescent probes are precedents for numerous applications for instances target-binding affinity, chemical reactivity, and subcellular localization of the probe (Valeur et al., 2011). In addition, fluorescence can be used for visual analyte detection and high-throughput analysis for screening applications that include the creation of optics for environmental and medical monitoring (Zhang et al., 2015). Fluorescent molecules respond to the analyte by various mechanisms, including photoinduced electron transfer, intramolecular charge transfer, aggregation-induced emission, and Forster resonance energy transfer (Georgiev et al., 2023). The importance of sensing biologically and environmentally relevant species is well-known chemosensor research. These fluorescent molecules along with fluorometry is notably beneficial for measuring analytes because of their rapid, nondestructive, and sensitive nature. Multiple scientists working in this sector have provided immense theoretical and practical knowledge for the invention, synthesis, and use of probes, expediting the production of new fluorescence probes. One of the most widely used techniques for observing live cells and determining the structure of biomolecules in tissues and cells is fluorescence microscopy, which enables in situ research on these molecules without the need for hazardous or time-consuming staining procedures. To make precise measurements of the preserved sample, they can be fixed before the addition of a fluorophore, or cell dynamics can be studied in living samples with a level of accuracy and sensitivity that allows one to follow the journey of a single protein over a course of lifetime. Modern fluorophore probes' high degree of specificity makes them feasible to emphasize and highlight particular proteins or other biological components, which when combined with confocal imaging results in a clear three-dimensional internal structure of the cell. The variety and specificity of the available probes enable the simultaneous identification of several overlapping structures within the same sample. These probes are sensitive enough to distinguish between these features clearly; they only need a few dozen active fluorophores to produce a detectable signal in a cubic micrometer of space, and this number will

DOI: 10.1201/9781032706245-7

continue to rise as more effective contrast enhancement techniques are used.

There are various fluorophore scaffolds that are commonly used for the preparation of fluorescent probes. These scaffolds are developed very vigorously for selectivity and specificity; also, the previously developed scaffolds are modified routinely to improve their detection. Some common scaffolds include coumarins, napthalimides, fluorescein, rhodamine, BODIPYs, and cyanine (Fu et al., 2018). Each of these has various advantages over the other with some limitations. These limitations are overcome by researchers to enhance the quality of results (Lin et al., 2002).

Chemically, cyanines consist of the conjugated moiety between two nitrogen atoms and possess required properties for acting as a fluorophore for designing new sensors (Figure 7.1). The first member of cyanine dye was synthesized by Greville Williams over 100 years ago with high intensity and selective toward the analyte (He et al., 2015). A low quantum yield of generally less than 0.25 and a high extinction coefficient make it an impressive fluorophore. Cyanine-based probes are very popular and show fluorescence in the region of red or near-infrared (NIR). This moiety has the limitation of easy oxidation under various conditions, which makes it difficult to study, but now the newly developed fluorescent probes show increased quantum yield and improved oxidation resistance (Kiyose et al., 2008). The focus of analytical and biological research has recently shifted to cyanine dyes, which are long-wavelength (NIR area at about 650–900 nm), analyte-responsive fluorescence probes. They have been successfully employed in in vivo fluorescence imaging due to their minimal photodamage to biological materials, deep tissue penetration, and minimal interference from background autofluorescence by biomolecules in live systems (Yuan et al., 2012; Abd-El-Aziz et al., 2015).

7.2 VARIOUS APPLICATIONS OF CYANINE FLUOROPHORE

7.2.1 Glutathione

Glutathione (GSH) plays an imperative part in normal body functioning. The imbalance of GSH disturbs the various redox reactions in the human body resulting in various pathological conditions (Herzenberg et al., 1997). Also, glutathione acts as an endogenous antioxidant and protects the cells from the damage caused by reactive oxygen species. Different methodologies were developed to detect glutathione by various researchers, including Capillary Zone Electrophoresis (CZE), HPLC, spectrometry, voltammetry, and electrochemistry (Chen et al., 2010). The cyanine-based fluorescent probe is one of the various probes that are used to detect thiols while some are selective for glutathione (Sheng et al., 2016).

Yin et al. designed and synthesized probes 1 and 2 (Figure 7.2) by treating IR-780 and piperazine in DMF, which was further reacted with sulfonyl chloride in the presence of a base to give probes 1 and 2. The first probe contains sulfonamide with a nitro group. It detects the thiols within our body including glutathione, cysteine, and homocysteine. It is highly cell permeable making it easy for detection in a living being. When the cell was treated with probe 1, there appeared a bright red fluorescence wherever thiols were present and when it was treated with 1 mM of N-methyl maleimide (NMM) for around 20 minutes, quenching of the fluorescence appeared showing that the action of the probe can be reversed by using NMM. When

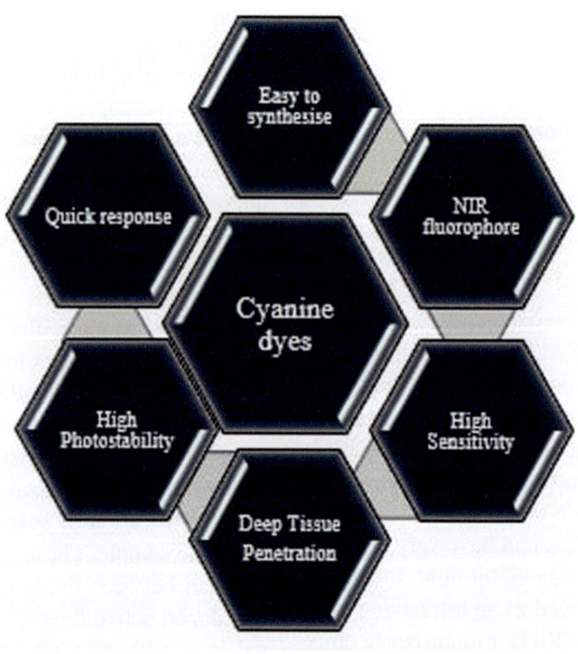

FIGURE 7.1 Properties of cyanine as a fluorophore.

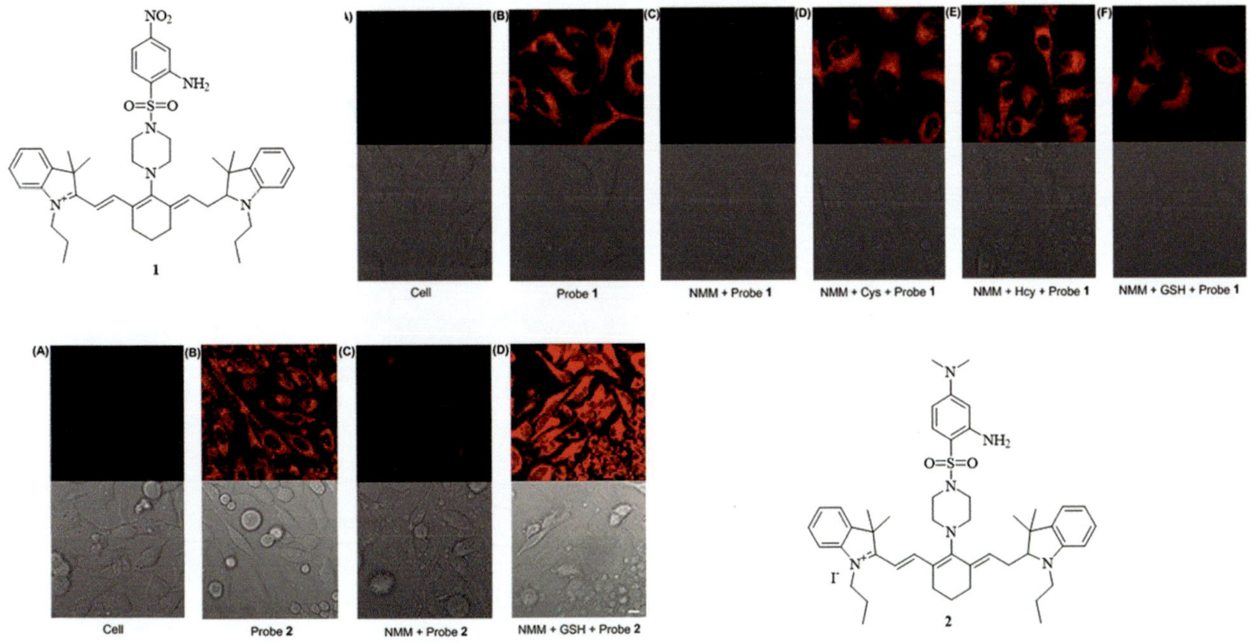

FIGURE 7.2 Structures of probes 1 and 2 and confocal microscope images following the addition of probes 1 and 2 to HeLa cells. Reprinted (adapted) with permission from Yin et al. (2014). Copyright: American Chemical Society, 2014.

the UV-visible spectrum of probe 1 was recorded, it showed emission wavelength at 635 nm, and when it was treated with thiols, it showed an increase in absorbance at 730 nm. To study the selectivity of probe 1, it was treated with the other amino acids proving that probe 1 didn't exhibit any change in absorbance. So, it clears up that the designed probe 1 is selective for the detection of thiols only (Figure 7.2.). Whereas probe 2 in which the nitro group was replaced by the dimethyl amino group showed fluorescence selective of glutathione paving for better detection. The emission of the probe increases with glutathione but not with cysteine and homocysteine was proved in a time-dependent study. But in the case of the absorbance with different amino acids, it didn't show any change. In the titration study of glutathione and probe 2, an increase in concentration increases the fluorescence intensity in the time up to 20 minutes. The action of probe 2 is reversed just as probe 1 by adding NMM. This probe is viewable in near IR fluorescence making it easier to identify in a living cell (Figure 7.2). Probe 2 is injected intravenously into the mouse and after 20 minutes, it shows the bright fluorescence in the organ where a large amount of the glutathione is present, i.e., liver, kidney, lung, and spleen. Here, the reduction in the concentration of glutathione in mice is recorded using acetaminophen. This proves that the designed probes 1 and 2 show better activity by the off-on mechanism (Yin et al., 2014).

Sheng et al. designed and synthesized probe 3 (Figure 7.3.) containing the amino and thiol group, which may selectively detect glutathione, known to be present in many fluorescent probes that are used to selectively detect glutathione. It is synthesized by reacting IR-780 with 4-aminobenzenethiol in DMF. Probe 3 was studied against various analytes, such as Boc-D-Glu, GSH, Ala, Boc-D-Tyr, Lys, Glu, Ser, His, Arg, Gly, Hcy, Cys, H_2S, SO_3^{2-}, and GSH (100 μmol. L^{-1}), in which only GSH gave the increase in fluorescence intensity while the absorption study showed no change in the absorbance or wavelength. The kinetic study of probe 3 with thiol-containing amino acid was carried out with GSH where its emission peak increased with time and showed maximum at 818 nm in 40 minutes while the other biothiols like Cys and Hcy didn't respond at the same time interval. Fluorescence titration of the probe showed that an increase in concentration increases the fluorescence intensity. The effect of pH on the probe was also recorded, which showed the highest fluorescence at pH around 6–11. To determine the selectivity of the probe toward GSH, HepG2 cells were incubated at 37°C along with the probe and different laser confocal fluorescence microscopy (LCFM) images were taken in which the GSH-containing cells gave the bright red fluorescence. The cytotoxicity study of probe 3 was also performed to show that increasing the concentration does not show any toxic effect on a cell. The detailed structure of the probe and the product formed with GSH was characterized by using the time-dependent density functional theory including the complexity of the probe. So, by performing various studies, researchers showed that probe 3 is selective toward GSH (Sheng et al., 2016).

As we discussed the various probes that selectively detect the GSH level in the body, the greatest disadvantage lies wherein these probes have low quantum yield of usually less than 0.25 (Xu et al., 2013). Another researcher designed probe 4 (Figure 7.3) that had a high quantum yield for which they selected the rhodamine moiety that generally shows the high quantum yield in the various probes that were previously reported but it has the major disadvantage that it exhibits emission in the visible region, which isn't

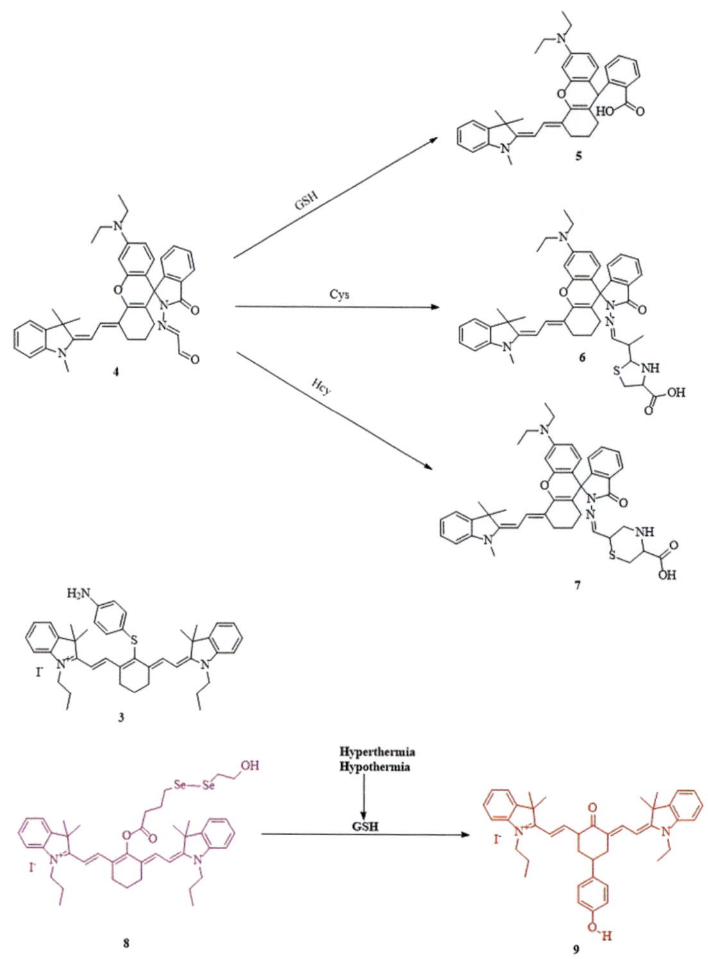

FIGURE 7.3 Probes for detection of GSH.

much efficient and is also nonselective or autofluorescent in the living cell. In this case, rhodamine is attached to the cyanine moiety, which is a NIR fluorophore for designing the novel high-quantum-yield NIR probes.

Spectroscopic properties of the synthesized probe studies showed that there was no absorption by the probe and when GSH was added, it showed an absorption band at 698 nm with a colorimetric change from yellow to green. Also, in the presence of another amino acids, it does not show any significant absorption. In the fluorescence emission study, the probe shows the turn-on fluorescence at 735 nm with the GSH and this response increases with the concentration of the GSH. When the effect of the pH on the probe was studied, it showed intense fluorescence at a physiological pH range of 6–8. The time-dependent kinetic study of probe with GSH shows a plateau after 3 minutes showing that it reacts within 3 minutes thereby causing the ring to open in the presence of GSH.

Probe 4 was reacted with GSH, Cys, and Hcy and the products formed are characterized by various spectroscopic techniques, and the possible products formed are shown below (Figure 7.3, products 5, 6, and 7), which are characterized using ^{1}H NMR, ^{13}C NMR, and mass spectra. Further, the fluorescence imaging performed in the living cell showed that when GSH is present in the cell, it shows bright red fluorescence. In summary, the synthesized probe appears in NIR with a high quantum yield and is selective toward GSH (Xie et al., 2016). GSH plays a very important role as an antioxidant in the human body, and its level affects the body's functioning as the temperature affects the level of the GSH. When the body temperature is above 39–50°C (hyperthermia), an excessive generation of ROS results in oxidative damage to the various proteins present in the body. Determining the level of GSH in hypothermia and hyperthermia conditions is important and is difficult to track due to diverse changes in the concentration in both conditions.

Han et al. designed probe 8 (Figure 7.3) that easily detects the intracellular concentration of GSH in various physiological conditions. They developed this probe based on the selenium-sulfur exchange reaction. This determines the GSH level qualitatively and quantitatively in various conditions. The designed probe consists of cyanine, bis(2-hydroxyethyl) diselenide, and D-galactose. The effect of the probe in the HepG2 and HL-7702 cells at hypothermia and hyperthermia conditions was studied as concentration of GSH in various cells changes with temperature. Liver cells are known to be stronger and fight against temperature due to which no such change in the concentration of GSH was observed; on the other hand, cancer cells can't fight against temperature because of which the concentration of GSH in these cells vary abnormally. Probe 8 reacts with GSH to produce product 9 (Figure 7.3) (Han et al., 2017).

7.2.2 Nitric Oxide

The production of NO, an essential signaling molecule, is involved in the regulation of numerous physiological and pathological systems happens in vivo by the NO synthase, which transforms L-arginine into L-citrulline. Numerous disorders associated with impaired NO signaling have been reported (Yun et al., 1997; Napoli et al., 2006). Therefore, techniques for visualizing NO would be effective tools to investigate the NO signaling pathways in-depth and eventually could be helpful for diagnosis (Nagano et al., 2002).

Sasaki et al. designed a moiety comprising tricarbocyanine, which acts as NIR fluorophore, and *o*-phenylenediamine as NO sensing element (Figure 7.4: probe 10). The detection of NO is based on the modification of o-phenylenediamine that has ability to donate electrons followed by selective NO-mediated conversion of diamine directly into triazole under aerobic conditions, hypothesizing that o-phenylenediamine would quench the fluorescence of tricarbocyanine because of electron transfer from o-phenylenediamine to fluorophore (Figure 7.4). However, as the triazole ring does not have enough electron-donating capacity for PET to occur, so the corresponding triazole synthesis will restore its NIR fluorescence. The spectral properties of the probe showed an increase in the fluorescence ability of the probe concerning the concentration of NO. The efficiency of probe 10 is 53 times more than the standard probe used for the detection of NO (DAF-2). At last, probe 10 was injected into the rats' kidney to know if the probe was working on ex vivo biological system.

It was found that there was an increase in the fluorescence after NO (NOC13—NO donor) administration suggesting that probe 10 works in ex vivo systems for biological determination. NO-detecting probes were proved to be relevant in both cellular and in vivo NO imaging due to the rapid response rate of DACs Probe 10 with NO and their detection in the NIR region, most importantly being less sensitive to be interfered by other biological substances (Sasaki et al., 2005).

7.2.3 Detection of Cations

7.2.3.1 Mercury

Various human activities that include manufacturing, energy production, or consumption generate enormous amount of waste. This waste results in pollution as there is discharge of heavy metals into all the basic natural resources like atmosphere, water, and soil thereby entering our bodies. Heavy metals are arsenic, cerium, cobalt, lead, mercury, and platinum. Mercury is very toxic and easily accumulates in human beings and causes various odd effects on physiological activities that cause brain-related problems, kidney damage, immunity dysfunction, and movement disorder (Miller et al., 1996). So, detecting this ion outside the body and inside the human being is very critical as this is one of the most dangerous heavy metals in this world (Yardim et al., 2003). Mercury contamination is prevalent, and it is thought that every year, between 4 to 7 tons of mercury are discharged into the environment globally (Tchounwou et al., 2003). It can be found in a variety of forms, including metallic and ionic, as well as organic and inorganic compounds and salts (Boening et al., 2000). One of the most stable inorganic forms of mercury, dissolved inorganic mercuric ion (Hg^{2+}), is a toxic and carcinogenic substance with significant cellular toxicity (Ziemba et al., 2005). Additionally, bacteria in aqueous environments can convert inorganic mercury into methylmercury, which can accumulate in seafood and enter the food chain (Harris et al., 2003). Methylmercury causes several serious diseases, including sensory, endocrine, and neurological damage, as well as acute and chronic toxicity, when it is absorbed into biological systems. As a result, worries about

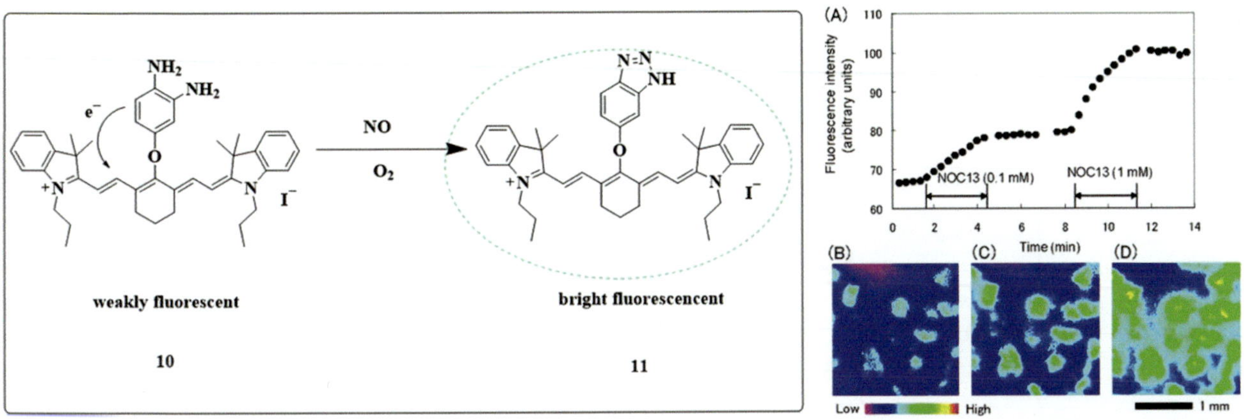

FIGURE 7.4 Turn on phenomenon for detection of nitric oxide (A) Increase of the fluorescence intensity of Probe 10 in rat kidney upon administration of NOC13.(B) The captured NIR fluorescence image of a part of a rat kidney after the loading of Probe 10. (C) The image after the administration of NOC13(0.1 mM). (D) The image after the administration of NOC13 (1 mM).Reprinted (adapted) with permission from ref. Sasaki et al. (2005). Copyright, ACS, 2005.

hazardous mercury have inspired to develop highly focused and sensitive technologies to monitor mercury. These methods are particularly beneficial for highlighting mercury in live cells and certain organisms (Clarkson et al., 2003).

Li et al. developed a cyanine-based NIR fluorescent probe 14 (Figure 7.5) in which thioether contains the ligand that is known to selectively bind only to mercury. The absorption spectra of probe 14 showed the shift in the presence of nickel and mercury and didn't show any shift with other ions. During fluorescence study, fluorescence intensity of the probe increased in the presence of mercury. Probe 14 was evaluated for mercury in different pH to study the stability of the probe-mercury complex and at physiological pH, it showed the highest fluorescence intensity. The UV and fluorescence spectra of most of the metal cations had negligible interference for the probe while mercury showed a drastic change. Then the absorption and emission spectra of the probe at different concentrations of mercury were recorded, and fluorescence intensity of the probe gradually increased with an increase in the concentration of mercury. Stoichiometric study performed on this probe 14 showed 1:1 stoichiometry through the Job's plot analysis. This study proves that the developed probe selectively detects mercury outside the body and within the body. The result obtained from the MTT assay showed very low cytotoxicity. Then the probe was incubated with U87 cells that showed no fluorescence but when mercury was present within the cell, it showed bright red fluorescence. From all the experiments, the designed cyanine-based probe containing disulfide and dicarboxylic acid moieties generates NIR fluorescence, which selectively detects mercury in a 1:1 ratio and shows turn-on fluorescence (Li et al., 2020).

Three tricarbocyanine dyes 16–18 (Figure 7.5) (IR-897, IR-877, and IR-925) with various thiourea substituents have been developed by Guo et al. that serve as chemo dosimeter units and have been developed as a quick indicator paper

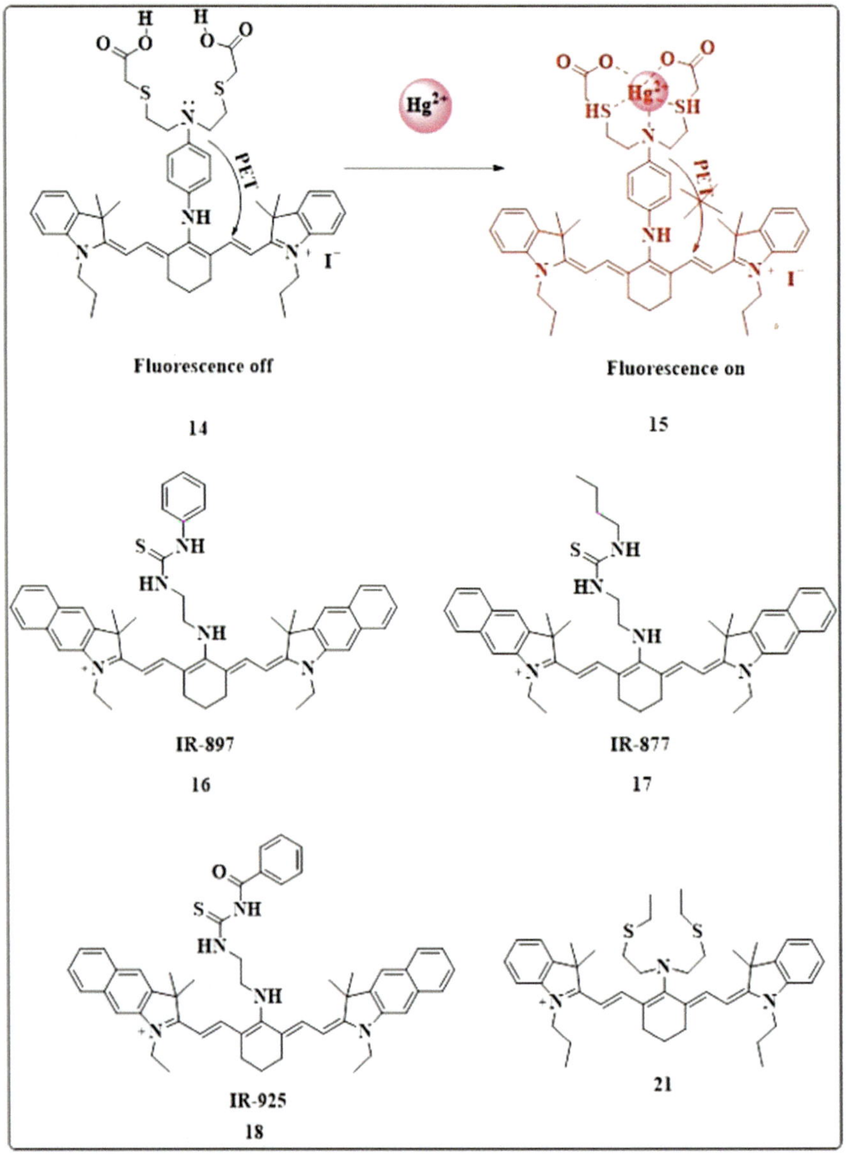

FIGURE 7.5 Probes for detection of Mercury.

for the detection of $MeHg^+$ and Hg^{2+} ions. The advantages of IR-897 and IR877 over other Hg^{2+}-selective chemo dosimeters include facile synthesis, very long wavelengths in the near-infrared (NIR) region (650–900 nm) with high molar extinction coefficients, a ratiometric response, and relatively little disturbance from Ag^+ and Cu^{2+} ions. They show massive red shifts that produce a distinct color change from deep blue to pea green that may be observed with a normal eye and are easily detected by an indicator paper. They exhibit a characteristic turn-off mode at 780 nm and a turn-on mode at 830 nm in emission spectra when Hg^{2+} ions are titrated. Remarkably, the S/N ratio with other thiophilic metal ions (Ag^+ and Cu^{2+}) is greatly elevated with ratiometric measurement of two channels: excitation spectra mode (I810 nm/I670 nm, monitored at 830 nm) and emission spectra mode (I830 nm/I780 nm), isosbestic absorption point at 730 nm. The unique response is influenced by the thiourea substituents' ability to donate electrons. Additionally, research with active SW1116 cells demonstrates that these three low-toxicity tricarbocyanine dyes can display unique properties that help detect intracellular Hg^{2+} and $MeHg^+$ ions in biological systems, including excellent membrane permeability, little interference from biological sample absorption and fluorescence, low scattering, and deep tissue penetration (Guo et al., 2010).

A new derivative of cationic cyanine dyes, probe 19 (Figure 7.6), which was designed and developed by Gao et al., contains the thiosemicarbazone unit that acts as ionophore for the detection of heavy metals. The photophysical study of probe 19 showed the best absorbance changes in pH 3–6. Further, the effect of concentration studied showed an increase in the concentration of mercury decreases the absorbance linearly. A variety of heavy, transition, and main group metal ions were evaluated for their capacity to bind with probe 19 to examine the specificity of probe 19-Hg (Hg $(CyL)_2$) complex formation in the presence of

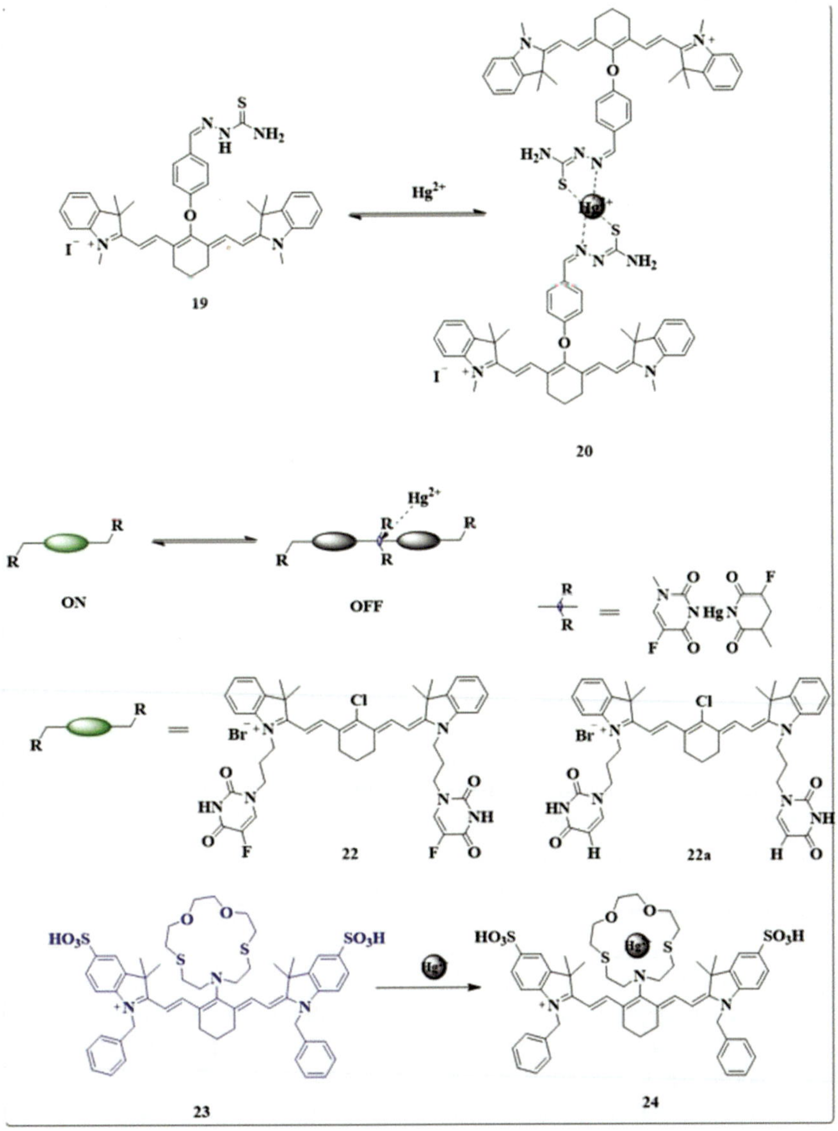

FIGURE 7.6 Structures of probes 19, 22, and 23 and their mechanism of sensing of mercury ions.

environmentally relevant metal ions. It was made clear that with the addition of 1 equiv. of the tested ions, such as Li^+, Na^+, K^+, Ca^{2+}, Mg^{2+}, Ba^{2+}, Co^{2+}, Cr^{3+}, Cu^{2+}, Cd^{2+}, Fe^{3+}, Ni^{2+}, Mn^{2+}, Zn^{2+}, and Pb^{2+}, probe 19 showed only a little drop in absorbance at 760 nm. A notable decrease in absorbance at 760 nm and a new absorption at a shorter wavelength of 666 nm were produced only by the addition of Hg^{2+}. Based on the Jobs' plot, the mechanism of sensing Hg^{2+} is proposed where 2:1 probe 19:Hg^{2+} complex (Figure 7.6) is confirmed by the mass spectroscopy. The investigated technique suggested probe 19 as a novel Hg^{2+} sensor for detection by absorption spectrometry in aqueous solution, and it may be employed as a good substitute for the existing Hg^{2+} sensors (Gao et al., 2023).

The 3,9-dithia-6-monoazaundecane-tricarbocyanine probe 21 (Figure 7.6) was developed by Tang et al. where 3,9-dithia-6-monoazaundecane serves as a highly selective binding site for metal ions and tricarbocyanine with two propyl groups functions as the NIR fluorophore. Tricarbocyanine's cyclohexane bridgehead's vinyl chlorine is reactive and may be substituted by dimethyl amine (DMA), making it simple to synthesize the probe in a single step under normal conditions. Probe 21 was studied in PBS buffer, and λ_{max} of 685 nm for excitation and 763 nm for emission was observed. Further, the selectivity study of probe 21 in the presence of different metal ions showed the probe is highly selective toward mercury over the other even in high concentration. The stoichiometry of probe 21 with mercury is determined using the Job's plot, which showed 1:1 probe:mercury stoichiometry. The limit of detection (LOD) was found to be 1.39×10^{-8} M. This outcome proved that the probe was capable of both qualitative and quantitative Hg^{2+} detection. It was found that the apparent dissociation constant, K_d, is 7.90×10^{-7} M. The confocal fluorescence microscopic analysis was done in zebra fish for detecting mercury proving that the probe may be beneficial for investigating the toxic effects of Hg^{2+} in living things (Tang et al., 2008).

One of the primary sites where mercury ions are localized is mitochondria, where substantial mercury-induced toxicity has been well established. Two heptamethine cyanine dyes 22 and 22a (Figure 7.6) were developed that have thymines attached to the N-alkyl chains and can particularly chelate with mercury ions to form a T-Hg-T complex, allowing us to detect mercury ions both in aqueous solutions and mitochondria of living cells with high sensitivity. Further research showed that adding fluorine atoms to thymine will considerably improve the binding capacity of T-Hg-T, offering suggestions for creating more precise and sensitive Hg^{2+} sensors (Long et al., 2017).

There are various chemosensors developed previously containing tricarbocyanine, which determined that the change by the fluorometric method depends on the instrumental method but here Zu et al. discuss how probe 23 (Figure 7.6) showed the colorimetric changes in the presence of metal ions. Detection by the naked eye has various advantages, for instance, easy detection, economical, and do not require a trained individual. To design the colorimetric probe selective toward mercury over several interfering metal ions, the dithia-dioxa-monoaza crown ether moiety was attached to cyanine fluorophore and to provide the photostability, tricarbocyanine with benzyl group was selected. A large red shift of 122 nm and color change from blue to colorless was observed after the addition of mercury; this change is reversed by the addition of an excess of EDTA. The fluorescent intensity of the probe decreases on titration with increasing concentration of Hg^{2+}. Selectivity of the probe with mercury is studied by various interfering cations like Fe^{3+}, Co^{2+}, Cu^{2+}, Zn^{2+}, Cd^{2+}, Pb^{2+}, Li^+, Na^+, K^+, Ca^{2+}, and Mg^{2+}; significantly, quenching of fluorescence is observed only with mercury. The mechanism of formation of the complex is seen by using the ^{1}H NMR spectrum (Figure 7.6: 24). The mechanism suggests the formation of a complex by downfield the ppm value between nitrogen and sulfur as they are involved in forming a coordinate bond with mercury. Hence, probe 23 shows high selectivity and sensitivity toward mercury (Zhu et al., 2008).

7.2.3.2 Zinc

Zinc is widely found in a variety of biological fluids, various tissues, and the environment. In the body, it is stored in the vesicles and released as an ionic signal through the gated membrane channel. Endogenous zinc activates the excitatory neurotransmitter, i.e., glutamate and γ-aminobutyric acid (GABA) (Fredreickson et al., 2005). The concentration of zinc above the limit enhances the formation of $A\beta$ amyloid (more than 300 nM) by destabilizing the soluble $A\beta_{1-40}$, which results in Alzheimer's disease (Bush et al., 1994). As discussed, zinc in high concentration is toxic, so determining zinc by the selective method is required, which detects zinc in low concentration.

Guo et al. designed a highly selective and sensitive cyanine-based fluorescent probe 25 (Figure 7.7) that detects endogenous zinc in cells and organelles. Probe 25 contains three pyridine rings to have more binding affinity toward zinc and polymethine chain attached to TMPA, which increases the conjugation and results in an increase in fluorescence. When probe 25 is treated with the zinc ion it suddenly changes its color from blue to light red, which is visually detected. Also, in the UV-visible spectroscopy, probe 25 shows a slight red shift in absorption maximum and a hypsochromic shift in both absorption and emission from 730 nm to 590 nm. The observed hypsochromic shift is due to the disruption of the highly conjugated system by the deprotonation of amine to form imine; this transfer results in the interruption of the pull-push in the conjugated system. Due to this, the conjugation in polymethine cyanine gets shortened and causes a large hypsochromic shift. In proton NMR, by addition of zinc in a 1:1 ratio into probe 25 resulted in the disappearance of N-H proton of amine (Figure 7.7: 26). Apart from this, probe 25 has a major application in the detection of zinc in live organelles. Here, the author treated a zebra fish with probe 25 at various stages of development in 2 μM concentration. The bright red fluorescence was seen in various organelles of the fish where zinc

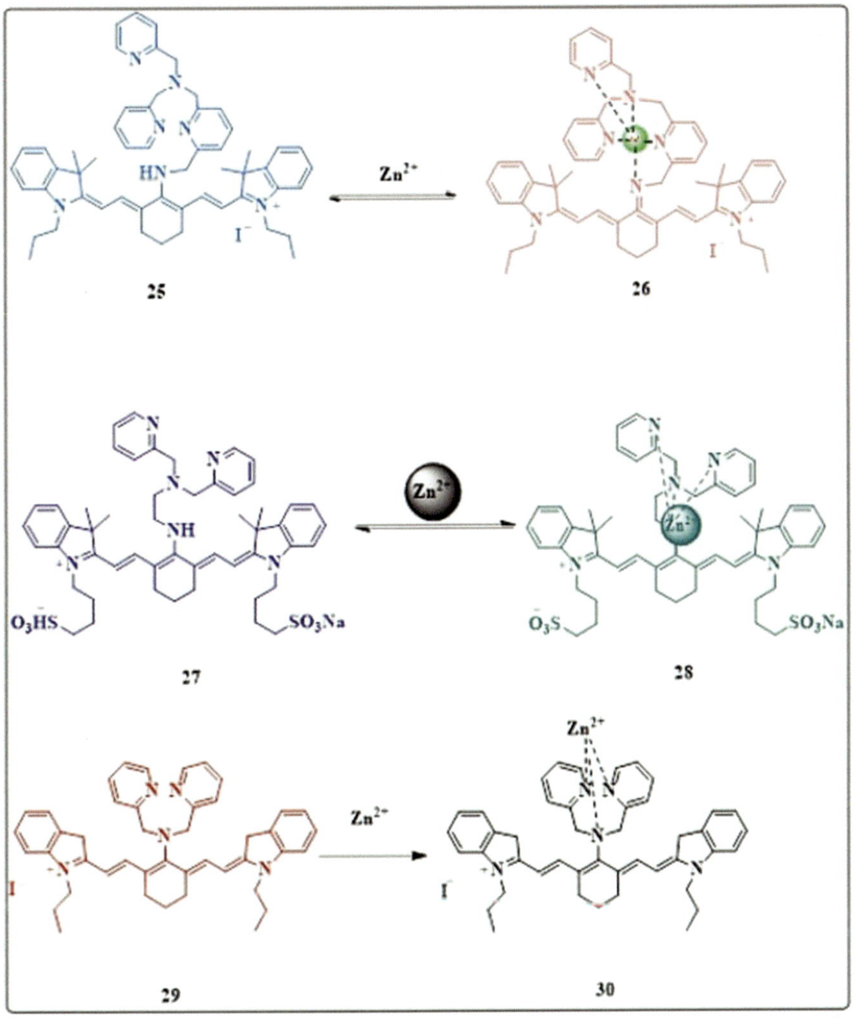

FIGURE 7.7 Probe for detection of Zinc.

is present. That the red spot is due to zinc is confirmed by pretreating it with the N,N,N′,N′-tetrakis(2-pyridylmethyl) ethylenediamine (TPEN), which results in no bright red spot. From this, it is clear that probe 25 is very efficient in determining endogenous zinc in live cells and also it has a bright future for zinc in biology (Guo et al., 2012).

Kiyose et al. reported probe 27 (Figure 7.7) consisting of amine-substituted tricarbocyanine fluorophore, which has excellent molar extinction coefficient value of 7.0 × 10 M^{-1} cm^{-1} with large Stokes shift. It contains the dipicolylethylenediamine (DPEN) moiety that behaves as metal chelator. The DPEN moiety's electron density will drop when dipicolylamine (DPA) coordinates to a metal ion (Figure 7.7; 28), which will reduce the amine's capacity to donate electrons to the fluorophore. It would be appropriate to use the IR-783 chromophore, which contains extremely water-soluble sulfonates, as the mother structure for determining the presence of metal ions in aqueous solution. The color of the solution changed from blue to teal because probe 27 reacts with Zn^{2+} and when TPEN, a potent chelator of Zn^{2+}, was added, the color returned to its original state. ESI-MS spectrometry was used to demonstrate the formation of a 1:1 complex of probe 27-Zn^{2+}. A 44 nm red shift of the absorption maximum was noted at a high enough concentration of Zn^{2+}, suggesting that probe 27 might function as a ratiometric probe for Zn^{2+}. The ring dipicolylamine acts as metal chelator for Mn^{2+}, Fe^{2+}, Co^{2+}, $Ni^{2+,}$ Cu^{2+}, Ag^{+}, Zn^{2+}, Cd^{2+}, Hg^{2+}, and Pb^{2+}, but here the designed probe containing the same moiety shows a significant change only with zinc and gives the selectivity toward the zinc ions only. The maximum absorption was red shifted by 62 nm when there was a 20-fold excess of Cu^{2+} present, but the fluorescence was fully suppressed. Additionally, the shifting of absorption band progressively vanished when TPEN was introduced (Kiyose et al., 2006).

Tang and coworkers designed and synthesized a sensitive and selective NIR probe 29 (Figure 7.7.) toward zinc. Based on PET quenching mechanisms, the most prevalent form of fluorescent probe for metal ions is used. Two components make up a PET probe: an ion-specific receptor with a fluorophore. The author used DPA to chelate zinc and tricarbocyanine with two propyl groups as the NIR fluorophore since it may pass through cellular membranes without any alteration.

Transition metal coordinates with amines or protons inhibiting PET is a frequently seen method for fluorescence amplification. Spectral properties were studied for the synthesized probe, which showed absorption maximum at 606 nm and emission toward 800 nm. Significant Stokes shift was observed in acetonitrile solvent of 194 nm. Employing a PET mechanism, probe 29 detects zinc with a 20-fold increase in turn-on response, while other metal ions quenched the fluorescent intensity of the probe. Probe 29 has a significant Stokes shift and thus is very selective for zinc. This probe is advantageous as NIR fluorescence probe for detecting zinc because it is cell permeable and has a rapid response to zinc. According to the Job's plot analysis, the point of intersection for probe 29 and zinc in solution appeared at 0.5 mole fraction indicates a 1:1 stoichiometry. Study of probe 29 on biological systems showed red fluorescence but on treatment with excess chelator TPEN, quenching was observed proving that the action of probe 29 is reversible (Tang et al., 2006).

7.2.3.3 Silver

Silver is an important catalyst in various industrial processes including in the production of various molecules, and sometimes it is directly thrown into the environment, which results in water pollution ending up in the contamination of agricultural crop (Kursunlu et al., 2016). Various studies are done to evaluate the effect of silver, which suggests that it is very harmful beyond the limit where it acts as a strong heavy metal pollutant. So, a selective detectable fluorescent probe was needed to be designed that has various advantages over the traditional methods.

Probe 31 developed by Zhang et al. (Figure 7.8) was evaluated against silver for its selectivity and specificity in ethanol and PBS solvent where its absorption band appeared at 619 nm and fluorescence emission at 760 nm. An increase in concentration of silver ions causes a decrease in the fluorescence intensity; when the concentration of silver reaches 7.0 μM, the fluorescence is quenched no further change is observed with increasing the concentration of silver. This change is also seen visually in the change in color from blue to red. Probe 31 shows a linear response with increasing silver concentration. Selectivity of probe 31 toward silver was studied by using Ag^+, Fe^{2+}, K^+, Na^+, Ca^{2+}, Zn^{2+}, Mg^{2+}, Pb^{2+}, Ni^{2+}, Mn^{2+}, Cd^{2+}, Cu^{2+}, and Hg^{2+}. Fluorescence quenching was observed only with silver ions while the other metal ions did not interfere with probe fluorescence intensity. A time-dependent study of probe 31 with silver was performed to determine the time required for the probe to bind with silver. The result showed that within 1 minute, silver formed a complex with probe 31 and caused quenching in fluorescence. The further detection limit of probe 31 toward silver ion is determined, which was found to be 0.03 μM. Further, stoichiometry was determined by measuring the maximum fluorescence quenching at mole fraction 0.5 as predicted by the Job's plot analysis confirming 1:1 stoichiometry. Probe 31 shows a turn-off phenomenon with selectivity and specificity (Zhang et al., 2019).

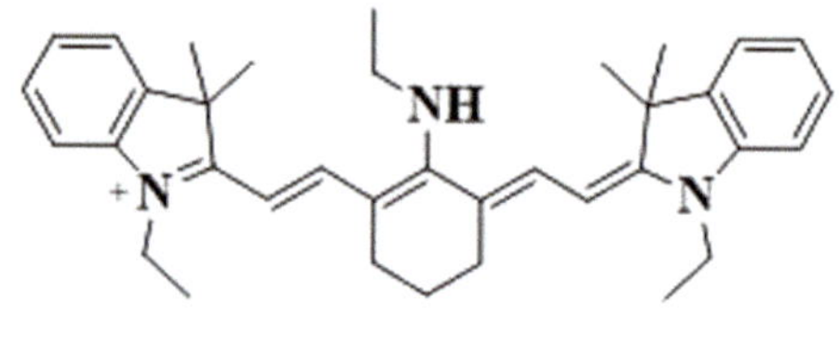

FIGURE 7.8 Probes for detection of Silver.

7.2.3.4 Copper

All life systems require this trace element (Prohaska et al., 2004). Many enzymes, including the respiratory enzyme cytochrome oxidase (COX), copper/zinc superoxide dismutase, and metallothionein, depend on copper's redox activity (Cobine et al., 2006). However, excess copper is toxic to living beings as it causes oxidative damage to proteins, lipids, DNA, and other biomolecules. Menkes and Wilson's disorders, familial amyotrophic lateral sclerosis, Alzheimer's disease, and prion illnesses are linked to disruption of copper homeostasis (Cobine et al., 2006). Consequently, there has been a lot of interest in developing methods for copper detection in biological systems (Tapiero et al., 2003). To track copper ions, a wide range of techniques including absorption spectrophotometry, mass spectrometry, and inductively coupled plasma atomic emission spectroscopy have been developed (Humbert et al., 1993). For high sensitivity and ease of use of making desired fluorescence sensing, great progress has been made in the creation of fluorescent probes that are selective for the oxidized form of copper, Cu^{2+} (Chew et al., 2000). In contrast, there aren't many fluorescent probes that are specific to copper's reduced form, Cu^+ (Zeng et al., 2006). Some fluorescent Cu^+ probes are based on fluorescent dyes, such as BODIPY and fluorescein, which only have visible light emission and absorption wavelengths.

Based on a tricarbocyanine scaffold, Lin et al. have synthesized a novel NIR fluorescent turn-on Cu^+ probe 32 (Figure 7.9) that is appropriate for imaging endogenous Cu^+ ions in living cells. For detection of Cu^{2+}, bis(2-((2-(ethylthio) ethyl)-thio) ethyl) amine (BETA) moiety is selected since it has a high affinity toward copper. The designed probe was synthesized by alkylation of tricarbocyanine with BETA. After the synthesis of this probe, the spectral properties were studied that showed an absorption peak at 696 nm and weak emission at 792 nm. After the addition of copper, a change was observed in both the absorption and emission spectra. A significant red shift was observed in the absorption peak from 696 to 750 nm after the addition of copper in the absorption spectrum. Addition of copper resulted in fluorescence turn-on. Titration of copper stated the most fluorescence intensity was observed up to 1 equivalent of copper. The Job's plot analysis utilizing the continuous

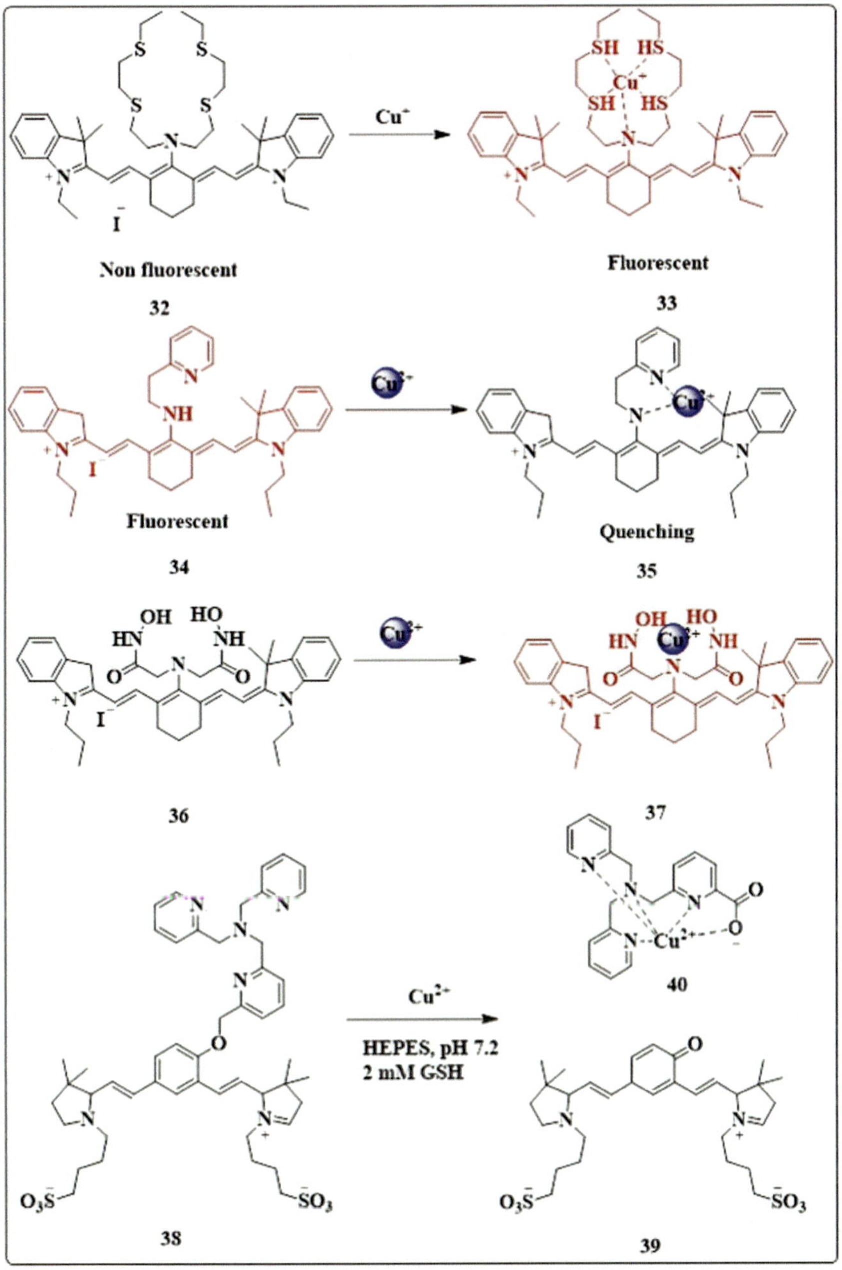

FIGURE 7.9 Probes for detection of Copper.

variations method proved that the fluorescence amplification is caused by the 1:1 probe/Cu^{2+} complex (Figure 7.9: 33). Detection limit for copper was obtained to be 4.1 × 10^{-13} M. Copper selectivity was determined in the presence of K^+, Na^+, Ca^{2+}, Mg^{2+}, Mn^{2+}, Ni^{2+}, Zn^{2+}, Co^{2+}, Fe^{2+}, Cd^{2+}, Pb^{2+}, Hg^{2+}, Cu^{2+}, and Cu^+. Addition of copper increases fluorescence intensity by 10-fold, while other metal enhanced the fluorescence by one to two fold. Observed data suggest selectivity toward Cu^+ over the other metal ion including Cu^{2+}. To study the stability of the fluorescent probe over the pH change is important for the biological study of any molecule, for probe 32 suggests the insignificant change with respect to pH range of 5–8. In acidic conditions, increase in fluorescence and in basic conditions, decrease in fluorescence are observed due to precipitation of copper. Study of copper detection endogenously was done in MG63 cells where they are incubated first with probe that shows no fluorescence. Incubation with copper gave the bright red fluorescence. The change is reversed by addition of a chelating agent in live cells that were incubated with EDTA resulting in quenching of fluorescence intensity (Cao et al., 2012).

Han et al. reported 2-(2-aminoethyl) pyridine-tricarbocyanine fluorescent probe 34 (Figure 7.9) for detection of Cu^{2+} selectively. Absorbance maxima of 640 nm and emission maxima of 670–800 nm were observed for probe 34. The absorbance study of probe shows the decrease in intensity after the addition of copper with no shift in the absorbance wavelength. Using a pH range of 5.0–9.0 and a fixed Cu^{2+} concentration of 5 mol/L, the effects of pH on the fluorescence intensity of probe 34 in the presence of Cu^{2+} were examined. The protonation of probe without binding with the metal ion may be the reason of decrease in fluorescence intensity of probe 34 at lower pH levels. On the other hand, $Cu(OH)_2$ precipitated when the pH was too high. The selectivity of synthesized probe was studied by using various interfering metal ions including Zn^{2+}, Cd^{2+}, Mn^{2+}, Ca^{2+}, Mg^{2+}, K^+, Na^+, Li^+, Al^{3+}, Co^{2+}, Ni^{2+}, Pb^{2+}, Hg^{2+}, Fe^{3+}, Ag^+, and Cu^{2+}. Probe 34 shows fluorescence quenching with copper ions by chelating with nitrogen atoms (probe 35) (Figure 7.10) while other metal ions don't involve in any significant change in fluorescent intensity (Han, 2013).

Li et al. synthesized NIR probe 36 (Figure 7.9) for detection of copper based on the photoinduced electron transfer mechanism. Copper selective moiety 2,2'-azanediyl bis(N-hydroxyacetamide) was incorporated in tricarbocyanine fluorophore (Figure 7.10: probe 37). They studied the spectral properties of probe 36 with and without addition of copper. The absorption maximum was observed at 750 nm and emission at 800 nm. As copper was added into the probe 36 solution, it resulted in a blue shift by 30 nm to 770 nm. The addition of 5 μM copper metal in solution of probe 36 gave the maximum fluorescence increase up to 10-fold. Probe 36 shows pH-dependent behavior; in pH range 5.5–8.5, weak variation in fluorescence intensity was observed. The limit of detection was found to be 5.0×10^8 with dissociation constant of 2.7 μM and good linear regression coefficient of 0.9935. The interference study showed an increased probe 36 fluorescence in presence of Cu^{2+} only, whereas other metal ions just slightly altered the probe 36 fluorescence. Even when Na^+, K^+, Mg^{2+}, and Ca^{2+} ions were present at micromolar quantities, the effect of coexisting cations on Cu^{2+} measurement was minimal (Li et al., 2011).

A Cu specific, water-soluble switch-on NIR fluorescence probe 38 (Figure 7.10) has been synthesized by Maity et al. Probe 38 can effectively detect submicromolar concentrations of Cu by employing NIR-emitting Cy-quinone dye generated during the metal ion-catalyzed process (Figure 7.9). In general, probe 38 can be utilized as a noninvasive method for in vivo imaging applications and the NIR fluorescence detection of Cu pool in biological fluids. This research finding might open the way for a fresh approach to the development of reaction-based NIR molecular probes for other metal ions and analytes (Maitey et al., 2015).

7.2.3.5 Cadmium

Cadmium poses a huge environmental risk as well as a substantial health issue to people because of its widespread use in batteries, fertilizers, and other products (Taylor, 1997; Mendes et al., 2006). Environmental cadmium exposure in human beings may increase the risk of developing endometrial, breast, prostate, or lung cancer (Byber et al., 2016). Therefore, there is a pressing need to create excellent techniques for monitoring cadmium in both environment and in vivo (U.S. EPA., 2005).

Yang et al. designed and synthesized two cyanine-based fluorescent probe 40 (Figure 7.10; CYP 1 and CYP 2) based on the commonly observed mechanism of photo-induced electron transfer. Probe 40 consists of three moieties selective toward receptor and fluorophore attached by covalent bond (probe 41) (Figure 7.10). The probes are cell permeable and functionalized with two sulfonate groups, which improve solubility and self-aggregation in water. Fluorescence intensity studied by titration shows increase in the intensity with an increase in Cd^{2+} concentration. When the probes were saturated with Cd^{2+}, the fluorescence emission spectra were altered by adding a significant quantity of EDTA, demonstrating that the sensing action was reversible. The detection limit for CYPs was also discovered to be 3.1 μM for CYP-1 and 2.3 μM for CYP-2. The selectivity of the probe was evaluated against Pb^{2+}, Hg^{2+}, Ag^+, Zn^{2+}, Cu^{2+}, Mn^{2+}, Co^{2+}, Ca^{2+}, Fe^{3+}, Li^+, Ni^{2+}, Ba^{2+}, Cr^{3+}, K^+, Fe^{2+}, and Mg^{2+}, which showed no significant change except with cadmium. The Job's plot analysis demonstrated the Cd^{2+}-CYP-2 stoichiometry is 2:1. The living cell imaging technique using HeLa cells provided more evidence of its relevance in real-world biological system applications as addition of Cd^{2+} to probe 40 causes increase in fluorescence (Yang et al., 2011).

7.2.3.6 Lithium

The release of harmful metals into the environment results from a variety of processes mostly industrial, though agricultural sources and sewage disposal also play a role. These contaminants can accumulate in large quantities and are primarily released into the environment as solutes. Legal

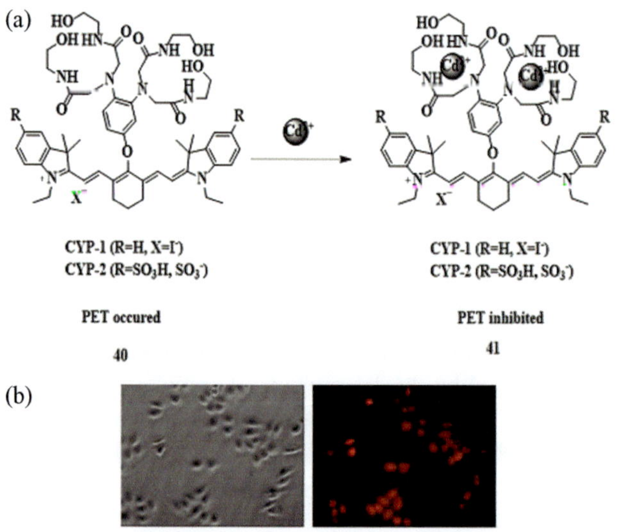

FIGURE 7.10 Fluorescent images of Cd^{2+} in HeLa cells for probe 40 (3 μM). (a) Bright-field transmission image of HeLa cells incubated with probe 40 after adding Cd^{2+} (500 μM). (b) Fluorescence image of HeLa cells incubated with probe 40 after adding Cd^{2+}. Reprinted (adapted) with permission from Yang et al. (2011). Copyright, American Chemical Society, 2011.

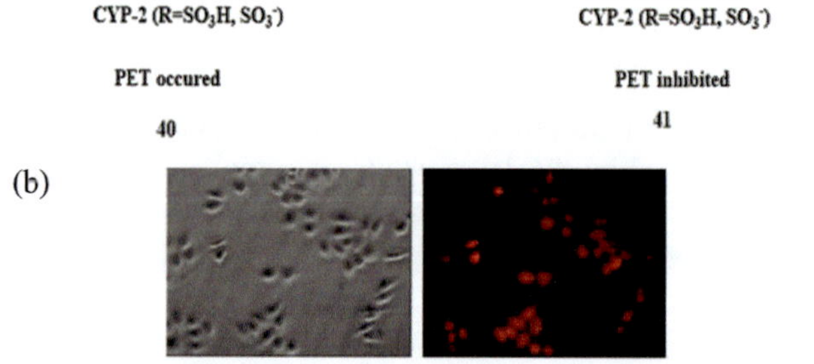

restrictions on the permitted concentrations are tightening up as people become more aware of the negative impacts (Kushner, 1993). The substitution of lithium for other cations (Na^+, K^+) involved in electrochemical gradients is what causes its hazardous effects. Additionally, toxicity develops because of cellular hemostatic balance being upset (Khayachi et al., 2021). Damage to the kidneys, cardiovascular system, central nervous system, and neuromuscular system may develop when serum lithium concentrations are above 1.6 meq/L. Since renal lithium clearance rises during pregnancy, therefore careful monitoring of lithium levels is necessary. Lithium tolerance is set at 0.25 mg/m^3 (Poels et al., 2018).

To overcome this, Tarazi et al. designed a crown ether containing cyanine dye, which improves the selectivity of probe toward metal ion. Probe 42 (Figure 7.11) was synthesized by reacting IR-786 iodide with 4'-aminobenzo-15-crown-5. The absorption and emission spectra of probe 42 gave maximum absorption at 780 nm and emission maximum at 795 nm in acetonitrile solvent. Stability of any molecule is an important aspect for the various studies, so the stability of probe with respect to time is studied over 3 hours. This shows the intensity of sample placed in dark condition gave constant result while in the presence of light the results fluctuated proving light sensitivity of probe 42. Within the examined pH range, probe 42 absorbance and fluorescence intensity were determined, which proved to be pH independent. Furthermore, the absorption or emission maxima did not change when the pH changed. The selectivity of probe 42 with various metals ions showed that the intensity was quenched significantly only by lithium ions. Further, limit of detection for lithium was found to be 0.00743 ppb. This suggests that an ICT from the donor moiety (amino group) to the acceptor moiety (N^+ of the pyridymerine ring) is the main cause of the fluorescence emission of probe 42. Lithium interferes with the lone pair's ability to transfer charge to nitrogen, lowering the fluorophore's quantum yield as a result. The crown in probe 42 may also serve as a second electron donor, although its capacity to do so is diminished in the presence of a complex cation. It was discovered that probe 42 stability constant was higher than that of comparable compounds. Probe 42 also absorbs and fluoresces in the NIR spectrum. As a result, background interference is reduced, sensitivity is increased, and the dye is perfect for determining trace metals (Tarazi et al., 2002).

7.2.3.7 Iron

Iron is the most prevalent transition metal in cellular systems (Fakih et al., 2008) and because it is found in so many enzymes, proteins, and transcriptional events, it is essential to many biological processes (Domaille et al., 2008). Its absence or overproduction can interfere with cellular homeostasis and metabolism. Cellular harm could have subsequent consequences in numerous illnesses, including cancer, neurodegeneration, Parkinson's disease, Alzheimer's disease, and hepatitis (Zecca et al., 2004). Iron rapidly participates in one-electron oxidation-reduction processes between its ferric (Fe^{3+}) and ferrous (Fe^{2+}) forms in living organisms (Haliawell et al., 1999). Ferrous iron (Fe^{2+}), in particular, may promote the production of extremely reactive oxygen species, particularly the hydroxyl radical (Ma et al., 2004). A significant factor in determining a cell's future development may be the presence of Fe^{2+}. The reported Fe^{2+} fluorescent probes suffered from lesser sensitivity since they were mostly built using a fluorescence quenching mechanism. Furthermore, the UV/vis region shows fluorescence for Fe^{2+} sensing probes suffered from relatively strong background fluorescence in addition to being biologically harmed (Petrat et al., 2002). Even worse, in ambient conditions, Fe^{2+} is readily oxidized to Fe^{3+}. Fe^{2+} fluorescence probe development is limited by all these issues. As a result, it is still difficult to build a new Fe^{2+} fluorescent probe, especially one that could effectively overcome those drawbacks.

Li et al. designed the NIR fluorescent probe 43 (Figure 7.11) by taking superior optical qualities of fluorophores 4,4-difluoro-4-bora-3a,4a-diaza-s-indacene (BODIPY), borondipyrromethene, (BDP), and cyanine (Cy). The Fe^{2+} responsive segment, 4'-(aminomethylphenyl)-2,2',6',2''-terpyridine (Tpy),

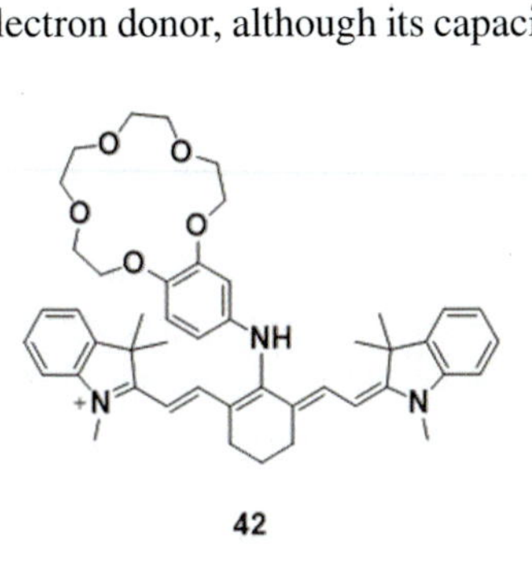

42

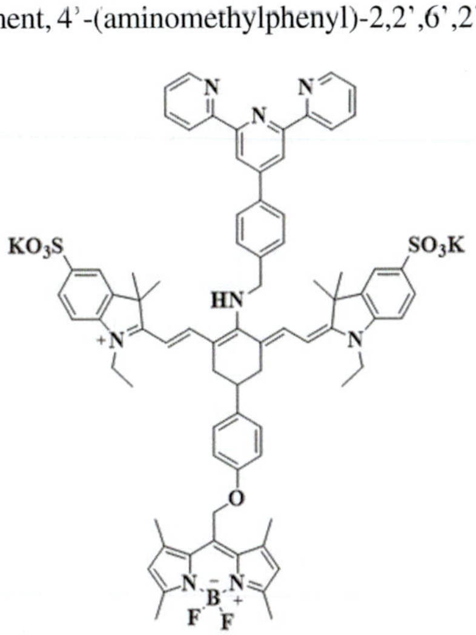

43

FIGURE 7.11 Probes for detection of Cadmium, Lithium, Iron.

and fragment of BDP were fused to the cyanine fluorophore. The new ratiometric fluorescent probe 43 featured two emission and excitation peaks. BDP and Cy-Tpy fragments' maximum emission wavelengths were 507 nm (λ_{ex} = 485 nm) and 635 nm (λ_{ex} = 569 nm), respectively. When iron was added to probe 43, fluorescence intensity of the Cy-Tpy fragment quenched while others did not exhibit any change. Fluorescence intensity shows linear relationship with increasing metal ion concentration. The LOD was found to be 12 nM, and the regression equation was R = 291.77 + 793.68 [Fe^{2+}] (μM, R = 0.9931). Evidently, probe 43 had the sensitivity to qualitatively and quantitatively detect Fe^{2+}. Also, the [Fe^{2+}]/[probe] complex's stoichiometry revealed that the ratio of Fe^{2+} to probe 43 was 1:1. For the selectivity determination, effect of other metal ion study results remained unchanged regarding fluorescence intensity. This shows that other metal ions do not interfere for determination of iron. Probe 43 proved reliable and could react quickly to changes in the Fe^{2+} concentration. Probe 43 also demonstrated a low toxicity and good cell membrane penetration ability. Confocal fluorescence ratiometric imaging experiments were used to verify probe 43 viability for intracellular Fe^{2+} imaging. The authors also stated that this is the first instance of a fluorescent ratiometric probe that may be utilized to effectively identify changes in intracellular and extracellular Fe^{2+} concentration in living cells (Li et al., 2011).

7.3 pH SENSING

Acid-base balance regulation is a capability of living things. However, this ability is compromised when the organisms are infected with cancerous diseases, which results in an acid-base imbalance and raises the pH of body fluids above the acceptable range (Yue et al., 2017). With changes in pH levels, certain dyes show different absorption or fluorescence properties (Xia et al., 2018). The inadequacy of existing methods for pH detection can be done by fluorescence pH measurement, which exhibits various optical signals with pH variations (Zhu et al., 2016). The fluorescence approach also offers the benefits of high detection sensitivity, ease of use, and mobility, making it appropriate for detecting real-time pH changes in cells and fluorescence biomolecular research.

The pH fluorescence probes can be split into those that detect neutral ranges and those that detect acidic environments. The pH range of the cytoplasm is roughly 6.8–7.4, whereas intracellular acidity is between 4.5 and 6.0. Intracellular pH levels are related to numerous physiological activities that occur normally in cells and organelles (Vegesna et al., 2014). Muscle spasm, cell proliferation, apoptosis, ion transport, homeostasis, multidrug resistance, malignancies, endocytosis, and Alzheimer's disease are all linked to changes in intracellular pH. As a result, the main in vivo study focus is on hydrogen ion change, and pH-sensitive NIR fluorescence probes offer a fresh way to track pH variations (Fang et al., 2017).

Wan et al. designed and synthesized the cyanine-based probe 44 a–d (Figure 7.12) by condensation of 9-chl oro-1,2,3,4-tetrahydro-10-methyl-acridinium iodide with Fisher's aldehyde to develop a pH-responsive probe where they first modified the chlorine group (44a) with tert-butyl(2-aminoethyl)carbamate by a substitution reaction (44b) followed by Boc deprotection (44c) and reacting with coumarin (44d). The intermediate probes were evaluated, and probe 44a gave low fluorescence quantum yield, i.e., 7%, while probes 44b, c, d gave 60%, 32%, and 35% quantum yield, respectively. In neutral and basic pH conditions, probes 44b and 44c exhibited intense fluorescence with the push-pull mechanism. However, the push-pull aspect of the probes is disrupted by protonation of the second amine at the 9-position under acidic conditions, which causes the fluorescence to be quenched as a result to give probe 45. Probe 44 showed increase in fluorescence because of the coumarin subgroup, although the fluorophores' fluorescence was completely quenched at acidic pH. So, probe 44 that was used to detect pH changes in HeLa cells, showed that it can specifically label mitochondria, which was further proved by in vivo fluorescence study using *Drosophila melanogaster* first-instar larvae (Wan et al., 2021).

A fluorescent probe 46 (Figure 7.12) was designed by Fang et al. using tetrastyrene donors and cyanide receptors that formed ether linkages in probe 46, which displayed aggregation-induced emission. They studied the effect of pH change on probe 46 that exhibits intense fluorescence at an acidic pH of 5.0. However, from 5.0 to 10.7, pH increase caused quenching in the fluorescence of both the cyanine acceptors causing fluorescence peak to slightly shifted toward blue as a result of the interaction between the TPE donor and the acceptor. Probe 46 becomes less aggregated in a basic aqueous solution due to charge repulsion among three negative phenolate ions, and its low fluorescence intensity at basic pH results from the energy loss is due to the rotation of the ether bond between the TPE donor and the cyanine acceptor and for the spinning of TPE four phenyl groups that was corroborated by dynamic light scattering measurements. Additionally, the cyanine acceptor's absorbance deteriorated. As the pH is increased from 5.0 to 10.7, the cassette's nanoparticle size is lessened. Further, the density functional theory was used to predict the exact mechanism involved in the excitation, which experimentally suggests that this is due to the HOMO/LUMO gap between the cyanine acceptor and the TPE donor. The cytotoxicity study of the probe was done using the MTT assay on HeLa cells that showed low toxicity with good compatibility up to 80% viability at 15 μM. In conclusion, the probe can identify intracellular pH fluctuations in HeLa cells and has dually visible and NIR excitation and emission capabilities (Fang et al., 2017).

The synthesis of a near-infrared rhodamine donor to a near-infrared cyanine acceptor (rhodamine cyanine pH) through stable C-N bonds via a nucleophilic substitution reaction facilitated the generation of a near-infrared fluorescent probe based on fluorescence resonance energy

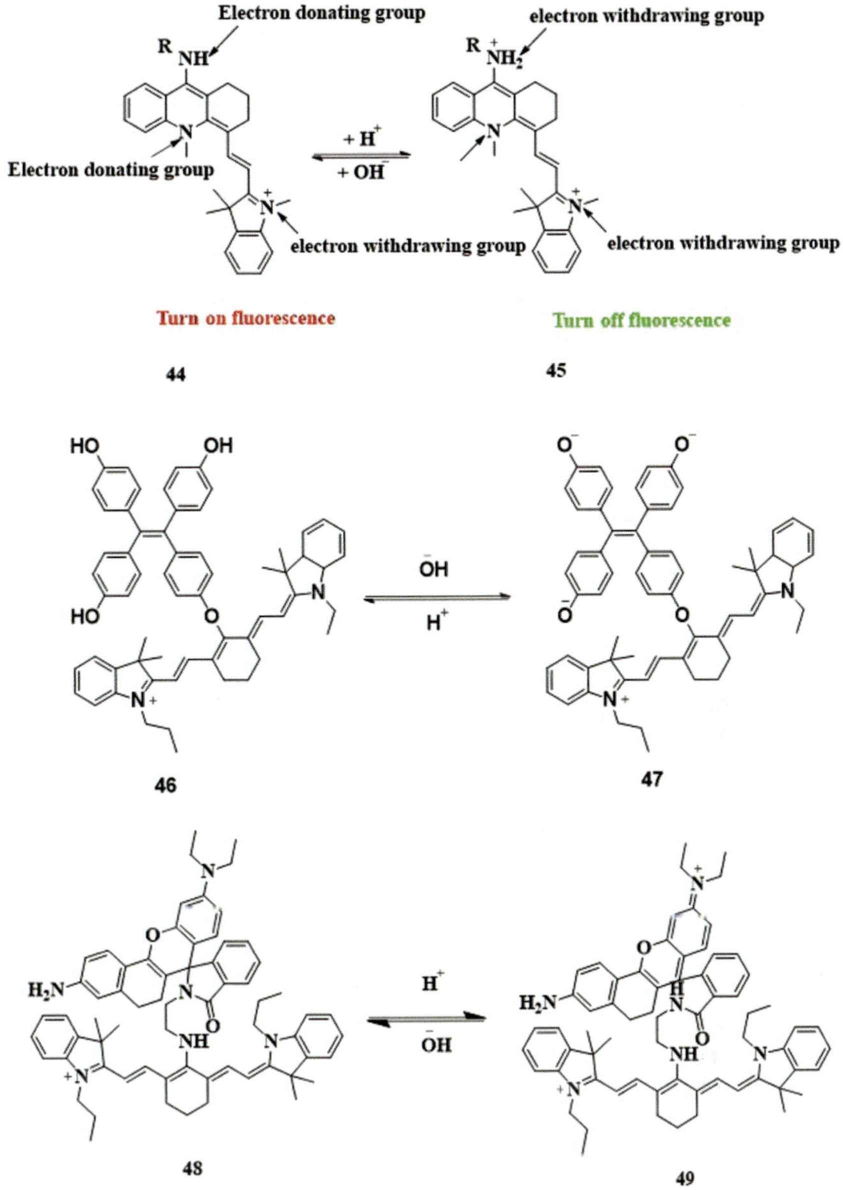

FIGURE 7.12 pH sensing probes.

transfer (FRET) for the double-checked sensitive detection of intracellular pH changes. Probe 48 (Figure 7.12) was developed by Zhang et al. by adding a rhodamine donor to a cyanine acceptor through a very short ethylene spacer to produce high efficiency of energy transfer from the donor to the acceptor to monitor pH changes and the presumed double-checked feature. The fluorescence peak of probe 48 was at 743 nm, and the weak absorption peak of the same was at 664 nm. However, as pH gradually lowers from 10.8 to 2.4, a new absorption peak at 413 nm is formed as well as a gradual increase in the fluorescence peaks at 616 nm and 743 nm under 450 nm of excitation. Acidic pH causes the central amine atom of the cyanine acceptor to protonate and increases the fluorescence of probe 48 through the spiro-lactam ring opening proving that cyanine acceptor excitation occurs at 645 nm and the fluorescence intensity of the cyanine moiety also increases upon pH de-escalation. When a strong C-N bond is added to the cyanine moiety, the cyanine acceptor absorbs with a blue shift at a wavelength of around 650 nm. This enables for effective FRET processes because the cyanine acceptor's absorption overlaps with the emission with the rhodamine donor.

Studies of this probe against different essential metal ions and amino acids was done, but the probe didn't show any significant changes in the presence of any anion or cation but with amino acid, it exhibited only little change in fluorescence. Theoretical calculations were performed based on the density functional theory, where the probe gets protonated to produce BH^{2+} due to three ES transitions which was reversible after transition. By incubating HeLa cells with probe 48 in various pH buffers containing 10 μM nigericin, they checked whether probe 48 could be utilized to detect intracellular pH changes with a double-checked feature. Under rhodamine donor stimulation at 440 nm, a gradual drop in intracellular pH from 9.00 to 3.01 resulted in increase in deep-red fluorescence in channel 1 and near-infrared fluorescence in channel 2. Additionally, the same steady pH reduction also resulted in the rise of intracellular near-infrared fluorescence in channel 3 under cyanine stimulation at 635 nm, showing that probe 48 is capable of monitoring intracellular processes (Zhang et al., 2018).

7.4 SIMULTANEOUS DETECTION OF TWO CATIONS

7.4.1 Copper and Zinc

Hu et al. synthesized a cyanine-based probe 50 (Figure 7.13) consisting of dipicolylamine moieties. They studied the various optical properties that caused pH-dependent change. When the probe was in an alkaline medium (pH 10), it showed green color change from azure and in an acidic medium (pH 4), it showed a dark blue color. The UV absorption study showed a broad band in the NIR region with absorption maxima at 742 nm. As the pH increases, the absorbance of probe 50 decreases. They also studied the effect of different metals on probe 50 for which color change was seen in the case of the copper and zinc from green to blue. Many probes of the zinc and copper complex were used to detect the various anions present in the sample, and no significant change in the absorbance spectra was observed in the case of the zinc-probe complex, but the copper-probe complex showed selectivity toward the sulfide ions because of change in absorbance and the red shift in the wavelength. Color change from blue to green in HEPES was seen. The results of this investigation demonstrated the probe 50 potential as a highly selective chemosensor for the detection of sulfide ions in aqueous media (Hu et al., 2016).

7.4.2 Oxygen and Mercury

The NIR three-channel fluorescent probe 51 (Figure 7.13.) was developed by Wang et al. to detect O_2 and Hg^{2+} in dangerous mercury poisoning models in cells and animals. The three-channel responses proportional to the fluorescence signal removed any interference from an uneven load or

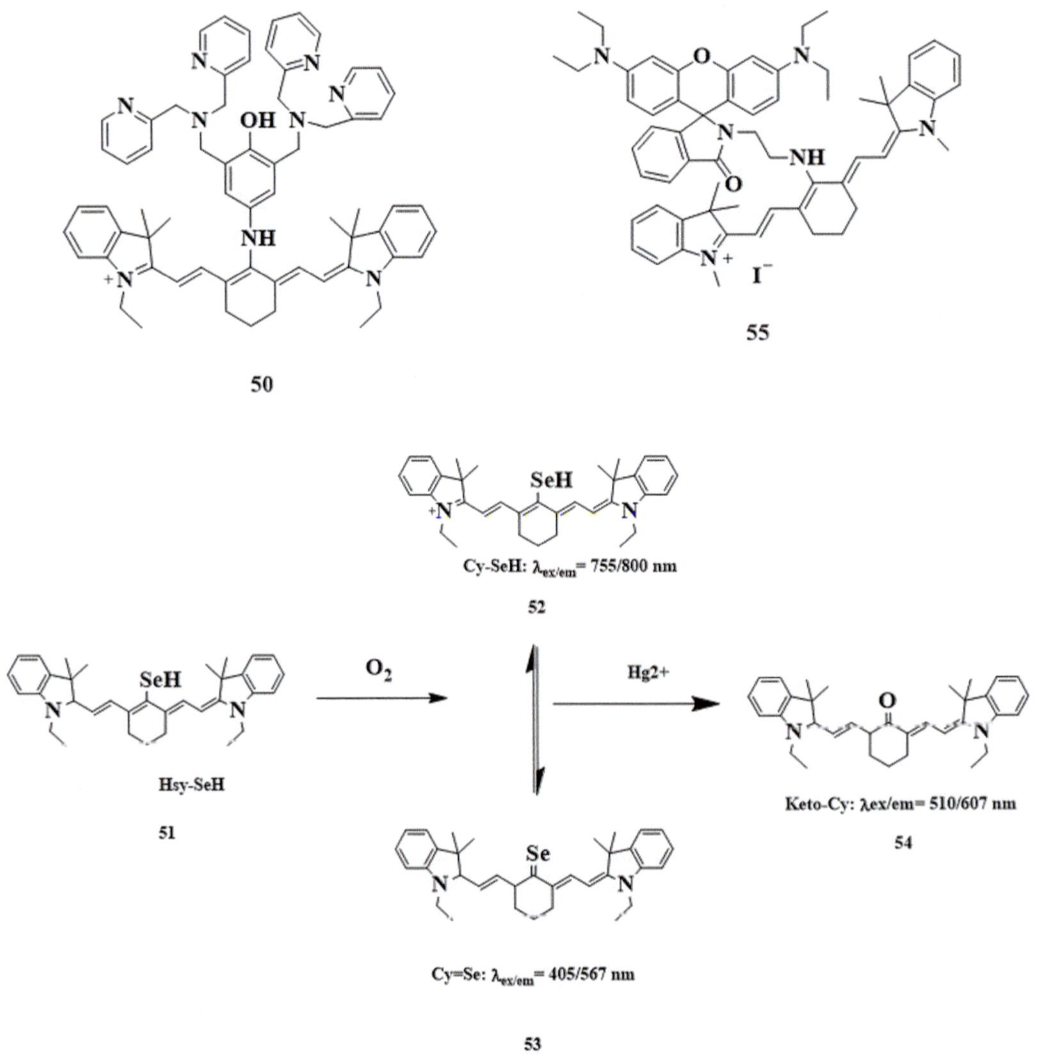

FIGURE 7.13 Probes that help in sensing two cations simultaneously.

uniform distribution during detection. The probe demonstrated good selectivity and sensitivity for the O_2 (probe 52, 53) and Hg^{2+} detection (probe 54) (Figure 7.13). The effectiveness of the probe was examined using the flow cytometry analysis and fluorescence imaging for the in situ detection of O_2 and Hg^{2+} in the HEK 293 cell model. The overproduction of O_2 by cells as a result of the accumulation of Hg^{2+} weakened their antioxidant defense system. The outcomes demonstrated probe 51 potential as a potent tool for the association detection of O_2 and Hg^{2+} in vitro and in vivo (Wang et al., 2018).

7.4.3 Copper and Iron

Li et al. designed and synthesized a novel rhodamine-cyanine fluorophore 55 (Figure 7.14) for the detection of the metal ions. It shows visual changes spectroscopically with two metals copper and iron that selectively responded to probe 55 but when carried out with different metal ions the probe showed the response at 650 nm with lesser deviation in the absorbance intensity while with Fe^{3+} it showed a significant blue shift to 514 nm and also showed a change in color from blue to red. The fluorescent response of probe 55 alone with metal ions resulted in a remarkable elevation in fluorescence while probe 55 with other metal ions did not show any fluorescence change UV titration experiment was performed to find out more about the interaction between probe 55 and Fe^{3+}. A new absorption band at 514 nm appeared as the concentration of Fe^{3+} ions in the solution increased, while the absorption band with at 650 nm vanished. This is likely due to the Fe^{3+}-induced spirocyclic opening reaction and activation of the PET process of cyanine to Fe^{3+} metal ion. A Job's plot analysis showed a concentration ratio of 0.50 for 1:1 stoichiometric relationship between probe 55 and Fe^{3+}. The detection limit of probe 55 for iron was found to be 0.737 μM. Also, probe 55 evinced the reversibility in the detection if high concentration of K_3PO_4 was added showing the color change from red to colorless but further metal addition converts it back to red. Probe 55 was also studied for the detection of copper for which it showed selectivity among the 19 selected metal ions. Probe 55 shows a strong absorption band at 628 nm due to the cyanine group but when copper was added into the probe 55 solution, it resulted in a hypochromic shift in absorption spectra and a change of color from blue to pink making it easy for detection. A Job's plot analysis corresponding to 0.67 suggested that the stoichiometric ratio between probe 55 and copper was 2:1. So, probe 55 selectively determines Fe^{3+} and copper in two different solvents, $MeOH/H_2O$ and $MeCN/H_2O$, respectively (Li et al., 2016).

7.5 SELECTIVE DETECTION OF ANIONS

7.5.1 Sulfide

Hydrogen sulfide (H_2S) and its ionized form, sulfide ion (S^{2-}), are known for potent reductant capabilities and direct consequences on human health (Whiteman et al., 2009). H_2S is a hazardous gas that is mostly produced by the extraction of natural gas or oil and through bacterial breakdown of organic waste. H_2S is widely known for its foul smell (Köhn et al., 2012). Despite its toxicity, H_2S, along with nitric oxide and carbon monoxide, is a crucial endogenous signaling molecule that has a significant impact on a variety of organs and tissues (Szabó, 2007). Due to its effects on vasorelaxation, hepatic circulation, inflammation, cell angiogenesis, and hippocampus memory formation, sulfide even in low quantities is crucial. Numerous disorders, including Alzheimer's disease, Down syndrome, diabetes, and liver cirrhosis, are associated with abnormal sulfide levels. It plays an antioxidant effect in fruits and is used to preserve food and drinks as sodium salt (Na_2S) (Fiorucci et al., 2005). Therefore, it is important to create quick, simple, and accurate S^{2-} detection methods. For the detection of S^{2-} to date, information about colorimetry, sulfide precipitation, electrochemical analysis, and gas chromatography methods has been published. However, these approaches have drawbacks due to complicated setup procedures, expensive instrumentation requirements, and lack of sensitivity. Because of its ease of use, high sensitivity, and real-time detection of S^{2-} in living biological systems, the fluorescence probe has emerged as one of the most effective tools for sensing and tracing samples (Hu et al., 2012). In both a physiological and pathological understanding, H_2S affects the neurological and cardiovascular systems. However, current data suggest that the real signaling substances in cells are hydrogen polysulfides H_2S_n (Yu et al., 2014). The main one electron that declines respiratory chain's mitochondrial electron transport is superoxide anion ($O_2^{\bullet}$). It acts as the principal source of other ROS (Chen et al., 2011). The cross-talk process between $O_2^{\bullet}$ and H_2S_n involves signal transduction, homeostatic regulation, oxidative stress, and antioxidant healing in time in living cells. Combined-response fluorescence probes for in situ detection of $O_2^{\bullet}$ and H_2S_n are urgently needed. A single probe's integration of many responses from various species can better illustrate the redox homeostasis process (Lin et al., 2013).

Two near-infrared fluorescent probes 56a and 56b were successfully designed by Huang et al. for the detection of $O_2^{\bullet}$ and H_2S_n in living organisms and cells (Figure 7.14). Probe 56a serves as the test's representative and displays exceptional sensitivity and selectivity for the detection of endogenous O_2 (probe 57) and H_2S_n (probe 58). The bioassays performed on RAW2647 cells and HUVECs have conclusively shown that mitochondrial oxidative stress can stimulate the production of H_2S_n. Tests using flow cytometry for apoptosis demonstrate the significance of H_2S_n in antioxidant systems. By directly scavenging O_2, H_2S_n can shield cells from mitochondrial oxidative stress. The use of in vivo NIR fluorescence imaging demonstrates that probes have the ideal depth of penetration for imaging O_2 and H_2S_n. Probe 56b can target and assess $O_2^{\bullet}$ and H_2S_n in malignant tissue, as shown by fluorescent imaging of BALB/c mice

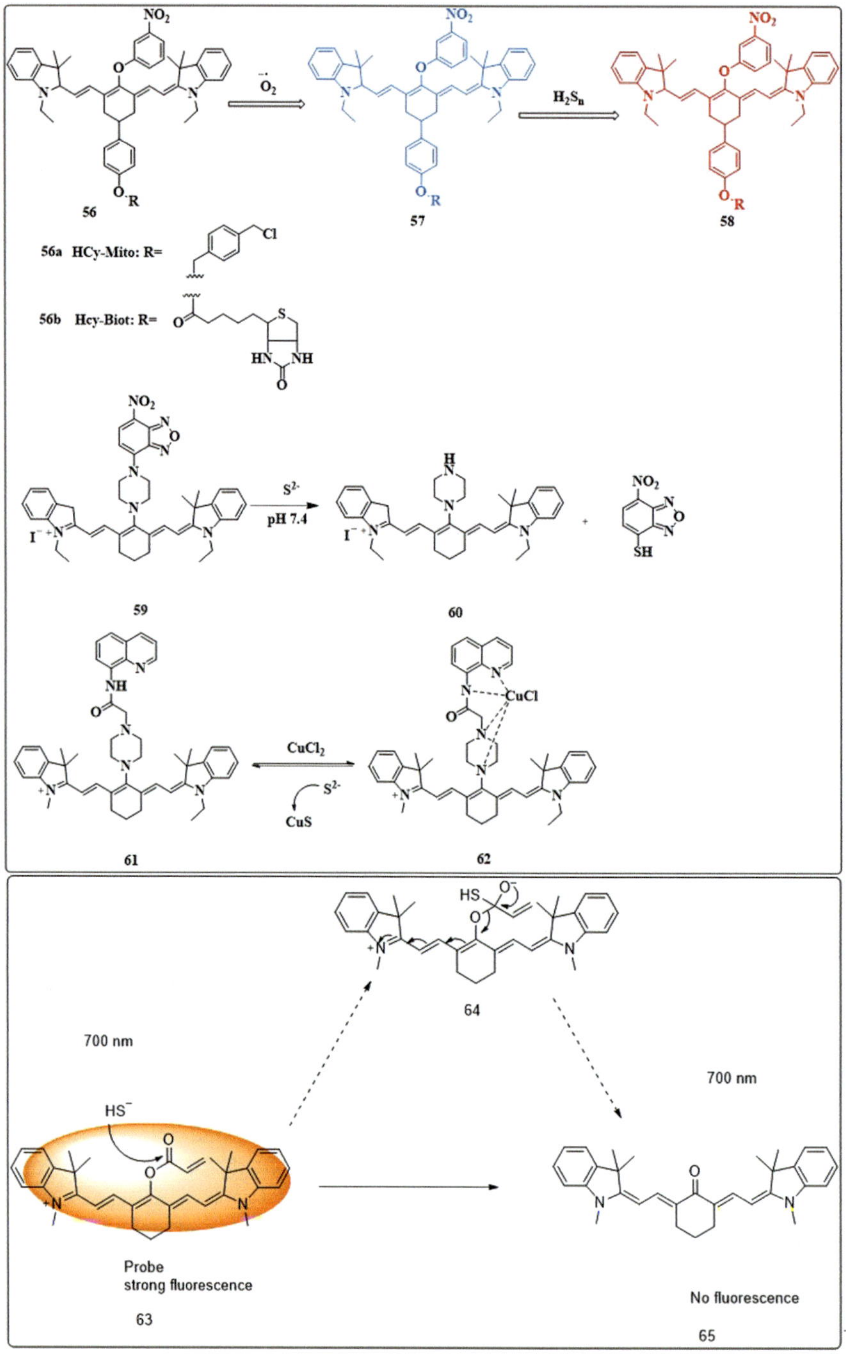

FIGURE 7.14 Detection of Sulphides.

carrying the murine sarcoma S180 tumor. Moreover, the use of these probes in cells and in living organisms will disclose the biological functions of $O_2^{\bullet}$ and H_2S_n in normal and abnormal processes (Huang et al., 2016).

Huang et al. designed a fluorescent probe 59 (Figure 7.14) consisting of heptamethine cyanine as a fluorophore with 7-nitro-1,2,3-benzoxadiazole (NBD) amine as a metal sensor. It showed a significant increase in fluorescence emission after the addition of sulfide than other metal ions with significant color change from dark blue to light blue, which acts as a novel turn-on fluorescent probe for sulfide. The designed probe was synthesized by using piperazine and cyanine. The intensity of the absorption decreased with the addition of sulfide and resulted in blue shift from 780 to 700 nm. The probe IR1800-DPII showed fluorescence intensity based on the viscosity of the solution due to the restriction of rotation in high viscous solution. When a study was conducted on nonsticky medium fluorescence, quenching was observed due to the rotation of meso-substituted NBD. With increasing concentration of sodium sulfide, the fluorescence intensity increases. This increase is observed up to the concentration of 20 μM. The detection limit for sulfide ions was found to be 0.15 μM. Several reductive and potentially interfering biological substances, including GSH, Cys, $S_2O_3^{-2}$, NO_2^-, SO_3^{-2}, Zn^{2+}, Cu^{2+}, Fe^{3+}, H_2O_2, ClO^-, CH_3COO^-, F^-, and I^-, were tested to study the probe IR800-DPII's selectivity toward S^{2-}. The probe shows sensitivity to biological pH range 6–8. In alkaline pH range, it shows a decrease in fluorescent intensity due to instability of the probe. In acidic condition, the probe changes its color due to change in the structure of chromophore due to protonation. Using confocal fluorescence microscopy, the probe's ability to detect S^{2-} in HeLa cells was evaluated. Without the addition of Na_2S, the HeLa cells treated with probe showed no fluorescence. Bright fluorescence images were seen after the Na_2S treatment in HeLa cells. Mechanism of sensing sulfide ion showed fragment at *m/z* 561.5 in mass spectrum which is formed due to removal of nitrobenzofurazen (Figure 7.14: 60). Through a nucleophilic attack on the probe's NBD group, which causes the group to be removed, the S^{2-} triggers the reactions. Because of this gradual cleavage reaction, the probe is appropriate for measuring the accumulated S^{2-} (Huang et al., 2019).

Cao et al. constructed a probe based on cyanine fluorophore 61 (Figure 7.14). Probe 61 shows an absorption band at 694 nm at pH 7.0 in HEPES buffer. The probe 61, which on coordination with sulfide anion formed a stable complex and behaved as a turn-on probe for sensing sulfide ions (Figure 7.14: 62). The linear relationship was observed with increasing concentration of sulfide anion, and the detection limit was found to be 280 nM. To study the selectivity, the copper complex with probe is studied by interfering various anions of 2,000 equiv. with F^-, Cl^-, Br^-, NO_3^-, NO_2^-, N_3^-, SO_4^{2-}, SO_3^{2-}, CO_3^{2-}, PO_4^{3-}, and CH_3COO^-, and 10 equiv. of I^- and CN^- didn't show any change in fluorescent intensity enhancement but the addition of sulfide anions resulted in a marked increase in fluorescence intensity, showing the high selectivity for sulfide ions. To get detailed knowledge about the mechanism sensing NMR and MS studies were done. Probe 61 was firstly treated with copper ion, and ^1H spectrum observation shows a broad peak due to complexation with copper. Further addition of sulfide anions resulted in a resolved peak. The peak at m/z 844.4 vanished because of the addition of sulfide anions, which also caused the creation of a new peak at 745.5. A transition-metal-based displacement strategy could pave the way for the creation of NIR fluorescent sensors for a variety of anion targets as strong NIR fluorescence turn-on response (Cao et al., 2011).

Jiang and his team designed a novel cyanine-based fluorescent probe 63 (Figure 7.14B). The response mechanism of the heptamethine cyanine-based near-infrared (NIR) fluorescent probe 63 for sodium hydrosulfide (NaHS) detection is based on the nucleophilic substitution of the acryloyl moiety by HS^-. Selective sensing of NaHS is a color change from blue green to pink color. The addition of 10 equiv. of NaHS gives rapid and intensive response within a few seconds. As the sulfide is being added, absorbance at 770 nm decreases causing the formation of a new band at 544 nm with increase in the absorption intensity. The response studied against the different time intervals after 10 minutes shows that most quenching occurred at 700 nm while at 544 nm, there was an increase in fluorescence intensity. The selectivity of probe 63 toward sulfide is determined in the presence of various metal ions, including F^-, Cl^-, Br^-, I^-, NO_3^-, NO_2^-, AcO^-, OH^-, CO_3^{2-}, HCO_3^-, HSO_3^-, $H_2PO_4^-$, HPO_4^{2-}, SO_4^{2-}, HSO_4^-, SCN^-, $S_2O_3^{2-}$, and biothiols (Cys and GSH) but there is no significant change observed in case of absorption and fluorescent intensity of the probe except sulfide. It demonstrated a speedy response (within tens of seconds) and a low detection limit (about 0.33 μM) based on nucleophilic substitution by HS^-. Probe 63 also demonstrated its capacity to image living cells and detect HS^- in FBS samples. These studies demonstrated that probe 63 could be a superb tool for the quick and highly selective detection of HS^-, offering an exciting way to study pathological and physiological processes concerning H_2S and holding tremendous promise for future biological applications (Jiang et al., 2017).

7.5.2 Singlet Oxygen

A crucial part of the pathological and physiological processes is played by intracellular singlet oxygen ($^1O_2$), a biological oxidant (Kang et al., 2002) oxidizing a variety of biological substances, including proteins, DNA, and lipids, resulting in degenerative disorders or destroying cancerous cells as a cancer treatment (Weishaupt et al., 1976). However, singlet oxygen has been suggested to take part in the transmission of cell signaling as well as the activation of gene expression. Due to a dearth of current measurement methods for intracellular $^1O_2$, the precise chemical mechanism of singlet oxygen as a signaling molecule in vivo is not yet fully understood (Umezawa et al., 1999). The reactive anthracene moiety was incorporated into a xanthene ring to provide some sensitive fluorescence probes for $^1O_2$. These probes have some drawbacks, including high natural fluorescence, light damage, weak light penetration, and pH sensitivity (Tanaka et al., 2001).

To overcome the limitations faced by other sensors, Xu et al. designed a cyanine-based fluorescent probe 66 (Figure 7.15) that shows weak fluorescence intensity due to electron transfer at 20 mM in tris-HCL buffer at 7.4 pH, where the oxidized probe 67 shows an increase in fluorescence intensity and this oxidation is confirmed by mass spectroscopy. 3.0 mM of probe 67 and various $^1O_2$ concentrations were used to test the susceptibility of probe 71 to singlet oxygen. Fluorescence intensity increased in a concentration-dependent fashion. The detection limit for $^1O_2$ is 7.9 nM, demonstrating a sensitive characteristic, and is computed as the concentration equivalent to three standard deviations of the background signal. Study of probe 67 in the presence of reactive oxygen species, reactive nitrogen species, and biological antioxidants such as glutathione and ascorbic acid does not involve much more changes in the fluorescence intensity. In further study regarding the effect of probe 67 on the living cells, RAW 264.7 cells gave weak fluorescence, but bright fluorescence was observed with the addition of phorbol myristate acetate. To confirm the

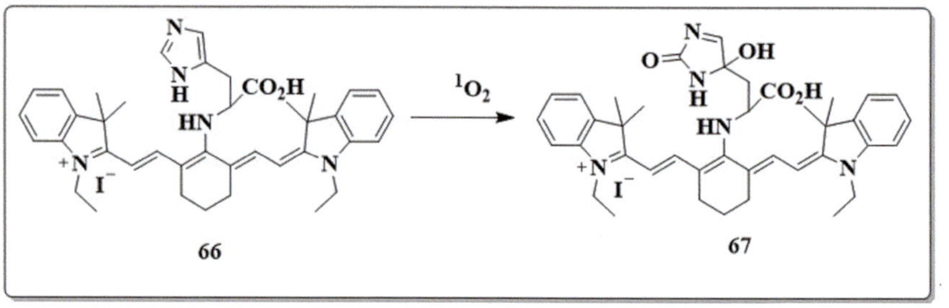

FIGURE 7.15 Sensing mechanism for singlet oxygen.

changes that took place by the addition of oxygen, study by adding histidine ($^1O_2$ scavenger) into the treated cell resulted in a quenching of fluorescence intensity. MTT assay showed low toxicity of the probe (Xu et al., 2011).

7.5.3 Cyanide

Various metal ions play essential part in the body where their levels if increasing or decreasing affect the body's functioning. But some metals are very toxic to our body even at very low amount. Cyanide metal is one example, which is very poisonous to our body and is also life-threatening. For this reason, the WHO fixed the amount of cyanide in drinking water to be less than 500 μM (Shan et al., 2004). To determine its concentration, various methods are developed that include titrimetric, spectroscopic, and other instruments but each has some disadvantages. On the other hand, the fluorescent probes can be developed for selective detection of ions (Safavi et al., 2004).

Gosi et al. designed a cyanine-based probe 12 (Figure 7.16) consisting of indolium and indole moieties and was synthesized in two steps. Probe 12 exhibited selectivity toward the cyanide ion through various visual and spectroscopic techniques. When different ions were treated with this probe 12, a visual color change from orange to colorless was seen only with the cyanide and not with the others. The spectroscopic study of probe 12 showed the UV-Vis absorbance spectra change only with cyanide in which the fluorescence study proved the selectivity is only toward the cyanide ion where it showed a turn-on fluorescence phenomenon with cyanide and the intensity of fluorescence increases as the concentration of cyanide increases while pH does not have any effect on probe 12 and CN^- complex (Figure 7.16). It was characterized by using the NMR spectroscopy, which showed the proton of N-methyl going upfield from 4.01 to 2.74 ppm due to the diminishing indolium ion and olefin proton shift from 7.19 to 6.17 and 8.70 to 7.20 ppm. Probe 12 also gives the paper strip test due to which one can easily determine the presence of cyanide visually without any need for an economical instrument (Gosi et al., 2021).

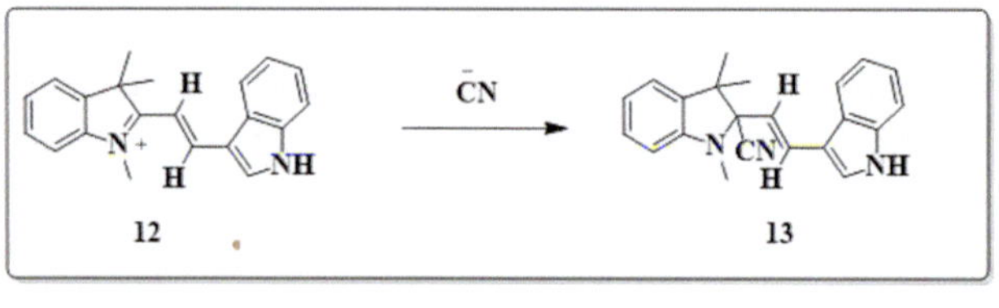

FIGURE 7.16 Probes for detection of Cyanide.

7.6 CONCLUSION

Cyanine dyes are extensively used in biological and biochemical studies. They are used in photodynamic therapy, which is used in cancer treatment where some cyanine dyes can be activated by light to produce reactive oxygen species, making them potential agents in cancer therapy. Cyanine dyes can also be suitable for staining live cells due to their low cytotoxicity, making them valuable for studying cellular dynamics in real time and for live cell imaging for detecting various metal ions, amino acids, proteins, etc. Cyanine dyes can be conjugated to antibodies or proteins for tracking and studying protein interactions and localization in various biological systems.

BIBLIOGRAPHY

Abd-El-Aziz, A. S., Strohm, E. A., & Okasha, R. M. (2015). Design and spectroscopic characterization of novel series of near infrared indocyanine dyes. *Journal of Molecular Structure*, *1091*, 228–235.

Boening, D. W. (2000). Ecological effects, transport, and fate of mercury: A general review. *Chemosphere*, *40*(12), 1335–1351.

Bush, A. I., Pettingell, W. H., Multhaup, G., d Paradis, M., Vonsattel, J. P., Gusella, J. F., ... Tanzi, R. E. (1994). Rapid induction of Alzheimer Aβ amyloid formation by zinc. *Science*, *265*(5177), 1464–1467.

Byber, K., Lison, D., Verougstraete, V., Dressel, H., & Hotz, P. (2016). Cadmium or cadmium compounds and chronic kidney disease in workers and the general population: a systematic review. *Critical Reviews in Toxicology*, *46*(3), 191–240.

Cao, X., Lin, W., & He, L. (2011). A near-infrared fluorescence turn-on sensor for sulfide anions. *Organic Letters*, *13*(17), 4716–4719.

Cao, X., Lin, W., & Wan, W. (2012). Development of a near-infrared fluorescent probe for imaging of endogenous Cu+ in live cells. *Chemical Communications*, *48*(50), 6247–6249.

Chen, X., Tian, X., Shin, I., & Yoon, J. (2011). Fluorescent and luminescent probes for detection of reactive oxygen and nitrogen species. *Chemical Society Reviews*, *40*(9), 4783–4804.

Chen, X., Zhou, Y., Peng, X., & Yoon, J. (2010). Fluorescent and colorimetric probes for detection of thiols. *Chemical Society Reviews*, *39*(6), 2120–2135.

Chew, L. T., Bradley, D. A., Mohd, A. Y., & Jamil, M. M. (2000). Zinc, lead and copper in human teeth measured by induced coupled argon plasma atomic emission spectroscopy (ICP-AES). *Applied Radiation and Lsotopes*, *53*(4–5), 633–638.

Clarkson, T. W., Magos, L., & Myers, G. J. (2003). The toxicology of mercury—current exposures and clinical manifestations. *New England Journal of Medicine*, *349*(18), 1731–1737.

Cobine, P. A., Pierrel, F., Bestwick, M. L., & Winge, D. R. (2006). Mitochondrial matrix copper complex used in metallation of cytochrome oxidase and superoxide dismutase. *Journal of Biological Chemistry*, *281*(48), 36552–36559.

Cobine, P. A., Pierrel, F., & Winge, D. R. (2006). Copper trafficking to the mitochondrion and assembly of copper metalloenzymes. *Biochimica et Biophysica Acta (BBA)-Molecular Cell Research*, *1763*(7), 759–772.

Domaille, D. W., Que, E. L., & Chang, C. J. (2008). Synthetic fluorescent sensors for studying the cell biology of metals. *Nature Chemical Biology*, *4*(3), 168–175.

Fakih, S., Podinovskaia, M., Kong, X., Collins, H. L., Schaible, U. E., & Hider, R. C. (2008). Targeting the lysosome: fluorescent iron (III) chelators to selectively monitor endosomal/lysosomal labile iron pools. *Journal of Medicinal Chemistry*, *51*(15), 4539–4552.

Fang, M., Adhikari, R., Bi, J., Mazi, W., Dorh, N., Wang, J., ... Liu, H. (2017). Fluorescent probes for sensitive and selective detection of pH changes in live cells in visible and near-infrared channels. *Journal of Materials Chemistry B*, *5*(48), 9579–9590.

Fang, M., Xia, S., Bi, J., Wigstrom, T. P., Valenzano, L., Wang, J., ... Liu, H. (2018). A cyanine-based fluorescent cassette with aggregation-induced emission for sensitive detection of pH changes in live cells. *Chemical Communications*, *54*(9), 1133–1136.

Fiorucci, S., Antonelli, E., Mencarelli, A., Orlandi, S., Renga, B., Rizzo, G., ... Morelli, A. (2005). The third gas: H2S regulates perfusion pressure in both the isolated and perfused normal rat liver and in cirrhosis. *Hepatology*, *42*(3), 539–548.

Frederickson, C. J., Koh, J. Y., & Bush, A. I. (2005). The neurobiology of zinc in health and disease. *Nature Reviews Neuroscience*, *6*(6), 449–462.

Fu, Y., & Finney, N. S. (2018). Small-molecule fluorescent probes and their design. *RSC Advances*, *8*(51), 29051–29061.

Gao, X., Shu, Z., Liu, X., Lin, J., & Zhang, P. (2023). Manipulating the monomer–dimer transformation of a heptamethine cyanine ligand: near infrared chromogenic recognition of biothiols. *Analytical Sciences*, *39*(8), 1215–1222.

Georgiev, N. I., Bakov, V. V., Anichina, K. K., & Bojinov, V. B. (2023). Fluorescent probes as a tool in diagnostic and drug delivery systems. *Pharmaceuticals*, *16*(3), 381.

Gosi, M., Marepu, N., & Sunandamma, Y. (2021). Cyanine-based fluorescent probe for cyanide ion detection. *Journal of Fluorescence*, *31*(5), 1409–1415.

Guo, Z., Kim, G. H., Shin, I., & Yoon, J. (2012). A cyanine-based fluorescent sensor for detecting endogenous zinc ions in live cells and organisms. *Biomaterials*, *33*(31), 7818–7827.

Guo, Z., Zhu, W., Zhu, M., Wu, X., & Tian, H. (2010). Near-infrared cell-permeable Hg2+-selective ratiometric fluorescent chemodosimeters and fast indicator paper for MeHg+ based on tricarbocyanines. *Chemistry–A European Journal*, *16*(48), 14424–14432.

Haliawell, B., Gutreridge, J. M. C., Huberman, E., & Barr, S. H. (1999). Book reviews fee radicals in biology and medicine carcinogenesis. *Journal of Free Radicals in Biology & Medicine*, *1*(116), 331–334.

Han, X., Song, X., Yu, F., & Chen, L. (2017). A ratiometric fluorescent probe for imaging and quantifying anti-apoptotic effects of GSH under temperature stress. *Chemical Science*, *8*(10), 6991–7002.

Han, Z. (2013). A new heptamethine cyanine-based near-infrared fluorescent probe for divalent copper ions with high selectivity. *Advances in Materials Physics and Chemistry*, *3*(08), 314.

Harris, H. H., Pickering, I. J., & George, G. N. (2003). The chemical form of mercury in fish. *Science*, *301*(5637), 1203–1203.

He, L., Lin, W., Xu, Q., Ren, M., Wei, H., & Wang, J. Y. (2015). A simple and effective "capping" approach to readily tune the fluorescence of near-infrared cyanines. *Chemical Science*, *6*(8), 4530–4536.

Herzenberg, L. A., De Rosa, S. C., Dubs, J. G., Roederer, M., Anderson, M. T., Ela, S. W., ... Herzenberg, L. A. (1997). Glutathione deficiency is associated with impaired survival in HIV disease. *Proceedings of the National Academy of Sciences*, *94*(5), 1967–1972.

Hu, L. Y., Hu, S. L., Wu, J., Li, Y. H., Zheng, J. L., Wei, Z. J., ... Zhang, H. (2012). Hydrogen sulfide prolongs postharvest shelf life of strawberry and plays an antioxidative role in fruits. *Journal of Agricultural and Food Chemistry*, *60*(35), 8684–8693.

Hu, Y., Yin, J., & Yoon, J. (2016). A multi-responsive cyanine-based colorimetric chemosensor containing dipicolylamine moieties for the detection of Zn (II) and Cu (II) ions. *Sensors and Actuators B: Chemical*, *230*, 40–45.

Huang, X., Liu, H., Zhang, J., Xiao, B., Wu, F., Zhang, Y., ... Jiang, Y. (2019). A novel near-infrared fluorescent hydrogen sulfide probe for live cell and tissue imaging. *New Journal of Chemistry*, *43*(18), 6848–6855.

Huang, Y., Yu, F., Wang, J., & Chen, L. (2016). Near-infrared fluorescence probe for in situ detection of superoxide anion and hydrogen polysulfides in mitochondrial oxidative stress. *Analytical Chemistry*, *88*(7), 4122–4129.

Humbert, R., Adler, D. A., Distechе, C. M., Hassett, C., & CEF, C. J. O. (1993). Fusion of the dominant negative transcription regulator CHOP with a novel gene FUS by translocation (12; 16) in malignant liposarcoma. *Nature Genetics*, *3*, 73–96.

Jiang, Y., Jin, D., Li, Y., Yan, X., & Chen, L. (2017). A near-infrared fluorescent probe for rapid and selective detection of hydrosulfide and imaging in live cells. *Research on Chemical Intermediates*, *43*, 2945–2957.

Kang, P., & Foote, C. S. (2002). Formation of transient intermediates in low-temperature photosensitized oxidation of an 8-13C-guanosine derivative. *Journal of the American Chemical Society*, *124*(17), 4865–4873.

Khayachi, A., Ase, A., Liao, C., Kamesh, A., Kuhlmann, N., Schorova, L., ... Milnerwood, A. (2021). Chronic lithium treatment alters the excitatory/inhibitory balance of synaptic networks and reduces mGluR5–PKC signalling in mouse cortical neurons. *Journal of Psychiatry and Neuroscience*, *46*(3), E402–E414.

Kiyose, K., Kojima, H., & Nagano, T. (2008). Functional near-infrared fluorescent probes. *Chemistry–An Asian Journal*, *3*(3), 506–515.

Kiyose, K., Kojima, H., Urano, Y., & Nagano, T. (2006). Development of a ratiometric fluorescent zinc ion probe in near-infrared region, based on tricarbocyanine chromophore. *Journal of the American Chemical Society, 128*(20), 6548–6549.

Köhn, C., Dubrovska, G., Huang, Y., & Gollasch, M. (2012). Hydrogen sulfide: Potent regulator of vascular tone and stimulator of angiogenesis. *International Journal of Biomedical Science: IJBS, 8*(2), 81.

Kursunlu, A. N., Ozmen, M., & Guler, E. (2016). Novel magnetite nanoparticle based on BODIPY as fluorescent hybrid material for Ag (I) detection in aqueous medium. *Talanta, 153*, 191–196.

Kushner, D. J. (1993). Effects of speciation of toxic metals on their biological activity. *Water Quality Research Journal, 28*(1), 111–128.

Li, G., Guan, Y., Ye, F., Liu, S. H., & Yin, J. (2020). Cyanine-based fluorescent indicator for mercury ion and bioimaging application in living cells. *Spectrochimica Acta Part A: Molecular and Biomolecular Spectroscopy, 239*, 118465.

Li, P., Duan, X., Chen, Z., Liu, Y., Xie, T., Fang, L., ... Tang, B. (2011). A near-infrared fluorescent probe for detecting copper (II) with high selectivity and sensitivity and its biological imaging applications. *Chemical Communications, 47*(27), 7755–7757.

Li, P., Fang, L., Zhou, H., Zhang, W., Wang, X., Li, N., ... Tang, B. (2011). A new ratiometric fluorescent probe for detection of Fe 2+ with high sensitivity and its intracellular imaging applications. *Chemistry–A European Journal, 38*(17), 10520–10523.

Li, S., Zhang, D., Xie, X., Ma, S., Liu, Y., Xu, Z., ... Ye, Y. (2016). A novel solvent-dependently bifunctional NIR absorptive and fluorescent ratiometric probe for detecting Fe3+/Cu2+ and its application in bioimaging. *Sensors and Actuators B: Chemical, 224*, 661–667.

Lin, V. S., Lippert, A. R., & Chang, C. J. (2013). Cell-trappable fluorescent probes for endogenous hydrogen sulfide signaling and imaging H2O2-dependent H2S production. *Proceedings of the National Academy of Sciences, 110*(18), 7131–7135.

Lin, Y., Weissleder, R., & Tung, C. H. (2002). Novel near-infrared cyanine fluorochromes: synthesis, properties, and bioconjugation. *Bioconjugate Chemistry, 13*(3), 605–610.

Long, L., Tan, X., Luo, S., & Shi, C. (2017). Fluorinated near-infrared fluorescent probes for specific detection of Hg 2+ in an aqueous medium and mitochondria of living cells. *New Journal of Chemistry, 41*(17), 8899–8904.

Ma, Y., Luo, W., Quinn, P. J., Liu, Z., & Hider, R. C. (2004). Design, synthesis, physicochemical properties, and evaluation of novel iron chelators with fluorescent sensors. *Journal of Medicinal Chemistry, 47*(25), 6349–6362.

Maitey, D., Raj, A., Karthigeyan, D., Kundu, T. K., & Govindaraju, T. (2015). A switch-on near-infrared fluorescence-ready probe for Cu (I): live cell imaging. *Supramolecular Chemistry, 27*(9), 589–594.

Mendes, A. M. S., Duda, G. P., Nascimento, C. W. A. D., & Silva, M. O. (2006). Bioavailability of cadmium and lead in a soil amended with phosphorus fertilizers. *Scientia Agricola, 63*, 328–332.

Miller, J. R., Rowland, J., Lechler, P. J., Desilets, M., & Hsu, L. C. (1996). Dispersal of mercury-contaminated sediments by geomorphic processes, Sixmile Canyon, Nevada, USA: Implications to site characterization and remediation of fluvial environments. *Water, Air, and Soil Pollution, 86*, 373–388.

Nagano, T., & Yoshimura, T. (2002). Bioimaging of nitric oxide. *Chemical Reviews, 102*(4), 1235–1270.

Napoli, C., de Nigris, F., Williams-Ignarro, S., Pignalosa, O., Sica, V., & Ignarro, L. J. (2006). Nitric oxide and atherosclerosis: an update. *Nitric Oxide, 15*(4), 265–279.

Petrat, F., Weisheit, D., Lensen, M., de GROOT, H., Sustmann, R., & Rauen, U. (2002). Selective determination of mitochondrial chelatable iron in viable cells with a new fluorescent sensor. *Biochemical Journal, 362*(1), 137–147.

Poels, E. M., Bijma, H. H., Galbally, M., & Bergink, V. (2018). Lithium during pregnancy and after delivery: A review. *International Journal of Bipolar Disorders, 6*(1), 26.

Prohaska, J. R., & Gybina, A. A. (2004). Intracellular copper transport in mammals. *The Journal of Nutrition, 134*(5), 1003–1006.

Safavi, A., Maleki, N., & Shahbaazi, H. R. (2004). Indirect determination of cyanide ion and hydrogen cyanide by adsorptive stripping voltammetry at a mercury electrode. *Analytica Chimica Acta, 503*(2), 213–221.

Sasaki, E., Kojima, H., Nishimatsu, H., Urano, Y., Kikuchi, K., Hirata, Y., & Nagano, T. (2005). Highly sensitive near-infrared fluorescent probes for nitric oxide and their application to isolated organs. *Journal of the American Chemical Society, 127*(11), 3684–3685.

Shan, D., Mousty, C., & Cosnier, S. (2004). Subnanomolar cyanide detection at polyphenol oxidase/clay biosensors. *Analytical Chemistry, 76*(1), 178–183.

Sheng, X., Chen, D., Cao, M., Zhang, Y., Han, X., Chen, X., ... Yin, J. (2016). A near infrared cyanine-based fluorescent probe for highly selectively detecting glutathione in living cells. *Chinese Journal of Chemistry, 34*(6), 594–598.

Szabó, C. (2007). Hydrogen sulphide and its therapeutic potential. *Nature Reviews Drug Discovery, 6*(11), 917–935.

Tanaka, K., Miura, T., Umezawa, N., Urano, Y., Kikuchi, K., Higuchi, T., & Nagano, T. (2001). Rational design of fluorescein-based fluorescence probes. Mechanism-based design of a maximum fluorescence probe for singlet oxygen. *Journal of the American Chemical Society, 123*(11), 2530–2536.

Tapiero, H., Townsend, D. Á., & Tew, K. D. (2003). Trace elements in human physiology and pathology: Copper. *Biomedicine & pharmacotherapy, 57*(9), 386–398.

Tang, B., Cui, L. J., Xu, K. H., Tong, L. L., Yang, G. W., & An, L. G. (2008). A sensitive and selective near-infrared fluorescent probe for mercuric ions and its biological imaging applications. *ChemBioChem, 9*(7), 1159–1164.

Tang, B., Huang, H., Xu, K., Tong, L., Yang, G., Liu, X., & An, L. (2006). Highly sensitive and selective near-infrared fluorescent probe for zinc and its application to macrophage cells. *Chemical Communications, 34*, 3609–3611.

Tarazi, L., Choi, H., Mason, J. C., Sowell, J., Strekowski, L., & Patonay, G. (2002). Characterization of a novel crown ether-bearing near-infrared heptamethine cyanine dye. A study of fluorescence quenching by lithium. *Microchemical Journal, 72*(1), 55–62.

Taylor, M. D. (1997). Accumulation of cadmium derived from fertilisers in New Zealand soils. *Science of the Total Environment, 208*(1–2), 123–126.

Tchounwou, P. B., Ayensu, W. K., Ninashvili, N., & Sutton, D. (2003). Environmental exposure to mercury and its toxicopathologic implications for public health. *Environmental Toxicology: An International Journal, 18*(3), 149–175.

Umezawa, N., Tanaka, K., Urano, Y., Kikuchi, K., Higuchi, T., & Nagano, T. (1999). Novel fluorescent probes for singlet oxygen. *Angewandte Chemie International Edition, 38*(19), 2899–2901.

U.S. EPA. (2005, March). Ecological soil screening levels for manganese, interim final. *OSWER Dir 92857–70*, 242.

Valeur, B., & Berberan-Santos, M. N. (2011). A brief history of fluorescence and phosphorescence before the emergence of quantum theory. *Journal of Chemical Education*, *88*(6), 731–738.

Vegesna, G. K., Janjanam, J., Bi, J., Luo, F. T., Zhang, J., Olds, C., ... Liu, H. (2014). pH-activatable near-infrared fluorescent probes for detection of lysosomal pH inside living cells. *Journal of Materials Chemistry B*, *2*(28), 4500–4508.

Wan, S., Xia, S., Medford, J., Durocher, E., Steenwinkel, T. E., Rule, L., ... Liu, H. (2021). A ratiometric near-infrared fluorescent probe based on a novel reactive cyanine platform for mitochondrial pH detection. *Journal of Materials Chemistry B*, *9*(25), 5150–5161.

Wang, Y., Gao, M., Chen, Q., Yu, F., Jiang, G., & Chen, L. (2018). Associated detection of superoxide anion and mercury (II) under chronic mercury exposure in cells and mice models via a three-channel fluorescent probe. *Analytical Chemistry*, *90*(16), 9769–9778.

Weishaupt, K. R., Gomer, C. J., & Dougherty, T. J. (1976). Identification of singlet oxygen as the cytotoxic agent in photo-inactivation of a murine tumor. *Cancer Research*, *36*(7_Part_1), 2326–2329.

Whiteman, M., & Moore, P. K. (2009). Hydrogen sulfide and the vasculature: A novel vasculoprotective entity and regulator of nitric oxide bioavailability?. *Journal of Cellular and Molecular Medicine*, *13*(3), 488–507.

Wu, D., Sedgwick, A. C., Gunnlaugsson, T., Akkaya, E. U., Yoon, J., & James, T. D. (2017). Fluorescent chemosensors: The past, present and future. *Chemical Society Reviews*, *46*(23), 7105–7123.

Xia, S., Wang, J., Bi, J., Wang, X., Fang, M., Phillips, T., ... Liu, H. (2018). Fluorescent probes based on π-conjugation modulation between hemicyanine and coumarin moieties for ratiometric detection of pH changes in live cells with visible and near-infrared channels. *Sensors and Actuators B: Chemical*, *265*, 699–708.

Xie, J. Y., Li, C. Y., Li, Y. F., Fei, J., Xu, F., Ou-Yang, J., & Liu, J. (2016). Near-infrared fluorescent probe with high quantum yield and its application in the selective detection of glutathione in living cells and tissues. *Analytical Chemistry*, *88*(19), 9746–9752.

Xu, K., Qiang, M., Gao, W., Su, R., Li, N., Gao, Y., ... Tang, B. (2013). A near-infrared reversible fluorescent probe for real-time imaging of redox status changes in vivo. *Chemical Science*, *4*(3), 1079–1086.

Xu, K., Wang, L., Qiang, M., Wang, L., Li, P., & Tang, B. (2011). A selective near-infrared fluorescent probe for singlet oxygen in living cells. *Chemical Communications*, *47*(26), 7386–7388.

Yang, Y., Cheng, T., Zhu, W., Xu, Y., & Qian, X. (2011). Highly selective and sensitive near-infrared fluorescent sensors for cadmium in aqueous solution. *Organic Letters*, *13*(2), 264–267.

Yardim, M. F., Budinova, T., Ekinci, E., Petrov, N., Razvigorova, M., & Minkova, V. (2003). Removal of mercury (II) from aqueous solution by activated carbon obtained from furfural. *Chemosphere*, *52*(5), 835–841.

Yin, J., Kwon, Y., Kim, D., Lee, D., Kim, G., Hu, Y., ... Yoon, J. (2014). Cyanine-based fluorescent probe for highly selective detection of glutathione in cell cultures and live mouse tissues. *Journal of the American Chemical Society*, *136*(14), 5351–5358.

You, L., Zha, D., & Anslyn, E. V. (2015). Recent advances in supramolecular analytical chemistry using optical sensing. *Chemical Reviews*, *115*(15), 7840–7892.

Yu, F., Han, X., & Chen, L. (2014). Fluorescent probes for hydrogen sulfide detection and bioimaging. *Chemical Communications*, *50*(82), 12234–12249.

Yuan, L., Lin, W., Zhao, S., Gao, W., Chen, B., He, L., & Zhu, S. (2012). A unique approach to development of near-infrared fluorescent sensors for in vivo imaging. *Journal of the American Chemical Society*, *134*(32), 13510–13523.

Yue, Y., Huo, F., Lee, S., Yin, C., & Yoon, J. (2017). A review: The trend of progress about pH probes in cell application in recent years. *Analyst*, *142*(1), 30–41.

Yun, H. Y., Dawson, V. L., & Dawson, T. M. (1997). Nitric oxide in health and disease of the nervous system. *Molecular Psychiatry*, *2*(4), 300–310.

Zecca, L., Youdim, M. B., Riederer, P., Connor, J. R., & Crichton, R. R. (2004). Iron, brain ageing and neurodegenerative disorders. *Nature Reviews Neuroscience*, *5*(11), 863–873.

Zeng, L., Miller, E. W., Pralle, A., Isacoff, E. Y., & Chang, C. J. (2006). A selective turn-on fluorescent sensor for imaging copper in living cells. *Journal of the American Chemical Society*, *128*(1), 10–11.

Zhang, W., Kang, J., Li, P., Wang, H., & Tang, B. (2015). Dual signaling molecule sensor for rapid detection of hydrogen sulfide based on modified tetraphenylethylene. *Analytical Chemistry*, *87*(17), 8964–8969.

Zhang, Y., Bi, J., Xia, S., Mazi, W., Wan, S., Mikesell, L., ... Liu, H. (2018). A near-infrared fluorescent probe based on a FRET rhodamine donor linked to a cyanine acceptor for sensitive detection of intracellular pH alternations. *Molecules*, *23*(10), 2679.

Zhang, Y., Ye, A., Yao, Y., & Yao, C. (2019). A sensitive near-infrared fluorescent probe for detecting heavy metal Ag+ in water samples. *Sensors*, *19*(2), 247.

Zhu, H., Fan, J., Du, J., & Peng, X. (2016). Fluorescent probes for sensing and imaging within specific cellular organelles. *Accounts of Chemical Research*, *49*(10), 2115–2126.

Zhu, M., Yuan, M., Liu, X., Xu, J., Lv, J., Huang, C., ... Zhu, D. (2008). Visible near-infrared chemosensor for mercury ion. *Organic Letters*, *10*(7), 1481–1484.

Ziemba, S. E., McCabe Jr, M. J., & Rosenspire, A. J. (2005). Inorganic mercury dissociates preassembled Fas/CD95 receptor oligomers in T lymphocytes. *Toxicology and Applied Pharmacology*, *206*(3), 334–342.

8 Fluorescein-Based molecules
Design, Synthesis, and Applications

Yashashvi Dhamapurkar, Mareechika Gaddam and Abha Sharma
National Institute of Pharmaceutical Education and Research,
Raebareli, Lucknow, UP, India

LIST OF ABBREVIATION

AD	Alzheimer's disease
PET	Photoinduced electron transfer
CHEF	Chelation-enhanced fluorescence
FRET	Fluorescence resonance energy transfer
FIA	Flow injection analysis
LOD	Limit of detection
EDTA	Ethylenediaminetetraacetic acid
TLC	Thin layer chromatography
WHO	World Health Organization
API	Active pharmaceutical ingredient
CNS	Central nervous system
PPB	Parts per billion
DTPA	Diethylenetriaminepentaacetic acid
ROS	Reactive oxygen species
CO	Carbon monoxide
NO	Nitric oxide
ATP	Adenosine triphosphate
Pi	Inorganic phosphate

8.1 INTRODUCTION

Fluorescein is a widely used fluorophore that has gained significant attention in various applications, particularly in the field of bioimaging and sensors. It was first synthesized by Bayer in 1871 and has subsequently become popular due to its remarkable properties, including a high molar extinction coefficient, high quantum yield, and strong emission (Ueno et al., 2004; Setsukinai et al., 2003). In recent years, there has been a growing interest in detecting and quantifying biologically and environmentally significant species, and a variety of methods have been developed for this purpose. These methods include liquid chromatography-mass spectrometry (LC-MS), electrochemistry, titrimetric analysis, conductometry, chemiluminescence, and spectrophotometry. However, among these techniques, fluorescent sensors have emerged as one of the best approaches due to their high sensitivity and specificity, ease of handling, and low limit of detection (LOD). Fluorescent sensors also enable real-time detection of analytes in biological sensing applications (Shi et al., 2023). Fluorescent sensors typically consist of a fluorophore, linker, and recognition moiety. These sensors interact with the target analyte through binding to enzymes, receptors, or chemical modifications (Piriya V.S et al., 2017). A wide range of fluorophores have been used as chemosensors, including BODIPY (Niu et al., 2012), xanthene (Chen et al., 2012), quinolone (Patil et al., 2014), phthalimide (Patil et al., 2019), and coumarin (Ngororabanga et al., 2020). Among them, xanthene analogs, which include fluorescein, are commonly employed due to their unique properties such as high absorption coefficient, high quantum yield, excellent photostability, and longer emission wavelengths. The xanthene moiety in fluorescein consists of two benzene rings connected by a methylene group and an oxygen atom. This structural arrangement contributes to the enhanced fluorescent properties observed in fluorescein and makes it a favorable choice for fluorescent sensing applications.

8.2 SYNTHESIS OF FLUORESCEIN

Fluorescein was initially synthesized by Von Bayer through the condensation of phthalic anhydride and phenol under acidic conditions. In more recent methods, fluorescein can be prepared using a Friedel-Crafts reaction (Grimm et al., 2011). This synthesis involves the mixing of phthalic anhydride and resorcinol with a catalyst such as zinc chloride or methane sulfonic acid (Figure 8.1). When using zinc chloride as the catalyst, the reaction typically requires the addition of concentrated hydrochloric acid (HCl) and high temperature in the range of 170–180°C. The presence of zinc chloride and the elevated temperature promote the reaction between phthalic anhydride and resorcinol, leading to the formation of fluorescein. However, it is mentioned that this synthesis method may result in the loss of a significant amount of initial material due to sublimation, making it less suitable for small-scale reactions. An alternative catalyst, methane-sulfonic acid, can be used in the Friedel-Crafts reaction for fluorescein synthesis. In this case, methane-sulfonic acid serves both as an appropriate solvent and an acid catalyst. It was noted that using methane-sulfonic acid under milder reaction conditions can yield better results (Anwar et al., 2018).

DOI: 10.1201/9781032706245-8

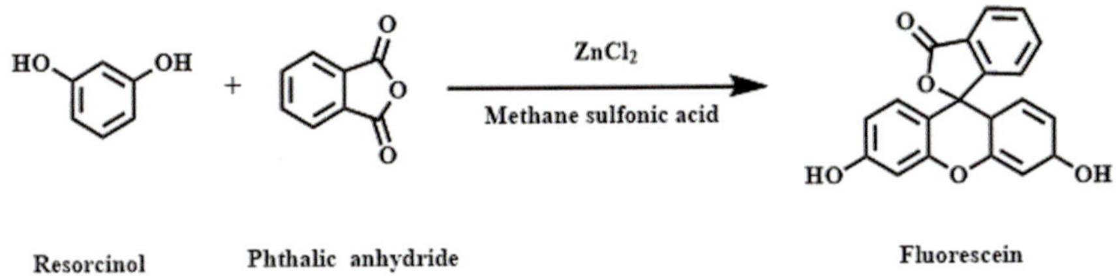

FIGURE 8.1 Synthetic pathway of fluorescein.

8.3 SPECTRAL PROPERTIES OF FLUORESCEIN

The fluorescence properties of fluorescein are based on the formation of a conjugated system in the xanthene moiety. Fluorescein exists in equilibrium between a "closed" lactone form and an "open" quinoid form, which leads to pH-dependent absorption and emission properties in the range of 5–9. At acidic pH levels ranging from 2 to 4, fluorescein predominantly exists in the "closed" lactone form. Under mild acidic pH conditions of 5–7, a significant portion of fluorescein may still be in the protonated, nonfluorescent form and in the less fluorescent monoanionic form. In contrast, under physiological conditions with a pH of 7.4, fluorescein exists predominantly as a highly hydrophilic dianion (Figure 8.2). Overall, the pH-dependent properties of fluorescein make it a versatile tool for various applications in bioimaging and sensing, as its fluorescence can be modulated by changes in pH levels (Rajasekar et al., 2021).

8.4 FLUORESCEIN SCAFFOLD AS A FLUOROPHORE

Fluorescein is a yellowish-green xanthene dye synthesized by Von Bayer in 1871 using phthalic anhydride and resorcinol through cyclo-degradation or the Friedel-Crafts acylation reaction. Fluorescein exists in two forms: the open carboxylic acid ring, which is fluorescent, and the closed spirocyclic lactone ring, which is nonfluorescent. The spiro-lactam structure of fluorescein, in its nonfluorescent and colorless form, is significant due to its ability to respond to different environments in an "open and close" manner. The spiro-lactam form of fluorescein is nonfluorescent until it interacts with a specific analyte, causing the ring to open and induce bright emission with high quantum yield and molar absorptivity. This change in fluorescence color is often observed as well. Additionally, fluorescein exhibits pH-sensitive behavior as it exists in an open-close

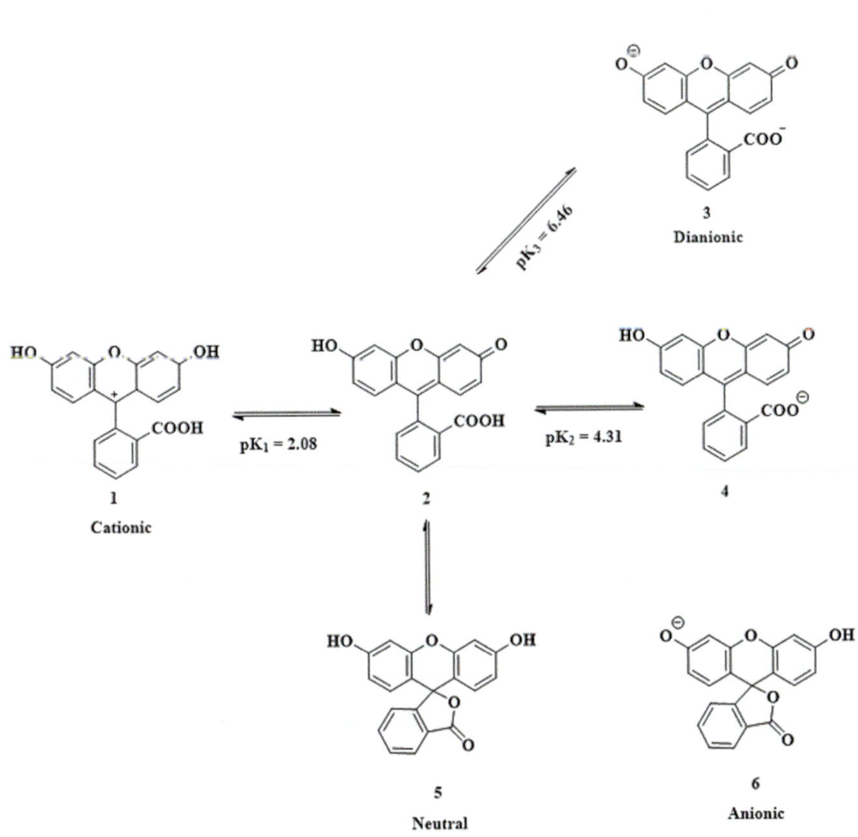

FIGURE 8.2 pH-dependent behavior of fluorescein.

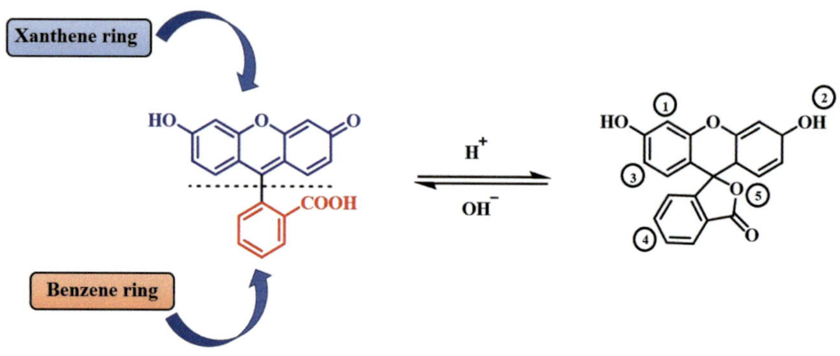

FIGURE 8.3 Structure of fluorescein with its modification sites.

equilibrium state, allowing its fluorescence properties to be modulated by changes in pH (Figure 8.2) (Le Guern et al., 2020). Based on the structure of fluorescein, it is possible to modify the xanthene and benzene rings, which are positioned orthogonally to each other. This structural modification at the five sites (positions) of fluorescein enables the development of a wide range of fluorescein-based probes (Figure 8.3). These modified probes influence the characteristic properties of fluorescein to sense-specific analytes by detecting the open and closed equilibrium states of fluorescein. Overall, the unique properties of fluorescein and its ability to be modified make it a valuable tool for the development of fluorescent probes and sensors for various applications, including bioimaging, chemical sensing, and analytical assays.

At **site 1** of the xanthene ring, fluorescein aldehydes are synthesized via the Reimer-Tiemann reaction, which can be further modified by connecting with amino groups based on the Schiff base reaction. This results in probes with amino groups that have lone pairs of electrons that can induce fluorescence quenching during the PET mechanism within the molecule. There are two types of probes available: bonding signaling type and reactive fluorescein probes. Fluorescein probes of the bonding-signaling type produce an increase or decrease in fluorescence intensity due to electron transfer from amino group-containing compounds to the xanthene ring. Ions with an empty orbit can hinder electron transfer, resulting in fluorescence recovery. Reactive fluorescein probes can facilitate fluorescein spiro-lactam ring opening and increase molecule rigidity due to the active Schiff base. The goal of detection can be accomplished by inhibiting C=N isomerization and promoting the chelation-enhanced fluorescence (CHEF) effect.

Fluorescein probes are synthesized by modifying hydroxyl group of the xanthene moiety (**site 2**) by esterifying two phenolic hydroxyl groups of fluorescein and connecting them with small organic molecules. Initially, fluorescein probes are colorless and nonfluorescent because of breaking in pi-conjugation. Ester bonds are broken and small organic molecules are removed when a specific ion is added in an irreversible chemical reaction. The recovery of pi-conjugation results in obvious color change with changes in fluorescence (Zheng et al., 2013).

At **site 3**, fluorescein-based probes are modified with halogen by the substitution reaction by connecting small organic molecules. This results in an increase in fluorescence properties with a high complex ability to selectively detect metal ions. Many probes are synthesized based on bonding signaling type of probes.

At **site 4**, the benzene moiety in fluorescein is modified with the connection of gold nanoparticles or organic fluorophores for the detection of particular analytes based on the FRET mechanism. The connection of fluorescein with gold nanoparticles results in the quenching of fluorescence via FRET mechanism. This quenching is caused due to overlap between the absorption and emission spectra of gold nanoparticles at plasmon resonance wavelength. But in the case of some analytes, the FRET process is blocked because of the high binding affinity of analytes, which causes fluorescence recovery.

Fluorescein probes are also synthesized by modifying the carboxyl group of the benzene moiety (**site 5**) via a fluorescein hydrazine-based Schiff base reaction by connecting aldehyde-containing compounds. These probes are synthesized by using fluorescein as the signaling unit and aldehyde-containing compounds as the binding unit based on noncovalent supramolecular interaction. The Schiff bases have high activity, which provides additional binding sites that help to enhance the sensitivity for analyte detection (Yan et al., 2017).

8.5 FLUORESCEIN-BASED SENSOR APPLICATIONS

There are several applications of fluorescein-based fluorescent probes, which have been extensively studied and depicted in Figure 8.4.

8.5.1 Detection of Metal Ions

Fluorescent sensors have gained significant attention for the detection of metal ions due to their high sensitivity and selectivity. When these sensors interact with specific metal ions, they change color and fluorescence intensity, allowing for their detection. Fluorescein has been widely utilized as a core scaffold for developing fluorescent probes to detect

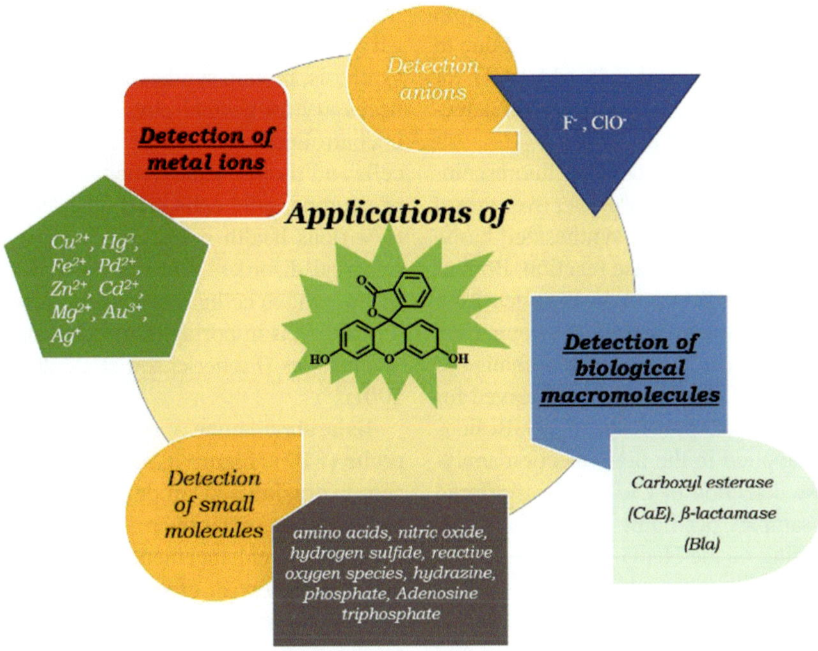

FIGURE 8.4 Applications of fluorescein.

various metal ions. Numerous probes based on fluorescein have been reported for the detection of metal ions such as copper, mercury, zinc, silver, gold, iron, palladium, cadmium, lead, magnesium, etc.

8.5.1.1 Detection of Copper Ions

Copper (Cu^{2+}) is an essential metal ion in the human body, playing a vital role in various processes such as gene expression and protein functioning. However, excessive intake and accumulation of Cu^{2+} can lead to health issues such as Wilson's disease and Alzheimer's disease. Therefore, it is crucial to monitor and maintain the proper levels of Cu^{2+} in the body. The recommended limit capacity for Cu^{2+} ion concentration is 31.48 μM. It is important to have accurate methods for monitoring the levels of Cu^{2+} to ensure both healthcare and environmental safety. Fluorescent sensors based on fluorescein and other fluorescent probes have been extensively explored for the detection and monitoring of Cu^{2+} ions. These sensors utilize the specific binding affinity of Cu^{2+} ions with the probe molecules, resulting in changes in fluorescence properties. These changes can be correlated with the concentration of Cu^{2+} ions present in the sample, allowing for accurate detection and quantification (Cakic et al., 2013; Strausak et al., 2001; Stelmashook et al., 2014; Kumar et al., 2012).

Mondal et al. synthesized a ratiometric fluorescent probe (**FP1**) consisting of a fluorescein acceptor and a phenanthroline donor for the detection of copper ions. This probe is synthesized based on a Schiff base reaction between fluorescein hydrazine and phenanthroline. For copper ion detection, most of the probes are developed based on the copper-induced spiro-ring opening of rhodamine but only some probes are based on the FRET phenomenon. This probe can detect copper based on the FRET "off-on" mechanism (Figure 8.5a). In the absence of copper ions, no fluorescence emission is observed. On the other hand, upon the addition of copper ion, the spirocyclic ring of the fluorescein probe opens, and new green emission develops at 506 nm, which indicates the FRET-on mechanism. This probe showed a good linear relationship between the fluorescence intensity and concentration of copper ion (0.002–0.119 mM) with the LOD of 15.47 μM. In addition, this sensor can detect Cu^{2+} ions through the naked eye. Also, it can be applicable for copper ion detection in aqueous media and living cells (Mondal et al., 2017).

Rathod et al. based on the Schiff base reaction had synthesized the fluorescein-based fluorometric and colorimetric sensor (**FP2**) (Figure 8.5b) for selective determination of copper ion metal. This probe upon reaction with copper resulted in a color change from colorless to deep yellow color in ACN solvent because of molecular interaction. Significant fluorescence quenching was observed with this probe at 519 nm. The limit of detection of copper ion was found to be 1.25 μM. Also, this probe showed efficient binding for copper ions in aqueous media (Rathod et al., 2016).

Using photochromic chemical diarylethene to improve the sensitivity of the sensor toward metal ions, Xue et al. synthesized a fluorescein-based probe (**FP3**) that is selective toward copper ions. This probe showed enhancement in fluorescence emission because of the change in the pirocyclic ring to open ring, upon being treated with copper. Due to the FRET mechanism, fluorescence quenching was also observed. This probe showed a fast response toward copper with LOD of 2.86 μM (Xue et al., 2017).

Fan et al. synthesized a fluorescein-based sensor (**FP4**) (Figure 8.5d) for the selective detection of copper by using phthalic anhydride and benzothiazole derivative. This sensor showed absorption maxima at 559 nm; upon the addition of copper, in the linear range in copper concentration from 0.03 to 2 μM, decrease in absorbance was observed due to spiro cyclization. This sensor showed 7.48 nM LOD for cupric ions. Also, it can be applicable for copper ion detection in water and tea samples (Fan et al., 2015).

Tachapermpon and his coworkers devised fluorescein-based probes (**FP5**) (Figure 8.5c) that are selective toward the copper ion metal. These probes are synthesized based on the Birch reduction and the Schiff base reaction. Both of these probes showed binding with copper by the quenching of fluorescence. Additionally, the sensing of the copper ion was observed with the visual eye detection by a change in color from yellow to orange. Detection limits observed for MF3 and DF3 were 0.23 and 0.43 ppb, respectively. Besides these, MF3 sensor was applied in the flow injection analysis (FIA) system for the detection of Cu^{2+} where it offered rapid detection of 32 samples per hour. Also, this sensor MF3 was used for imaging of the HepG-2 cellular system (Tachapermpon et al., 2017).

Helal et al. synthesized a fluorescent probe (**FP6**) (Figure 8.5e) based on the conjugation of fluorescein and imidazole. The probe is synthesized based on a Schiff base reaction between N-methyl imidazole and fluorescein hydrazine. Due to electron transfer from the imidazole ring to fluorescein, this probe showed an emission at 525 nm that gives green fluorescence. After the addition of copper, fluorescence gets quenched due to the inhibition of electron transfer between fluorescein and imidazole. As the electron-donating group is present on the imidazole ring, electron-donating capacity increases, making this probe more selective toward copper ions. This probe showed a 37 nM limit of detection (LOD) for copper ions. Moreover, this probe can be used in the detection of Cu^{2+} in biological samples (Helal et al., 2016).

Based on the FRET mechanism, Reza Hormozi-Nezhad et al. reported a probe for copper ion detection. It was synthesized using an acceptor-donor structure with gold nanoparticles and fluorescein isothiocyanate. Initially, this probe showed a turn-off fluorescence because of adsorption to the surface of gold nanoparticles. Moreover, the fluorescence was turn-on upon the addition of D-penicillamine (D-PC) to the probe solution. By the displacement of the probe with D-PC on the surface of nanoparticles, the probe gets released. By this method, the strong interaction of copper was observed with the chelating agent of D-PC. Here two linear calibrations were obtained within the range of 1–9 nM and 10–40 nM. LOD of 0.3 nM was found. This probe can be applied for the determination of copper content in water and food samples (Hormozi-Nezhad et al., 2015).

8.5.1.2 Detection of Zinc Ions

Zinc is an essential metal ion in the human body and plays a crucial role in various biological processes. It is involved in numerous cellular functions, including signal transmission, gene transcription, enzymatic reactions, and immune system regulation. It is also essential for proper growth, development, and reproduction. It serves as a cofactor for many enzymes and transcription factors, enabling them to carry out their biological functions. It is involved in DNA synthesis, protein synthesis, and cell division. While zinc is necessary for normal physiological functions, excessive or uncontrolled levels of zinc can be toxic and detrimental to cells and tissues. Zinc toxicity can result in the destruction of neurons, glial cells, and other cell types. This can lead to various health issues and impairments, including neurological disorders and tissue damage. Understanding the role of zinc in cellular processes, as well as its implications in toxicity, is important for both basic research and clinical applications (Frederickson et al., 2005; Beyersmann et al., 2001).

Balasubramanian Vidya et al. synthesized a fluorescent probe (**FP7**) (Figure 8.5f) using a combination of fluorescein hydrazine and imidazole moiety based on the Schiff base reaction. After the reaction of this probe with zinc ion, fluorescence enhancement with a change in color from colorless to yellowish green was observed. Detection of zinc ion with this probe is reversible, which can recover the fluorescence by EDTA. The detection limit was 0.54 nM for the determination of zinc ions. In addition, this probe was applicable for zinc ion imaging in HEK 293 cells (Vidya et al., 2016).

Khwan chanok Chantalakana et al. synthesized the probe (**FP8**) (Figure 8.5g) based on the PET mechanism with the combination of ionophore and fluorescein moiety. Ionophore incorporation helps in the enhancement of fluorescence with high selectivity toward zinc (Zn^{+2}) ions. It showed a ratiometric response for zinc ion detection with 1:1 binding in the pH range of 7–9. It showed potential for the detection of zinc ions in biological systems, specifically in fluorescence imaging. The detection limit was 1.2 μM for the determination of zinc ions (Chantalakana et al., 2016).

Based on the Schiff base reaction, Wang et al. synthesized a fluorescent probe (**FP9**) (Figure 8.5h) with fluorescein hydrazine and dihydroxy benzaldehyde. Interaction of probe with zinc ion causes bathochromic shift nearly about 50 nm from 346 nm to 400 nm. Upon binding with zinc, this probe showed the CHEF effect with induced molecular rigidity. It showed an increase in fluorescence with the change in fluorescence color from colorless to yellow. The detection limit was found to be 10 μM for the determination of zinc ions. The detection of this probe for zinc is reversible, which can recover using EDTA. In addition, it showed potential for imaging and sensing of zinc ions in living cells (Wang et al., 2014).

Serkan Erdemir et al. developed a fluorescein-benzothiazole-based sensor (**FP10**) (Figure 8.5i) for the selective detection of zinc ions. This sensor recognized zinc ions by spiro-lactam ring opening with a significant increase in fluorescence enhancement. Also, it showed the detection of zinc ions by the paper strip method. The LOD was found to be 5.64 μM for the detection of zinc ions (Erdemir and Tabakci, 2017).

Das et al. synthesized a cost-effective fluorescent sensor (**FP11**) (Figure 8.5j) based on the Schiff base reaction with the combination of fluorescein hydrazine and phenol moiety. The interaction of this sensor with zinc ions showed a significant increase in fluorescence due to the inhibition of C=N isomerization, PET, and CHEF. This probe showed the ability to detect zinc ions in different solvent media and also by the TLC paper with a fast response. The detection limit was found to be 1.59 μM for the determination of zinc ions. In addition, this probe is applicable for imaging of zinc ions in the intracellular region of Vero cells (Das et al., 2019).

Jun-mei An et al. reported a probe (**FP12**) (Figure 8.5k) by a combination of fluorescein and coumarin moiety. This sensor was highly selective for zinc metal ions, which showed enhancement in fluorescence (33 fold) with the presence of zinc ions. This change was observed due to the spiro-lactum ring opening power of zinc. The detection limit was found to be 6.54 ppb for the determination of zinc ions (An et al., 2013).

8.5.1.3 Detection of Magnesium Ions

Magnesium is an essential mineral and the fourth most abundant cation in the human body. It plays a vital role in various physiological functions. Some of its important functions include skeletal development, DNA synthesis, enzymatic reactions, muscle contraction, nerve function, and the functioning of the immune system and central nervous system. Proper levels of magnesium are necessary for maintaining overall health. Magnesium deficiency can lead to several health issues, including coronary heart disease, diabetes, osteoporosis, and muscle disorders. It is crucial to monitor and maintain adequate levels of magnesium in the body to support optimal health and well-being. The development of magnesium ion sensors is important for both research and clinical applications (Svagzdiene, 2015; Rude et al., 2004; Janning et al., 2010; Martinez-Jimenez et al., 2015).

Chao-rui Li et al. reported a fluorescein-based fluorescent probe (**FP13**) (Figure 8.5l) that contains chromone moiety for the detection of magnesium ions. This sensor's reaction with magnesium ions induces ring opening with enhancement in fluorescence at 504 nm. The probe binding ratio to Mg^{2+} is 2:1, which is distinct compared to other fluorescein hydrazine probes. In addition, this sensor could be used to detect Mg^{2+} in HeLa cells (Ge et al., 2015).

Fengyan Ge et al. developed a fluorescein-based probe, which is formed by linking morpholine with amino fluorescein using triazine chloride and has proven to be an effective method for sensing magnesium ions (Mg^{2+}) in biological systems. The choice of triazine chloride as the linking agent is due to its ability to easily undergo substitution, allowing for the attachment of the morpholine and amino fluorescein moieties. The morpholine ring contains a nitrogen atom that can act as a Lewis base and coordinate with metal ions, forming stable complexes. When Mg^{2+} ions bond with the carbonyl group of the probe, a fluorescent enhancement occurs. This means that the fluorescence of the probe is increased or intensified upon binding with Mg^{2+}. This change in fluorescence can be easily detected and quantified, providing a means for sensitive detection of Mg^{2+} ions in biological systems. The effectiveness of a probe for sensing Mg^{2+} in biological systems indicates its potential utility in various applications (Li et al., 2018).

8.5.1.4 Detection of Silver Ions

Silver has traditionally been considered a kind of precious metal that is useful to humans. It is widely used specifically in the photographic, imaging, and electronic industries. This metal has long been known to have the ability to destroy harmful bacteria. The excessive intake of this metal leads to various side effects on human health. The World Health Organization (WHO) has set a permissible limit for silver ions in drinking water at 0.9 μM (approximately 0.1 mg/L). Higher concentrations of silver ions can cause cardiac enlargement, growth retardation, and degenerative changes in the liver. Therefore, detecting this metal ion is very needful nowadays.

Sun et al. reported a fluorescein-based sensor (**FP14**) (Figure 8.5m) based on the Schiff base reaction. The sensor was synthesized by using allyl bromide, fluorescein hydrazine, and 2-thiazole carboxaldehyde. This sensor forms the copolymer due to the polymerization of allyl ether. In the presence of silver ions, enhancement in fluorescence was observed at 505 nm. Also, a good linear relationship was observed between fluorescence intensity and concentration of silver ions in the range of 10–60 μM. The limit of detection was found to be 20 μM for the detection of silver ions in the water sample (Sun et al., 2016).

Based on the Schiff base reaction, Shen et al. developed a reactive fluorescein-based sensor. It is synthesized using fluorescein hydrazine and 2-formylthiophene. This sensor showed the sensing ability for mercury as well as silver ions. In the presence of mercury ions, it showed a colorimetric response with an increase in fluorescence enhancement. It showed LOD of 0.21 μM for the mercury ions. Also, this sensor showed high selectivity for silver ions with a low limit of detection in the nanomolar range, i.e., 9 nM. In addition, this sensor can detect mercury and silver ions in L-02 human liver cells (L-02) (Shen et al., 2016).

8.5.1.5 Detection of Gold Ions

Since earlier in history, gold is an important metal and has been used as a material for jewelry, coinage, and other arts. Gold's distinct chemical and physical attributes have also led to numerous applications such as electronic materials, catalysts, sensing materials, medicines, and so on. Gold ions have been used as catalysts for the activation of C-C triple bonds. Some diseases have also been treated with gold-based molecules, including malaria, asthma, brain lesions, rheumatoid arthritis, tuberculosis, cancer, and HIV. In spite of their various applications, gold ions are extremely harmful due to their ability to strongly bind to DNA, causing cleavage, and may lead to causing harm to the kidneys, liver, and nervous system. Thus, developing efficient techniques to detect gold ions in both environmental and biological samples is extremely important (Zhang et al., 2016; Gu et al., 2015).

Srinivasulu Kambam et al. reported a fluorescein-based sensor (**FP15**) (Figure 8.5o) derived from fluorescein dicarboxyaldehyde. This sensor caused enhancement in fluorescence at 523 nm in the presence of gold ions. Also, it showed a linear relationship between fluorescence intensity and concentration of gold ions in the range of 0–60 μM. The detection limit was found to be 0.07 μM for the detection of gold ions. The fluorescence enhancement is due to the irreversible C=N bond hydrolysis that happened in the presence of gold ions with this probe, resulting in the fluorescent 4,5-fluorescein dicarboxyaldehyde. In addition, imaging experiments revealed that this sensor is cell permeable and can be applied to detect gold ions within living cells (Kambam et al., 2015).

Anna et al. reported a probe (**FP16**) (Figure 8.5n) by using diphenyl phosphino group linked with fluorescein using a linker o-phenylene group. This sensor showed a unique response toward gold ions based on the reaction and coordination mechanism. It showed a linear relationship in the range of 0–500 nM with an increase in fluorescence at 542 nm in the presence of gold ions. The LOD was found to be 76 nM for the detection of gold ions (Christianson et al., 2016).

8.5.1.6 Detection of Palladium Ions

Palladium metal has an important role as a catalyst in synthetic chemistry because of its selectivity, versatility, etc. This metal is used widely in the synthesis of active pharmaceutical ingredients (APIs). But the by-products of palladium are generally difficult to remove and also this is very toxic. So, it is very necessary to detect the presence of palladium ions (Yusop et al., 2011; Wiseman et al., 2009; Saleem et al., 2014).

Mithun Santra et al. reported fluorescein-based propargyl ether derivative (**FP17**) (Figure 8.5p) to detect palladium ions at room temperature. Notably, this sensor detects palladium species in all of the typical oxidation states (0, +2, and +4) without the use of any other reagents. This probe senses palladium ions due to the cleavage of propargyl ethers by depropargylation reaction. In addition, it can detect palladium ions in zebra fish (Santra et al., 2010).

Most of the probes showed some limitations like background fluorescence. To overcome this issue, Weston R. Kitley et al. reported a fluorescein-based probe (**FP18**) (Figure 8.5r) with a combination of allyl bromide and fluorescein. This probe is synthesized based on an etherification reaction. In the presence of palladium ions, this sensor showed fluorescence enhancement at 529 nm due to the cleavage of allyl ether. The LOD was found to be 4.6 nM for the detection of palladium ions with the linear range observed between 0 and 235 nm (Kitley et al., 2015).

Guo Wei et al. reported a novel fluorescein-based sensor (**FP19**) (Figure 8.5q) for the detection of palladium ions. This sensor in the presence of other metal ions shows very weak fluorescence because of protected hydroxyl groups by alkyl groups. But in the presence of palladium ions, this protected -OH group hydrolyzes and results in the change in fluorescence. This sensor showed high selectivity and sensitivity only in the presence of copper, which acts as synergic trigger to detect palladium ions (Kitley et al., 2015; Wei et al., 2012).

8.5.1.7 Detection of Mercury Ions

Mercury is regarded as a major toxic metal because it can harm the CNS, kidneys, brain, endocrine system, etc. Even a trace amount of this metal ion is toxic, so the detection of this metal ion in a biological system is very needful. Previously, most of the probes were reported based on rhodamine, napthalimide, and BODIPY. But these probes are unsuitable due to some limitations like water solubility is not good. To overcome these limitations, water-soluble fluorescein-based probes were synthesized.

Pornthip et al. reported a sensor based on fluorescein dithia-cyclic skeleton for the detection of mercury ions. This probe showed a fluorescence on-off mechanism toward the mercury ion. It showed the selectivity for mercury ions with fluorescence quenching and a change in color from yellow to orange. Also, it showed high sensitivity and selectivity for mercury ions over other different metal ions. The 1.5 ppb detection limit was observed for mercury ions. This sensor can be applied for the detection of mercury ions in drinking water samples (Piyanuch et al., 2016).

Wang et al. reported a fluorescein-based probe by introducing thiol-DNA-functionalized gold nanoparticles. Upon interaction with mercury ion, this sensor formed the hairpin-like structure of ssDNA due to thymine-mercury-thymine coordination. Also, fluorescence quenching was observed because of the FRET mechanism between the acceptor gold nanoparticles and donor fluorescein. The limit of detection was found to be 8 nM for mercury ions. This probe can be applied for mercury ion detection in tap water samples (Dong et al., 2018).

Feng et al. synthesized a novel water-soluble thiooxo fluorescein-based fluorescent probe for the detection of mercury ions. In the presence of mercury ions, this probe showed fluorescence enhancement due to the transformation of the spiro-lactum ring form of fluorescein to the ring-opened amide form. Also, this probe can detect mercury ions through the naked eye. The detection limit was found to be 39 nM. Additionally, this probe had the ability to detect mercury ions in vivo (Feng et al., 2017).

8.5.1.8 Detection of Iron Ions

Iron plays a very important role in various biological processes. Its deficiency causes different diseases, namely sickle cell anemia, iron deficiency anemia, hemochromatosis, and thalassemia. So, the detection of iron ions is very needful. Yan et al. reported a fluorescein-based sensor based on the Schiff base reaction between fluorescein hydrazine and benzothiazole. It displayed the response for iron ions via the naked eyes. This sensor opens up the ring

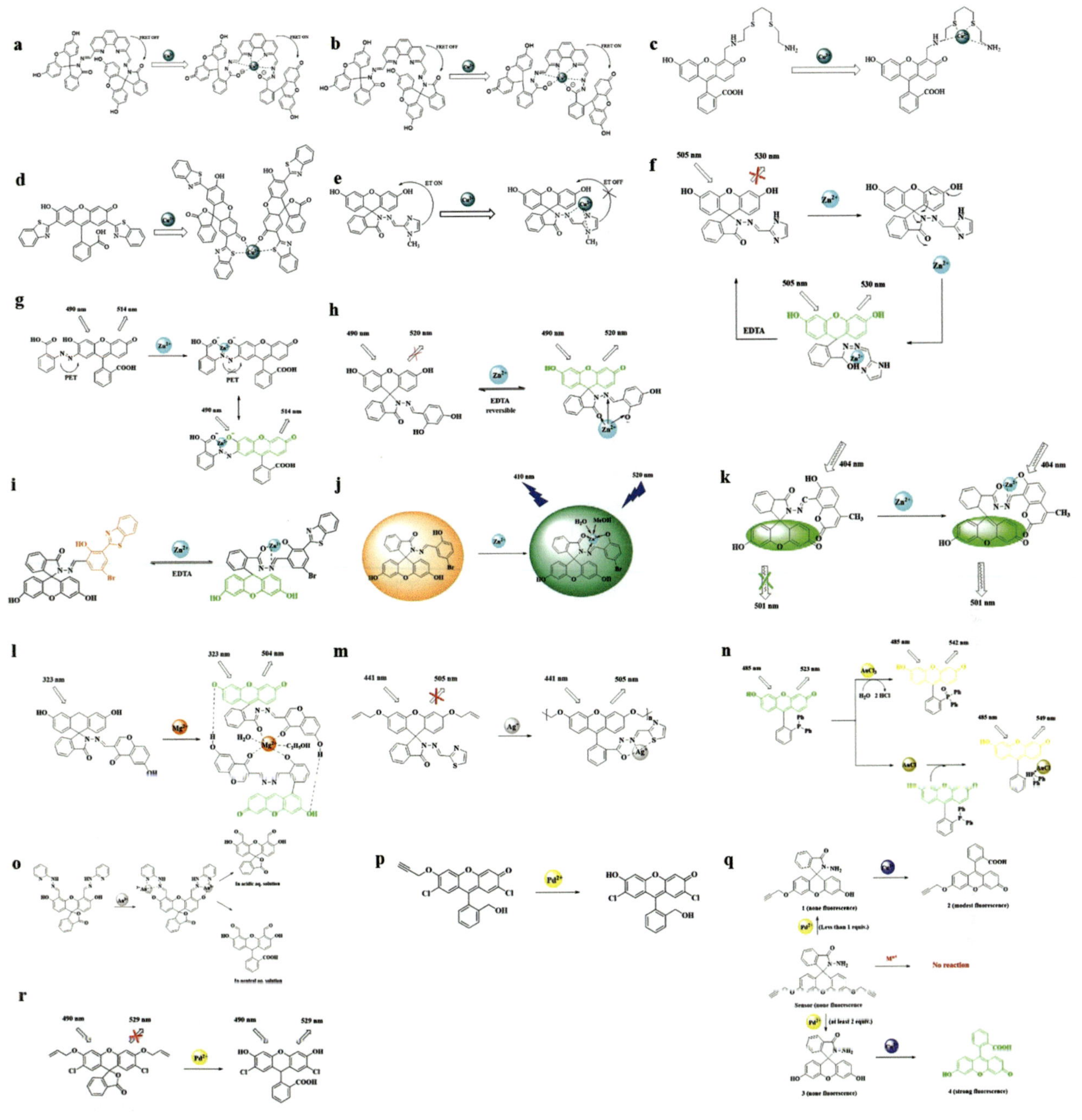

FIGURE 8.5 Structure of probes for detection of metal ions.

in the presence of iron ions with an increase in fluorescence enhancement. The LOD was found to be 7.4 nM for the detection of iron ions. In addition, this probe showed a low cytotoxic nature. Due to its good cell membrane permeability, it has the capacity to detect iron ions in Hep G2 living cells (Gao et al., 2016).

8.5.2 Detection of Anions

8.5.2.1 Detection of Fluoride Ions

Fluoride ions play an important role in a wide variety of applications such as chemical, biological, industrial, and environmental activities. Its optimum level is very essential to prevent osteoporosis and dental caries. On the other hand, its high level results in kidney damage and also gastric problems. According to the WHO, the optimum limit of F^- in drinking water for human health is 1.5 mg L^{-1} (79 µmol L^{-1}) (Liu et al., 2013).

Sharad Kumar et al. reported a reaction-based fluorescein sensor (**FP20**) (Figure 8.6a) for the selective detection of F^- ions. This sensor consists of signaling unit fluorescein and masking unit 4-nitrobenzene sulfonyl chloride. Sensing of fluoride ions is caused by two processes, firstly by addition of F^- ions, which deprotects N-bromo succinimide (NBS), which results in the monoionic form of fluorescein that again gets deprotonated by the addition of F^-, which results in the dionic form of fluorescein of high fluorescence. This sensor showed high selectivity for F^- over thiols or sulfides with detection limits of 4.6×10^{-7} M and 2.75×10^{-8} M determined by UV and fluorescence spectroscopy, respectively (Asthana et al., 2014).

Hong Yeong Kim et al. developed a chromogenic and fluorogenic sensor (**FP21**) (Figure 8.6b) for the selective detection of F^- based on a fluorescein scaffold. This sensor is synthesized from dichlorofluorescein with protecting group dimethyl phosphinyl that detached easily by the addition of fluoride ions. Sensing of F^- was achieved due to the cleavage of the protecting group upon the addition of F^- with the regeneration of dichloro fluorescein, which results in the enhancement in fluorescence. The interference of Hg^{2+} and Cu^{2+} is suppressed by the masking agent diethylene triamine pentascetic acid (DTPA). This sensor has the potential to be useful as a dual signaling system for the detection of F^-. The limit of detection was found to be 9.8 nM for F^- (Kim et al., 2015).

Shan Jiao et al. synthesized the fluorescent sensor (**FP22**) (Figure 8.6c) by combining fluorescein and coumarin moiety to detect F^-. In the designing of the probe, tert-butyldiphenylsilyl was selected as a functional unit as it shows the selectivity for F^- ions. The detection was caused by the desilyation reaction (i.e., breaking of O-Si bond) that happened in the presence of F^- ions. A good linear relationship was observed between the fluorescence intensity and concentration in the range of 0–20 µmol L^{-1}. The low detection limit was observed with 0.025 µmol L^{-1} for the detection of F^- ions. In addition, this sensor showed the potential to detect F^- ions in zebra fish and living cells (Jiao et al., 2018).

8.5.2.2 Detection of Hypochlorite Anion

At volumes in the range of 102–105 mol/L, hypochlorite is extensively utilized for disinfection and as a bleaching agent. Its protonated form (i.e., hypochlorous acid, HClO) is a biologically significant reactive oxygen species (ROS) that serves as an antimicrobial agent for water treatment and in organisms. The excessive presence of these ions (HClO) can cause several diseases such as neuron degeneration, atherosclerosis, hepatic ischemia-reperfusion, rheumatoid arthritis, and cancer. So, sensitive and selective probes are needed for hypochlorite (ClO^-) ion detection (Taheri et al., 2019; Samanta et al., 2019; Xu et al., 2013).

Ning Wang et al. developed a fluorescein-based probe (**FP23**) (Figure 8.6d) for the detection of OCl^-. This probe was designed by combining two moieties, such as fluorescein and carbazole. In the presence of OCl^-, C=N bond breaks and results in the opening of a ring structure. The recognition occurs by an oxidation reaction favored by OCl^-. The NH_2 group present in the probe oxidized to the imino group, which further results in enhancement in fluorescence. The probe showed a good linear relationship with the low LOD of 0.056 µmol/L. Moreover, the probe showed low toxicity and cell permeability. So, the probe showed an ability to identify OCl^- endogenously and exogenously in RAW264.7 cells. Also, good compatibility was detected for OCl^- in zebra fish organs (Wang et al., 2020).

To overcome the background fluorescence, Jing Lv et al. reported two fluorescein-based probes, **FP24 and FP25**, respectively (Figure 8.6e). These probes showed rapid response toward $HOCl/OCl^-$ due to the breakage of C=N bond and PET process inhibition between diaminonapthalene and fluorescein moieties. This process causes enhancement in fluorescence. Also, a good linear relationship was observed between fluorescence intensity and concentration of $HOCl/OCl^-$ in the range of 0–10 µM. The probes showed low detection limits of 0.21 and 0.23 µM, for **FP24 and FP25**, respectively. In addition, these sensors showed cell permeability with the detection of $HOCl/OCl^-$ in living cells (Lv, 2017).

In the year 2013, Xilang Jin et al. synthesized two probes (**FP26**) (Figure 8.6f) based on a fluorescein scaffold and evaluated them for OCl^-. Both the probes have shown sensing ability for OCl^- based on the oxidation reaction. The linear relationship was determined between fluorescence intensity and concentration in the range of 1–10 µM. Also, the detection limit was found to be 20 nM. These sensors detect OCl^- qualitatively and quantitively in drinking and tap water samples. Moreover, sensing was also identified for OCl^- in *Rhodobacter ferroxidans* sp. SW2 (Jin, 2013).

Xiaohong Cheng et al. reported a probe (**FP27**) (Figure 8.6g) consisting of fluorescein dye and protecting group oxime as a reactive moiety. This probe showed high selectivity and sensitivity over other ions. The sensing was confirmed for OCl^- through deimination reaction (removal of C=N bond) with the development of turn-on fluorescence. This sensor showed the change in color from colorless to light yellow via naked eye, while in a normal UV lamp, fluorescence change was observed with nonfluorescent to strong green color. Also, this sensor showed potential for detecting OCl^- in living cells and in vivo (Cheng et al., 2011).

Lei Jin et al. synthesized the probe (**FP28**) (Figure 8.6h) for specific detection of OCl^-. The recognition mechanism was achieved through observing change in color with the addition of OCl^-. Color change from colorless to light yellow and colorless to pink was detected by UV lamp at 365 nm and naked eye, respectively. It showed low LOD with 0.023 µM. The monitoring of OCl^- was successfully identified in HeLa cells by this probe (Jin et al., 2013).

Yi Zhou et al. developed three fluorescein-based sensors (**FP29**) (Figure 8.6j). Among them, sensor **FP29** showed sensing ability toward HOCl based on oxidation reaction. It showed a fast response within 20 minutes with the enhancement in fluorescence (1,643.4 fold). Also, a lower detection limit was found with 0.71 ppb for the detection of HOCl. In addition, this sensor could detect HOCl in living cells and zebra fish organs (Zhou et al., 2012).

To detect hypochlorite ions, Fang-Jun Huo et al. reported a fluorescein-based colorimetric and fluorescent sensor. Upon the addition of ClO^-, the sensor forms a complex that occurs due to the oxidization of Cu^+ to Cu^{2+}. This complex results in strong green fluorescence. This was the first fluorescein-based sensor that was reported for the detection of HOCl or ClO^- in the aqueous system. In addition, this sensor could detect ClO^- in tap water (Huo et al., 2012).

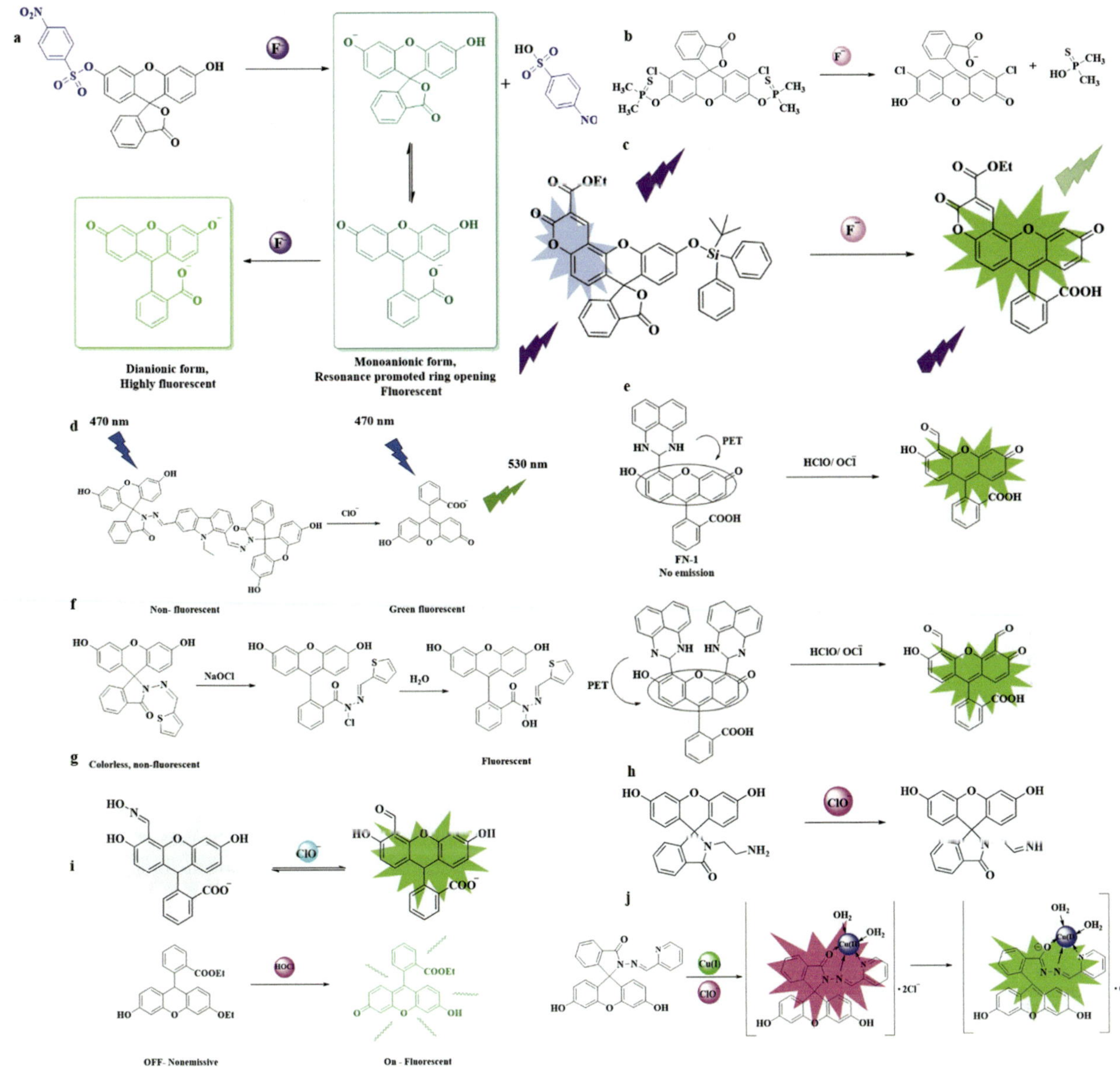

FIGURE 8.6 Structure of probes for detection of anions.

8.5.3 Detection of Two Cations

8.5.3.1 Detection of Cu^{2+} and Al^{3+}

For rapid detection and low cost, a single fluorescent sensor with multianalysis has gained much attention nowadays. Only a few fluorescein-based sensors are available that give simultaneous detection of two cations. Lingjie Hou et al. reported a colorimetric and fluorometric fluorescein-based probe (**FP30**) (Figure 8.7a) for the detection of both ions, i.e., Cu^{2+} and Al^{3+}. This sensor shows a selective response to these ions in two solvent systems with different binding modes and fluorescent signals. This sensor when exposed to Cu^{2+} results in a change in color from pink to orange-yellow with significant enhancement in fluorescence at 520 nm with the binding ratio of 1:2, while with exposure of Al^{3+}, color change from pink to yellow with enhancement in fluorescence at 440 nm with the binding ratio of 1:1 was detected. This sensor shows the limits of detection of 1.47 × 10^{-8} M and 7.32 × 10^{-8} M for Cu^{2+} and Al^{3+}, respectively. Moreover, this sensor can detect Cu^{2+} and Al^{3+}in living cells (Hou et al., 2017).

8.5.3.2 Detection of Zn^{2+} and Al^{3+}

Gang Zhao et al. developed a multiple turn-on fluorescein-based probe for the simultaneous identification of Zn^{2+} and Al^{3+}. This probe is synthesized using fluorescein, cyanuric chloride, hydrazone, and salicylaldehyde. In the presence of Zn^{2+}, a sensor with nonfluorescent properties shows strong green fluorescence while with Al^{3+}, blue fluorescence was detected. The detection limits for Zn^{2+} and Al^{3+} were found to be 7.90 × 10^{-8} M and 5.37 × 10^{-8} M, respectively. Furthermore, this probe shows low toxicity, so it can be applicable for quantitative detection of Zn^{2+} and Al^{3+} (Zhao et al., 2020).

8.5.3.3 Detection of Cd^{2+} and Co^{2+}

Kaur et al. developed a novel fluorescent probe (**FP31**) (Figure 8.7b) for detecting Cd (II) and Co (II) metal ions using a chemosensor called 1,2,3-triazole-linked fluorescein. This sensor exhibited a remarkable "turn-off" response with Cd^{2+} ions and a "turn-on" response with Co (II) ions. The motivation behind developing this sensor was the high toxicity of Cd (II) and Co (II) ions in water as well as the challenge of differentiating between them in a mixture of metal ions using the naked eye in an aqueous solution. The 1,2,3-triazole-linked fluorescein probe, which incorporates a fluorescein backbone, demonstrated excellent selectivity and specificity. The selectivity and specificity of the **FP31** probe can be attributed to a mechanism called photo electron transfer (PET). This mechanism involves the transfer of electrons between the **FP31** probe and the metal ions, resulting in distinct "on-off" responses for Cd^{2+} and Co^{2+}ions. The sensor shows the detection limit for Cd^{2+} and Co^{2+} within 10 μM (Kaur et al., 2019).

8.5.3.4 Detection of Zn^{2+} and Hg^{2+}

Serkan Erdemir et al. reported a novel fluorescent sensor (**FP32**) (Figure 8.7c) based on phenolphthalein-fluorescein that enables the detection of Zn^{2+} and Hg^{2+} ions through different binding modes. This sensor exhibited selective binding to Zn^{2+} and Hg^{2+}, leading to significant fluorescence enhancement at distinct emission wavelengths. This fluorescence enhancement demonstrated the potential of the sensor as a selective and sensitive fluorescent sensor for Zn^{2+} and Hg^{2+} ions. Based on a Job's plot analysis, the coordination mode between **FP32** and Zn^{2+} and Hg^{2+} was proposed to be in a 1:2 ratio. The association constants for the binding of **FP32** with Zn^{2+} and Hg^{2+} were determined to be $6.45 \times 10^{10}\ M^{-2}$ and $2.11 \times 10^{10}\ M^{-2}$, respectively. The practical application of the probe for detecting Zn^{2+} and Hg^{2+} ions in water samples demonstrated its high accuracy and potential. This suggests that this sensor holds significance in analysis and detection, which indicates its promising future applications (Erdemir and Kocyigit, 2017).

8.5.4 Detection of Both Anions and Cations

8.5.4.1 Detection of Al^{3+} and F^-

Diao et al. reported fluorescein-based probe (**FP33**) (Figure 8.8) consisting of fluorescein monoaldehyde and picolylamine. The picolylamine group does not contain a loan pair, which is responsible for PET mechanism in the molecule. This probe showed quenching in fluorescence with F^- ions due to electron transfer, while in the presence of Al^{3+}, the fluorescence intensity of molecules gets reconstructed. In addition, this probe can detect F^- and Al^{3+} in HeLa cells (Diao et al., 2016).

8.5.4.2 Detection of Hg^{2+} and F^-

Shen et al. reported a ratiometric fluorescent probe based on a fluorescein scaffold to detect both Hg^{2+} and F^-. This probe forms a complex with Hg^{2+} on thiourea moiety that leads to fluorescence and absorbance properties. The probe causes ring opening structure with the cleavage of the silyl ether bond due to FRET transfer from fluorescein to the

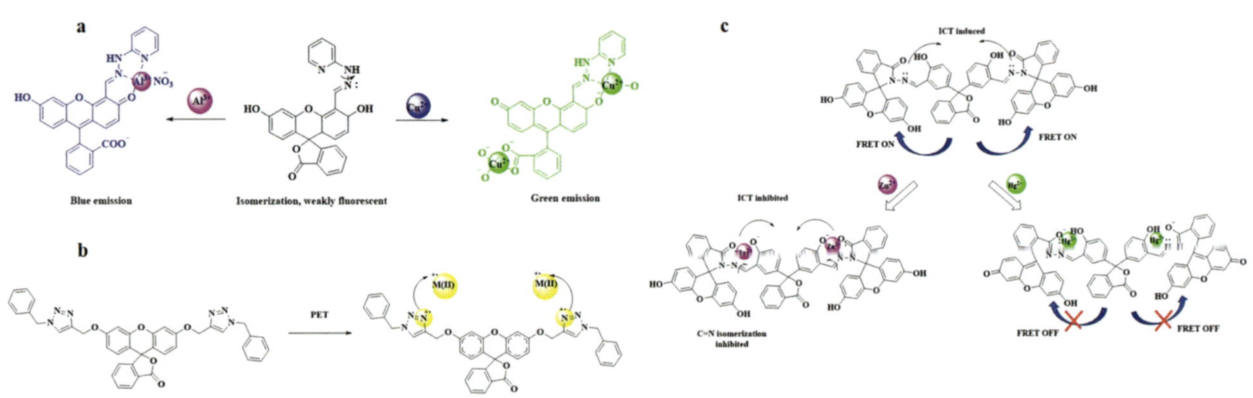

FIGURE 8.7 Structure of probes for detection of two cations.

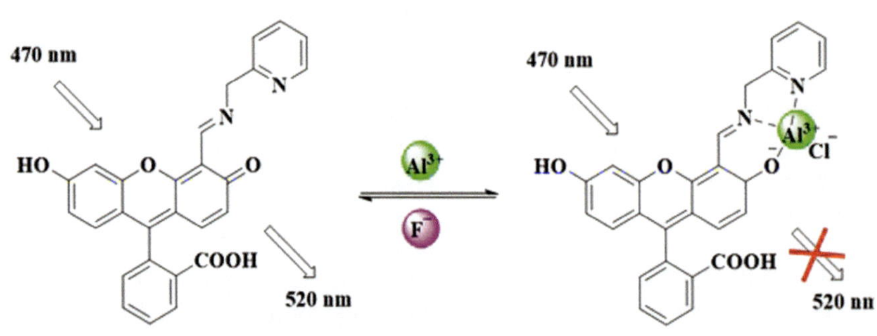

FIGURE 8.8 Structure of probes for detection of cations and anions.

rhodamine group. The LODs were found to be 51 and 5.4 nM, respectively, for F^- and Hg^{2+}. Furthermore, this can detect Hg^{2+} and F^- in W138 normal lung fibroblast cells (Chereddy et al., 2015).

8.5.5 Detection of Small Molecules

The alkylation and esterification with fluorescein increases the lipophilicity of probes based on fluorescein. The molecules, namely nitric oxide, ROS, hydrazine, hydrogen sulfide, and amino acids, can break ester bonds and allow the conjugation of molecules that further cause changes in the fluorescence. In comparison to ions, the detection of small biomolecules is very complex. Because of the highly complex structure and their intrinsic properties, various strategies are developed to detect these biomolecules. It aids in the comprehension of various physiological and toxicological properties of molecules.

8.5.5.1 Detection of Sulfur Group-Containing Analytes

8.5.5.1.1 Detection of Cysteine

To detect cysteine, Meng et al. reported a ratiometric fluorescent probe (**FP34**) (Figure 8.9a) based on a fluorescein scaffold. The probe consists of donor and acceptor, acryloyl and BODIPY, respectively. The probe itself has emission due to the BODIPY group present on it. Upon interaction with cysteine, deacylation reaction caused due to FRET results in green fluorescence. A good linear relationship was obtained in the range of 10–600 mM. The detection limit for cysteine was found to be 0.1 mM (Ma et al., 2015).

By combining coumarin and fluorescein moiety, Peng et al. developed a ratiometric and chromogenic probe (**FP35**) (Figure 8.9b) for the selective detection of cysteine. With the presence of cysteine, fluorescence gets turned on with the changes in color from colorless to yellow. The LOD for cysteine was 0.084 μM. Also, the probe can be applied for the detection of cysteine in HepG2 cells and human blood plasma (Figure 8.9(ii)) (Fu et al., 2017).

By modifying fluorescein with bromoacetyl ester and aldehyde, Lee et al. synthesized a fluorescent probe (**FP36**) (Figure 8.9c) to specifically detect cysteine. The interaction of the probe with cysteine causes cleavage of the ester bond of nonfluorescent bromoacetyl group resulting in a strong fluorescent compound. The reaction of the aldehyde group with cysteine causes the generation of oxazolidine. The LOD for cysteine was found to be 0.51 μM. Moreover, this probe could detect cysteine in cellular regions (Lee et al., 2008).

To trace thiol group, Meng et al. synthesized fluorescein-based sensor (**FP37**) (Figure 8.9d). This probe was synthesized by reaction between 4-nitrobenzenesulfonyl (DNBS) and fluorescein methyl ester. The probe upon reaction with RSH showed strong fluorescence at 521 nm. The LOD was found to be 0.16 μM. Also, it showed potential for bioimaging of -SH (Liu et al., 2016).

8.5.5.1.2 Detection of Hydrogen Sulfide

Hydrogen sulfide (H_2S) is the third most important gasotransmitter. It has many roles, i.e., gastrointestinal functions. It is also used in signaling molecule in cells, cardiovascular system, and nervous system. But due to its toxic nature, it is very harmful to humans. It damages the functioning of body organs and causes diseases such as inflammation, cardiovascular diseases, asthma, etc. So, the detection of H_2S is very essential (Saha et al., 2015).

Zhang et al. developed a fluorescein-based reactive probe (**FP38**) (Figure 8.9e) with the incorporation of 2,4-dinitrobenzene sulfonyl chloride and fluorescein dicarboxaldehyde for the detection of H_2S. Upon addition of H_2S, ring opening is caused due to nucleophilic reaction between H_2S and benzene sulfonate group that results in enhancement in fluorescence. This probe turns yellow in the presence of H_2S.

A good linear relationship was detected in the range of 0–10 μM with LOD of 0.024 μM. In addition, this probe could detect H_2S in HeLa cells (Lv et al., 2017).

Wu et al. developed a probe (**FP39**) (Figure 8.9g) based on a substitution reaction between 2-thiophenecarbonyl chloride and -OH group of fluorescein. It detects H_2S selectively in the linear range of 20–100 μM. Moreover, it is able to detect H_2S in the biological bovine plasma system (Jin et al., 2016).

8.5.5.2 Detection of Carbon Monoxide

Carbon monoxide (CO) is a molecule that was traditionally considered unimportant until it was recently identified as a mediator in cellular functions and recognized for its biological importance. It has been found to play a significant role in inflammatory and cardiovascular diseases, leading to its utilization in the field of drug design. The Pd (palladium)-mediated Tsuji-Trost reaction mechanism is commonly observed in the development of CO sensors. Due to the challenges associated with directly using CO molecules for studies, a CO-releasing molecule called CO-RM-3 is employed. CO-RM-3 is designed to release controlled amounts of carbon monoxide in a precise manner, allowing researchers to study the effects of CO in various biological systems (Heinemann et al., 2014; Romao et al., 2012).

In a study conducted by Feng et al., they developed a fluorescent probe (**FP40**) (Figure 8.9f) called FL-CO-1 (F) utilizing allyl chloroformate as a mediator. $PdCl_2$ was included as an additive trap for CO. FL-CO-1 is initially nonfluorescent in the presence of Pd^{2+} but exhibits fluorescence when Pd^{2+} is reduced to Pd^0. This reduction is facilitated by the presence of CO. The reduction of Pd^{2+} to Pd^0 by CO promotes the Tsuji-Trost reaction, a palladium-catalyzed reaction. As a result of this reaction, a highly fluorescent molecule called fluorescein is released. This fluorescence can be observed and measured, providing a means to detect and study the presence and activity of CO. The development of FL-CO-1 and its fluorescent properties with the reduction of Pd^{2+} to Pd^0 by CO offers a valuable tool for investigating CO-mediated processes and studying

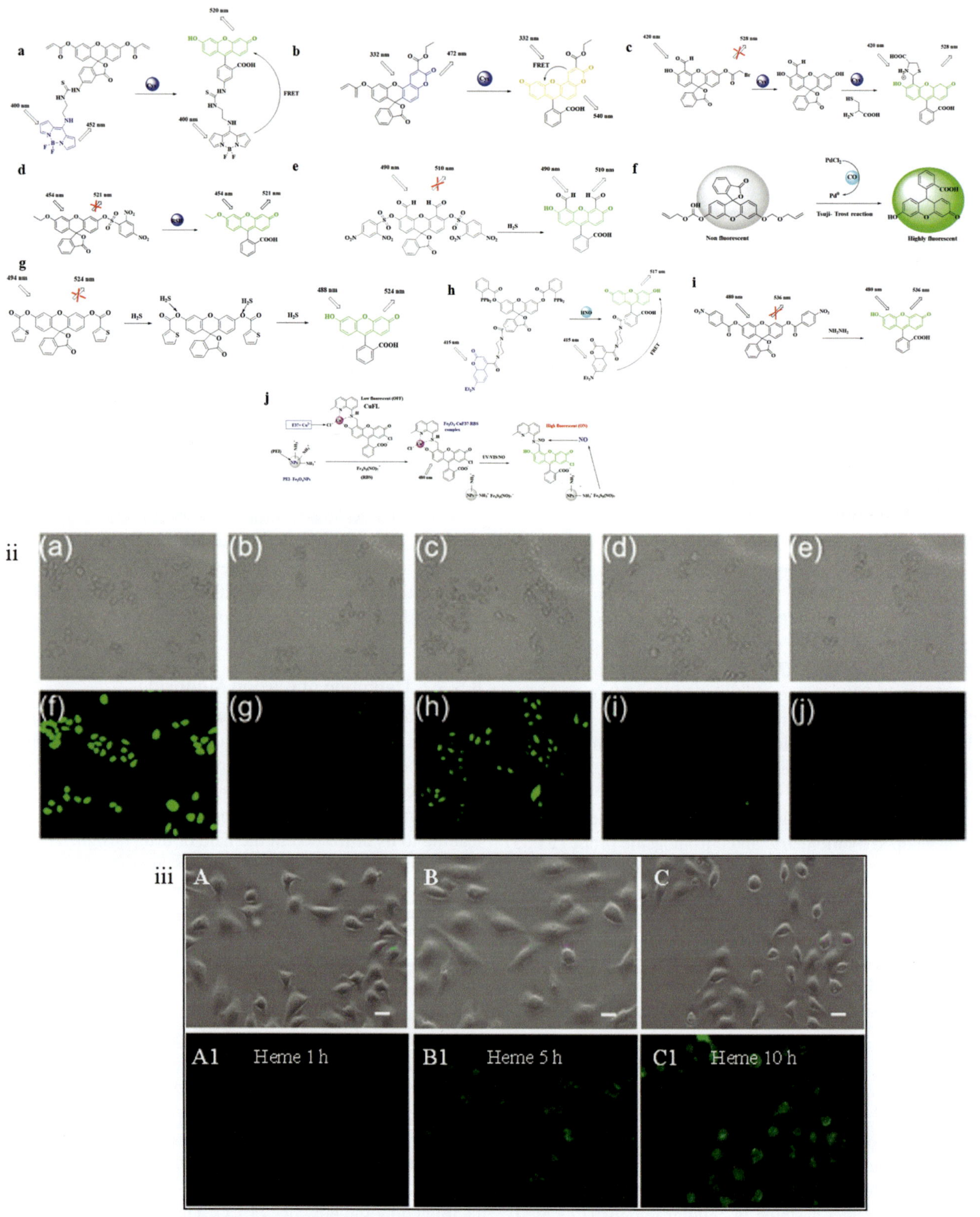

FIGURE 8.9 (i) Structure of probes for detection of sulfur group-containing compounds, carbon monoxide and nitric oxide. (ii) HepG2 cell images: (a) bright field and (f) fluorescence images of HepG2 cells with FP46 (10.0 μM) for 20 minutes; (b) bright field and (g) fluorescence images of HepG2 cells incubated with NEM (50.0 μM) for 30 minutes, and then incubated with FP46 (10.0 μM) for 20 minutes; (c, d, e) bright field and (h, i, j) fluorescence images of HepG2 cells incubated with NEM (50.0 μM) for 30 minutes, then incubated with Cys, Hcy, and GSH (100.0 μM) for 30 minutes, and finally incubated with probe 2 (10.0 μM) for 20 minutes. (iii) Fluorescent imaging of CO produced via heme stimulation in HeLa cells using the probe system (FP51 + $PdCl_2$). Reprinted (adapted) with permission from Fu et al. (2017). Copyright 2017, American Chemical Society.Reprinted (adapted) with permission from Feng et al. (2016). Copyright 2016, American Chemical Society.

the mechanisms underlying CO-related biological functions. Since it was reported that heme can stimulate more hemeoxygenase (HO)-derived CO in living cells, the capability of the probe to detect the endogenous produced CO by heme treatment was studied in HeLa cells (Figure 8.9 (iii)) (Feng et al., 2016).

In response to the instability of the allylcarbonate protection group in FL-CO-1, Shumin Feng et al. made further modifications to enhance the stability of the probe. They replaced the allylcarbonate group with an allyl ether group, resulting in the development of probes. By incorporating the allyl ether protection group, the stability of the probe was improved. Similar to FL-CO-1, the allyl group in probes a and b can still be removed through a Pd^0-mediated Tsuji-Trost reaction. This reaction enables the release of the desired fluorescent molecule, fluorescein, upon reduction of Pd^{2+} to Pd^0 by CO. The use of the allyl ether protection group addresses the stability issues associated with the allylcarbonate group, allowing for improved performance and reliability of the probes in studying CO-mediated processes and detecting CO activity (Feng et al., 2017).

8.5.5.3 Detection of Nitric Oxide

Nitric oxide is synthesized in endothelial cells and serves an important role in various biological processes. It plays an important role as a γ -sensitizer that kills neoplastic tissues. Recently, researchers focused on detecting NO sources for the storage of NO. Based on the need for the detection of NO, Jiang et al. synthesized a fluorescein-based probe (**FP41**) (Figure 8.9j). For the preparation of **FP41** probe, Fe_3O_4-CuFl-RBS nanocomposites and polyethyleneimine-stabilized Fe_3O_4 magnetic nanoparticles were electrostatically assembled with light-sensitive NO donors, $Fe_4S_3NO_7$, RBS, and Cu-fluorescein complex (Cu-Fl). This complex (Cu-Fl) has turn-off fluorescence property while when reacted with NO, it results in a significant increase in fluorescence. This change occurs due to nitrosation of secondary amine on ligand CuFl that yields N-nitrosated ligand (Fl-NO). Under 480 nm light, the prepared nanocomposites exhibit weak fluorescence and barely release NO, while under 365 nm light, these nanocomposites release NO via RBS photolysis. This released NO then conjugates with CuFl and forms Fl-NO complex, which was confirmed with the increase in fluorescence of Fl-NO. It showed a good linear relationship in the range of 0.01–0.25 μM with LOD 6 nM (Jiang et al., 2015).

To detect nitroxyl (HNO), Huatang Zhang et al. synthesized a ratiometric fluorescent probe (**FP42**) (Figure 8.9h). This probe is synthesized based on a reaction between 5(6)-carboxyfluorescein, coumarin piperazine, and 2-diphenylphosphanyl benzoic acid. Here fluorescein is used as an acceptor while coumarin as a donor moiety. Firstly, the reaction between HNO and phosphine leads to the formation of aza-ylide intermediate. Then, aza-ylide attacks the adjacent electrophilic ester moiety causing intramolecular interaction that leads to release of alcohol with turn-on fluorescence at 517 nm. Also, the color change was observed from blue to green after the reaction of the probe with HNO. A good linear relationship was detected in the range of 0–100 μM. In addition, this probe can be applied for bioimaging of HNO in living cells (Zhang et al., 2015).

8.5.6 Detection of Reactive Oxygen Species

Reactive oxygen species (ROS) such as O_2, $OH^\bullet$, and H_2O_2, which constitute significantly for transmitting signals, and are utilized to regulate a variety of physiological processes. The excessive concentration of these species leads to oxidative stress. By modifying sites on fluorescein moiety, it is possible to use fluorescein derivatives for the detection of ROS.

8.5.6.1 Detection of O_2

Zhang et al. reported a probe (**FP43**) (Figure 8.10) by connecting aryl triflate with fluorous substituted fluorescein to detect O^{2-}. Upon adding O^{2-}, ring opening is caused by the hydrolysis of the ester bond. This process leads to enhancement in fluorescence.

The LOD was found to be 23 nM with good linear relationship in the range of 0–12 μM. Moreover, this probe can detect O_2 in zebra fish organs (Hu et al., 2015).

8.5.6.2 Detection of Hydrogen Peroxide

Hydrogen peroxide has a short lifespan, and it is very difficult to detect this species due to its low concentration. To accurately detect this species, Qiao et al. develop an

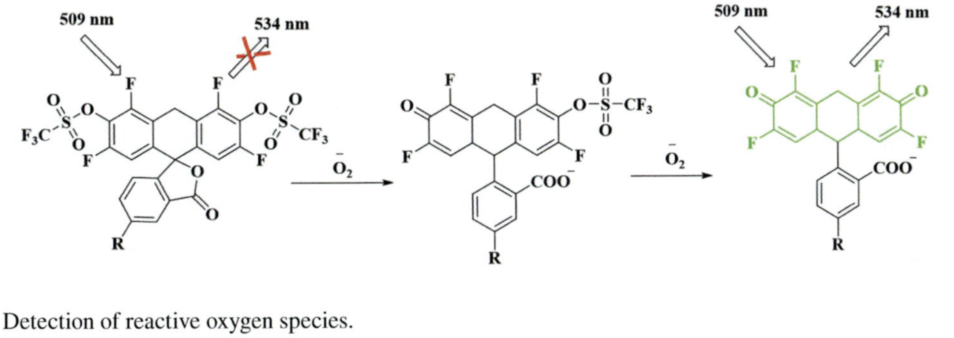

FIGURE 8.10 Detection of reactive oxygen species.

efficient ratiometric probe. This probe consists of tetraphenylethene as a donor, fluorescent boronate as an acceptor, self-assembled polymer micelles as a carrier, and triphenylphosphonium as a mitochondria-targeted moiety. With the interaction of H_2O_2, the probe transfers the boronate into phenol that causes lactone ring opening, which results in a strong green fluorescence product. The LOD to detect H_2O_2 was found to be 0.95 μM. Furthermore, it can be applicable to detect H_2O_2 in living cells exogenously and endogenously (Qiao et al., 2015).

8.5.7 Detection of Hydrazine

Only few fluorescent probes are available based on fluorescein for the detection of hydrazine hydrate. It can be absorbed easily through inhalation, oral, and dermal exposure. Fluorescein-based probes detect hydrazine due to hydrolysis of ester bonds that leads to changes in fluorescence. As the nature of NH_2NH_2 is toxic, detection is very needful. So, Li et al. reported a probe (**FP44**) (Figure 8.11a and 8.9i) by connecting p-nitrobenzoic acid and fluorescein. This probe showed great sensitivity and selectivity for hydrazine. Also, the enhancement in fluorescence with the color change was observed from colorless to yellow in the presence of NH_2NH_2. The LOD for NH_2NH_2 was found to be 46 nM. This probe can also be applied to detect NH_2NH_2 in living cells (Li et al., 2017).

8.5.8 Detection of Adenosine Triphosphate

As a major transporter of chemical energy in living organisms, adenosine triphosphate (**ATP**) plays a role as a mediator of chemo-sensing transmission in the central nervous system. Aptamers are very short, artificial single-stranded DNA fragments that have excellent affinity for target molecules and are thought to be effective in the detection and diagnostic tools. To detect this major transporter, Wen et al. developed a probe (**FP45**) (Figure 8.11b) based on graphene platform. The probe consists of graphene, fluorescein isothiocyanate, and Exo-III. In the presence of ATP, the probe binds with aptamer and forms a complex with it. Then, Exo-III specifically catalyzes the removal of mononucleotides from a complex of an aptamer that results in the removal of the fluorescein isothiocyanate-labeled ATP aptamer probe. Then this cycle gets repeated and causes continuous cleavage of aptamer targets. Here, graphene is used for fluorescence quenching. A good linear relationship was observed in the range of 10–1000 nM. The LOD to detect ATP was calculated to be 5 nM. In addition, this sensor could detect ATP in human serum samples (Wen et al., 2015).

8.5.9 Detection of Inorganic Phosphate (Pi)

Inorganic phosphate is ubiquitous in biological systems. Its extreme level leads to abnormal renal functioning. To detect the presence of Pi, Meng et al. reported a fluorescent probe that consists of fluorescein monoaldehyde and aminopyridine. When the probe forms the complex with Fe^{3+} ions at the oxygen site of fluorescein and the nitrogen site of aminopyridine, fluorescence gets quenched due to the nature of Fe^{3+}, i.e., paramagnetic. In the presence of Pi, fluorescence gets enhanced. A good linear relationship was obtained in the range of 0–2.5 μM. Furthermore, this probe can also detect Pi in living cells and aqueous systems (Meng et al., 2015).

8.5.10 Detection of Biomolecules

In biological systems, macromolecules such as carboxylesterase and *β*-lactamase play a vital role. But their insufficient or excess level leads to physiological problems. So, monitoring the level of these macromolecules is very needful. To monitor the level of carboxylesterase, Wu et al. designed and synthesized the fluorescein-based ratiometric probe (**FP46**) (Figure 8.12a). The presence of this enzyme is confirmed through the electrostatic interaction between TPE-N+ dots (cationic) and negatively charged fluorescein molecules, which results in the FRET mechanism. The reaction of a probe with carboxylesterase leads to enhancement in fluorescence at 520 nm. The detection limit was found to be 0.26 U/L with a linear range observed within 0–50 U/L (Wu et al., 2015).

To detect the level of *β*-lactamase, Chen et al. reported a fluorescent probe (**FP47**) (Figure 8.12 b) consisting of Tokyo green fluorescein and cephalosporin. Recognition of *β*-lactamase occurred through the hydrolysis of cephalosporin. Also, the *β*-lactam ring opening causes fragmentation that results in the breakage of the ether bond with the release of Tokyo green in free form that exhibited the emission at 525 nm. The presence of *β*-lactamase was confirmed through the recovery of Tokyo green. In the real

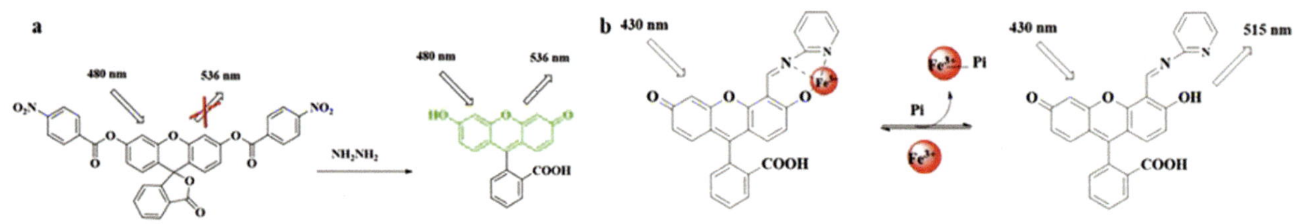

FIGURE 8.11 Detection of hydrazine, ATP, Pi.

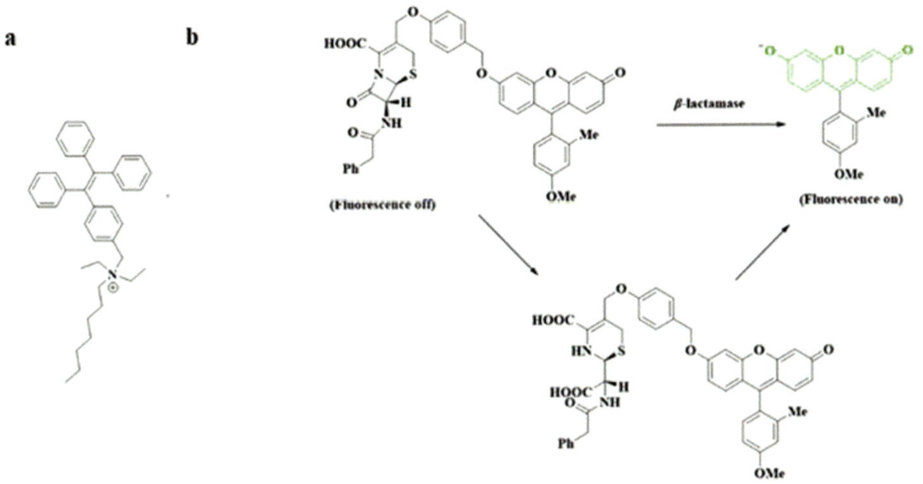

FIGURE 8.12 Detection of small molecules.

milk samples, this fluorescent probe can detect β-lactamase in 15 minutes. The detection limit provided by this probe was 10–50 U/mL. This probe detected β-lactamase with high accuracy and efficiency.

REFERENCES

An, J. M.; Yan, M. H.; Yang, Z. Y.; Li, T. R.; Zhou, Q. X. A Turn-on Fluorescent Sensor for Zn(II) Based on Fluorescein-Coumarin Conjugate. *Dyes and Pigments* **2013**, *99* (1), 1–5. https://doi.org/10.1016/j.dyepig.2013.04.018.

Anwar, M.; Nadeem, F.; Faiz, S. *A Review of Synthesis of Fluorescein Based Advanced Materials*; 2018; Vol. 14. www.iscientific.org/Journal.html.

Asthana, S. K.; Kumar, A.; Neeraj; Upadhyay, K. K. A Reaction Based Chromofluorogenic Turn-on Probe for Specific Detection of Fluoride over Sulfide/Thiols. *Tetrahedron Lett* **2014**, *55* (43), 5988–5992. https://doi.org/10.1016/j.tetlet.2014.09.051.

Beyersmann, D.; Haase, H. *Functions of Zinc in Signaling, Proliferation and Differentiation of Mammalian Cells*, Springer 2001; Vol. 14.

Cakic, M.; Mitic, Z.; Nikolic, G.; Savic, I.; Savic, I. M. Design and Optimization of Drugs Used to Treat Copper Deficiency. *Expert Opinion on Drug Discovery*, October **2013**, 1253–1263. https://doi.org/10.1517/17460441.2013.825245.

Chantalakana, K.; Choengchan, N.; Yingyuad, P.; Thongyoo, P. A Highly Selective "turn-on" Fluorescent Sensor for Zn2+ Based on Fluorescein Conjugates. *Tetrahedron Lett* **2016**, *57* (10), 1146–1149. https://doi.org/10.1016/j.tetlet.2016.01.106.

Chen, X.; Pradhan, T.; Wang, F.; Kim, J. S.; Yoon, J. Fluorescent Chemosensors Based on Spiroring-Opening of Xanthenes and Related Derivatives. *Chemical Reviews*, March 14, **2012**, 1910–1956. https://doi.org/10.1021/cr200201z.

Chen, Y.; Xianyu, Y.; Wu, J.; Zheng, W.; Rao, J.; Jiang, X. Point-of-Care Detection of β-Lactamase in Milk with a Universal Fluorogenic Probe. *Anal Chem* **2016**, *88* (11), 5605–5609. https://doi.org/10.1021/acs.analchem.6b01122.

Cheng, X.; Jia, H.; Long, T.; Feng, J.; Qin, J.; Li, Z. A "Turn-on" Fluorescent Probe for Hypochlorous Acid: Convenient Synthesis, Good Sensing Performance, and a New Design Strategy by the Removal of CN Isomerization. *Chemical Communications* **2011**, *47* (43), 11978–11980. https://doi.org/10.1039/c1cc15214a.

Chereddy, N. R.; Nagaraju, P.; Niladri Raju, M. V.; Saranraj, K.; Thennarasu, S.; Rao, V. J. A Two Fluorophore Embedded Probe for Collective and Ratiometric Detection of Hg2+ and F- Ions. *Dyes and Pigments* **2015**, *112*, 201–209. https://doi.org/10.1016/j.dyepig.2014.07.004.

Christianson, A. M.; Gabbaï, F. P. Synthesis and Coordination Chemistry of a Phosphine-Decorated Fluorescein: "Double Turn-On" Sensing of Gold(III) Ions in Water. *Inorg Chem* **2016**, *55* (12), 5828–5835. https://doi.org/10.1021/acs.inorgchem.6b00080.

Das, B.; Jana, A.; Mahapatra, A. Das; Chattopadhyay, D.; Dhara, A.; Mabhai, S.; Dey, S. Fluorescein Derived Schiff Base as Fluorimetric Zinc (II) Sensor via 'Turn on' Response and Its Application in Live Cell Imaging. *Spectrochim Acta A Mol Biomol Spectrosc* **2019**, *212*, 222–231. https://doi.org/10.1016/j.saa.2018.12.053.

Diao, Q.; Ma, P.; Lv, L.; Li, T.; Sun, Y.; Wang, X.; Song, D. A Water-Soluble and Reversible Fluorescent Probe for Al3+ and F- in Living Cells. *Sens Actuators B Chem* **2016**, *229*, 138–144. https://doi.org/10.1016/j.snb.2016.01.136.

Dong, K.; Zhou, J.; Yang, T.; Dai, S.; Tan, H.; Chen, Y.; Pan, H.; Chen, J.; Audit, B.; Zhang, S.; Xu, J. Sensitive Hg2+ Ion Detection Using Metal Enhanced Fluorescence of Novel Polyvinyl Pyrrolidone (PVP)-Templated Gold Nanoparticles. *Appl Spectrosc* **2018**, *72* (11), 1645–1652. https://doi.org/10.1177/0003702818775704.

Erdemir, S.; Kocyigit, O. A Novel Dye Based on Phenolphthalein-Fluorescein as a Fluorescent Probe for the Dual-Channel Detection of Hg2+ and Zn2+. *Dyes and Pigments* **2017**, *145*, 72–79. https://doi.org/10.1016/j.dyepig.2017.05.053.

Erdemir, S.; Tabakci, B. Selective and Sensitive Fluorescein-Benzothiazole Based Fluorescent Sensor for Zn2+ Ion in Aqueous Media. *J Fluoresc* **2017**, *27* (6), 2145–2152. https://doi.org/10.1007/s10895-017-2153-8.

Fan, C.; Luo, S.; Liu, R. Optimization of an Analytical Method for the Spectrophotometric Determination of Copper in Tea and Water Samples after Ultrasonic Assisted Cloud Point Extraction Using a Benzothiazole Fluorescein Derivative Complexing Agent. *RSC Adv* **2015**, *5* (80), 65321–65327. https://doi.org/10.1039/c5ra11212e.

Feng, S.; Liu, D.; Feng, W.; Feng, G. Allyl Fluorescein Ethers as Promising Fluorescent Probes for Carbon Monoxide Imaging in Living Cells. *Anal Chem* **2017**, *89* (6), 3754–3760. https://doi.org/10.1021/acs.analchem.7b00135.

Feng, W.; Liu, D.; Feng, S.; Feng, G. Readily Available Fluorescent Probe for Carbon Monoxide Imaging in Living Cells. *Anal Chem* **2016**, *88* (21), 10648–10653. https://doi.org/10.1021/acs.analchem.6b03073.

Feng, Y.; Kuai, Z.; Song, Y.; Guo, J.; Yang, Q.; Shan, Y.; Li, Y. A Novel "Turn-on" Thiooxofluorescein-Based Colorimetric and Fluorescent Sensor for Hg2+ and Its Application in Living Cells. *Talanta* **2017**, *170*, 103–110. https://doi.org/10.1016/j.talanta.2017.03.099.

For Fluoride Concentration in Drinking Water for the Prevention of Dental Caries Services Federal Panel on Community Water Fluoridation; Vol. 130.

Frederickson, C. J.; Koh, J. Y.; Bush, A. I. The Neurobiology of Zinc in Health and Disease. *Nature Reviews Neuroscience,* June **2005**, 449–462. https://doi.org/10.1038/nrn1671.

Fu, Z. H.; Han, X.; Shao, Y.; Fang, J.; Zhang, Z. H.; Wang, Y. W.; Peng, Y. Fluorescein-Based Chromogenic and Ratiometric Fluorescence Probe for Highly Selective Detection of Cysteine and Its Application in Bioimaging. *Anal Chem* **2017**, *89* (3), 1937–1944. https://doi.org/10.1021/acs.analchem.6b04431.

Gao, Y.; Liu, H.; Liu, Q.; Wang, W. A Novel Colorimetric and OFF-ON Fluorescent Chemosensor Based on Fluorescein Derivative for the Detection of Fe3+ in Aqueous Solution and Living Cells. *Tetrahedron Lett* **2016**, *57* (17), 1852–1855. https://doi.org/10.1016/j.tetlet.2016.03.050.

Ge, F.; Yang, C.; Cai, Z. Fluorescence Sensor Performance of a New Fluorescein Derivate: [2-Morpholine-4-(6-Chlorine-1,3,5-s-Triazine)-Amino]Fluorescein. *Bull Korean Chem Soc* **2015**, *36* (11), 2703–2709. https://doi.org/10.1002/bkcs.10551.

Grimm, J. B.; Lavis, L. D. Synthesis of Rhodamines from Fluoresceins Using Pd-Catalyzed c-n Cross-Coupling. *Org Lett* **2011**, *13* (24), 6354–6357. https://doi.org/10.1021/ol202618t.

Gu, J.; Hu, D.; Wang, W.; Zhang, Q.; Meng, Z.; Jia, X.; Xi, K. Carbon Dot Cluster as an Efficient "off-on" Fluorescent Probe to Detect Au(III) and Glutathione. *Biosens Bioelectron* **2015**, *68*, 27–33. https://doi.org/10.1016/j.bios.2014.12.027.

Heinemann, S. H.; Hoshi, T.; Westerhausen, M.; Schiller, A. Carbon Monoxide – Physiology, Detection and Controlled Release. *Chemical Communications* **2014**, *50* (28), 3644–3660. https://doi.org/10.1039/c3cc49196j.

Helal, A.; Kim, H. S.; Yamani, Z. H.; Nasiruzzaman Shaikh, M. Fluorescein-N-Methylimidazole Conjugate as Cu2+ Sensor in Mixed Aqueous Media Through Electron Transfer. *J Fluoresc* **2016**, *26* (1), 1–9. https://doi.org/10.1007/s10895-015-1713-z.

Hormozi-Nezhad, M. R.; Taghipour, M. An Ultrasensitive and Selective Turn-off Fluorescent Nanoprobe for the Detection of Copper Ions. *Analytical Methods* **2015**, *7* (12), 5067–5073. https://doi.org/10.1039/c5ay00766f.

Hou, L.; Feng, J.; Wang, Y.; Dong, C.; Shuang, S.; Wang, Y. Single Fluorescein-Based Probe for Selective Colorimetric and Fluorometric Dual Sensing of Al3+ and Cu2+. *Sens Actuators B Chem* **2017**, *247*, 451–460. https://doi.org/10.1016/j.snb.2017.03.027.

Hu, J. J.; Wong, N. K.; Ye, S.; Chen, X.; Lu, M. Y.; Zhao, A. Q.; Guo, Y.; Ma, A. C. H.; Leung, A. Y. H.; Shen, J.; Yang, D. Fluorescent Probe HKSOX-1 for Imaging and Detection of Endogenous Superoxide in Live Cells and in Vivo. *J Am Chem Soc* **2015**, *137* (21), 6837–6843. https://doi.org/10.1021/jacs.5b01881.

Huo, F. J.; Zhang, J. J.; Yang, Y. T.; Chao, J. Bin; Yin, C. X.; Zhang, Y. Bin; Chen, T. G. A Fluorescein-Based Highly Specific Colorimetric and Fluorescent Probe for Hypochlorites in Aqueous Solution and Its Application in Tap Water. *Sens Actuators B Chem* **2012**, *166–167*, 44–49. https://doi.org/10.1016/j.snb.2011.11.081.

Janning, C.; Willbold, E.; Vogt, C.; Nellesen, J.; Meyer-Lindenberg, A.; Windhagen, H.; Thorey, F.; Witte, F. Magnesium Hydroxide Temporarily Enhancing Osteoblast Activity and Decreasing the Osteoclast Number in Peri-Implant Bone Remodelling. *Acta Biomater* **2010**, *6* (5), 1861–1868. https://doi.org/10.1016/j.actbio.2009.12.037.

Jiang, Y.; Zhang, W.; Huang, S.; Gui, R. Assembled Magnetic Nanoparticles for Photosensitive Nitric Oxide Release and Turn-on Fluorescence Detection in Situ. *Mater Chem Phys* **2015**, *167*, 231–235. https://doi.org/10.1016/j.matchemphys.2015.10.037.

Jiao, S.; Wang, X.; Sun, Y.; Zhang, L.; Sun, W.; Sun, Y.; Wang, X.; Ma, P.; Song, D. A Novel Fluorescein-Coumarin-Based Fluorescent Probe for Fluoride Ions and Its Applications in Imaging of Living Cells and Zebrafish in Vivo. *Sens Actuators B Chem* **2018**, *262*, 188–194. https://doi.org/10.1016/j.snb.2018.01.186.

Jin, L.; Xu, M.; Jiang, H.; Wang, W.; Wang, Q. A Simple Fluorescein Derived Colorimetric and Fluorescent "off-on" Sensor for the Detection of Hypochlorite. *Analytical Methods* **2018**, *10* (37), 4562–4569. https://doi.org/10.1039/c8ay01489b.

Jin, X.; Hao, L.; Hu, Y.; She, M.; Shi, Y.; Obst, M.; Li, J.; Shi, Z. Two Novel Fluorescein-Based Fluorescent Probes for Hypochlorite and Its Real Applications in Tap Water and Biological Imaging. *Sens Actuators B Chem* **2013**, *186*, 56–60. https://doi.org/10.1016/j.snb.2013.05.079.

Jin, X.; Wu, S.; She, M.; Jia, Y.; Hao, L.; Yin, B.; Wang, L.; Obst, M.; Shen, Y.; Zhang, Y.; Li, J. Novel Fluorescein-Based Fluorescent Probe for Detecting H2S and Its Real Applications in Blood Plasma and Biological Imaging. *Anal Chem* **2016**, *88* (22), 11253–11260. https://doi.org/10.1021/acs.analchem.6b04087.

Kambam, S.; Wang, B.; Wang, F.; Wang, Y.; Chen, H.; Yin, J.; Chen, X. A Highly Sensitive and Selective Fluorescein-Based Fluorescence Probe for Au3+ and Its Application in Living Cell Imaging. *Sens Actuators B Chem* **2015**, *209*, 1005–1010. https://doi.org/10.1016/j.snb.2014.12.085.

Kaur, P.; Lal, B.; Kaur, N.; Singh, G.; Singh, A.; Kaur, G.; Singh, J. Selective Two Way Cd(II) and Co(II) Ions Detection by 1,2,3–Triazole Linked Fluorescein Derivative. *J Photochem Photobiol A Chem* **2019**, *382*. https://doi.org/10.1016/j.jphotochem.2019.05.010.

Kim, H. Y.; Im, H. G.; Chang, S. K. Colorimetric and Fluorogenic Signaling of Fluoride Ions by Thiophosphinated Dichlorofluorescein. *Dyes and Pigments* **2015**, *112*, 170–175. https://doi.org/10.1016/j.dyepig.2014.06.030.

Kitley, W. R.; Santa Maria, P. J.; Cloyd, R. A.; Wysocki, L. M. Synthesis of High Contrast Fluorescein-Diethers for Rapid Bench-Top Sensing of Palladium. *Chemical Communications* **2015**, *51* (40), 8520–8523. https://doi.org/10.1039/c5cc02192h.

Kumar, M.; Kumar, N.; Bhalla, V.; Sharma, P. R.; Kaur, T. Highly Selective Fluorescence Turn-on Chemodosimeter Based on Rhodamine for Nanomolar Detection of Copper Ions. *Org Lett* **2012**, *14* (1), 406–409. https://doi.org/10.1021/ol203186b.

Le Guern, F.; Mussard, V.; Gaucher, A.; Rottman, M.; Prim, D. Fluorescein Derivatives as Fluorescent Probes for Ph Monitoring along Recent Biological Applications. *International Journal of Molecular Sciences*. MDPI AG December 1, 2020, pp 1–23. https://doi.org/10.3390/ijms21239217.

Lee, K. S.; Kim, T. K.; Lee, J. H.; Kim, H. J.; Hong, J. I. Fluorescence Turn-on Probe for Homocysteine and Cysteine in Water. *Chemical Communications* **2008**, No. 46, 6173–6175. https://doi.org/10.1039/b814581d.

Li, C.; Li, S.; Wang, G.; Yang Z.. Spectroscopic Properties of a Chromone-Fluorescein Conjugate as Mg2+ "Turn on" Fluorescent Probe. *J Photochem Photobiol A Chem* **2018**, *356*, 700–707. https://doi.org/10.1016/j.jphotochem.2016.03.001.

Li, G.; Liu, Y.; Song, J.; Ye, Y. A Fluorescein-Based Colorimetric and Fluorescent Probe for Hydrazine and Its Bioimaging in Live Cells. *J Fluoresc* **2017**, *27* (1), 323–329. https://doi.org/10.1007/s10895-016-1961-6.

Liu, J. M.; Lin, L. P.; Wang, X. X.; Jiao, L.; Cui, M. L.; Jiang, S. L.; Cai, W. L.; Zhang, L. H.; Zheng, Z. Y. Zr(H2O)2EDTA Modulated Luminescent Carbon Dots as Fluorescent Probes for Fluoride Detection. *Analyst* **2013**, *138* (1), 278–283. https://doi.org/10.1039/c2an36055a.

Liu, Y.; Xiang, K.; Tian, B.; Zhang, J. A Fluorescein-Based Fluorescence Probe for the Fast Detection of Thiol. *Tetrahedron Lett* **2016**, *57* (23), 2478–2483. https://doi.org/10.1016/j.tetlet.2016.04.068.

Lv, J.; Wang, F.; Qiang, J.; Ren, X.; Chen, Y.; Zhang, Z.; Wang, Y.; Zhang, W.; Chen, X. Enhanced Response Speed and Selectivity of Fluorescein-Based H2S Probe via the Cleavage of Nitrobenzene Sulfonyl Ester Assisted by Ortho Aldehyde Groups. *Biosens Bioelectron* **2017**, *87*, 96–100. https://doi.org/10.1016/j.bios.2016.08.018.

Lv, J.; Wang, F.; Wei, T.; Chen, X. Highly Sensitive and Selective Fluorescent Probes for the Detection of HOCl/OCl- Based on Fluorescein Derivatives. *Ind Eng Chem Res* **2017**, *56* (13), 3757–3764. https://doi.org/10.1021/acs.iecr.7b00381.

Ma, D. H.; Kim, D.; Akisawa, T.; Lee, K. H.; Kim, K. T.; Ahn, K. H. An FITC-BODIPY Fret Couple: Application to Selective, Ratiometric Detection and Bioimaging of Cysteine. *Chem Asian J* **2015**, *10* (4), 894–902. https://doi.org/10.1002/asia.201403073.

Martínez-Jiménez, M. I.; García-Gómez, S.; Bebenek, K.; Sastre-Moreno, G.; Calvo, P. A.; Díaz-Talavera, A.; Kunkel, T. A.; Blanco, L. Alternative Solutions and New Scenarios for Translesion DNA Synthesis by Human PrimPol. *DNA Repair (Amst)* **2015**, *29*, 127–138. https://doi.org/10.1016/j.dnarep.2015.02.013.

Meng, Q.; Wang, Y.; Yang, M.; Zhang, R.; Wang, R.; Zhang, Z. A New Fluorescent Chemosensor for Highly Selective and Sensitive Detection of Inorganic Phosphate (Pi) in Aqueous Solution and Living Cells. *RSC Adv* **2015**, *5* (66), 53189–53197. https://doi.org/10.1039/c5ra08712k.

Mondal, S.; Manna, S. K.; Maiti, K.; Maji, R.; Ali, S. S.; Manna, S.; Mandal, S.; Uddin, M. R.; Mahapatra, A. K. Phenanthroline-Fluorescein Molecular Hybrid as a Ratiometric and Selective Fluorescent Chemosensor for Cu2+ via FRET Strategy: Synthesis, Computational Studies and in Vitro Applications. *Supramol Chem* **2017**, *29* (8), 616–626. https://doi.org/10.1080/10610278.2017.1301452.

Ngororabanga, J. M. V.; Tshentu, Z. R.; Mama, N. A New Highly Selective Colorimetric and Fluorometric Coumarin-Based Chemosensor for Hg2+. *J Fluoresc* **2020**, *30* (5), 985–997. https://doi.org/10.1007/s10895-020-02542-x.

Niu, L. Y.; Guan, Y. S.; Chen, Y. Z.; Wu, L. Z.; Tung, C. H.; Yang, Q. Z. BODIPY-Based Ratiometric Fluorescent Sensor for Highly Selective Detection of Glutathione over Cysteine and Homocysteine. *J Am Chem Soc* **2012**, *134* (46), 18928–18931. https://doi.org/10.1021/ja309079f.

Patil, P.; Sehlangia, S.; Patil, A.; Pradeep, C.; Sahoo, S. K.; Patil, U. A New Phthalimide Based Chemosensor for Selective Spectrophotometric Detection of Cu(II) from Aqueous Medium. *Spectrochim Acta A Mol Biomol Spectrosc* **2019**, *220*. https://doi.org/10.1016/j.saa.2019.05.034.

Patil, S. R.; Nandre, J. P.; Patil, P. A.; Bothra, S.; Sahoo, S. K.; Klasek, A.; Rodríguez-López, J.; Mahulikar, P. P.; Patil, U. D. Quinolone Based Chemosensor for the Naked-Eye and Spectrophotometric Detection of Cu2+in Aqueous Media. *Inorg Chem Commun* **2014**, *49*, 59–62. https://doi.org/10.1016/j.inoche.2014.09.021.

Piriya V.S, A.; Joseph, P.; Daniel S.C.G., K.; Lakshmanan, S.; Kinoshita, T.; Muthusamy, S. *Colorimetric Sensors for Rapid Detection of Various Analytes: Materials Science and Engineering C*. Elsevier Ltd September 1, 2017, pp 1231–1245. https://doi.org/10.1016/j.msec.2017.05.018.

Piyanuch, P.; Watpathomsub, S.; Lee, V. S.; Nienaber, H. A.; Wanichacheva, N. Highly Sensitive and Selective Hg2+-Chemosensor Based on Dithia-Cyclic Fluorescein for Optical and Visual-Eye Detections in Aqueous Buffer Solution. *Sens Actuators B Chem* **2016**, *224*, 201–208. https://doi.org/10.1016/j.snb.2015.09.131.

Qiao, J.; Liu, Z.; Tian, Y.; Wu, M.; Niu, Z. Multifunctional Self-Assembled Polymeric Nanoprobes for FRET-Based Ratiometric Detection of Mitochondrial H2O2 in Living Cells. *Chemical Communications* **2015**, *51* (17), 3641–3644. https://doi.org/10.1039/c4cc09120e.

Rajasekar, M. *Recent Development in Fluorescein Derivatives: Journal of Molecular Structure*. Elsevier B.V. January 15, 2021. https://doi.org/10.1016/j.molstruc.2020.129085.

Rathod, R. V.; Bera, S.; Singh, M.; Mondal, D. A Colorimetric and Fluorometric Investigation of Cu(II) Ion in Aqueous Medium with a Fluorescein-Based Chemosensor. *RSC Adv* **2016**, *6* (41), 34608–34615. https://doi.org/10.1039/c6ra03021a.

Recent Progress of Fluorescent Probe for Selective Detection of Silver. **2002**. https://doi.org/10.24327/ijcar.2018.

Romão, C. C.; Blattler, W. A.; Seixas, J. D.; Bernardes, G. J. L. Developing Drug Molecules for Therapy with Carbon Monoxide. *Chem Soc Rev* **2012**, *41* (9), 3571–3583. https://doi.org/10.1039/c2cs15317c.

Rude, R. K.; Gruber, H. E. Magnesium Deficiency and Osteoporosis: Animal and Human Observations. *J Nutr Biochem*, December **2004**, 710–716. https://doi.org/10.1016/j.jnutbio.2004.08.001.

Saha, T.; Kand, D.; Talukdar, P. Performance Comparison of Two Cascade Reaction Models in Fluorescence Off-on Detection of Hydrogen Sulfide. *RSC Adv* **2015**, *5* (2), 1438–1446. https://doi.org/10.1039/c4ra13086c.

Saleem, F.; Rao, G. K.; Kumar, A.; Kumar, S.; Singh, M. P.; Singh, A. K. Palladium(II) Complexes Bearing the 1,2,3-Triazole Based Organosulfur/ Selenium Ligand: Synthesis, Structure and Applications in Heck and Suzuki-Miyaura Coupling as a Catalyst via Palladium Nanoparticles. *RSC Adv* **2014**, *4* (99), 56102–56111. https://doi.org/10.1039/c4ra09574j.

Samanta, S.; Halder, S.; Manna, U.; Das, G. Specific Detection of Hypochlorite: A Cyanine Based Turn-on Fluorescent Sensor. *J Chem Sci* **2019**, *131* (5). https://doi.org/10.1007/s12039-019-1612-y.

Santra, M.; Ko, S. K.; Shin, I.; Ahn, K. H. Fluorescent Detection of Palladium Species with an O-Propargylated Fluorescein. *Chem Commun* **2010**, *46* (22), 3964–3966. https://doi.org/10.1039/c001922d.

Setsukinai, K. ichi; Urano, Y.; Kakinuma, K.; Majima, H. J.; Nagano, T. Development of Novel Fluorescence Probes That Can Reliably Detect Reactive Oxygen Species and Distinguish Specific Species. *J Biol Chem* **2003**, *278* (5), 3170–3175. https://doi.org/10.1074/jbc.M209264200.

Shen, W.; Wang, L.; Wu, M.; Bao, X. A Fluorescein Derivative FLTC as a Chemosensor for Hg2 + and Ag+ and Its Application in Living-Cell Imaging. *Inorg Chem Commun* **2016**, *70*, 107–110. https://doi.org/10.1016/j.inoche.2016.05.001.

Shi, Y.; Zhang, W.; Xue, Y.; Zhang, J. Fluorescent Sensors for Detecting and Imaging Metal Ions in Biological Systems: Recent Advances and Future Perspectives. *Chemosensors*. MDPI April 1, 2023. https://doi.org/10.3390/chemosensors11040226.

Stelmashook, E. V.; Isaev, N. K.; Genrikhs, E. E.; Amelkina, G. A.; Khaspekov, L. G.; Skrebitsky, V. G.; Illarioshkin, S. N. *Role of Zinc and Copper Ions in the Pathogenetic Mechanisms of Alzheimer's and Parkinson's Diseases. Biochemistry (Moscow)*. Maik Nauka Publishing / Springer SBM 2014, pp. 391–396. https://doi.org/10.1134/S0006297914050022.

Strausak, D.; Mercer, J. F. B.; Dieter, H. H.; Stremmel, W.; Multhaup, G. *Copper in Disorders with Neurological Symptoms: Alzheimer's, Menkes, and Wilson Diseases*; 2001.

Sun, H.; Lai, J. P.; Lin, D. S.; Huang, X. X.; Zuo, Y.; Li, Y. L. A Novel Fluorescent Multi-Functional Monomer for Preparation of Silver Ion-Imprinted Fluorescent on-off Chemosensor. *Sens Actuators B Chem* **2016**, *224*, 485–491. https://doi.org/10.1016/j.snb.2015.10.052.

Švagždiene, M.; Širvinskas, E.; Baranauskiene, D.; Adukauskiene, D. Correlation of Magnesium Deficiency with C-Reactive Protein in Elective Cardiac Surgery with Cardiopulmonary Bypass for Ischemic Heart Disease. *Medicina (Lithuania)* **2015**, *51* (2), 100–106. https://doi.org/10.1016/j.medici.2015.03.003.

Tachapermpon, Y.; Chaneam, S.; Charoenpanich, A.; Sirirak, J.; Wanichacheva, N. Highly Cu2+-Sensitive and Selective Colorimetric and Fluorescent Probes: Utilizations in Batch, Flow Analysis and Living Cell Imaging. *Sens Actuators B Chem* **2017**, *241*, 868–878. https://doi.org/10.1016/j.snb.2016.11.009.

Taheri, M.; Mansour, N. Functionalized Silicon Nanoparticles as Fluorescent Probe for Detection of Hypochlorite in Water. *J Photochem Photobiol A Chem* **2019**, *382*. https://doi.org/10.1016/j.jphotochem.2019.111906.

Ueno, T.; Urano, Y.; Setsukinai, K. I.; Takakusa, H.; Kojima, H.; Kikuchi, K.; Ohkubo, K.; Fukuzumi, S.; Nagano, T. Rational Principles for Modulating Fluorescence Properties of Fluorescein. *J Am Chem Soc* **2004**, *126* (43), 14079–14085. https://doi.org/10.1021/ja048241k.

Vidya, B.; Sivaraman, G.; Sumesh, R. V.; Chellappa, D. Fluorescein-Based ""Turn On"" Fluorescence Detection of Zn2+ and Its Applications in Imaging of Zn2+ in Apoptotic Cells. *ChemistrySelect* **2016**, *1* (13), 4024–4029. https://doi.org/10.1002/slct.201600863.

Wang, D.; Xiang, X.; Yang, X.; Wang, X.; Guo, Y.; Liu, W.; Qin, W. Fluorescein-Based Chromo-Fluorescent Probe for Zinc in Aqueous Solution: Spirolactam Ring Opened or Closed? *Sens Actuators B Chem* **2014**, *201*, 246–254. https://doi.org/10.1016/j.snb.2014.05.019.

Wang, N.; Xu, W.; Song, D.; Ma, P. A Fluorescein-Carbazole-Based Fluorescent Probe for Imaging of Endogenous Hypochlorite in Living Cells and Zebrafish. *Spectrochim Acta A Mol Biomol Spectrosc* **2020**, *227*. https://doi.org/10.1016/j.saa.2019.117692.

Wei, G.; Wang, L.; Jiao, J.; Hou, J.; Cheng, Y.; Zhu, C. Cu 2+ Triggered Fluorescence Sensor Based on Fluorescein Derivative for Pd 2+ Detection. *Tetrahedron Lett* **2012**, *53* (27), 3459–3462. https://doi.org/10.1016/j.tetlet.2012.04.108.

Wen, C.; Huang, Y.; Tian, J.; Hu, K.; Pan, L.; Zhao, S. A Novel Exonuclease III-Aided Amplification Assay Based on a Graphene Platform for Sensitive Detection of Adenosine Triphosphate. *Analytical Methods* **2015**, *7* (9), 3708–3713. https://doi.org/10.1039/c5ay00354g.

Wiseman, C. L. S.; Zereini, F. Airborne Particulate Matter, Platinum Group Elements and Human Health: A Review of Recent Evidence. *Sci Total Environ*, April 1, **2009**, 2493–2500. https://doi.org/10.1016/j.scitotenv.2008.12.057.

Wu, Y.; Huang, S.; Zeng, F.; Wang, J.; Yu, C.; Huang, J.; Xie, H.; Wu, S. A Ratiometric Fluorescent System for Carboxylesterase Detection with AIE Dots as FRET Donors. *Chem Commun* **2015**, *51* (64), 12791–12794. https://doi.org/10.1039/c5cc04771d.

Xu, Q.; Lee, K. A.; Lee, S.; Lee, K. M.; Lee, W. J.; Yoon, J. A Highly Specific Fluorescent Probe for Hypochlorous Acid and Its Application in Imaging Microbe-Induced Hocl Production. *J Am Chem Soc* **2013**, *135* (26), 9944–9949. https://doi.org/10.1021/ja404649m.

Xue, D.; Zheng, C.; Qu, S.; Liao, G.; Fan, C.; Liu, G.; Pu, S. A Highly Selective Fluorescent Chemosensor for Cu2+: Synthesis and Properties of a Rhodamine B-Containing Diarylethene. *Luminescence* **2017**, *32* (4), 652–660. https://doi.org/10.1002/bio.3234.

Yan, F.; Fan, K.; Bai, Z.; Zhang, R.; Zu, F.; Xu, J.; Li, X. *Fluorescein Applications as Fluorescent Probes for the Detection of Analytes: TrAC - Trends in Analytical Chemistry*. Elsevier B.V., December 1, 2017, pp. 15–35. https://doi.org/10.1016/j.trac.2017.08.013.

Yusop, R. M.; Unciti-Broceta, A.; Johansson, E. M. V.; Sánchez-Martín, R. M.; Bradley, M. Palladium-Mediated Intracellular Chemistry. *Nat Chem* **2011**, *3* (3), 239–243. https://doi.org/10.1038/nchem.981.

Zhang, H.; Liu, R.; Tan, Y.; Xie, W. H.; Lei, H.; Cheung, H. Y.; Sun, H. A FRET-Based Ratiometric Fluorescent Probe for Nitroxyl Detection in Living Cells. *ACS Appl Mater Interfaces* **2015**, *7* (9), 5438–5443. https://doi.org/10.1021/am508987v.

Zhang, R.; Yan, F.; Huang, Y.; Kong, D.; Ye, Q.; Xu, J.; Chen, L. *Rhodamine-Based Ratiometric Fluorescent Probes Based on Excitation Energy Transfer Mechanisms: Construction and Applications in Ratiometric Sensing: RSC Advances*. Royal Society of Chemistry, 2016, pp. 50732–50760. https://doi.org/10.1039/c6ra06956h.

Zhao, G.; Wei, G.; Yan, Z.; Guo, B.; Guang, S.; Wu, R.; Xu, H. A Multiple Fluorescein-Based Turn-on Fluorophore (FHCS) Identified for Simultaneous Determination and Living Imaging of Toxic Al3+ and Zn2+ by Improved Stokes Shift. *Anal Chim Acta* **2020**, *1095*, 185–196. https://doi.org/10.1016/j.aca.2019.10.025.

Zheng, H.; Zhan, X. Q.; Bian, Q. N.; Zhang, X. J. Advances in Modifying Fluorescein and Rhodamine Fluorophores as Fluorescent Chemosensors. *Chem Commun* **2013**, *49* (5), 429–447. https://doi.org/10.1039/c2cc35997a.

Zhou, Y.; Li, J. Y.; Chu, K. H.; Liu, K.; Yao, C. Fluorescence Turn-on Detection of Hypochlorous Acid via HOCl-Promoted Dihydrofluorescein-Ether Oxidation and Its Application in Vivo. *Chem Commun* **2012**, *48* (39), 4677–4679. https://doi.org/10.1039/c2cc30265a.

9 Quinoline-Based Fluorescent Probe for Various Applications

Kisan Pawar, Somidi Srinu, Mareechika Gaddam and Abha Sharma
National Institute of Pharmaceutical Education and Research,
Raebareli, Lucknow, UP, India

LIST OF ABBREVIATIONS

OLEDs	Organic light-emitting diodes
ATP	Adenosine triphosphate
ROS	Reactive oxygen species
RNS	Reactive nitrogen species
RSS	Reactive sulfur species
GSH	Glutathione
Cys	Cysteine
Hcy	Homocysteine
AD	Alzheimer's disease
ICT	Intramolecular charge transfer
PDT	Photodynamic treatment
DCFH	Dichloro-dihydro-fluorescein
DCF	Dichlorofluorescein
CBS	Cystathionine β-synthase
3-MST	Mercaptopyruvate sulfurtransferase
PET	Positron emission tomography
CT	Computed tomography
MRI	Magnetic resonance imaging
CHEF	Chelation enhanced fluorescence
PET	Photoinduced electron transfer
EDTA	Ethylenediaminetetraacetic acid
TPM	Two-photon microscopy
OPM	One-photon microscopy
LOD	Limit of detection
LDs	Lipid droplets

9.1 INTRODUCTION

In the realm of molecular sciences, the quest for versatile tools to unravel intricate biological phenomena and catalyze technological advancements is unending. Among these, fluorescent probes stand out as invaluable assets, offering a window into the microscopic world with their ability to emit light upon excitation. Among the myriad compounds explored for this purpose, quinoline derivatives have emerged as a class of paramount importance. In this chapter, we embark on a journey through the illuminating realm of quinoline derivatives, exploring their diverse roles as fluorescent probes in various applications. Quinoline, a heterocyclic compound composed of a benzene ring fused to a pyridine ring, possesses inherent fluorescence properties that can be finely tuned through chemical modification. This versatility forms the foundation for its widespread use as a fluorescent probe. By harnessing the unique photophysical properties of quinoline derivatives, researchers have been able to design probes tailored to specific applications, ranging from biomedical research to materials science and beyond. The biomedical field has witnessed a surge in the utilization of quinoline-based fluorescent probes for imaging and sensing purposes. One of the key areas of application lies in the detection and monitoring of biomolecules within living systems. By conjugating quinoline derivatives with targeting moieties such as peptides or antibodies, researchers have developed probes capable of selectively labeling biomolecules of interest, from proteins and nucleic acids to metabolites and ions. The exquisite sensitivity and specificity of these probes have revolutionized our ability to visualize cellular processes in real time, shedding light on the intricate dynamics underlying various physiological and pathological conditions. Moreover, quinoline derivatives have found utility in the field of drug discovery and development. Their ability to interact with specific biological targets, coupled with their fluorescent properties, renders them invaluable tools for high-throughput screening assays. By incorporating quinoline-based probes into screening platforms, researchers can expedite the identification and optimization of lead compounds, thereby accelerating the drug discovery process. Furthermore, the development of responsive probes capable of reporting on enzymatic activity or intracellular events has opened new avenues for probing drug mechanisms and elucidating pharmacological pathways. Beyond the biomedical realm, quinoline derivatives have made significant inroads into materials science and nanotechnology. Their facile synthesis and tunable photophysical properties make them ideal candidates for applications in sensing, imaging, and optoelectronic devices. For instance, quinoline-based fluorescent sensors have been employed for the detection of environmental pollutants, metal ions, and explosives, offering rapid and selective detection methods with potential applications in environmental monitoring and homeland security. Additionally, the integration of quinoline derivatives into organic light-emitting diodes (OLEDs) has led to the development of efficient emissive

DOI: 10.1201/9781032706245-9

materials with promising prospects for next-generation display technologies. Quinoline derivatives stand as versatile luminaries in the realm of fluorescent probes, offering a myriad of opportunities for exploration and innovation. From their pivotal role in biomedical imaging and drug discovery to their burgeoning applications in materials science and beyond, quinoline-based probes continue to illuminate our understanding of the microscopic world and catalyze transformative discoveries across diverse domains. Join us on a journey through the captivating world of quinoline derivatives, where fluorescence meets innovation, and possibilities abound.

9.2 QUINOLINE-BASED FLUORESCENT PROBE FOR BIOLOGICAL APPLICATION

Quinoline-based probes are reported for various biological applications including mitochondrial imaging, nucleic acid imaging, lysosomal imaging, nucleus imaging, etc. The quinoline-based probes are also used to detect various other environmental pollutants like metal ions and toxic chemicals.

9.2.1 Quinoline-Based Fluorescent Probe for Mitochondrial Imaging

Mitochondria are "power factory" in cells; mitochondria that exist in most eukaryotic cells are highly important. Mitochondria are involved in many biological processes such as signal transmission, cell differentiation, growth, and death. For example, mitochondria play a central role in the production of adenosine triphosphate (ATP). Accordingly, the function of mitochondria is realized based on many important biomolecules, including various metal ions, reactive oxygen/nitrogen/sulfur species (ROS/RNS/RSS), and so on. Therefore, it is of considerable significance to accurately detect these important biomolecules for studying the mitochondrial function (McBride et al., 2006).

9.2.1.1 Detection of Homocysteine and Cysteine in Mitochondria

Small molecular biothiols, such as glutathione (GSH), homocysteine (Hcy), and cysteine (Cys), are essential for maintaining the redox balance and metabolism of cells in organisms (Reddie et al., 2008; Robaczewska et al., 2016). Abnormal biothiol concentration has been linked to a number of illnesses. Reduced Cys levels, for instance, are closely linked to infantile dysplasia symptoms such as delayed growth, muscle atrophy, and discolored hair (Udha et al., 2002; Zhang et al., 2004). Moreover, Cys and Hcy function as antioxidants to eliminate ROS from mitochondria because ROS oxidative properties might lead to apoptosis (Schwarzländer et al., 2013). Therefore, it is critical to identify Cys/Hcy in mitochondria. Fluorescent probes are chosen as some of the best monitoring tools due to their advantages of high sensitivity, selectivity, and real-time detection (Liu et al., 2014; Xie et al., 2018; Xu et al., 2019).

Yating Yu et al. (2020) developed the rhodamine π-conjugated system by integrating quinoline moiety into xanthene to create a novel near-infrared rhodamine dye RhQ (Figure 9.1). A novel dye based on rhodamine exhibits emission in the near-infrared spectrum with a high quantum yield. To take the advantage of the spirocyclization open-ring states of the rhodamine platform, the

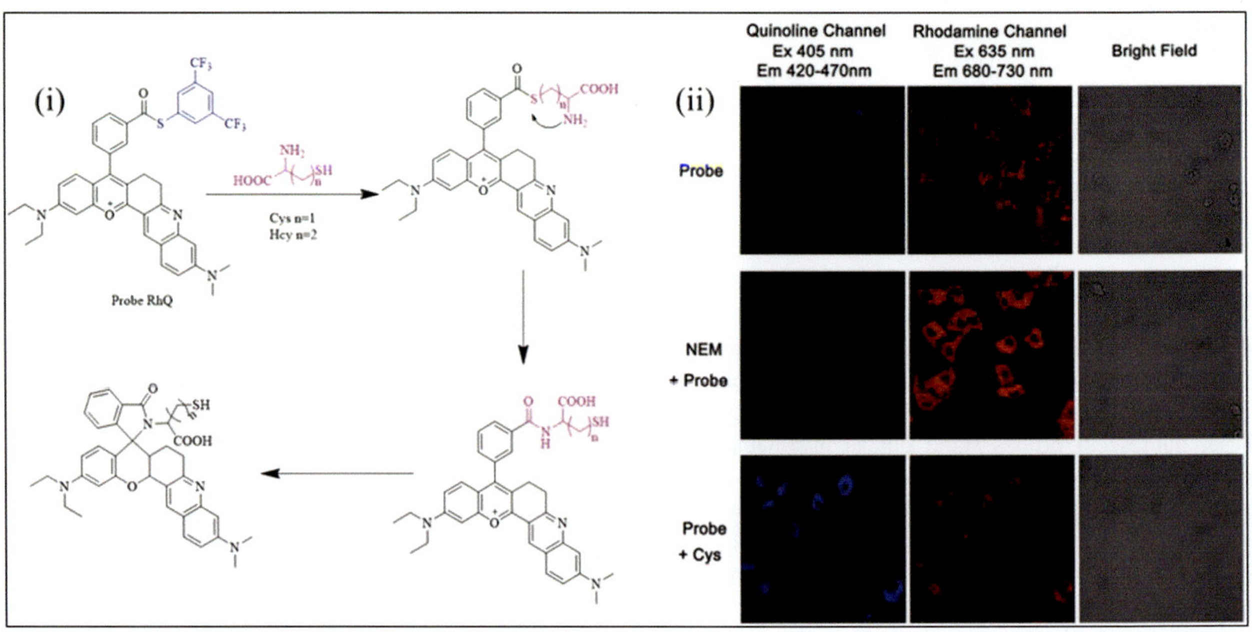

FIGURE 9.1 (i) The proposed sensing mechanism of RhQ to cysteine and homocysteine. (ii) Cellular fluorescence images of probe RhQ (10 μM) incubated with Caov3 cells in the absence and presence of Cys (100 μM) or under NEM (1 mM) treatment. Reprinted (adapted) with permission from Yu et al. (2020). Copyright (2020): Elsevier.

3,5-ditrifluormethylthiophenol moiety is introduced into the dye through a thioester linkage. The probe shows an impressive dual-channel fluorescence response in aqueous solutions between visible and near-infrared emission while exposed to Cys/Hcy. This is caused by the disruption of the molecular π-conjugation system by the creation of a closed spiro-lactam structure (Figure 9.1). The RhQ response in Caov3 cells in the presence and absence of Cys is shown in Figure 9.1 (ii).

9.2.1.2 Detection of Mitochondrial Analyte

Fluorescent probes targeted at mitochondria are crucial for obtaining information on mitochondrial function (McBride et al., 2006). The existing mitochondria-targeted fluorescent probe's accuracy is inadequate for the two reasons listed below. In the first instance, certain probes that are always targeted to mitochondria would react with the analytes outside of mitochondria and induce mitochondria to release a fluorescent signal, which causes a false favorable outcome (Liu et al., 2018; Zhu et al., 2016). In other case, certain probes may diffuse from mitochondria to other organelles in reaction to the analytes in mitochondria, resulting in a false-negative result. These two issues would result in present probes' low ability to detect mitochondrial analytes. A solid-state fluorophore named 2-(2'-hydroxyphenyl)-4(3H)-quinazolinone (HPQ) has been used to image active species in situ (Liu et al., 2017; Zhou et al., 2016). Furthermore, quinoline compounds lack the strong mitochondria-targeting ability and high water solubility of quinoline salts (Xu et al., 2016). Here, Zhe et al. (2021) developed and synthesized HQPQ, a new solid-state and water-insoluble fluorophore, to accurately detect mitochondrial analytes by utilizing the characteristics of 2-(2′-hydroxyphenyl)-4(3H)-quinazolinone (HPQ) and quinolone. They developed and synthesize a probe HQPQ-B (Figure 9.2) for H_2O_2 detection as a proof of concept by altering quinoline to create quinoline salts, which precipitate in situ after reacting with analytes. Because quinoline salts and quinone have differing mitochondria-targeting capacities, HQPQ loses its capacity to target mitochondria after interacting with analytes outside of mitochondria, preventing a false-positive result. On the contrary, HQPQ would precipitate and remain in mitochondria without diffusing to other sites when the probe initially localized in mitochondria and then reacted with analytes, preventing a false-negative result. Thus, the precise detection of mitochondrial analytes is made possible by HQPQ (Figure 9.2).

9.2.1.3 Quinoline-Based Fluorescent Probe for Monitoring Mitochondrial Viscosity

Viscosity is a crucial component of cellular membranes, acting as one of the most significant intracellular environments and influencing the exchange of mass and signals as well as protein-protein interactions (Luby-Phelps, 2013; Yang et al., 2014). It has been demonstrated that variations in intracellular viscosity can serve as a marker for a number of conditions, including diabetes, Alzheimer's disease, atherosclerosis, and cancer (Deliconstantinos et al., 1995; Humphreys, 1976; Zubenko et al., 1999). Even within the same type of cell, different organelles have quite varying viscosities. Detecting viscosity changes in individual organelles rather than the entire cytoplasm is likely more important for better understanding the implications of these changes in cells (Luby-Phelps, 1994). One of the most common organelles in cells, mitochondria play a crucial role in aerobic respiration, which supplies energy for a variety of cellular functions. There is evidence linking certain neurodegenerative illnesses, like Parkinson's disease and Huntington's disease, due to abnormalities in mitochondrial dynamics (Züchner et al., 2006). Mitochondrial viscosity is closely related to cellular respiration, as it can affect cellular metabolite diffusion and apoptosis and can regulate mitochondrial metabolism. Therefore, it is vital to build a molecular probe that is suitable for visualizing mitochondrial viscosity (Nie et al., 2005). The conventional methods for measuring viscosity are time-consuming, limited to measuring the viscosity of macroscopic liquids and unable to detect the viscosity of organelles. Small molecular fluorescent probes, which have produced a lot of attention, were created as a solution to these issues. They convert recognition information into fluorescent signals and can be used to visualize biological microenvironments such as pH, amino acid, viscosity, and polarity (Li et al., 2020). Near-infrared emission could minimize the interference by cell autofluorescence and the photodamage to living cells and organisms. Some intrinsic benefits of a big Stokes shift include minimal background fluorescence, noninvasiveness, and high sensitivity. Fluorescence confocal imaging is considered one of the most significant techniques for tracking biological processes because of its benefits, which include quick reaction times and the capacity to see data in real time (Sasaki et al., 2016).

Zhang et al. (2021) designed and synthesized a novel near-infrared fluorescent probe to visualize variations in viscosity. A quinidine derivative cation was employed to

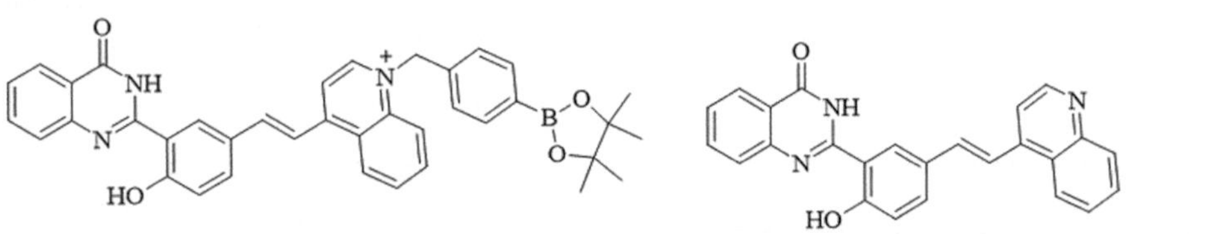

HQPQ-B HQPQ

FIGURE 9.2 Structure of HQPQ-B and HQPQ.

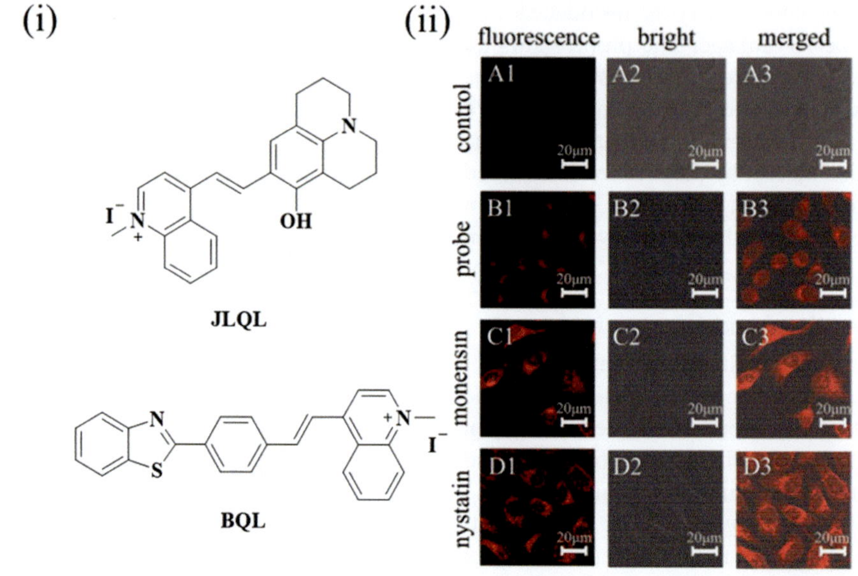

FIGURE 9.3 (i) Structures of JLQL and BQL. (ii) Fluorescence images of A549 cells incubated with monensin and nystatin. Fluorescence (A1) and bright-field (A2) images of control cells, and their merged image (A3). Fluorescence (B1) and bright-field (B2) images of A549 cells incubated with only 15 μM JLQL, and their merged image (B3). Fluorescence (C1) and bright-field (C2) images of A549 cells incubated with 6 μM monensin for 20 min, followed by 15 μM JLQL for another 30 min, and their merged image (C3). Fluorescence (D1) and bright-field (D2) of A549 cells incubated with 6 μM nystatin for 20 min, followed by 15 μM JLQL for another 30 min, and their merged image (D3). Reprinted (adapted) with permission from Zhang et al. (2021). Copyright (2021): Elsevier.

target mitochondria, and the modified julolidine group was selected as the fluorophore to raise the emission wavelength as close to the near-infrared wavelength as possible. Conveniently, JLQL (Figure 9.3), a novel organic water-soluble molecule composed of a fluorophore and a mitochondrial-targeting moiety as an acceptor group, joined by a double bond was synthesized. Its maximal emission wavelength was 734.6 nm, and its Stokes shift was 134.6 nm. The augmentation in viscosity in a PBS-glycerol system caused the fluorescence intensity of the sensor to increase by a factor of 175. It may be possible to lessen the interference from other microenvironments, such as pH, polarity, and other interference analytes. JLQL was able to distinguish between various levels of mitochondrial viscosity brought on by nystatin or monensin with sensitivity and selectivity (Figure 9.3). Additionally, the probe might offer a desirable means of tracking viscosity changes in real time during mitophagy, with the aforementioned characteristics. JLQL has the potential to be an extremely helpful tool for detecting mitochondrial viscosity.

Zhang et al. (2023) developed a novel fluorescent chemosensor called BQL (Figure 9.3), which is made up of a quinoline derivative, benzothiazole group and a benzene ring incorporated in the centre to expand the conjugation. The quinoline cation targets mitochondria and acts as an electron-withdrawing group in addition to the benzothiazole group, which emits fluorescence and donates electrons. Together, these functional groups form a D-π-A structure for viscosity detection. A molecular rotor and fluorophore together typically allow the relaxation process to occur by nonradiative transition, resulting in a considerable dimming of fluorescence. This is because the molecular rotor can rotate (relative to the whole molecule) in low-viscosity solvents, Earlier studies states that high-viscosity solvents restrict nonradiative energy transitions, which increases fluorescence intensity. Due to its low cytotoxicity and superior spatiotemporal resolution, BQL has a wide range of potential applications in biology, chemistry, medicine, and other domains.

9.2.1.4 Quinoline-Based Mitochondrial-Specific Fluorescent Probe for Assessment of Cell Viability

An efficient technique for assessing cell viability and toxicity is crucial for biomedical research, including cell biology, pathology, and drug discovery (Kepp et al., 2011). In these studies, cell viability tests are crucial in forecasting the in vivo therapeutic efficacy and unfavorable effects (Steffes et al., 2017). Great efforts have been made in the last few decades to create cytotoxicity assays, and some trustworthy techniques have been established. A single-channel integrated microfluidic chip, for instance, was proposed based on the cell detachment assay to increase the accuracy of cell viability assessment by measuring the adhesion strength of cells, which is more accurate than that evaluated using the cell counting assay (Chen et al., 2021). Due to its inherent advantages, MTT, a colorimetric test that quantifies the reduction of yellow 3-(4,5-dimethythia zol-2-yl)-2,5-diphenyl tetrazolium bromide by mitochondrial succinate dehydrogenase, has been determined to be the most practical approach. Nevertheless, the majority of them have some limitations, such as limited sensitivity, a complicated procedure, variable results brought on by cell extract interference, and time-consuming (Wang et al., 2011).

Fluorescence imaging is a powerful technology that may visualize cell activity in real time. It can also provide vital biological information that is needed for illness diagnosis and therapy monitoring. Nevertheless, nonspecific fluorescent background is frequently detrimental to its uses and is brought on by auto-fluorescence, light scattering, or low fluorescent labeling efficiency (He et al., 2014; Mouhib et al., 2019). Consequently, a great deal of work has gone into improving functional fluorescence imaging, with organelle-specific probes in particular greatly expanding its application scope. As a result, tracking mitochondria may offer a useful method for keeping an eye on cell fates. Numerous fluorescent probes for tracking dynamic changes related to mitochondria have been developed recently.

Chen et al. (2021) developed a novel class of fluorescent probes known as coumarin-quinoline conjugates (CQCs, Figure 9.5), which are made of 7-diethylaminocoumarin conjugated with a quinoline moiety and are intended to be used for the selective monitoring of changes in mitochondrial dynamics to assess cell viability. In diseased situations, CQCs could detect the fissions of mitochondria with great sensitivity and demonstrated a stronger affinity for them. Specifically, in contrast to the MTT test, CQCs could precisely determine the viability of the cells in less than 10 minutes, enabling the quick assessment of treatment effects with great sensitivity and accuracy. Therefore, for the study of cellular viability in a wide range of biomedical applications, CQCs offer a proof-of-concept test that is suitably practical either alone or in conjunction with other assays.

9.2.1.5 Quinoline-Based Probe for Monitoring Mitochondrial pH

Mitochondria are the sites of oxidative metabolism in eukaryotic organisms and are crucial to cell metabolism. They produce adenosine triphosphate through the respiratory chain, regulate membrane potential and programmed cell death, store calcium ions to maintain the dynamic balance of calcium concentration in cells, produce ROS, control cell proliferation, and synthesize heme and cholesterol (Wu et al., 2015). pH is a crucial component of cells, and the pH of most organelles ranges from weakly acidic to neutral; pH of mitochondria is slightly alkaline and only changes within an extremely narrow range (≈8.0). Serious diseases including neurodegenerative and mitochondrial genetic disorders, sideroblastic anemia, diabetes, and obesity, are brought on by disruptions to the pH homeostasis of the mitochondria. Thus, it is crucial to monitor variations in mitochondrial pH in real time in order to investigate the physiological and pathological characteristics of mitochondria as well as their function in cellular activities (Jiang et al., 2020).

Based on the intramolecular charge transfer (ICT) process, Chen et al. developed (2022) the switching-type weakly alkaline fluorescence probe (**Cys-pH,** Figure 9.5), which could target mitochondria and identify minute pH variations in them. The fluorescent structure of **Cys-pH** is characterized by a massive π-conjugated link produced by quinoline and vanillin. The recognition group of this bond is the hydroxyl of the phenol unit, the recognition mechanism is deprotonation, and the mitochondrial targeting group is quinolinium salt. **Cys-pH** has a large Stokes shift (142 nm) and acid dissociation constant (pKa = 8.53). In the fluorescence spectrum, **Cys-pH** has a very good linear relationship between fluorescence intensity and pH (6.50–10.00) with high selectivity and sensitivity. **Cys-pH** has been successfully used to monitor changes in mitochondrial pH in living cells.

9.2.2 Quinoline-Based Fluorescent Probe for Lysosomal Imaging

Acidic cellular organelles called lysosomes—also called "cellular debris"—serve as a center of control in eukaryotic cells. They are essential to several biological processes. The roles include macromolecule disintegration and lysosomal preservation, membrane trafficking facilitation, membrane integrity, and preservation of the ionic gradients.

9.2.2.1 Lysosome-Targeted Quinoline-BODIPY Photosensitizer

Treating malignant tumors with light irradiation and photosensitizer is a fast-emerging technique called photodynamic treatment (PDT) (Wang et al., 2020). When exposed to light, photosensitizers convert the triplet oxygen in the environment into highly energetic singlet oxygen, which has potential to destroy cancer cells. Due to its many benefits, including less trauma, quick treatment time, few side effects, and broad application, PDT has garnered significant interest from researchers (Usama et al., 2018). Here, the lysosome serves as an example; it is responsible for the breakdown of biological macromolecules like polysaccharides, proteins, and nucleic acids. Its membrane is essential to its function, and when the lysosomal membrane is destroyed, cells would eventually undergo necrosis and death (Li et al., 2017). Lysosome-targeted photosensitizers offer excellent application prospects to some degree. The following characteristics of an ideal photosensitizer should be present: high singlet oxygen output, low dark toxicity, high phototoxicity, good biocompatibility, and NIR absorption (Hou et al., 2016).

Bai et al. (2020) developed a quinoline-BODIPY fluorophore (**BOP-Lyso)** as a photosensitizer for PDT of cancer cells. Only a few photosensitizers with NIR absorption, good water solubility, low dark toxicity, and excellent colocation ability have been reported. They set out to create a photosensitizer based on the quinoline-BODIPY fluorophore with superior characteristics while keeping these issues in mind, and the creative modification was done, for instance, the addition of iodine at 2 and 6 positions of BODIPY increased the singlet oxygen efficiency. 1-(2-morpholinoethyl)-1H-indole-3-carbaldehyde was connected to the 3 position of BODIPY via a Knoevenagel reaction to achieve strong water solubility. **BOP-Lyso**, a NIR photosensitizer with a high singlet oxygen output,

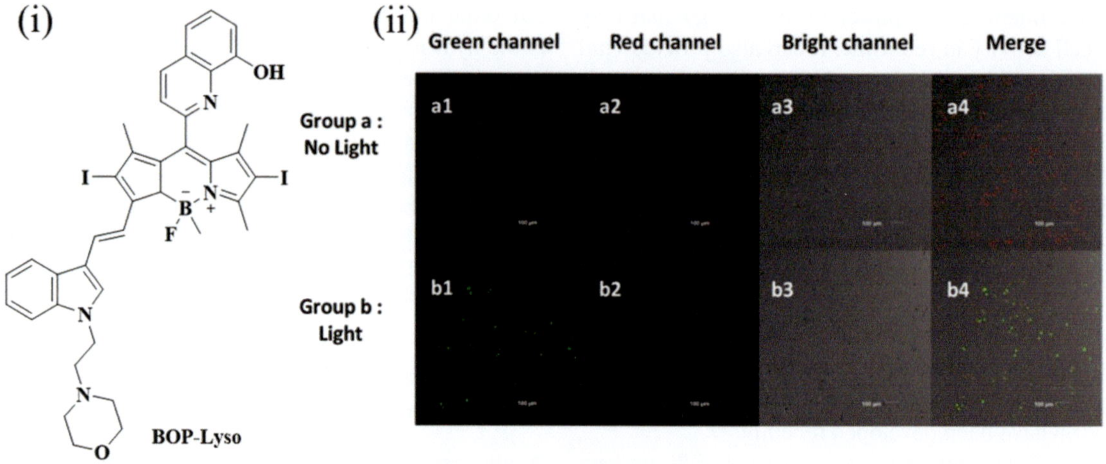

FIGURE 9.4 (i) Structure of BOP-Lyso. (ii) ROS generation in BOP-Lyso (1 μM, Ex: 635 nm, Em: 655–755 nm) treated SGC-7901 cells with the addition of DCFH-DA (2 μM, Ex: 488 nm, Em: 488–520 nm) in the absence and presence of NIR led light irradiation (660 nm, 10 min). Group a: BOP-Lyso without irradiation; Group b: BOP-Lyso with irradiation. Reprinted (adapted) with permission from Bai et al. (2020). Copyright (2020): Elsevier.

demonstrated good lysosomal localization ability, minimal dark toxicity, and good biocompatibility. The **BOP-Lyso** photosensitizer emission peak was at 709 nm, whereas its absorption peak was at 650 nm. Using methylene blue (MB) as the standard, the photosensitizer **BOP-Lyso** singlet oxygen yield under 660 nm NIR LED light irradiation reached 43.1%; ultra-low dark toxicity (cell viability >85%) and great cell compatibility were all present in the photosensitizer. In addition, the photosensitizer **BOP-Lyso** IC_{50} value was 0.24 μM (Figure 9.4). Reactive oxygen species convert nonfluorescent DCFH to green fluorescent DCF; the advent of green fluorescence is a proof that ROS were generated by BOP-Lyso in presence of NIR radiation.

9.2.2.2 Ratiometric Fluorescent Probe for Lysosomal pH Measurement

Intracellular pH values play vital roles in mediating many physiological processes such as cell metabolism, enzyme activity, and immunization (Roos et al., 1981). Lysosomes are acidic organelles (pH 4.0–5.5) in eukaryotic cells that maintain their pH value by the inward pumping of protons (H^+ ions) from the cytosol across the cell membrane via proton pumps. The lysosomes hydrolytic enzymes and proteins must have a low pH in order to work at their optimum (Mindell, 2012). For instance, for the digestive enzymes in lysosomes to break down ingested proteins, DNA, polysaccharides, lipids, viruses, RNA, and bacteria, the environment must be acidic, ranging from pH 4.0 to pH 5.5. Unusual pH levels in lysosomes can cause cellular malfunction, which in turn can cause a variety of serious lysosomal diseases, including mucolipidoses and lipid storage disorders (Platt, 2014). Health professionals must possess straightforward and efficient methods for the quick, accurate, and sensitive measurement of lysosomal pH levels in biological specimens due to the crucial role pH plays in lysosome function.

Liu et al. (2017) developed and synthesized a fluorescent probe (Figure 9.5: **CQ-Lyso**), which targets lysosomes and can measure the pH of lysosomes in living cells. **CQ-Lyso** was modified to include a morpholine moiety for lysosome targeting. The quinoline moiety of **CQ-Lyso** is expected to proportionately get protonated when treated with an increasing gradient of acidity, despite the probe's bright yellow fluorescence in neutral/basic media. Protonation could promote an improved ICT process, causing the absorption and emission spectra to both significantly red shift. Thus **CQ-Lyso** pH-sensitive emission spectra could be used as a ratiometric fluorescence probe to test the pH of lysosomes. In neutral or basic media, the probe exhibited a strong absorption band between 350 and 470 nm, maximum at 397 nm. The 397 nm band eventually vanished as the pH became more acidic, and a new band in the 410–580 nm region developed. The probe displayed a strong yellow florescence with λ_{em} = 560 nm in neutral/basic media (Φ = 0.26). In contrast, when the pH of the solution was down-regulated to more acidic values, the fluorescence gradually changed from yellow to red (λem = 613 nm, Φ = 0.12) under the same excitation wavelength. Thus, increasing acidity was accompanied by red shifts for both absorption and emission spectra of 104 nm and 53 nm, respectively. **CQ-Lyso** ratiometric fluorescent probe is based on chromoquinolines and has extended wavelength emission, great sensitivity, and excellent selectivity for lysosomal pH imaging and detection. **CQ-Lyso** having an exceptionally high Pearson's correlation coefficient (0.97) can selectively stain lysosomes.

Hande et al. (2020) developed a series of quinoline-pyrene probes (Figure 9.5: **6b**) in one step by oxonium ion-triggered alkyne carboamination sequence involving C-C, C-O, and C-N bond formation for intracellular pH sensing. At low pH, quinoline-pyrenes exhibited a noticeable red shift; the probe observations of fluorescence lifetime decay revealed a

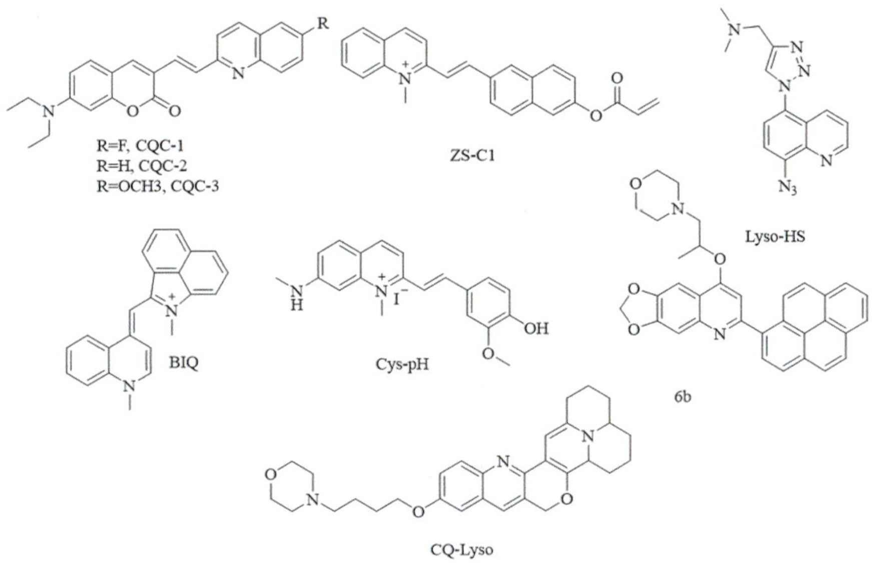

FIGURE 9.5 Structure of quinoline-based probes.

decrease in lifespan at pH 4. In the presence of different interfering agents, such as amino acids and cations/anions, the probes demonstrated outstanding selectivity. Additionally, the probes were discovered to be entirely reversible between pH 4 and 7. Using Lysotracker Deep Red DND-99 as a reference, a quinoline-pyrene probe based on morpholine effectively labeled lysosomes with high Pearson's correlation values (0.86). An investigation of the colocalization of the probe and lysotracker DND-99 revealed specific intracellular targeting as well as a shift in fluorescence emission resulting from an acidic lysosomal pH.

9.2.2.3 Lysosome-Targeted Probe for Detection of Hydrogen Sulfide in Living Cell

Enzymes involved in the endogenous generation of H_2S: cystathionine γ-lyase, cystathionine β-synthase (CBS), 3-cysteine aminotransferase and mercaptopyruvate sulfur transferase (3-MST) (Stipanuk, 2004). Recently, altered H_2S levels have been linked to several illnesses, including cancer, diabetes, and Alzheimer's disease (Eto et al., 2002). Monitoring H_2S levels and spatiotemporal patterns in living cells and organisms is therefore essential. The lysosome is a crucial organelle in numerous physiological processes, including cell death, membrane repair, metabolism, and so forth. Understanding some intracellular dynamic processes and developing treatments for associated disorders depend on observing reactive small molecules (RSMs) in lysosomes.

Dou et al. (2018) developed a fluorogenic probe based on 5-triazole quinoline (Figure 9.5: **Lyso-HS**); the probe was used to monitor H_2S in lysosomes in living cells. It demonstrated good turn-on capabilities with a 95-fold fluorescence enhancement. The lysosome-targeted probe **Lyso-HS** was synthesized via C-H direct azidation from 8-amino quinoline, which is capable of imaging endogenous H_2S in cells in real time. The H_2S treatment boosted the probe's fluorescence intensity by an amount of 95, based on the azidation bond's selective thiolysis. Under physiological conditions, **Lyso-HS** demonstrated great sensitivity, selectivity, and a low detection limit for H_2S in comparison to other physiologically relevant analytes, such as biothiols. It was also demonstrated to have minimal cytotoxicity and to be a useful tool for fluorescent imaging of H_2S in living cells. It may be particularly useful for identifying endogenous H_2S in cancer cells.

9.2.2.4 Simultaneous Detection of Lysosomal Viscosity and SO_2

Lysosomes are essential to the processes of cellular degradation since they are the digestive organelles that degrade biological polymers in live cells (Sun et al., 2020). Degradative enzymes in lysosomes keep fluid viscosity consistent; however in malfunctioning lysosomes, enzyme inactivation causes biomolecules to accumulate and aberrant viscosity augmentation (Wei et al., 2021). Conversely, sulfur dioxide (SO_2), found in aqueous solutions as HSO^{3-}/SO_3^{2-}, is linked to numerous physiological and pathological processes (Zhang et al., 2021). Nevertheless, research has shown that SO_2 concentration affects the quantity and activity of lysosomal enzymes, which may contribute to a number of illnesses, including cancer (Cheng et al., 2021). It is also linked to oxidative stress and metabolic dysfunction of lysosomes.

9.2.3 Quinoline-Based Fluorescent Probe for Detection of Cysteine

An intracellular biothiol called cysteine (Cys) is essential to many physiological and pathological processes (Paul et al., 2018). Cys, meanwhile, is engaged in a wide range of biological processes, such as detoxification, signal transduction, protein synthesis, and biocatalysis (Wlodek et al., 2001). A high concentration of this biothiol typically causes cancer, heart disease, liver damage, weakening, and lethargic behavior, among other illnesses (Hirai et al., 2013). Zhang et al. (2022) designed **ZS-C1** (Figure 9.5) fluorescent probe to provide a practical and effective tool for the early diagnosis of cancer by monitoring the physiological level of endogenous and exogenous Cys in real time. The most malignant brain tumor, glioblastoma (GBM), has a higher chance of survival if detected early. The clinical diagnostic methods for GBM, including PET (positron emission tomography), CT (computed tomography), MRI (magnetic resonance imaging), and biopsy tissue samples, have been substantially developed in recent years. The identification of cysteine levels as a novel GBM biomarker in human cells and in situ offers a new tool for the early detection and clinical surgical guiding of GBM. **ZS-C1** fluorescent probe uses a conjugate addition-cyclization reaction mechanism to detect Cys in vitro. Initially, the cationic quinoline group was chosen as the electron-accepting unit, giving **ZS-C1** the ability to penetrate the blood-brain barrier and aid in the early detection of GBM. Acrylate was then used as the acceptor unit so that Cys could be uniquely recognized by the probe **ZS-C1**. The probe demonstrated exceptional sensitivity and selective responsiveness in particular. Naturally, the fluorescent probe also functioned admirably in tissue imaging of 3D tumor spherical tissue and fluorescence imaging of live U87MG cells. In the pH range of 6–9, it can reliably detect Cys. Cys combines with the probe acrylate group to release ZS-C1-OH, which subsequently quenches fluorescence. **ZS-C1** demonstrated dose-dependent responses to the Cys level at low concentrations in living U87MG cells, and it was able to monitor both exogenous and endogenous Cys in these cells.

9.2.4 Quinoline-Based Fluorescent Probe for Detection of Nucleic Acids

9.2.4.1 Benzo[c,d]indole–Quinoline Monomethine Cyanine for Imaging of Nucleolar RNA in Living Cells

RNA-binding fluorescent probes have proven to be effective and adaptable tools for investigating RNA biological functions (Xia et al., 2017). This is because they have several benefits such as high space resolution, membrane permeability, and versatile photophysical properties. Small-molecular-weight probes appropriate for RNA analysis have not been as developed as protein-based and oligonucleotide-based fluorescent probes (Wang et al., 2018; Wilson et al., 1993). The primary reason for this is that tiny probes prefer to connect to DNA rather than RNA. Only SYTO RNA select is a commercially available probe that exhibits RNA selectivity. SYTO RNA select is nearly nonfluorescent, but when it binds to RNAs, its fluorescence intensity has increased by 100 times. The light-up feature of SYTO RNA select makes it easier to detect RNA with extreme sensitivity. Yoshino et al. (2019) developed a novel small probe based on an asymmetric monomethine cyanine scaffold (Figure 9.5: **BIQ**) to satisfy deep red emission and light-up response for RNA analysis simultaneously. Asymmetric monomethine cyanine dyes are commonly employed as fluorescent probes that bind nucleic acids. **BIQ** has emission maxima at 657 nm and exhibits a notable 105-fold light-up response and superior photostability compared to the RNA selective probe. BIQ was effectively used to image nucleolar RNAs fluoroscopically in living cells with minimal damage.

9.3 QUINOLINE-BASED FLUORESCENT PROBES FOR MISCELLANEOUS APPLICATIONS

9.3.1 pH-Sensing Probes

Numerous cellular processes, such as cell division, apoptosis, endocytosis, signal transduction, and ion transport, depend on the maintenance of the acid-base (pH) balance. For proper operation, cells have precise processes to control the intracellular pH. Hypoxia in the tumor microenvironment modifies the metabolism of cancer cells, hence influencing pH levels. Compared to normal cells, cancer cells often have greater intracellular pH and lower extracellular pH, which encourages invasion and proliferation while impeding apoptosis. Moreover, the effectiveness of pH-sensitive chemotherapy medications is impacted by this pH gradient. Cardiopulmonary and neurological problems might result from abnormal pH levels. Subcellular spaces such as mitochondria and lysosomes have different pH values that affect how they operate. Knowing the pH of living cells is crucial to comprehending physiological and pathological mechanisms.

Guria et al. (2021) have created fluorescent probes called **QNOH-MO** and **QNOH-OME** (Figure 9.6), which are based on quinoline. They have conducted a comprehensive analysis of the probes' photophysical characteristics. Both **QNOH-MO** and **QNOH-OME** can detect slight changes in pH using colorimetric and fluorescence techniques. Under conditions of low pH, the probes displayed red fluorescence in both living and preserved cells. As the pH level rises, the intensity of red fluorescence diminishes while the intensity of green fluorescence increases. The unique crimson fluorescence exhibited in cellular contexts is most likely a result of the combined influence of acidic pH and variables such as viscosity or hydrophobicity. This was verified using theoretical calculations and ^{1}H NMR titration tests. In addition, the process of quaternizing the morpholine unit in **QNOH-MO** gave it an inherent cationic feature. This led to the creation of the cationic probe **QNOH-MO-CA** (Figure 9.6), which showed higher water solubility and greater efficiency in detecting pH levels. A fluorescent microscope was subsequently used to capture images of the fixed cells. It was discovered from the experiment that the probe displayed emission up to

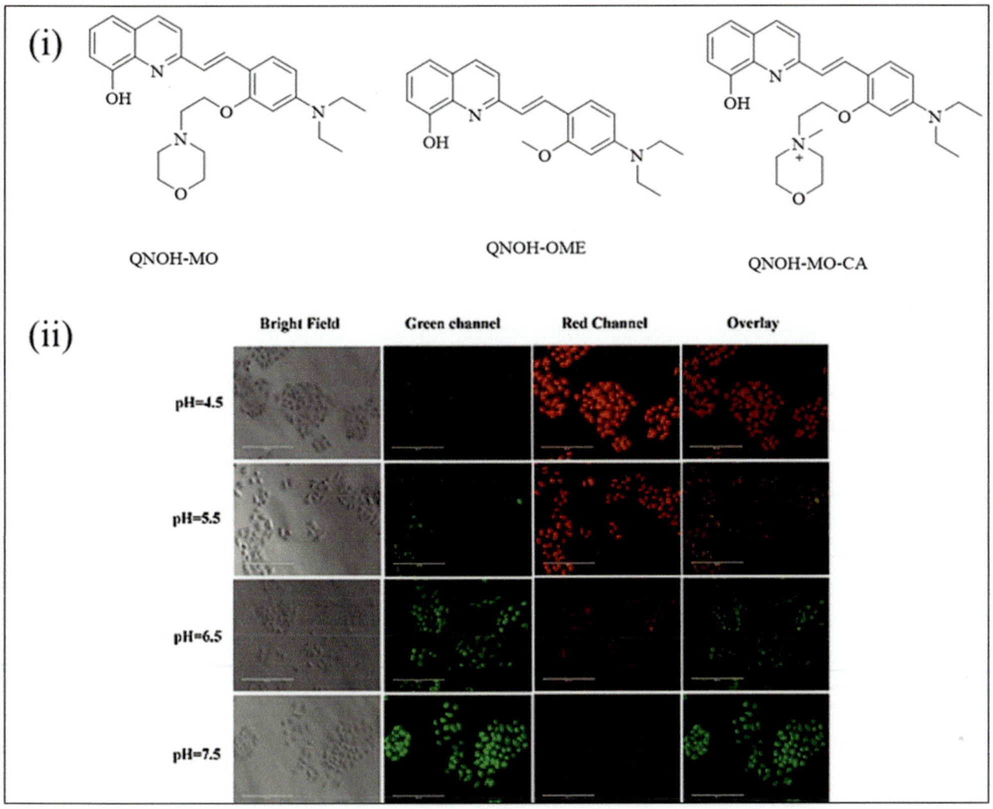

FIGURE 9.6 (i) Structures of QNOH-MO, QNOH-OME, and QNOH-MO-CA. (ii) Imaging of QNOH-MO-CA at different pH in live cells. Reprinted (adapted) with permission from Guria et al. (2021). Copyright (2021): Elsevier.

pH 4.6 in the red channel. The green channel's emission was minimal inside this pH range. In contrast, starting at pH 5, the fluorescence increases from the green channel and peaks at pH 7.5. At this pH level, no discernible red channel emission was found. These combined findings indicated that these probes might be utilized to quantify pH variations within cells, both living and fixed. In addition, **QNOH-OME** demonstrated exceptional efficacy as a marker for fluorescently tagging cholesterol molecules, which are crucial constituents of biological membranes and liposomal formulations used in drug administration (Guria et al., 2021).

9.3.2 Copper and Sulphur Sensing Probes

Copper (II) ions, which are the third most abundant transition metals in the human body, are essential for numerous physiological processes despite their diversity. On the other hand, illnesses like Menke's, Alzheimer's, and Wilson's, can be exacerbated by an excessive buildup of copper in nerve cells. Creating synthetic probes to find Cu^{2+} has garnered a lot of attention lately. Cu^{2+} magnetic characteristics usually cause a fluorophore to lose some of its fluorescence when it attaches to it, which is considered a restriction in Cu^{2+} detection. As an alternative, the subsequent restoration of the released fluorophore's fluorescence could provide an indirect means of identifying anions by fluorescence amplification if the dormant Cu^{2+}-fluorophore complex could serve as a highly selective sensor for anions by displacing Cu^{2+}. Sulfur is a common element in the environment and has several uses, such as being used in colors and cosmetics. On the other hand, prolonged exposure to elevated sulfide levels may result in a variety of physiological and biochemical problems, including irritation of mucous membranes, unconsciousness, and respiratory paralysis. A plethora of detection techniques, including spectroscopy, electrochemical methods, titration, ion chromatography, and chemiluminescence methods, have been developed in response to the growing interest in detecting sulfide anions in recent years. Among them, fluorescence spectroscopy has become more well liked because of its high sensitivity and user-friendliness. Tang et al. synthesized a new acylhydrazone probe **1** (Figure 9.7), which displayed selective sequential identification of Cu^{2+} and S^{2-} in HEPES-buffered solution (with 1% DMSO, HEPES 20 mM, pH = 7.4). Probe 1 interacts with Cu^{2+} in the solution with a 1:1 stoichiometry; the resultant 1-Cu^{2+} complex displays remarkable selectivity toward sulfide ions and a copper displacement mechanism-based "off-on" fluorescence response. Probe 1 is a potential sensor for detecting Cu^{2+} and S^{2-} in aqueous samples because of its special ability to recognize Cu^{2+} and S^{2-} sequentially (Tang et al., 2013).

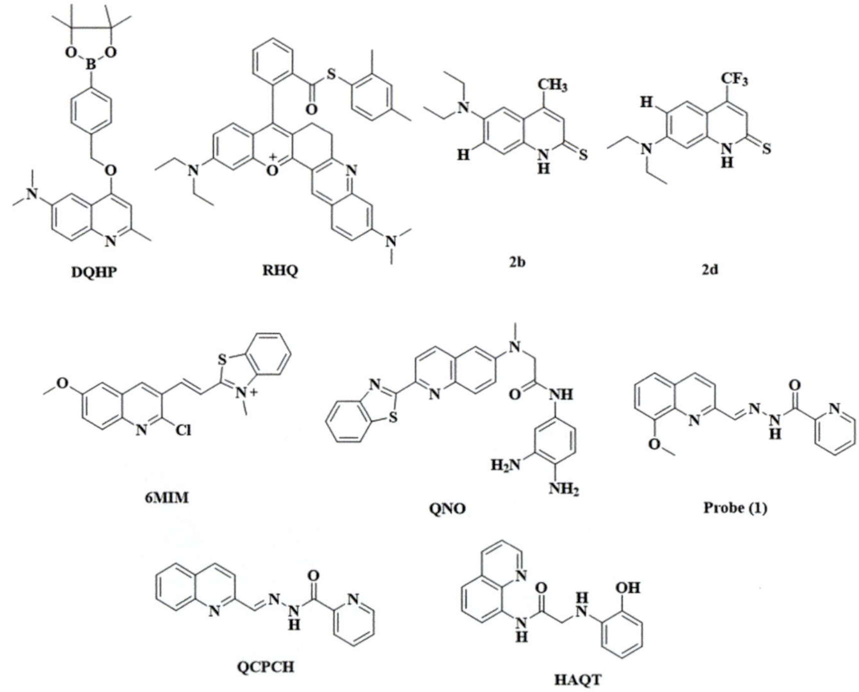

FIGURE 9.7 Structure of quinoline-based probes.

9.3.3 Zinc-Sensing Probes

Transition metal ions like zinc are important in life sciences, health, energy, and environmental chemistry. Zinc is an important micronutrient and the second most prevalent transition metal ion in the body. Neurotransmission, signal transduction, gene expression, cellular metabolism, and metalloenzyme control are its functions. Zinc shortage and excess may cause health problems, therefore maintaining the appropriate balance is important. Zinc deficiency may affect cognition, immunity, digestion, and development. Zinc overload may damage humans and the environment, especially plant metabolism and development. Thus, zinc levels must be carefully monitored in the body and environment to reduce hazards and maximize benefits. Understanding zinc's many functions is crucial to addressing zinc imbalance and using its potential for scientific development and environmental sustainability.

Wang et. al. used a simple one-step condensation procedure to effectively synthesize **QCPCH** (Figure 9.7), a quinoline-based probe. QCPCH exhibits remarkable performance qualities such as low detection limit of 72 nM for Zn^{2+}, a broad linear detection range, and excellent selectivity and sensitivity. The combined effects of chelation enhanced fluorescence(CHEF) and suppression of the photoinduced electron transfer (PET) process reveal the sensing mechanism of **QCPCH** toward Zn^{2+}. Comprehensive studies, such as Benesi-Hildebrand research, FT-IR spectrum analysis, Job's plot analysis, and ^{1}H NMR titration, are used to support this process.

Furthermore, **QCPCH** shows its versatility outside of lab environments. It exhibits adaptability for Zn^{2+} detection in actual water samples and is readily adaptable into test strips for quick, practical, and accurate Zn^{2+} measurement that is both qualitative and quantitative.

To sum up, this study's results have a substantial impact on the development of fluorescent chemosensors for Zn^{2+} detection, broadening the field's research and application potential. **QCPCH** strong performance and usefulness highlight its potential as a useful tool in a range of analytical and environmental monitoring scenarios (Wang et al., 2021).

A new era in analytical chemistry has been ushered in with the painstaking design and synthesis of **HAQT** (Figure 9.7), a revolutionary quinoline-based fluorescent Zn^{2+} chemosensor. It is noteworthy that **HAQT** exhibits exceptional sensitivity and selectivity for Zn^{2+}, along with a visually noticeable color shift from dark purple to dazzling yellow. This visual indication makes Zn^{2+} detection simple and quick, allowing for on-spot examination without the need for specialist tools.

Interference tests with other metal ions, especially Cd^{2+}, are intriguing since they clearly validate **HAQT** resilience against possible confounders and reinforce its dependability in Zn^{2+} sensing applications. Notably, EDTA may effectively be used to recover the probe, guaranteeing its reusability and reducing its environmental impact. It's exciting to note that **HAQT** has already shown to be useful in practical settings, having been effectively used to analyze water samples. This achievement highlights its potential as a useful instrument for environmental monitoring, providing a quick, sensitive, and affordable way to detect Zn^{2+} in a variety of aqueous matrices.

The unique design approach and exceptional photophysical characteristics of **HAQT** show promise for a wide range of biological research and environmental sample analysis applications. **HAQT** has the potential to transform Zn^{2+} detection and provide new avenues for study in environmental science, health monitoring, and other fields as scientists continue to test and improve it (Yue et al., 2015).

9.3.4 Nitric Oxide

Nitric oxide (NO) regulates many physiological and pathological processes in the cardiovascular, neurological, and immunological systems. Dysregulated NO production is linked to cancer, neurological disorders, and inflammation. NO reactivity, lipid solubility, and diffusion make detection difficult, despite its relevance. With various bioimaging fluorescent probes, real-time NO detection is promising. Most Nagano and Lippard probes are suited for one-photon microscopy (OPM), although they have short excitation wavelengths, small penetration depth, and cause photodamage. Two-photon microscopy (TPM) has been popular in bioimaging during the last decade because it overcomes these constraints. TPM-compatible NO probes have proved difficult to construct owing to a shortage of scaffolds. Only two TPM probes for intracellular NO have been published, stressing the need for more. **QNO** (Figure 9.7), a revolutionary efficient TPM probe for NO, was invented and prepared by Dong et al. A quinoline derivative is the fluorophore and o-phenylenediamine is the NO receptor coupled to glycinamide in **QNO**. The probe detects NO in living cells and tissues at depths beyond 100 μm without photobleaching or cytotoxicity. **QNO** is very sensitive to NO as shown by its impressive 12-fold rise in fluorescence in response to NO. It reacts with NO to provide a maximal TP action cross-section value of 52 GM at 810 nm, indicating excellent selectivity for NO. Using TPM, **QNO** can detect NO in living cells and tissues for prolonged periods of time at depths between 90 and 180 μm without photobleaching.

9.3.5 Sulfur Mustard

Since its creation in the 1800s, sulfur mustard (SM), often known as mustard gas, has been infamous for its employment as a chemical warfare weapon in World War I and later wars. Blisters, respiratory tract injuries, eye lesions, and bone marrow depression result from this very damaging chemical. As an alkylating chemical, SM reacts with DNA guanine nucleotides, causing carcinogenic and mutagenic consequences with extended exposure. SM symptoms appear 30 minutes to 8 hours after incubation, slowing response attempts and worsening damage.

Chemical weapons from previous wars still threaten lives and national security. In China, chemical weapons left by invaders, particularly the Japanese, still cause diseases and events like the 2003 Qiqihar City release, which killed 44 people. The simplicity of SM composition and ease of synthesis raise worries regarding terrorist usage, emphasizing the necessity for quick, reliable SM detection tools.

Using 2-sulfur quinolones and a thioamide moiety, four fluorescent probes have been created for the detection of sulfur mustard (SM) and its analogues, 2-chloroethyl ethyl sulfide (CEES) and nitrogen mustard hydrochloride (NH). When quinolones come into contact with KOH, the thioamide group functions as a reactive site and deprotonates to produce a highly nucleophilic thiophenol anion. These originally nonfluorescent probes undergo a sensing response to the three mustards, which causes them to react quickly and create D-A fluorescent molecules called thioethers. With quick reaction time (between 1 and 5 minutes), excellent selectivity, and great sensitivity even at nanomolar concentrations (LOD ~ nM level), 2b and 2d outperform the other four probes in terms of sensing capability. In order to detect CEES in the gas phase, test strips containing poly(ethylene oxide) and either 2b or 2d were also created. These test strips can detect SM vapor even below the acute exposure guideline levels threshold (0.6 ppm/min), as shown by their exceptional selectivity and sensitivity (0.2 ppm/4 minutes). As such, these two fluorescence probes have the potential for SM detection in gas phases as well as liquid solutions (Feng et al., 2021).

9.3.6 Glutathione

Glutathione (GSH), the most prevalent tripeptide among biological thiols, is vital for xenobiotic metabolism, intracellular redox control, gene expression, and signal transduction in both eukaryotic and mammalian cells. As a critical endogenous antioxidant, GSH is essential for cellular defense against toxins and free radicals. GSH imbalances have been linked to cancer, cardiac issues, aging, and other illnesses. Thus, it is essential to provide quick, easy, and precise techniques for identifying GSH and tracking changes in GSH levels in a physiological setting. **6MIM** (Figure 9.7), an intramolecular charge transfer (ICT) hemicyanine fluorescence probe containing quinoline, was produced and thoroughly investigated. Surprisingly, it demonstrates glutathione (GSH) preferential detection over homocysteine and cysteine, with a nanomolar detection limit. The usefulness of **6MIM** as a fluorescent probe for identifying GSH in live cells was further supported by cell research. Furthermore, real-time online GSH analysis was made possible by the probe's effective integration with a smartphone RGB color value application. This method's 120 nM detection limit demonstrated the probe's potential for useful applications in high-sensitivity and convenient GSH detection (Mani *et al.*, 2020).

9.3.7 Hydrogen Peroxide

Interest in studying intracellular hydrogen peroxide (H_2O_2) has increased with the development of fluorescence imaging technologies. This is because oxidative stress and serious health problems may be brought on by the buildup of H_2O_2, a consequence of oxygen metabolism, in cells. Recent research has also shown the function of H_2O_2 as a messenger in normal cellular functions. As a result, there is an increasing demand for accurate and focused fluorescent probes to identify intracellular H_2O_2, supporting research as well as possible therapeutic uses. Although several fluorescent probes have been used for H_2O_2 imaging, they have problems with selectivity against other reactive oxygen or nitrogen species (ROS/RNS). But new developments in design strategies, including benzil, boronate ester chemistry, or benzenesulfonyl ester chemistry, have produced probes with improved selectivity and sensitivity. However, since most of these probes depend on fluorescence intensity, which might be impacted by outside interference altering intracellular fluorescence, they could not provide the necessary precision for quantitative measurements. By enabling self-calibration using two selected emission wavelengths, ratiometric probes offer a viable remedy and guarantee precise and quantitative analysis. However, the development of highly sensitive ratiometric probes for H_2O_2 detection is desired, since presently there is a lack of such probes.

In order to detect H_2O_2, a novel fluorescent probe based on quinoline has been devised called **DQHP** (Figure 9.7). The monocationic condition of the probe causes it to produce fluorescence at a wavelength of 542 nm when it is submerged in water. The boronate cleavage that is caused by H_2O_2 is responsible for the drop in fluorescence at 542 nm and the noticeable shift in emission to 480 nm that occurs as a result of interaction with hypochlorous acid. As a result, a ratiometric measurement is accomplished by using a single excitation wavelength and dual emission wavelengths. This measurement is distinguished by a large blue shift in emission ($\Delta\ \lambda = 62$ nm) and considerable variations in the emission intensity ratio (F480 nm/F542 nm), with a maximum fold increase of 8.3. In addition, the probe has a high selectivity for detecting H_2O_2 in comparison to other reactive oxygen species that are present in biological systems (Qian *et al.*, 2012).

9.3.8 Lipid Droplets

Within the majority of eukaryotic cells, lipid droplets (LDs) perform the function of universal storage compartments for neutral lipids. They are also responsible for playing critical roles in the maintenance of energy balance and lipid metabolism. Additionally, LDs provide important activities inside cells, including functioning as buffers against stress, shielding mitochondria from damage during autophagy, and protecting against endoplasmic reticulum (ER) stress. These functions are in addition to their role as lipid storage carriers. Research that was done in the past has shed light on the close relationship that exists between the size and concentration of LDs and the physiological state of cells. It is important to note that changes in LDs may act as biomarkers for a variety of illnesses, such as insulin resistance, coronary artery disease, fatty liver diseases, and neurodegenerative disorders. Therefore, in vivo monitoring of LD-related behaviors is of major relevance for gaining knowledge of the biological roles of lipid droplets, gaining an awareness of the physiological condition of cells, and allowing early disease detection. Liet al. (2020) have designed bioprobes LDP1–LDP4 (Figure 9.8); these probes show solvatochromic behavior and biocompatibility in comparison to commercially available probes. LDP4 could stain LDs by collecting fluorescent signals from the green channel (500–600 nm) (Figure 9.10).

9.3.9 Hydrazine

Hydrazine is used extensively in pharmaceutical production as an intermediate, insecticide, and catalytic agent. It is also a popular option for propellants in rockets, satellites, and missiles due to its high-energy potential. Hydrazine has a lot of health and environmental risks, even with its usefulness. Along with being carcinogenic, it is acutely neurotoxic. The environment and living things may suffer greatly as a result of improper handling throughout the processes of manufacture, transportation, application, and disposal. Severe organ damage may result from hydrazine exposure by ingesting,

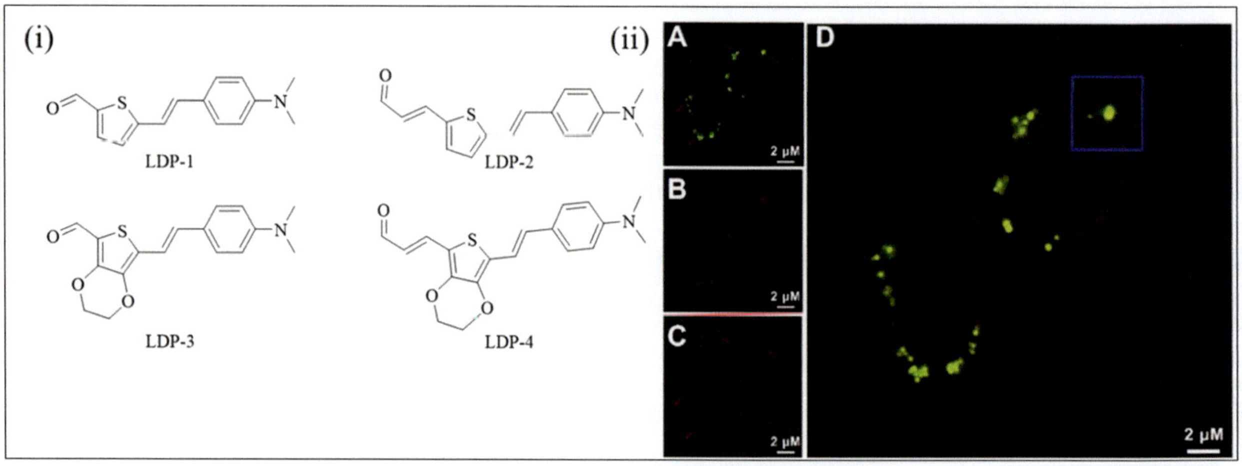

FIGURE 9.8 (i) Quinoline-based probes for labeling lipid droplets. (ii) Laser scanning confocal images of U87 cells stained with LDP4 (5 mM). (A) Green channel. (B) Red channel. (C) Bright-field image of U87 cells. (D) Merged image green and red channels. Reprinted (adapted) with permission from Zhang et al. (2020). Copyright (2020): Elsevier

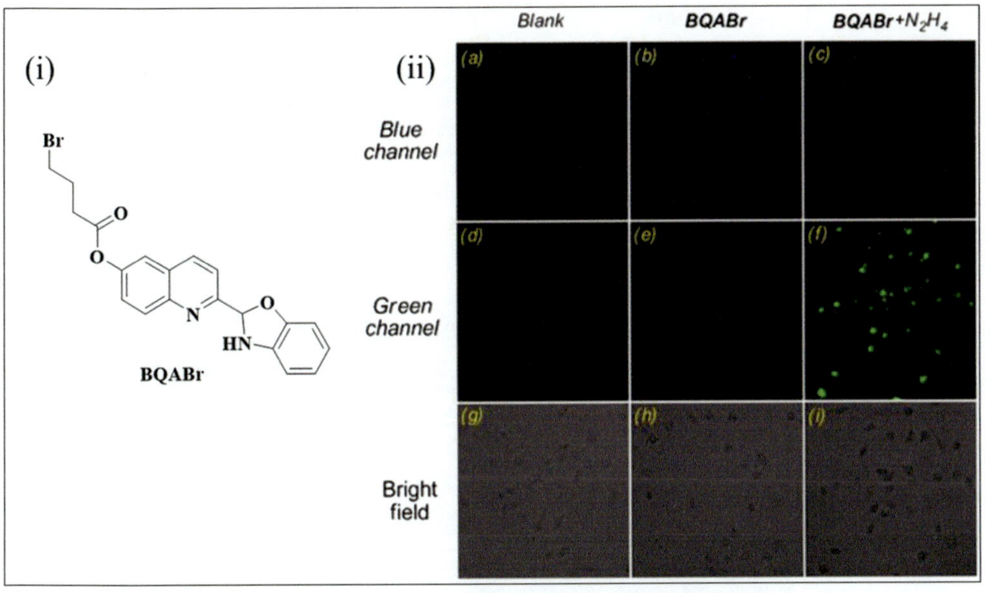

FIGURE 9.9 (i) Structure of BQABr. (ii) Fluorescence emission image of MDA-MB-231 cells; signals were collected from the blue channel (410–460 nm) or green channel (500–550 nm). Reprinted (adapted) with permission from Wu et al. (2019). Copyright (2019): Elsevier.

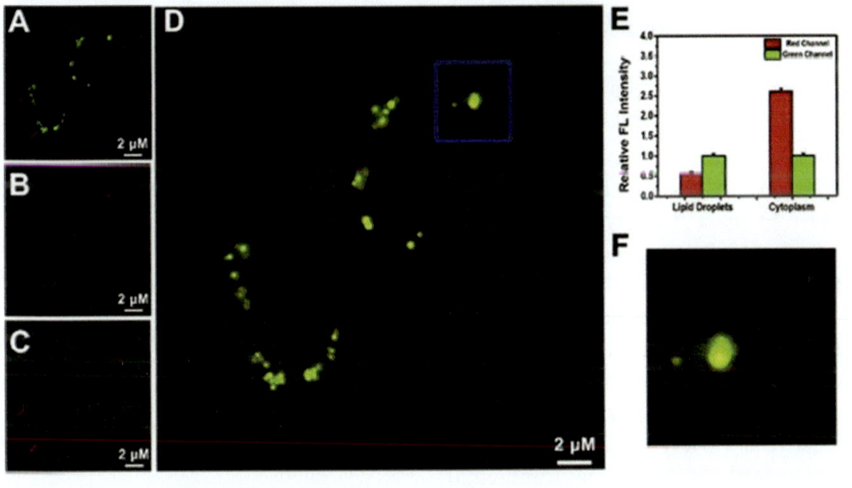

FIGURE 9.10 Laser scanning confocal images of U87 cells stained with LDP4 (5 mM). (A) Green channel. (B) Red channel. (C) The bright field of U87 cells. (D) The merge of green and red channels. (E) The semi-quantification of fluorescence signals in different channels. (F) The inset in D was zoomed in F. Note: The arrow points to the LDs. Reprinted (adapted) with permission from Li et al. Analytica Chimica Acta (2020),

inhalation, or oral and dermal contact. Because of the risks associated with hydrazine, regulatory organizations like the Environmental Protection Agency of the United States have established strict restrictions. The recommended threshold level of 10 parts per billion (ppb) for hydrazine in the environment highlights the need for stringent monitoring and control methods to lessen its detrimental impacts on ecological integrity and human health.

Using a special SN2 cyclization method, a new quinoline-derived fluorescent probe called **BQABr** (Figure 9.9) has been synthesized to selectively detect hydrazine in aqueous solutions. As a ratiometric fluorescent probe, **BQABr** performs very well and has a quick assay time of 30 minutes. **BQABr** has exceptional sensitivity and accuracy, exhibiting an emission shift of more than 130 nm and a signal response of more than 3,000 fold. It also features two separate emission peaks with little spectrum overlap, allowing for accurate identification. Sensitivity tests confirm the probe's effectiveness by demonstrating its very low detection limit of 5.8 nM (0.19 ppb) for quantitative hydrazine determination. Furthermore, test paper strips coated with **BQABr** have been effectively created and used to detect gaseous hydrazine, exhibiting discernible color shifts in fluorescence. Additionally, **BQABr** shows off its usefulness in environmental monitoring by exhibiting high levels of consistency between data discovered in actual environmental samples and reference data. The accuracy of **BQABr** for hydrazine detection in intricate environmental matrices is shown by this validation. Interestingly, investigations with cell imaging show that

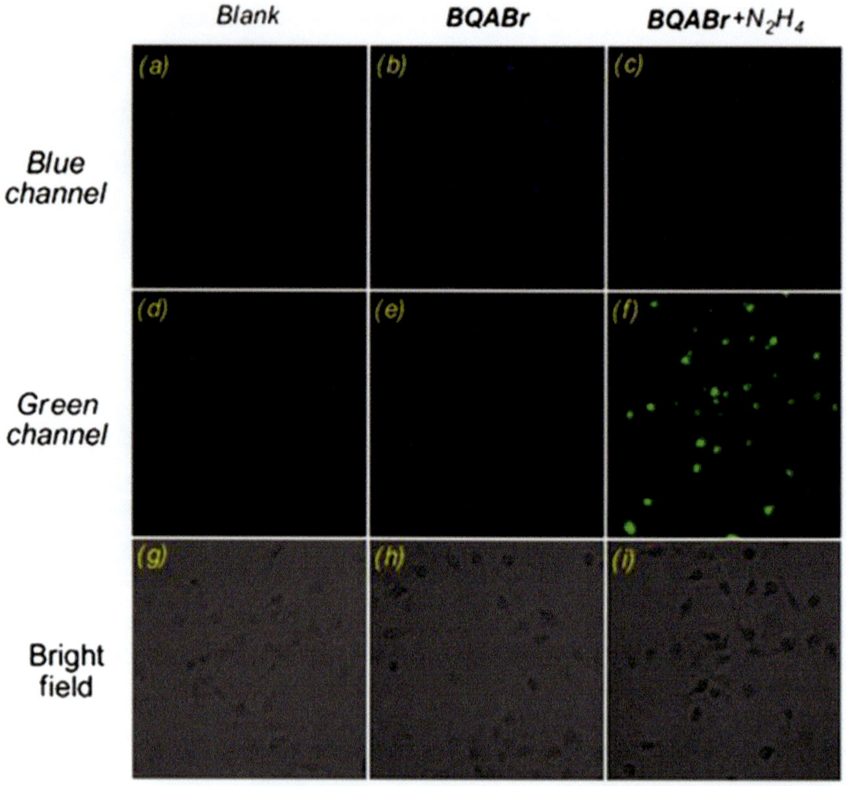

FIGURE 9.11 Confocal fluorescence emission image of MDA-MB-231 cells, signal was collected from blue channel (410–460 nm) or green channel (500–550 nm). Reprinted (adapted) with permission from Wu et al. (2019). Copyright (2019): Elsevier.

the probe is equally effective in biological environments. The capacity of **BQABr** to react with hydrazine in live cells efficiently is shown by the fluorescence color change of the cells from blue to green following hydrazine addition (Figure 9.11), allowing sensing in biological contexts. With unmatched sensitivity, precision, and adaptability, **BQABr** is a major improvement in the area of fluorescent probes for hydrazine detection. Its prospective uses promise revolutionary effects in a variety of fields, including biomedical research, industrial safety, and environmental monitoring (Wu et al., 2019).

REFERENCES

Bai, J., & Qian, Y. (2020). Construction of an NIR and lysosome-targeted quinoline-BODIPY photosensitizer and its application in photodynamic therapy for human gastric carcinoma cells. *Dyes and Pigments*, *181*, 108615.

Chen, J., Xu, Y., Gao, Y., Sun, L., Meng, X., Gu, K., Zhang, Y., & Ning, X. (2021). A mitochondria-specific fluorescent probe for rapidly assessing cell viability. *Talanta*, *221*, 121653.

Chen, Y. S., Zhou, B. Z., Liu, F. T., Miao, J. Y., Zhao, B. X., & Lin, Z. M. (2022). A quinoline-salt-based fluorescent probe for precise monitoring of pH changes on mitochondria and water. *Sensors and Actuators B: Chemical*, *373*, 132732.

Cheng, J., Li, Z., & Lin, W. (2021). Development of a one-step synthesized red emission fluorescent probe for sensitive detection of viscosity in vitro and *in vivo*. *Spectrochimica Acta - Part A: Molecular and Biomolecular Spectroscopy*, *258*, 119808.

Dou, Y., Gu, X., Ying, S., Zhu, S., Yu, S., Shen, W., & Zhu, Q. (2018). A novel lysosome-targeted fluorogenic probe based on 5-triazole-quinoline for the rapid detection of hydrogen sulfide in living cells. *Organic & Biomolecular Chemistry*, *16*(5), 712–716.

D'Emmanuele di Villa Bianca, R., Sorrentino, R., Maffia, P., Mirone, V., Imbimbo, C., Fusco, F., ... & Cirino, G. (2009). Hydrogen sulfide as a mediator of human corpus cavernosum smooth-muscle relaxation. *Proceedings of the National Academy of Sciences*, *106*(11), 4513–4518.

Deliconstantinos, G., Villiotou, V., & Stavrides, J. C. (1995). Modulation of particulate nitric oxide synthase activity and peroxynitrite synthesis in cholesterol enriched endothelial cell membranes. *Biochemical Pharmacology*, *49*(11), 1589–1600.

Eto, K., Asada, T., Arima, K., Makifuchi, T., & Kimura, H. (2002). Brain hydrogen sulfide is severely decreased in Alzheimer's disease. *Biochemical and Biophysical Research Communications*, *293*(5), 1485–1488.

Feng, W., Li, H., Xue, M. J., Zhang, Q. L., Liu, S. L., & Song, Q. H. (2021). Quinoline-2-thione-based fluorescent probes for selective and sensitive detections of mustard gas and its analogues. *Analytica Chimica Acta*, *1159*, 338440.

Guria, S., Ghosh, A., Mishra, T., Das, M. kumar, Adhikary, A., & Adhikari, S. (2021). X-ray structurally characterized quinoline based fluorescent probes for pH sensing: Application in intracellular pH imaging; DFT calculations and fluorescent labelling. *Journal of Photochemistry and Photobiology A: Chemistry*, *407*, 113704.

Hande, P. E., Mishra, M., Ali, F., Kapoor, S., Datta, A., & Gharpure, S. J. (2020). Design and expeditious synthesis of quinoline-pyrene-based ratiometric fluorescent probes for targeting lysosomal pH. *ChemBioChem*, *21*(10), 1492–1498.

He, L., Tan, C. P., Ye, R. R., Zhao, Y. Z., Liu, Y. H., Zhao, Q., Ji, L. N., & Mao, Z. W. (2014). Theranostic iridium(III) complexes as one- and two-photon phosphorescent trackers to monitor autophagic lysosomes. *Angewandte Chemie - International Edition, 53*(45), 12137–12141.

Hirai, T., Kanda, T., Sato, K., Takaishi, M., Nakajima, K., Yamamoto, M., Kamijima, R., DiGiovanni, J., & Sano, S. (2013). Cathepsin K is involved in development of psoriasis-like skin lesions through TLR-dependent Th17 activation. *The Journal of Immunology, 190*(9), 4805–4811.

Hou, W., Xia, F., Alves, C. S., Qian, X., Yang, Y., & Cui, D. (2016). MMP2-targeting and redox-responsive PEGylated Chlorin e6 nanoparticles for cancer near-infrared imaging and photodynamic therapy. *ACS Applied Materials and Interfaces, 8*(2), 1447–1457.

Humphreys, M. S. (1976). Relational information and the context effect in recognition memory. *Memory & Cognition, 4*(2), 221–232.

Jiang, X., Liu, Z., Yang, Y., Li, H., Qi, X., Ren, W. X., Deng, M., Lü, M., Wu, J., & Liang, S. (2020). A mitochondria-targeted two-photon fluorescent probe for sensing and imaging pH changes in living cells. *Spectrochimica Acta - Part A: Molecular and Biomolecular Spectroscopy, 224*, 117435.

Kepp, O., Galluzzi, L., Lipinski, M., Yuan, J., & Kroemer, G. (2011). Cell death assays for drug discovery. In *Nature reviews drug discovery* (Vol. 10, No. 3, pp. 221–237).

Li, L., Xu, Y., Chen, Y., Zheng, J., Zhang, J., Li, R., Wan, H., Yin, J., Yuan, Z., & Chen, H. (2020). A family of push-pull bio-probes for tracking lipid droplets in living cells with the detection of heterogeneity and polarity. *Analytica Chimica Acta, 1096*, 166–173.

Li, M., Tian, R., Fan, J., Du, J., Long, S., & Peng, X. (2017). A lysosome-targeted BODIPY as potential NIR photosensitizer for photodynamic therapy. *Dyes and Pigments, 147*, 99–105.

Li, Z., Ren, T.-B., Zhang, X.-X., Xu, S., Gong, X.-Y., Yang, Y., *et al.* (2021). *Precipitated fluorophore-based probe for accurate detection of mitochondrial analytes.* ACS Publications. Collection.

Liu, H. W., Chen, L., Xu, C., Li, Z., Zhang, H., Zhang, X. B., & Tan, W. (2018). Recent progresses in small-molecule enzymatic fluorescent probes for cancer imaging. In *Chemical society reviews* (Vol. 47, No. 18, pp. 7140–7180). Royal Society of Chemistry.

Liu, H. W., Hu, X. X., Li, K., Liu, Y., Rong, Q., Zhu, L., Yuan, L., Qu, F. L., Zhang, X. B., & Tan, W. (2017). A mitochondrial-targeted prodrug for NIR imaging guided and synergetic NIR photodynamic-chemo cancer therapy. *Chemical Science, 8*(11), 7689–7695.

Liu, Y., Yu, D., Ding, S., Xiao, Q., Guo, J., & Feng, G. (2014). Rapid and ratiometric fluorescent detection of cysteine with high selectivity and sensitivity by a simple and readily available probe. *ACS Applied Materials and Interfaces, 6*(20), 17543–17550.

Luby-Phelps, K. (1994). Physical properties of cytoplasm. *Current Opinion in Cell Biology, 6*, 3–9.

Luby-Phelps, K. (2013). The physical chemistry of cytoplasm and its influence on cell function: An update. In *Molecular Biology of the Cell* (Vol. 24, No. 17, pp. 2593–2596).

Mani, K. S., Rajamanikandan, R., Ilanchelian, M., Muralidharan, N., Jothi, M., & Rajendran, S. P. (2020). Smart phone assisted quinoline-hemicyanine based fluorescent probe for the selective detection of glutathione and the application in living cells. *Spectrochimica Acta - Part A: Molecular and Biomolecular Spectroscopy, 243*, 118809.

McBride, H. M., Neuspiel, M., & Wasiak, S. (2006). Mitochondria: More than just a powerhouse. *Current Biology, 16*(14), R551–R5560.

Mindell, J. A. (2012). Lysosomal acidification mechanisms. *Annual Review of Physiology, 74*, 69–86.

Mouhib, M., Antonucci, A., Reggente, M., Amirjani, A., Gillen, A. J., & Boghossian, A. A. (2019). Enhancing bioelectricity generation in microbial fuel cells and biophotovoltaics using nanomaterials. *Nano Research, 12*(9), 2184–2199. Tsinghua University Press.

Nie, G., Sheftel, A. D., Kim, S. F., & Ponka, P. (2005). *Overexpression of mitochondrial ferritin causes cytosolic iron depletion and changes cellular iron homeostasis, Blood (105)*, 2161–2167.

Paul, B. D., Sbodio, J. I., & Snyder, S. H. (2018). Cysteine metabolism in neuronal redox homeostasis. *Trends in Pharmacological Sciences, 39*(5), 513–524. Elsevier Ltd.

Platt, F. M. (2014). Sphingolipid lysosomal storage disorders. *Nature, 510*(7503), 68–75. Nature Publishing Group.

Qian, Y. Y., Xue, L., Hu, D. X., Li, G. P., & Jiang, H. (2012). Quinoline-based fluorescent probe for ratiometric detection of hydrogen peroxide in aqueous solution. *Dyes and Pigments, 95*(2), 373–376.

Reddie, K. G., & Carroll, K. S. (2008). Expanding the functional diversity of proteins through cysteine oxidation. *Current Opinion in Chemical Biology, 12*(6), 746–754.

Robaczewska, J., Kedziora-Kornatowska, K., Kozakiewicz, M., Zary-Sikorska, E., Pawluk, H., Pawliszak, W., & Kedziora, J. (2016). Role of glutathione metabolism and glutathione-related antioxidant defense systems in hypertension. *Journal of Physiology and Pharmacology, 67*, 331–337.

Roos, A., & Boron, W. F. (1981). Intracellular pH. *REVIEWS, 61*(2), 296 314.

Sasaki, S., Drummen, G. P. C., & Konishi, G. I. (2016). Recent advances in twisted intramolecular charge transfer (TICT) fluorescence and related phenomena in materials chemistry. *Journal of Materials Chemistry C, 4*(14), 2731–2743). Royal Society of Chemistry.

Schwarzländer, M., & Finkemeier, I. (2013). Mitochondrial energy and redox signaling in plants. *Antioxidants and Redox Signaling, 18*(16), 2122–2144.

Steffes, V. M., Murali, M. M., Park, Y., Fletcher, B. J., Ewert, K. K., & Safinya, C. R. (2017). Distinct solubility and cytotoxicity regimes of paclitaxel-loaded cationic liposomes at low and high drug content revealed by kinetic phase behavior and cancer cell viability studies. *Biomaterials, 145*, 242–255.

Stipanuk, M. H. (2004). Sulfur amino acid metabolism: Pathways for production and removal of homocysteine and cysteine. *Annual Review of Nutrition, 24*, 539–577.

Sun, H., Leng, H., Liu, J., Roy, G., Yan, J., & Zhang, L. (2020). A novel fluorescent protein chromophore analogue to simultaneously probe lysosome viscosity and β-amyloid fibrils. *Sensors and Actuators, B: Chemical, 305*, 127509.

Tang, L., Zhou, P., Qi, Z., Huang, Z., Zhao, J., & Cai, M. (2013). A new quinoline-based acylhydrazone for highly selective fluorescence recognition of Cu(II) and sulfide in aqueous solution. *Bulletin of the Korean Chemical Society, 34*(8), 2256–2260.

Udha, S., Eshadri, S., Lexa, A., Eiser, B., Acob, J., Elhub, S., Acques, A. F. J., Osenberg, R. H. R., Alph, R., D'a Gostino, B., Ilson, E. W. F. W., Hilip, P., & Olf, A. W. (2002). Plasma homocysteine as a risk factor for dementia and Alzheimer's disease a bstract background in cross-sectional studies, elevated. *New England Journal of Medicine, 346*(7), 476–483.

Usama, S. M., Thavornpradit, S., & Burgess, K. (2018). Optimized heptamethine cyanines for photodynamic therapy. *ACS Applied Bio Materials*, *1*(4), 1195–1205.

Wang, J., Liu, W., Luo, G., Li, Z., Zhao, C., Zhang, H., Zhu, M., Xu, Q., Wang, X., Zhao, C., Qu, Y., Yang, Z., Yao, T., Li, Y., Lin, Y., Wu, Y., & Li, Y. (2018). Synergistic effect of well-defined dual sites boosting the oxygen reduction reaction. *Energy and Environmental Science*, *11*(12), 3375–3379.

Wang, J. T., Pei, Y. Y., Yan, M. Y., Li, Y. G., Yang, G. G., Qu, C. H., Luo, W., Wang, J., & Li, Q. F. (2021). A fast-response turn-on quinoline-based fluorescent probe for selective and sensitive detection of zinc (II) and its application. *Microchemical Journal*, *160*, 105776.

Wang, S., Yu, H., & Wickliffe, J. K. (2011). Limitation of the MTT and XTT assays for measuring cell viability due to superoxide formation induced by nano-scale TiO 2. *Toxicology in Vitro*, *25*(8), 2147–2151.

Wang, Z., Ivanov, M., Gao, Y., Bussotti, L., Foggi, P., Zhang, H., Russo, N., Dick, B., Zhao, J., Di Donato, M., Mazzone, G., Luo, L., & Fedin, M. (2020). Spin–orbit charge-transfer intersystem crossing (ISC) in compact electron donor–acceptor dyads: ISC mechanism and application as novel and potent photodynamic therapy reagents. *Chemistry - A European Journal*, *26*(5), 1091–1102.

Wei, Y. F., Weng, X. F., Sha, X. L., Sun, R., Xu, Y. J., & Ge, J. F. (2021). Simultaneous imaging of lysosomal and mitochondrial viscosity under different conditions using a NIR probe. *Sensors and Actuators, B: Chemical*, *326*, 128954.

Wilson, W. D., Ratmeyer, L., Zhao, M., Strekowski, L., & Boykin, D. (1993). The search for structure-specific nucleic acid-interactive drugs: Effects of compound structure on RNA versus DNA interaction strength? *Biochemistry*, *32*, 4098–4104.

Wlodek, P. J., Iciek, M. B., Milkowski, A., & Smolenski, O. B. (2001). Various forms of plasma cysteine and its metabolites in patients undergoing hemodialysis a b a. *Clinica Chimica Acta*, *304*, 9–18.

Wu, M. Y., Li, K., Liu, Y. H., Yu, K. K., Xie, Y. M., Zhou, X. D., & Yu, X. Q. (2015). Mitochondria-targeted ratiometric fluorescent probe for real time monitoring of pH in living cells. *Biomaterials*, *53*, 669–678.

Wu, Q., Zheng, J., Zhang, W., Wang, J., Liang, W., & Stadler, F. J. (2019). A new quinoline-derived highly-sensitive fluorescent probe for the detection of hydrazine with excellent large-emission-shift ratiometric response. *Talanta*, *195*, 857–864.

Xia, Y., Zhang, R., Wang, Z., Tian, J., & Chen, X. (2017). Recent advances in high-performance fluorescent and bioluminescent RNA imaging probes. In *Chemical society reviews* (Vol. 46, No. 10, pp. 2824–2843). Royal Society of Chemistry.

Xie, X., Yin, C., Yue, Y., & Huo, F. (2018). Rational design of fluorescent probe for Cys dynamics imaging in living cell. *Sensors and Actuators, B: Chemical*, *267*, 76–82.

Xu, S., Zhou, J., Dong, X., Zhao, W., & Zhu, Q. (2019). Fluorescent probe for sensitive discrimination of Hcy and Cys/GSH in living cells via dual-emission. *Analytica Chimica Acta*, *1074*, 123–130.

Xu, W., Zeng, Z., Jiang, J., Chang, Y., & Yuan, L. (2016). Wahrnehmung der chemischen Prozesse in einzelnen Organellen mit niedermolekularen Fluoreszenzsonden. *Angewandte Chemie*, *128*(44), 13858–13902.

Yang, Z., Cao, J., He, Y., Yang, J. H., Kim, T., Peng, X., & Kim, J. S. (2014). Macro-/micro-environment-sensitive chemosensing and biological imaging. In *Chemical society reviews* (Vol. 43, Issue 13, pp. 4563–4601). Royal Society of Chemistry.

Yoshino, Y., Sato, Y., & Nishizawa, S. (2019). Deep-red light-up signaling of benzo [c, d] indole–quinoline monomethine cyanine for imaging of nucleolar RNA in living cells and for sequence-selective RNA analysis. *Analytical Chemistry*, *91*(22), 14254–14260.

Yu, Y., Wang, J., Xiang, H., Ying, L., Wu, C., Zhou, H., & Liu, H. (2020). A new near-infrared ratiometric fluorescent probe based on quinoline-fused rhodamine dye for sensitive detection of cysteine and homocysteine in mitochondria. *Dyes and Pigments*, *183*, 108710.

Yue, Y., Dong, Q., Zhang, Y., Sun, Y., & Gong, Y. (2015). A highly selective "turn-on" fluorescent chemosensor based on 8-aminoquinoline for detection of Zn2+. *Analytical Methods*, *7*(13), 5661–5666.

Zhang, Q., Li, Q., Yuan, L. C., Abbas, M., He, Z. X., Zhu, H. L., & Wang, Z. C. (2022). A novel quinoline-based fluorescent probe for real-time monitoring of Cys in glioma. *Analyst*, *147*(19), 4257–4265.

Zhang, S., Ong, C. N., & Shen, H. M. (2004). Critical roles of intracellular thiols and calcium in parthenolide-induced apoptosis in human colorectal cancer cells. *Cancer Letters*, *208*(2), 143–153.

Zhang, X., Yan, H., Huo, F., Chao, J., & Yin, C. (2021). Dual-emission NIR fluorescent probe sensitive response biological microenvironment and sulfur dioxide. *Sensors and Actuators, B: Chemical*, *344*, 130244.

Zhang, S., Zhang, Y., Zhao, L., Xu, L., Han, H., Huang, Y., ... Song, D. (2021). A novel water-soluble near-infrared fluorescent probe for monitoring mitochondrial viscosity. *Talanta*, *233*, 122592.

Zhang, S., Zhang, H., Zhao, L., Xu, L., Ma, P., Ren, P., & Song, D. (2023). A benzothiophene-quinoline-based targetable fluorescent chemosensor for detection of viscosity and mitochondrial imaging in live cells. *Spectrochimica Acta Part A: Molecular and Biomolecular Spectroscopy*, *284*, 121799.

Zhou, J., Shi, W., Li, L., Gong, Q., Wu, X., Li, X., & Ma, H. (2016). Detection of misdistribution of tyrosinase from melanosomes to lysosomes and its upregulation under Psoralen/Ultraviolet a with a melanosome-targeting tyrosinase fluorescent probe. *Analytical Chemistry*, *88*(8), 4557–4564.

Zhu, H., Fan, J., Du, J., & Peng, X. (2016). Fluorescent probes for sensing and imaging within specific cellular organelles. *Accounts of Chemical Research*, *49*(10), 2115–2126.

Zubenko, G. S., Kopp, U., Seto, T., & Firestone, L. L. (1999). Original investigation platelet membrane fluidity individuals at risk for Alzheimer's disease: A comparison of results from fluorescence spectroscopy and electron spin resonance spectroscopy. *Psychopharmacology*, *145*, 175–180.

Züchner, S., De Jonghe, P., Jordanova, A., Claeys, K. G., Guergueltcheva, V., Cherninkova, S., Hamilton, S. R., Van Stavern, G., Krajewski, K. M., Stajich, J., Tournev, I., Verhoeven, K., Langerhorst, C. T., De Visser, M., Baas, F., Bird, T., Timmerman, V., Shy, M., & Vance, J. M. (2006). Axonal neuropathy with optic atrophy is caused by mutations in mitofusin 2. *Annals of Neurology*, *59*(2), 276–281.

10 Phenanthiridine-Based Molecules
Design, Synthesis, and Applications

Kisan Pawar, Preeti Ashokkumar Chaudhran and Abha Sharma
National Institute of Pharmaceutical Education and Research,
Raebareli, Lucknow, UP, India

LIST OF ABBREVIATIONS

ICT	Intramolecular charge transfer
DNA	Deoxyribonucleic acid
DMSO	Dimethyl sulfoxide
UV Light	Ultraviolet light
GC-MS	Gas chromatography-mass spectroscopy
HPLC	High-performance liquid chromatography
MOF	Metal-organic framework
ESIPT	Excited-state intramolecular proton transfer
TICT	Twisted intramolecular charge transfer
THF	Tetrahydrofuran
DMF	Dimethylformamide
BA	Biological amine
ROS	Reactive oxygen species
WHO	World Health Organization
RNA	Ribonucleic acid
USEPA	United States Environmental Protection Agency
AAS	Atomic absorption spectroscopy
ICPMS	Inductively coupled plasma–mass spectrometry
FAAS	Flame atomic absorption spectrometry
PET	Photoinduced electron transfer
CHEF	Chelation-enhanced fluorescence
FRET	Fluorescence resonance energy transfer
GC	Gas chromatography
RPHPLC	Reversed phase high-performance liquid chromatography
IR	Infrared spectroscopy
TFA	Trifluoracetic acid
MAP	4-methoxy acetophenone
TBAB	Tetrabutylammonium bromide

10.1 INTRODUCTION

Phenanthridine is a nitrogen-containing heterocyclic compound similar to phenanthrene, except having nitrogen at position 10 (structure I). Phenanthridine was discovered by Ame Pictet and H. J. Ankersmit in 1891 and was first time synthesized by pyrolysis of the condensation product of benzaldehyde and aniline (Pictet et al., 1891). Phenanthridine shows fluorescent properties by the intramolecular charge transfer (ICT) mechanism where the phenanthridine moiety acts as a donor and the substituents on phenanthridine act as an acceptor and vice versa. Phenanthridine detects various cations, anions, biogenic amines, etc. Many macrocyclic dyes, such as Hoechst 33258, Hoechst 33342 (piperazine derivatives), acridine orange (acridine derivatives), propidium (phenanthridine derivatives), and ethidium bromide, bind to DNA bases; some of the dyes also bind to RNA (acridine orange). They are highly aromatic and planar molecules that show fluorescence by intercalation binding to DNA, i.e., they sandwich themselves between bases. Intercalation binding enhances the fluorescence, making the nucleic acid recognizable.

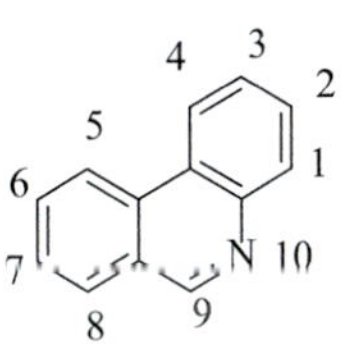

STRUCTURE I: Phenanthridine.

10.2 SYNTHETIC OVERVIEW ON PHENANTHRIDINE

By pyrolyzing the condensation product of benzaldehyde and aniline, Pictet and Ankersmit created phenanthridine for the first time at the end of the 19th century (Pictet et al., 1891). Additionally, Morgan and Walls changed the reaction conditions, conducting a reaction based on the cyclization of phenanthridine by dehydrative ring closure with phosphorus oxychloride in boiling nitrobenzene (Gilbert et al., 1931). Later in 20th century, this method was commonly used for the synthesis of phenanthridine. However, in the last decades, an increased interest resulted in the development of new and many more synthetic processes to improve yield, time, and other parameters of the reactions (Lysén et al., 2002).

An approach for the synthesis of 6-aryl phenanthridine derivatives using benzotriazole derivatives of thioxanthene, diarylmethane, xanthene, acridine, etc. was reported, where the steps involve the generation of a carbanion stabilized by benzotriazole, followed by oxidation using copper iodide to form a free radical. The elimination of nitrogen is followed by ring closure to form phenanthridine (scheme 1(i))

DOI: 10.1201/9781032706245-10

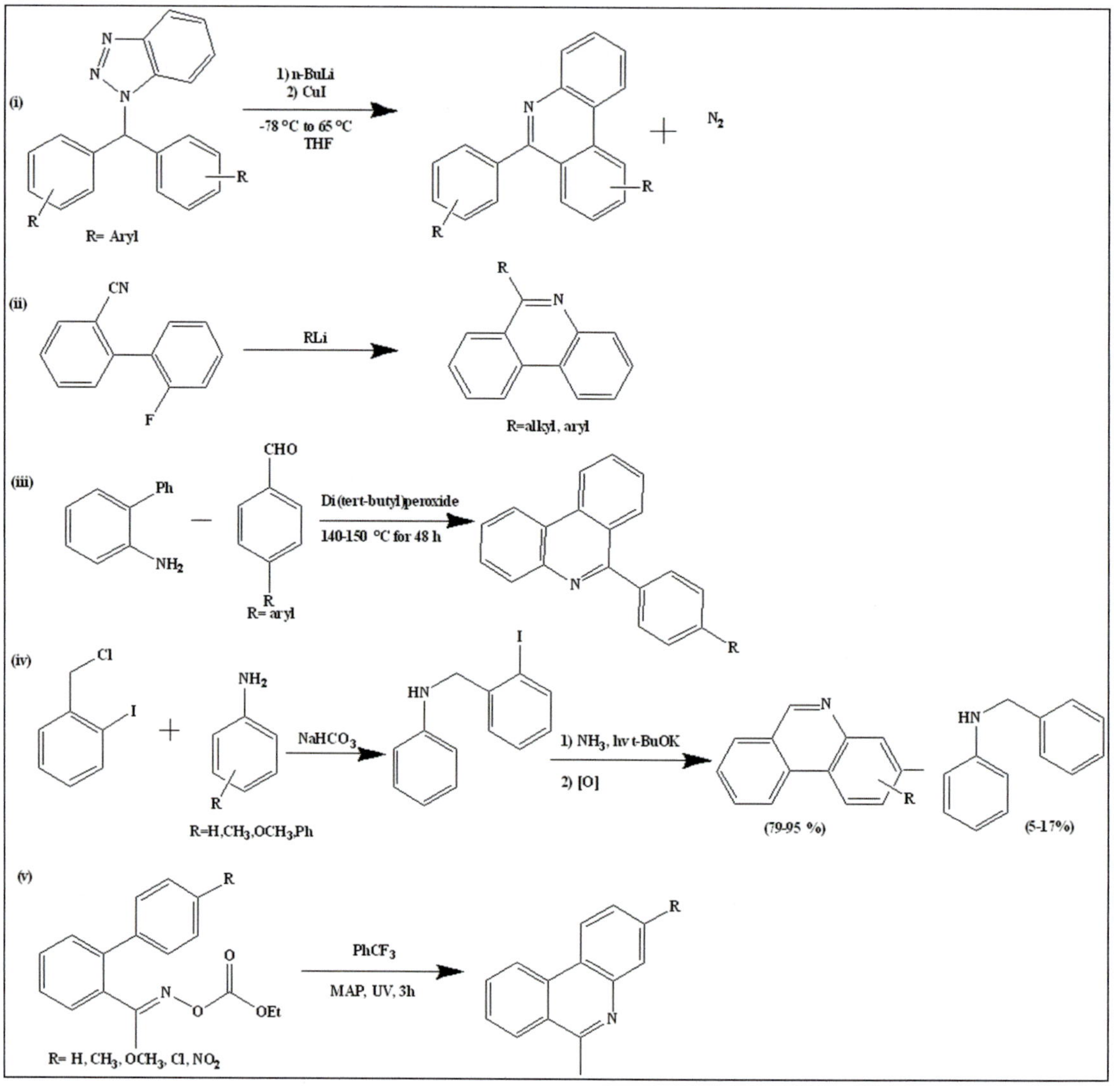

SCHEME 1 Synthetic methods for synthesis of phenanthridine derivatives.

(Katritzky et al., 1999; Katritzky et al., 1996). The Grignard reagent, alkyl or aryl lithium as a base, was used for the anionic ring closure reaction to synthesize alkyl-substituted phenanthridine compound (scheme 1(ii)) (Lysén et al., 2002; Mamalis et al., 1950). 6-aryl phenanthridine derivatives were prepared by the reaction of aromatic amine with alkyl substituted aldehyde and tert-butoxide used as a base, whereas the reaction proceeded through tert-butoxide free radical assisted the homolytic aromatic substitution mechanism (scheme 1(iii)) (Leardini et al., 1985; Bowman et al., 2012). The radical-based approach described here is one of the few that targets the synthesis of 5,6-substituted phenanthridines using N-(*o*-halo benzyl) arylamines as starting material, whereas *o*-haloaryl benzylamine obtained by the unimolecular radical nucleophilic substitution reaction (S_{RN}^{1}) of 2-iodobenzyl chloride and various anilines in NH_3 or DMSO as a solvent later acidified to obtain oxidized phenanthridine compound in good yields (95%) (scheme 1(iv)) (Budén et al., 2010). Various N-heterocycles prepared by McBurney et al., via oxime carbonates, are considered excellent precursors for iminyl radical generation using UV light with the help of standard photolysis conditions. The reaction produces a good yield (55%–99%) of 3-substituted 6-methyl phenanthridine (scheme 1(v)) (McBurney et al., 2011).

10.3 APPLICATIONS OF PHENANTHRIDINE-BASED FLUORESCENT PROBE

10.3.1 Phenanthridine-Based Fluorescent Probe for Aniline Detection

The sensitive and selective sensor development for chemical and biological agent detection is important in environmental, industrial, and biological systems (McDonagh et al., 2008; Li et al., 2017; Bhattacharjee, 2008). Organic amines specifically aniline derivatives are used widely in industries (Simenhoff et al., 1977; Pacquit et al., 2007). Aniline is toxic

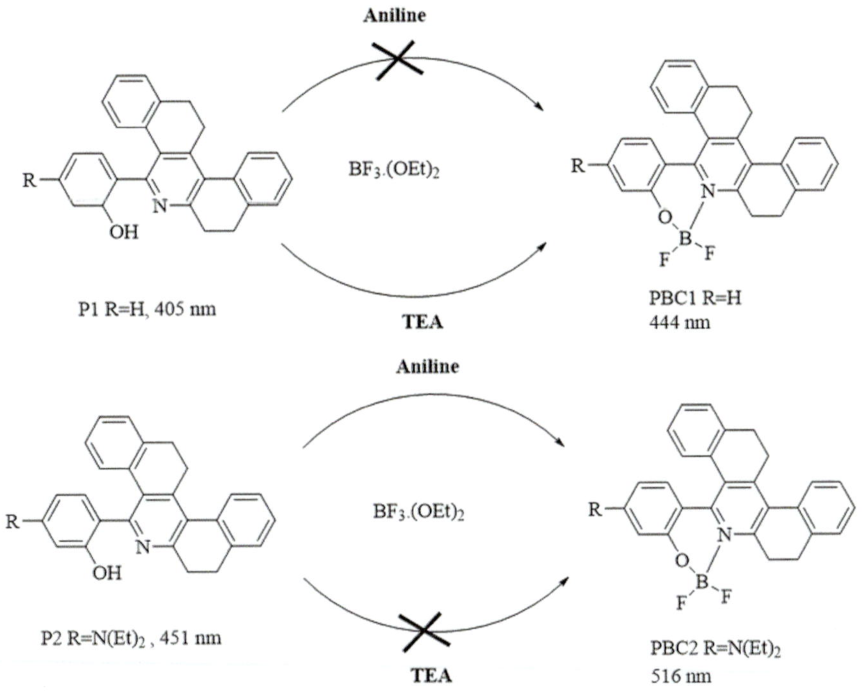

FIGURE 10.1 Phenanthridine boron complex of **PBC1** and **PBC2**.

in nature, which upon inhalation or ingestion causes toxicity in the body. Hence, the detection is crucial involving the conventional method for detection such as gas chromatography-mass spectrometry (GC-MS), HPLC, colorimetric arrays, etc. (Zhang et al., 2015; Wu et al., 2017). Recently, fluorescence-based detection has been commonly used, as it offers good selectivity as well as specificity, low cost, and easy operation (Huang et al., 2014; Wu et al., 2017). Most fluorescent probes are restricted to aliphatic amines because of their enormous molecular weights, and they are challenging to reshape in order to change the emission properties (Sekhar et al., 2014; Pinto et al., 2019). Few amine sensors, including poly(3-hexylthiophene) and graphene oxide, rhodamine B-Zn(II)-based metal-organic frameworks (MOFs), perylene-bridged cyclodextrins, lanthanide metal-organic frameworks, and 1,10-phenanthroline-azido-copper(II) complex, are being studied for aniline detection (Meng et al., 2014; Gimenez et al., 2022).

Balijapalli et al. (2019) reported P1 and P2 phenanthridine-based fluorophore that produced PBC1 and PBC2 boron combinations when $BF_3.(OEt)_2$ was present (Figure 10.1).

The fluorophore P1 shows λ_{abs} at 337 nm and low λ_{em} at 405 nm in tetrahydrofuran (THF), and the Stoke shift is calculated as 4,982 cm^{-1}. P2 shows λ_{abs} at 393, 260, and 292 nm, the lower energy bands red-shifted due to the charge transfer between NEt_2 group and phenanthridine moiety. P2 shows less emission in acetonitrile (ACN), methanol, ethyl acetate, toluene, chloroform, and THF. In dimethyl sulfoxide (DMSO) and dimethyl formamide (DMF), the emission intensity was maximum. Emission intensity depends upon solvent viscosity; in DMSO and DMF emission intensity was increased due to high viscosity of the solvent due to which the twist angle was reduced. P2 selectively detects aniline in the presence of $BF_3.OEt_2$. The aniline in THF showed a quench of the band at 333 nm, while the peak at 516 nm was increased as the concentration of aniline increased. P2 probe was found responding for Zn^{2+}, where the ESIPT phenomenon plays a role for metal ion detection. The absorption and emission spectra of P2 showed the spectra changes after interaction with Zn^{2+}, the band at 390 nm progressively disappeared and a new band at 425 nm forms as well as weak emission band at 516 nm by increasing concentration from 0 to 50 equivalent with the change in fluorescent color from blue to green was observed.

P1 containing hydroxyl group and pyridine nitrogen forms a planar six-membered ring by hydrogen bonding that leads to the ESIPT process; here the proton transfer process competes with the TICT effect. When molecules with electron donor and acceptor moieties connected by a single bond are photoexcited, an electron transfer process known as TICT takes place. P2 containing $-NEt_2$ as donor and phenanthridine as acceptor should reduce the ESIPT effect and induce TICT; subsequent research revealed that PBC2 is only formed when aniline is present, which enables the probe P2 to function as a fluorescent turn-on sensor for aniline detection.

10.3.2 Phenanthridine-Based Donor/Acceptor Fluorescent Dyes for Sensing of Biogenic Amines

The analytical method development for amines detection such as biological amines (BAs) is important as it serves as a critical analyte in agricultural, pharmaceutical, and food processing effluent (Garg et al., 2019). BAs are produced naturally in active monoamines or polyamines forms in the human cell. The presence of a large number of BAs leads to health risks, food poisoning, cancer, and bacterial infections. Hence, the detection of BAs is important for the diagnosis of disease. Several methods are used for the detection of BAs, for example, electrophoresis, electrochemical sensing, chromatography, and enzymatic assays (Landete et al., 2005). However, these methods have disadvantages, including low selectivity, low sensitivity, and a specialized instrument (Górski et al., 2003). In recent years, fluorescence-based detection has been used for biomedical, clinical testing, environmental monitoring, food testing, etc., and is also reported in other fields because of high selectivity and sensitivity (Zhang et al., 2019).

Trifluoromethyl-substituted ketone, hydrophobic porphyrin dyes, and some 1,3-diketones are used for BAs detection; however, these probes are used for sensing BAs in organic solvents, which is unfavorable for biological applications (Montañez et al., 2007; Rakow et al., 2005). Swaminathan and his group reported (2021) a large Stokes shift value and high extinction coefficient; ICT-based fluorescent probes PA1 and PA2 developed by using phenanthridine as an electron donor and the aldehyde functional group as an electron acceptor for sensing BAs (Figure 10.2).

The UV-visible and fluorescence spectroscopy were recorded in ACN solvent, where the probe exhibited longer absorption maxima generated due to π-π* and absorption at a shorter wavelength due to ICT between aldehyde and phenanthridine moiety. Both compounds emitted yellow light, where PA1 exhibited λ_{em} at 530 nm upon exciting at 340 nm while PA2 exhibited two emission bands at 467 nm and 536 nm. PA1 and PA2 in organic solvents of different polarity from 1,4-dioxane to ACN were observed for solvent effect. The PA1 blue-shifted by 57 nm from 1,4-dioxane to DMSO; similarly, PA2 also blue-shifted although with less wavelength shift (13 nm). PA1 has a higher quantum yield in a polar solvent (DMSO = 0.59, DMF = 0.43); however, in nonpolar solvent (1,4-dioxane), it exhibited a low quantum yield. PA1 has a higher quantum yield as compared to PA2, as PA2 showed lesser quantum yield in all the tested solvents.

The ACN solution of probes PA1 and PA2 was investigated for amine sensing by using UV-vis and fluorescence spectroscopy in the ACN solution. In this study, benzylamine and methylamine were used; the probe PA2 did not show spectral change, whereas PA1 showed absorption maxima at 360 nm due to ICT; however, upon interaction with the primary amine, the peak diminished and two bands at 325 nm and 420 nm appeared. The two isosbestic points were observed at 340 nm and 395 nm, confirming the formation of imine accompanied with a color change from yellow to blue.

The PA2 probe was investigated for detection of BAs such as serotonin, dopamine, glutamate, and epinephrine, where absorption spectra show no significant changes; however, fluorescence was found to be increased in the case of dopamine and serotonin. The PA2 probe was found to be an effective sensor for BAs, where the color changed from yellow to greenish-blue for primary amine by forming an imine group by reaction between aldehyde and amine functional group, whereas PA1 probe was also used for detection of BAs such as serotonin and dopamine. The amine attacks the aldehydic group of PA1 resulting in a new fluorescence peak, due to disturbance of ICT. The synthesized probes PA1 and PA2 have fast emission responses toward amines. The probes have excellent selectivity toward amine and no significant change with other interfering cations and anions.

10.3.3 Phenanthridine-Based Probes for Anion Detection

10.3.3.1 Detection of Hypochlorite Ions

Sodium hypochlorite (NaOCl) is a regularly used bleaching agent and disinfectant. The conjugated base ($^-$OCl) that generates upon dissociation is a reactive oxygen species (ROS). In biological systems, $^-$OCl is produced by the enzyme myeloperoxidase by the interaction of hydrogen peroxide (H_2O_2) and chloride (Cl) ion (Yuan et al., 2021; Jin et al., 2021). $^-$OCl ion is a powerful oxidizing agent, which potentially oxidizes different functional groups such as thiols, hydroxamic acid, nitroso groups, oxime, and acyl. The WHO has recommended a maximum residual limit 3.3 μM of $^-$OCl content in water because the high concentration causes harmful effects, such as oxidized lipid, RNA, DNA, cholesterol, and protein, and may also cause conditions

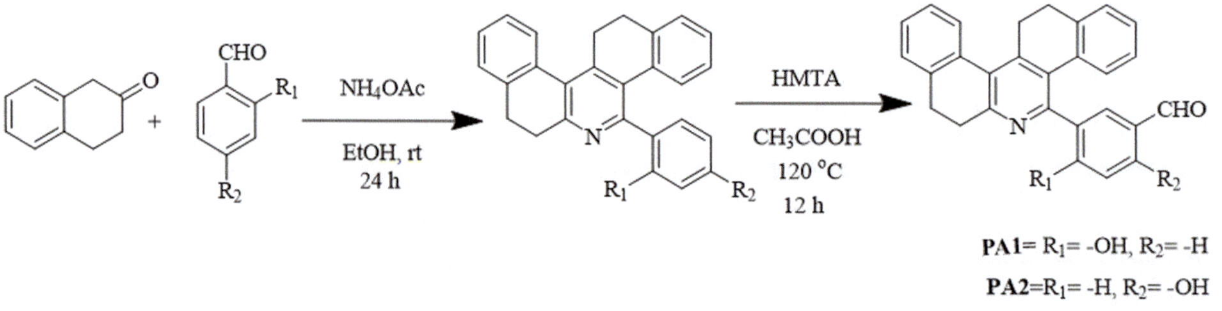

FIGURE 10.2 Synthesis of phenanthridine derivatives **PA1** and **PA2**.

like cancer, Alzheimer's disease, inflammatory diseases, and cardiovascular diseases. (Gao et al., 2017; Wang et al., 2018).

Various analytical methods are used for $^{-}$OCl ions detection like chromatography, spectrophotometry, potentiometric analysis, and radiolysis. These methods are time-consuming, less sensitive, and complex. However, fluorescent probes have various advantages to overcome this limitation, such as high sensitivity, selectivity, and simple operation (Lin et al., 2020; Singh et al., 2019). Over the past few years, various probes for $^{-}$OCl ions detection are discovered, such as triphenylamine, diaminomalenonitrile, thiocoumarin, pyrene, nopinone, rhodamine, benzimidazole, phenothiazine, and acridine are known.

Enbanathan et al. (2022a) designed an innovative oxime-based chemosensor PBO for the detection of $^{-}$OCl ions. The PBO was synthesized by a three-step reaction: step 1: a Mannich reaction between aldehyde and 2-tetralone in the presence of ammonium acetate to give PB (Figure 10.3). In step 2, the Suzuki coupling reaction between (4-formylphenyl) boronic acid and compound PB took place, which produced PBC. In step 3, hydroxyl ammonium chloride and PBC were dissolved in ethanol using triethylamine, for 3 hours under reflux at 70°C, then after the crude product was purified using column chromatography with mobile phase, n-hexane/ethyl acetate (Figure 10.3). During the study, it was found that the PBO probe has poor solubility in water; however, it has good solubility in system CH_3CN:H_2O (9:1) ratio.

The CH_3CN/H_2O (9:1) probe PBO solution was evaluated against anions, such as F^-, $^{-}$OCl, CH_3COO^-, CO_3^{2-}, $S_2O_3^-$, HCO_3^{2-}, and OH^-, and no change in absorption spectra was obtained for ions except $^{-}$OCl. $^{-}$OCl ion showed variation in the absorption peak; the absorption gradually decreased with the increased concentration of $^{-}$OCl ion from 0 to 2 equivalents. Furthermore, the color of the PBO + $^{-}$OCl solution changed to colorless, which could be observed by the naked eye. The fluorescence emission intensity at 455 nm shifted to 420 nm (blue shift) against $^{-}$OCl ion with color change from cyan to blue. The nucleophilic $^{-}$OCl ion attacks on the imine carbon, which leads to oxidation of the oxime group of PBO. Probe PBO (ϕ_{fl} = 0.23) showed higher fluorescence in contrast to PBO + ($^{-}$OCl), (ϕ_{fl} = 0.12), and it was found that PBO + ($^{-}$OCl) was weakly fluorescent than free PBO, hence showed a quenching response. If $^{-}$OCl is introduced to the PBO system, the intensity of the fluorescence emission declines in the 5–11 pH range, indicating that the detection of $^{-}$OCl ion by PBO probe can be used in a wide pH range of 5–11. The limit of detection (LOD) of PBO was discovered to be 8 nM, which is less than what had been reported before. This probe was also found effective in the detection of $^{-}$OCl ion in living cells. The practical application of the probe PBO is demonstrated by the paper strip method, where the color of the paper changes from cyan to strong blue, noticed under UV light at 365 nm, indicating the presence of $^{-}$OCl ion.

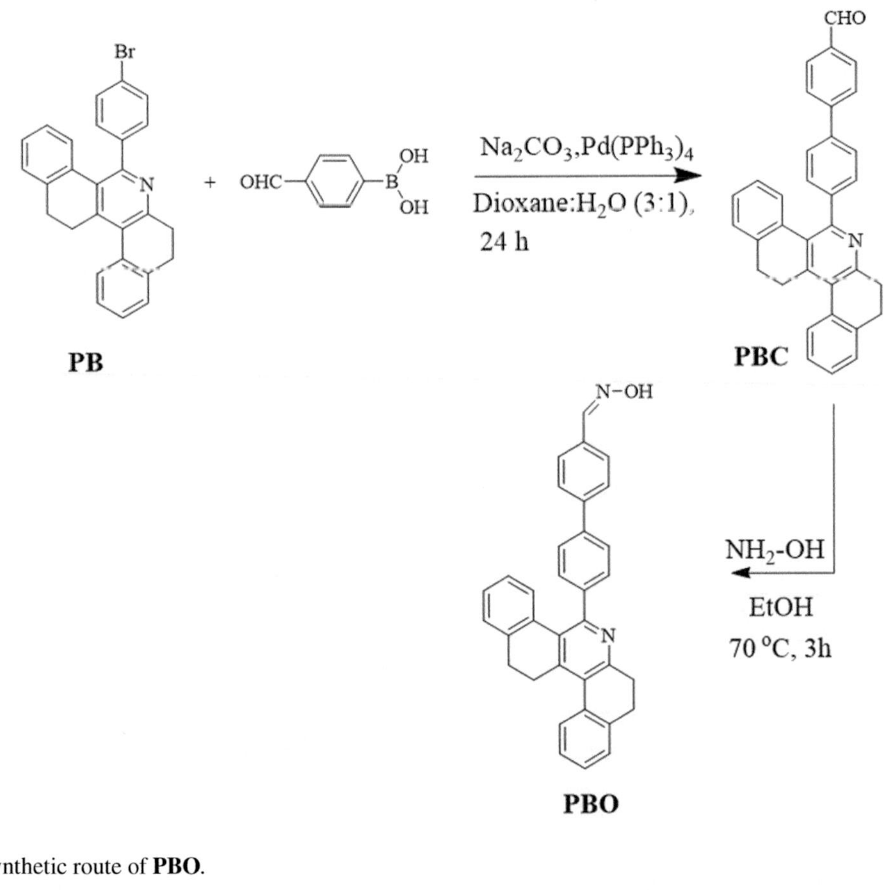

FIGURE 10.3 Synthetic route of **PBO**.

10.3.3.2 Detection of Cyanide Anion

Cyanide ion (CN^-) is considered the most toxic anion in environmental chemistry and also plays an important role in biology, medicine, and catalysis (Paul et al., 2001). Due to human activities, CN^- is a serious environmental pollutant in water, soil, and air. The trace amount of cyanide also accumulates in human and animal bodies, which can prove very toxic to the physiological system (Brugnara, 2003). Cyanide is very important in the synthetic industry for the synthesis of synthetic fibers, resins, and acrylic plastic and is also used in the extraction of gold (Zhang et al., 2014). The standard limit of CN^- in portable water is 1.9 μM set by the WHO (WHO guideline, 1996). The detection of CN^- is critical due to its toxicity; the traditional methods for CN^- detection such as voltametric, potentiometric, electrochemical, and optical methods were used (Jia et al., 2019). However, the optical methods based on spectroscopic properties are preferred because of easy detection and high sensitivity.

Manickam et al. (2020) developed the dicyanovinyl-based phenanthridine system containing dicyanovinyl moiety as a chelator for cyanide anion and phenanthridine moiety as a receptor. They reported a new colorimetric and turn-off fluorescent probe for CN^- in an aqueous medium. Probe **e** (Figure 10.4) consists of conjugated phenanthridine moiety and malononitrile group having the benefit of being highly reactive for the cyanide anions. The Suzuki coupling between 4-bromosubstituted phenanthridine (a) and 4-formylphenyl boronic acid (b) gave aldehyde-substituted compound (c) that undergoes Knoevenagel condensation with malononitrile (d) by using piperidine as base in THF:ethanol (3:1) as solvent, results into dicyanovinyl substituted product (e) (Figure 10.4).

The two absorption bands at 271 nm and 328 nm were observed for compound **c** at 20 μM CH_3CN solution. Compound **e** showed three absorptions at 267, 288, and 368 nm because of π-π * and ICT transitions and gave a yellow color solution in CH_3CN. After the addition of 100 μL of CN^- ion, the peak due to ICT decreased gradually and a new absorption band appeared at 325 nm; it was a blue-shifted absorption band, because of nucleophilic reaction. The compound **e** had a quantum yield of 21% and was found to be decreased by adding 12 equivalents of CN^- ions. The selectivity studies of compound **e** were performed with different anions (F^-, Cl^-, CH_3COO^-, CO_3^{2-}, OH^-, SO_4^-, $S_2O_3^-$, HCO_3^-, and PO_4^{3-}); however no distinct change was observed in the UV-vis absorption and emission spectrum. In the case of cyanide anion, color of the solution changed quickly from orange to colorless, which could be observed under a UV lamp at 365 nm, because of the interaction of the dicyanovinyl group of compound **e** with the cyanide anion; therefore, compound **e** acts as a highly selective and sensitive turn-off fluorescence probe for cyanide anion without any other anion interference. The emission intensity at 594 nm of compound **e** was instantly quenched and completed within 20 seconds, suggesting probe **e** could be an effective tool for the detection of cyanide anion. Probe **e** has an LOD of 39.3 nM, which is less than the standard limit of CN^- (1.9 μM).

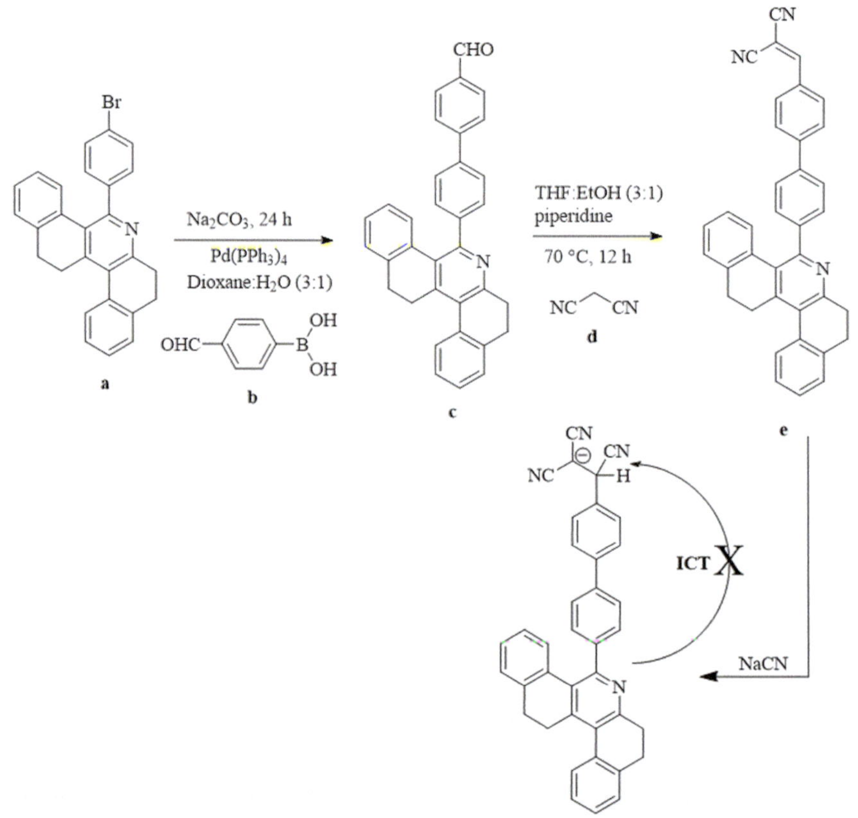

FIGURE 10.4 Synthetic route of compound **e** and its sensing mechanism.

10.3.4 Phenanthridine-Based Probes for Cations Detection

10.3.4.1 Detection of Cadmium Metal

Various toxic trace heavy metals are present in water as hazardous contaminants, for example, mercury (Hg), cadmium (Cd), thallium (Tl), lead (Pb), and arsenic (As). These heavy metals cause toxicity due to bioaccumulation in biological systems and also harm the environmental system (Peng et al., 2012). Cd^{2+} is a toxic heavy metal ion, produced from various natural activities such as forest fires, volcanic eruption, and manufacturing activities like burning of fossil fuels, wastewater, fertilizers, use of Cd-Ni batteries, mineral mining, and industrial effluent, are sources of Cd in the environment (Singh and Kushwaha, 2021; Xiang et al., 2021). Cd is widely used in agriculture and industry and is responsible for soil, air, and water pollution. Cd^{2+} is one of the 126 major contaminants listed by the US Environmental Protection Agency (EPA). The WHO states that 3 ppb of Cd^{2+} can be present in drinking water (Wang et al., 2019). Cd enters the living system through the contaminated food, water, or even cigarette smoke. The accumulation in body causes severe diseases or disorders like malignancies of the lungs, renal failure, prostate gland and kidney disorders, and calcium metabolic abnormalities (Krishnaveni et al., 2018).

There are numerous fluorescent probes accessible for Cd^{2+} detection, and the fluorescence sensing mechanism is based on PET, ICT, CHEF, and FRET. The metal ions determination by the D-π-A-based organic fluorescent probe is gaining interest; the detection of Cd^{2+} is very challenging as Zn^{2+} ion interferes with Cd^{2+}, so to selectively detect Cd^{2+} is quite difficult (Zhang et al., 2018). Enbanathan et al. (2022b) developed a fluorescent probe with the D-π-A system containing terpyridine as the receptor moiety that also assists extension of conjugation and phenanthridine as a signaling unit. The ratiometric fluorescence response of probe with Cd^{2+} ions was observed which led to the color changes from blue to orange. The ICT sensor shifts emission wavelength and changes the fluorescence intensity, resulting in ratiometric fluorescent responses.

The PBC (Figure 10.3) reacts with 2-acetylpyridine using the Krohnke pyridine synthesis to produce TBTP (Figure 10.5i). Absorption and emission titration was used to assess TBTP with different metal cations in $CH_3CN:H_2O$ (8:2 v/v). TBTP shows λ_{abs} at 277 nm and 321 nm, results of π-π* and n-π* transitions. Interference studies with different metal cations such as Pb^{2+}, Ag^{+}, Ru^{3+} Li^{+}, Ba^{2+}, Co^{2+}, Cu^{2+}, Fe^{2+}, Ce^{4+}, Hg^{2+}, Na^{+}, Mg^{2+}, Ca^{2+}, Mn^{2+}, Ni^{2+}, Al^{2+}, K^{+}, Fe^{3+}, Sr^{2+}, and Zn^{2+} show no significant shifting; however, the addition of Cd^{2+} resulted into appearance of new absorbance peak at 334 nm. TBTP has selectivity toward Cd^{2+}

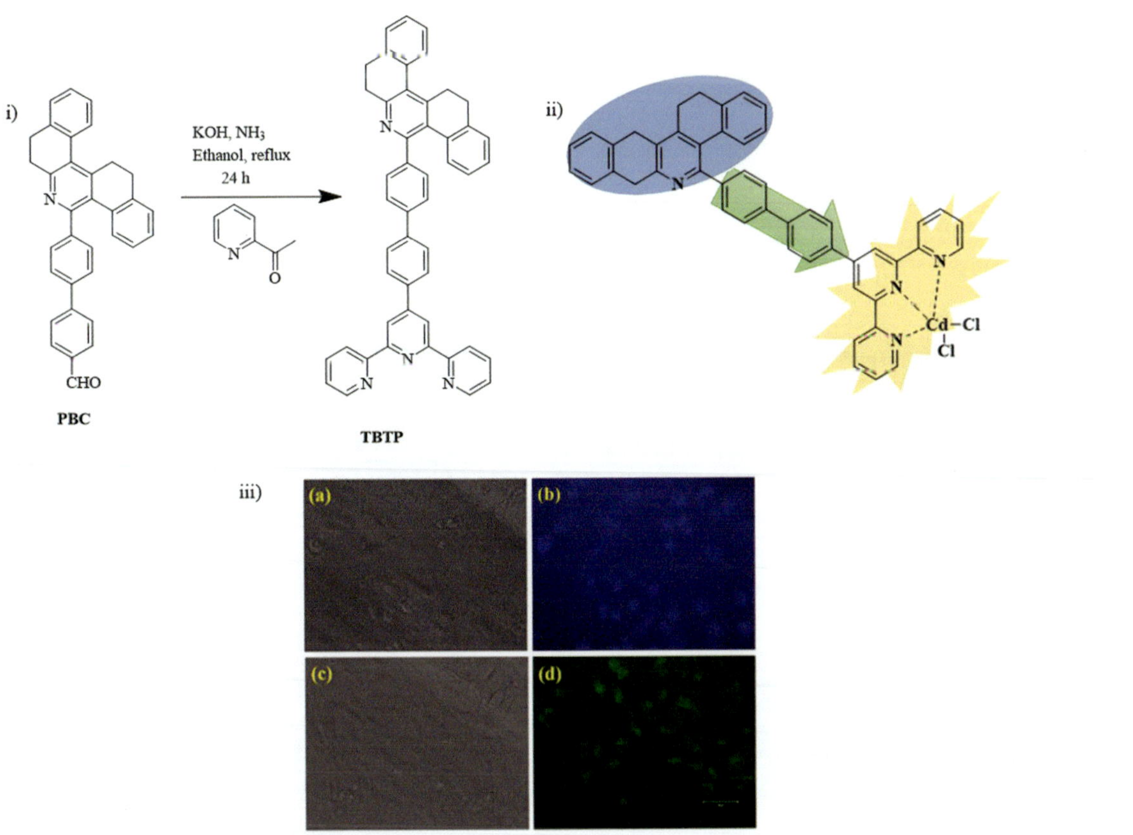

FIGURE 10.5 (i) Synthetic route of TBTP. (ii) The possible mechanism for the sensor TBTP in complex with Cd^{2+}. (iii) Fluorescence and bright-field pictures of HeLa cells that have been exposed to TBTP. In (a) and (c), HeLa cells were separately treated with TBTP and TBTP+Cd^{2+} (b and d): Fluorescence pictures of HeLa cells treated with TBTP and TBTP+Cd^{2+}. Reprinted with permission from Enbananthan et al. Copyright (2022) Elsevier.

than other metal ions. The absorbance at 321 nm gradually shifted toward a longer wavelength upon incorporating of Cd^{2+} ions (0–1 equivalent); however, the intensity reduces, and at 330 nm, an isosbestic point is observed, which indicates the adduct formation.

In the emission spectra of TBTP recorded in the $CH_3CN:H_2O$ (8:2 v/v) solvent system, the emission band appeared at 423 nm; the addition of other metal ions did not affect emission spectrum. However, the addition of Cd^{2+} ion generates new emission spectrum at 523 nm. Zn^{2+} also did not affect the emission spectra due to poor shielding of 4d electron of Cd^{2+} as compared to 3d electron of Zn^{2+}; the limit of detection of TBTP is 1.181×10^{-8} M. To see the translational potential of probe TBTP, authors prepared the paper strips by the dipping method and tested for Cd^{2+} ion where also a fluorescence color change was observed from blue to orange. The TBTP sensor differentiates Cd^{2+} from other metal ions by chelating Cd^{2+} ions via nitrogens of terpyridine moiety (Figure 10.5ii).

10.3.4.2 Detection of Mercury Metal

The environmental contamination by heavy metals is a very major concern, and the mercury ion (Hg^{2+}) is also a potential contaminant causing serious health problems due to accumulation in the body (Faye et al., 2012; Jiang et al., 2020). The well-known Minamata disease is cause due to accumulation of Hg^{2+} in the body because of the consumption of the mercury-contaminated food. Long-term accumulation of mercury causes DNA damage disruption of ligand-receptor interaction, organ dysfunction, etc. (Wu et al., 2019; Song et al., 2016). Older techniques for the detection of Hg^{2+} need complex procedures and sensitivity is also less, so the development of selective chemosensors is very important for the detection of Hg^{2+} (Kim et al., 2012; Soh, 2018).

Ponram et al. (2018) developed paper-based mono and tetra sulfur-containing phenanthridine chemosensors for the detection of mercury (Figure 10.6a). The tetradihydrobenzo phenanthridine is an important class of heterocycle-containing pyridine rings. It is widely present in natural products and also useful for metal chelation through a nitrogen atom. The addition of Hg^{2+} ions showed a significant change in the absorption spectra of probes 1 and 2. Probe 1 peak at 270 nm shifted to 280 nm and the peak at 326 nm also shifted to 348 nm; these changes were not shown in the case of other metals in the absorption spectra of probe 1. Similarly, in probe 2, upon the addition of Hg^{2+} ions, the peak at 255 nm shifted to 263 nm and the peak at 346 nm shifted to 330 nm (blue shift); however, these changes were not observed with other metal ions and at 287 nm isosbestic point was observed, indicating the interaction of Hg^{2+} ion

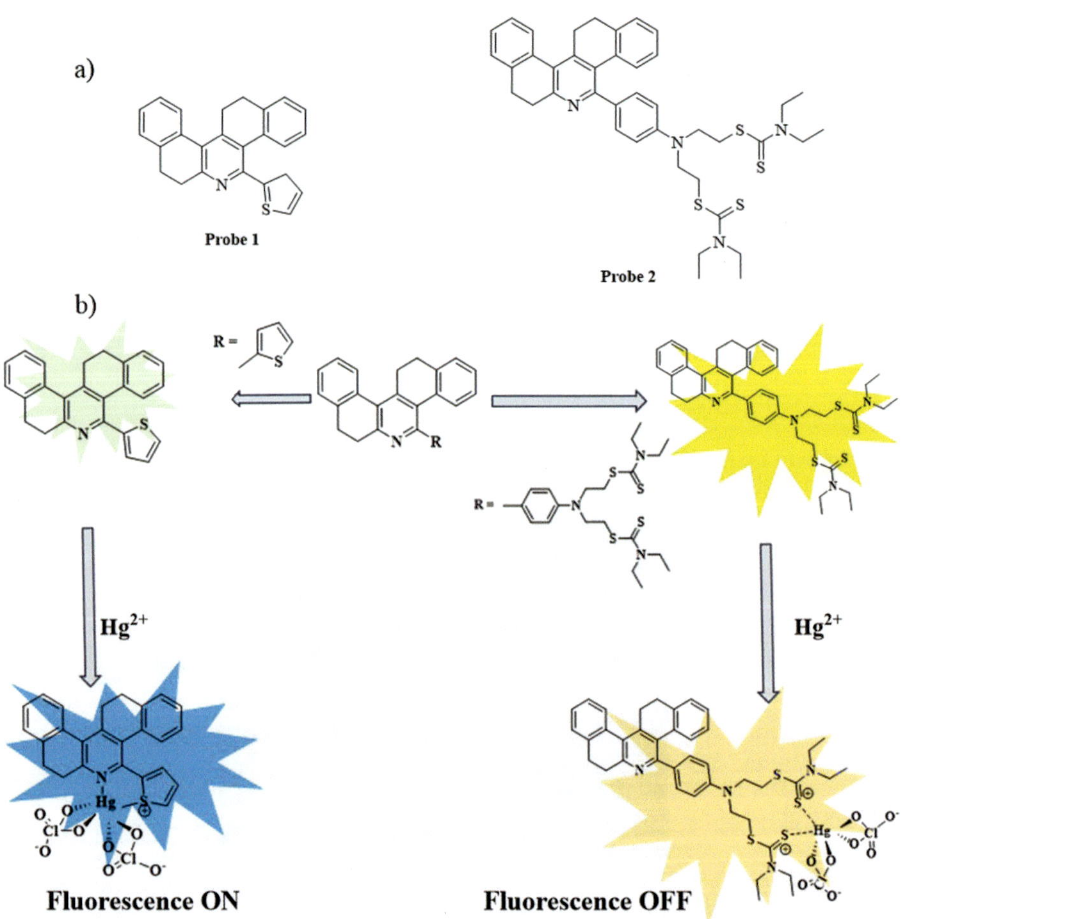

FIGURE 10.6 (a) Structure of probe 1 and probe 2. (b) Proposed mechanism for phenanthridine ligands.

with the probe. Emission spectra of probes 1 and 2 showed significant spectral changes with Hg^{2+} ion. Probe 1 showed λ_{max} at 408 nm, which after the addition of the Hg^{2+} ion shifted to 455 nm with an increase in high-energy emission bands at 375 and 400 nm. At the higher concentration of Hg^{2+} ion, the low-energy emission band was suppressed with the high-energy emission band and the emission color changed from pale blue to blue. In the case of probe 2, the emission spectral changes with Hg^{2+} ion were observed, reducing emission intensity at 456 nm that shifts toward longer wavelength, and also less intensity bands at 596 nm were observed. The color change is observed from blue to colorless; this quenching in emission spectra could be due to the TICT nature of probe 2. TCIT state returns to ground state by either nonradiative relaxation or red-shifted emission. In probe 2, nonradiative relaxation is responsible for quenching in emission spectra. In the paper strips study, probes 1 and 2 showed color change with an aqueous solution of Hg^{2+}; color change was not observed with other metal ions (Figure 10.6b).

Seenan et al. (2021) reported Hg^{2+} detection using a ratiometric fluorescent probe in an organic semi-aqueous medium based on phenanthridine pyridine moiety (Figure 10.7a). It is an ICT-based probe containing pyridine as a core moiety for binding with metal ions. The Hg^{2+} ion formed chelate with the nitrogen atom of pyridine, and the nitrogen atom of phenanthridine moiety takes part in spectrometric changes. The fluorescence color change from blue to green on the addition of Hg^{2+} was observed. The selectivity study of the probe was performed with other metal ions like Cd^{2+}, Cu^{2+}, Mg^{2+}, Bi^{2+}, Mn^{2+}, Ni^{2+}, Pb^{2+}, Sr^{2+}, Cr^{3+}, Al^{3+}, Fe^{2+}, and Hg^{2+}. The probe showed distinct spectral changes on the addition of Hg^{2+}, whereas no changes were observed on the addition of other metal ions. The probe has an absorption band at around 280 nm in the MeCN/water (9:1, v/v) system; the absorption intensity at 316 nm decreases upon the addition of Hg^{2+} with the bathochromic shift. After the progressive addition of Hg^{2+} ion, the red-shifted absorption band was observed at 373 nm with an isosbestic point at 296 nm. The red-shifted absorption band was observed due to the ICT mechanism between the probe and Hg^{2+} ion. The emission properties of the probe were investigated with alkali and alkaline earth metals, and the

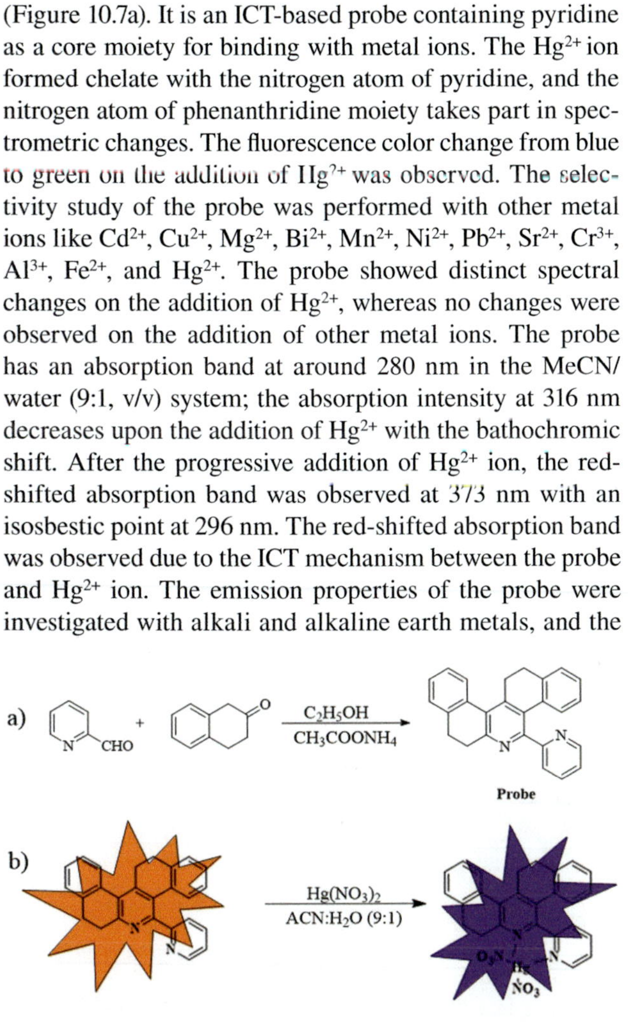

FIGURE 10.7 (a) Synthetic route for sensor probe. (b) The proposed binding mode of a probe with Hg^{2+} ions.

emission maxima of the probe was observed at 486 nm. In the fluorescence titration experiment of a probe with Hg^{2+}, the emission peak of a probe at 415 nm decreases with the new emission band at 480 nm and the isosbestic point at 453 nm indicates the formation of a complex between the probe and Hg^{2+} ion. The quantum yields of the probe and probe-Hg^{2+} complex were found to be 0.68 and 0.83, respectively; the increase in quantum yield also indicates the formation of a complex between the probe and Hg^{2+} ion. The pH sensitivity of the probe was also studied in the presence of Hg^{2+} ion; there was no change in fluorescence intensity when pH was less than 5, and fluorescence intensity was increased in the 6–11 pH range indicating the complex is more stable in these pH ranges. The probe detects Hg^{2+} ion without interference of moderately acidic, neutral, and moderately basic pH. The binding stoichiometry of the probe with Hg^{2+} ion was calculated using Job's plot, where the maximum mole fraction of the probe is 0.5, so it is indicated that the complex has 1:1 binding stoichiometry between analyte and probe (Figure 10.7b).

10.3.5 Imidazole-Phenanthridine-Based Probe for Ratiometric Detection of Methanol in Biodiesel

Recently, a continuous increase in industrialization has resulted in environmental contamination issues worldwide. Methanol is a regularly used chemical in biochemical and clinical labs, industries and also used in antifreeze, petroleum, etc. (Tephlyl, 1991). Methanol is a toxic compound that results in blindness or even death. The frequently encountered problem is contaminated beverages at a higher methanol content, such as fermented alcoholic drinks. Excess methanol in biodiesel can have detrimental effects like moisture absorption, flash point reduction, and metal corrosion (Zhang et al., 2014; Akbarinejad et al., 2016). The detection of ethanol and methanol in the system is challenging, and a more sensitive and selective method is required.

Sawminathan et al. (2021) developed ratiometric fluorescent probe imidazole-based compounds 7 and 9a, 9b, and 9c for the detection of methanol. They exhibit solvatochromic fluorescence and are strongly emissive in their solution state. The probes were synthesized via a multistep reaction: In step 1, the reactant 4-aminobenzaldehyde reacts with 2-tetralone in 2 equivalents in the presence of ammonium acetate and methanol as solvent at room temperature to give compound 3. In step 2, 9,10-phenanthrenequinone reacts with terephthaldehyde and amine (6 and 8) in the ammonium acetate's presence in refluxing acetic acid to provide 7 and 9a, 9b, and 9c (Gao et al., 2017; Reddy et al., 2016) as solid compounds in 68%–73% yields (Figure 10.8a). Compounds 7 and 9c, which contain aldehyde, demonstrated better ICT characteristics than compounds 9a and 9b. The imidazole ring's electron-withdrawing group, aldehyde, intensified the ICT effect. The developed probe has excellent methanol sensing and also good selectivity. The aldehyde group of probes forms hydrogen bonding with methanol and alters

ICT behavior of the probe, which makes the mechanism of detection.

In ACN solvent, compounds 7, 9a, 9b, and 9c displayed absorption peaks between 256 and 376 nm; absorption peaks are due to π–π* and n–π* transitions, respectively. Compound 7 has absorption maxima at 260 nm, and the compounds 9a, 9b, and 9c have absorption maxima at 256 nm. The molar absorption coefficient of compound 7 was less than the imidazole derivatives 9a, 9b, and 9c due to bulky phenanthridine ring. Methanol interferes with the binding site of imidazole and blocks the H-bonding, responsible for the ESIPT process, and low quantum yield was observed. The Stokes shift for compound 7 was observed around 4,000 cm^{-1}, and compounds 9a, 9b, and 9c have large Stokes shift at 6,000 cm^{-1}. In compounds 7, 9a, 9b, and 9c, a positive solvatochromism is displayed by the aldehyde group at the para position of the phenyl ring. In methanol, solvent compounds showed emission peaks in the range of 390–420 nm, and the selectivity of probes toward methanol was studied by using various metals, alcoholic solvents, and amino acids. After the addition of various alcoholic solvents, only methanol showed a selective peak at 510 nm and 398 nm shift of emission band due to intermolecular H-bonding between aldehyde group and methanol (Figure 10.8b). They also studied ratiometric detection of methanol in ACN solvent by changing the concentration of methanol.

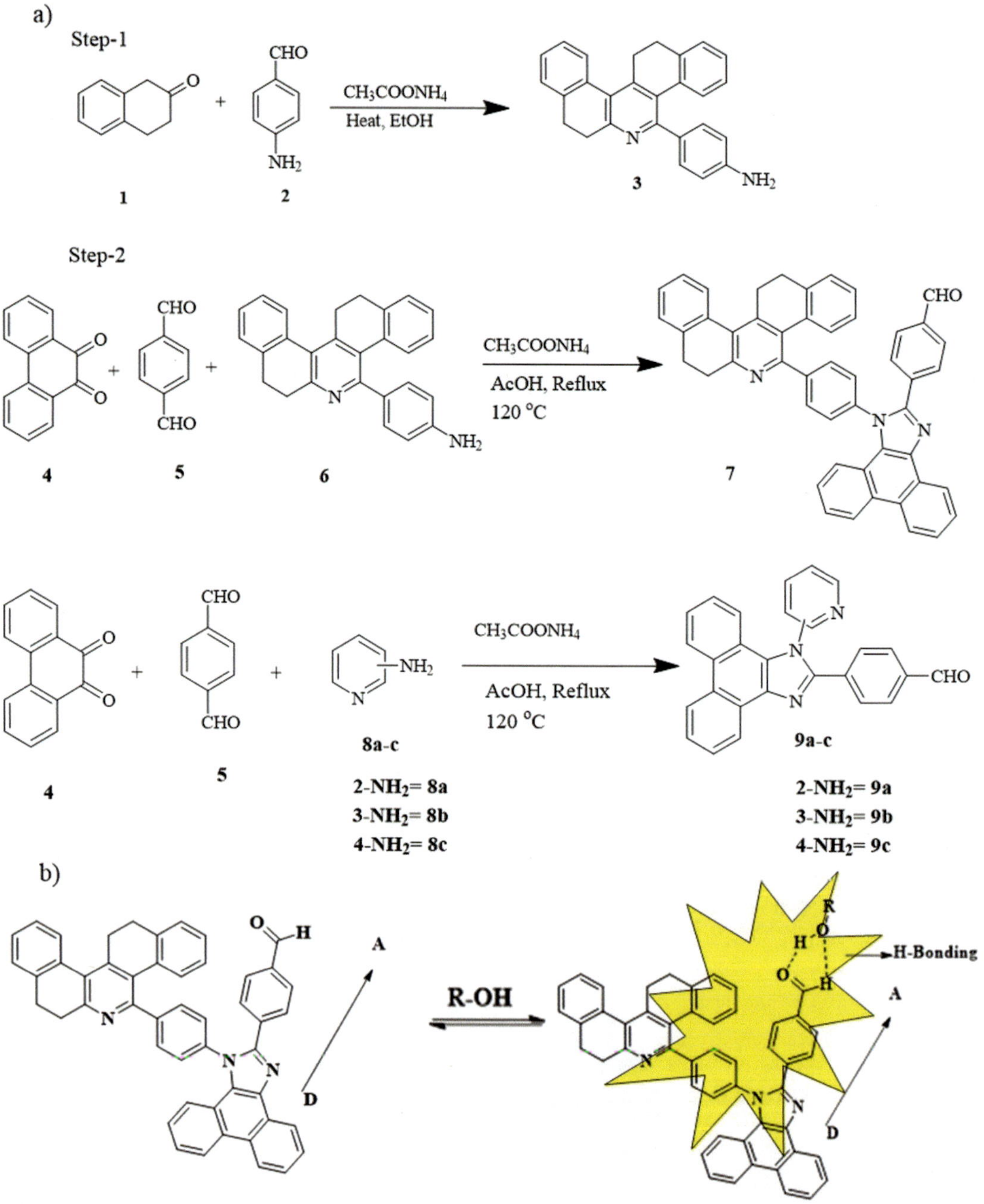

FIGURE 10.8 (a) Synthetic route for compounds **7**, **9a**, **9b,** and **9c.** (b) The intermolecular H-bonding of 7 with MeOH.

10.3.6 Phenanthridine-Pyrene-Based Fluorescent Probe for Detection of DNA/RNA and Inactive Mutant of Dipeptidyl Peptidase Enzyme

The small fluorescent molecules with the ability to specifically bind with various biomolecular structures, for instance, polynucleotides, proteins, DNA, RNA, etc., lead to changes in the fluorescence phenomenon, which can be a basis for the development of probes (Stojković et al., 2014). Pyrene derivatives are known fluorescent probes for biomolecules, as the large hydrophobic aromatic surface of pyrene is responsible for intercalation and minor groove binding with DNA/RNA base pairs. Pyrenes are sensitive to surrounding polarity and help to monitor conformational changes of protein (Newson et al., 2006; Lakowicz, 1999).

Matic et al. (2019) developed phenanthridine-pyrene-based fluorescent probes (Phen-Py-1 and Phen-Py-2) having differences in only linker length between pyrene and phenanthridine units. Prepared by alanine amino acid containing phenanthridine, first was deprotected in acidic condition (TFA-H_2O) and coupled with pyrene-carboxylic acid in the presence of N,N,N',N''-tetramethyl-*O*-(1H-benzotriazol-1yl)uronium hexafluorophosphate (HBTU) as coupling reagent and 1-hydroxybenotriazole (HOBT) as coupling additive and triethylamine (TEA) to give products (Figure 10.9).

In UV-vis spectra, a reduced absorption upon heating up to 90°C indicates intermolecular aggregation and stacking of compounds. Phenanthridine and pyrene chromophores show intermolecular contact assisted by a linker between them. The intermolecular interaction reduced the surface area exposed to water resulting in the Coulombic interaction between induced dipoles, and shielding of chromophores may result in hypochromism and a subsequent drop in the ε value.

Using cacodylate buffer, the emission spectra of Phen-Py-1 and Phen-Py-2 were measured at pH values of 5 and 7. In contrast to single units like phenanthridine and pyrene, Phen-Py-1 demonstrated excimer formation and was identified by a new fluorescence band at 475 nm that was considerably red-shifted. (λ_{max} = 400 and 378 nm), whereas Phen-Py-2 showed shoulder peak at 480 nm due to excimer formation. Excimer formation is pH dependent and formation was observed in water; however, it was not observed in methanol because of the lower polarity and dielectric constant, which affects intramolecular stacking. The small molecules bind to DNA/RNA polynucleotides, which usually affects the stability of the double helix, whereas it may stabilize or destabilize the polynucleotides. The stabilization or destabilization is indicated by a change in melting temperature (ΔT_m) value, i.e., the difference in melting temperature of free nucleotide and its complex with small molecules. For both DNA and RNA, the molecules Phen-Py-1 and Phen-Py-2 had less of an impact on the helix's stability (below 0.5°C).

The excimer signal of Phen-Py-1 at 470 nm was quenched by the addition of polynucleotide up to 30% and also showed the blue shift of emission maxima. With the exception of titration with poly dGdC, monomer emission at 400 nm remained unchanged. When Phen-Py-1 binds to polynucleotides, the intramolecular phenanthridine-pyrene dimer unstacks, quenching the excimer signal while leaving the fluorescence signal at 400 nm unchanged. Less thermal stabilization suggests no classical intercalation of phenanthridine moiety is possible whereas the possibility of partial intercalation of pyrene or phenanthridine moiety exist. The binding of Phen-Py-1 to the dipeptidyl peptidase (DPP III) inactive enzyme mutant E451A was also studied by fluorometric titrations, and it was found that Phen-Py-1 binds to protein with submicromolar affinity. This monozinc metalloexopeptidase enzyme hydrolyzes dipeptides at substrates' N-termini. This enzyme is also important in protein catabolism and has a function of pain modulation and oxidative stress, so it is a useful target for drug development. The excimer fluorescence band was quenched, and the emission band of pyrene was increased upon the addition of the enzyme. The hydrophobic unit, pyrene, is important for protein binding.

10.3.7 Styryl Tetrahydrodibenzo[*a,i*]phenanthridine-Based Naked-Eye Solvatochromic Probe for Acidochromic Application

Balijapalli et al. (2016) developed tetrahydrophenanthridines containing pyridine rings, mostly found in natural products and biologically active compounds. The research group has developed sensors containing donor-acceptor groups with extended conjugation (Figure 10.10a). They studied the effect of variation in substituents on optical

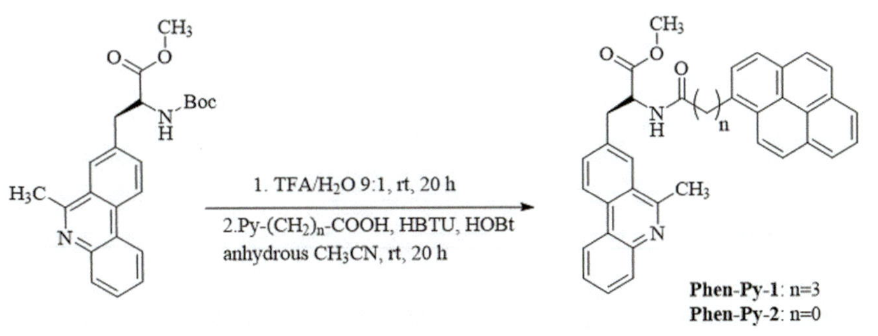

FIGURE 10.9 Synthesis of Phen-Py-1 and Phen-Py-2 by peptide coupling reaction.

activity and additionally examined the probe's sensitivity to pH and water-solvent mixture of different polarities. The change in color due to interaction with pH solutions and solvents is called acidochromism and solvatochromism, respectively.

They have developed styryl tetrahydrodibenzophenanthridine from cinnamaldehydes and 2-tetralone using suitable donor and acceptor groups (Figure 10.10a, dyes a–e). Rich optical characteristics are displayed by styryl phenanthridines in both excited and ground states. The styryl segment displayed a large impact on optical properties; λ_{max} lies in the range of 416–552 nm with good quantum yield. Dye D exhibited a large Stokes shift value of 7,265 cm^{-1} and more red-shifted, yellow emission at 552 nm than the others. The nature of the solvent does not influence the absorption spectra of dyes. The positive solvatochromism was observed in the emission spectra of dyes. The polar solvent stabilizes more polar excited states. The AIE is observed in dye **c**; fluorescence was triggered by the aggregation of molecules in a 10% THF-water mixture. The emission spectra of dye **c** significantly shifted with the change in pH; good linearity is observed in the emission intensity ratio in the pH range 4.1–6.8, so it is useful for the quantitative determination of pH in the acidic range. The binding of H^+ with N atom of pyridine was responsible for the enhancement of the ICT process and for the sensor's ratiometric reaction to an acidic pH. The phenanthridine unit was protonated as a result of the addition of TFA and was reorganized by the addition of a base like TEA and turned into acidochromism. Red shift of spectra due to protonation of the pyridine unit generates phenanthridine as an electron-withdrawing unit and triggers a facile ICT process; a ratiometric fluorescence enhancement due to enhancement of acid content in organic solvent due to formation of ionic fluorescent structure on protonation of dye **c** can be used to detect acid impurity in solvents (Figure 10.10b).

10.3.8 Fused Pyrazole-Phenanthridine-Based Probe for Live Cell pH Imaging

The stimuli-responsive organic compounds have attracted considerable attention in material, medicine, biology, and environmental science (Petruci et al., 2016; Chung et al., 2009). The acidochromic material has an emission response at a particular pH that can be useful for various applications in scientific research such as agriculture, biological processes, ecology, and environmental samples analysis. The pH measurement is essential to study physical and chemical processes in cells. The development of chemosensors for pH detection in cells has many advantages like high selectivity, sensitivity, as well as a cost-effective method (Antonijević et al., 2005). Several fluorophores, including cyanine, BODIPY, phthalimide, and rhodamine, are used as pH probes (Li et al., 2019; Aigner et al., 2013).

Muthukumar et al. (2019) developed a fused pyrazole-phenanthridine-based probe (Figure 10.11a), containing both pyrazole and pyridine as donor moiety, which showed weak fluorescence; however after undergoing protonation, it resulted in the formation of pyridinium ion to get D-π-A type probes with ICT process in which pyridinium ion act as an acceptor and pyrazole as a donor moiety. Trifluoroacetic acid-induced greenish-yellow fluorescence can be reversed by incorporating triethylamine. When metal ions and other competitive moiety exist, the synthesized probes demonstrate high selectivity toward H^+. The probes have excellent cell permeability useful for monitoring pH fluctuation in

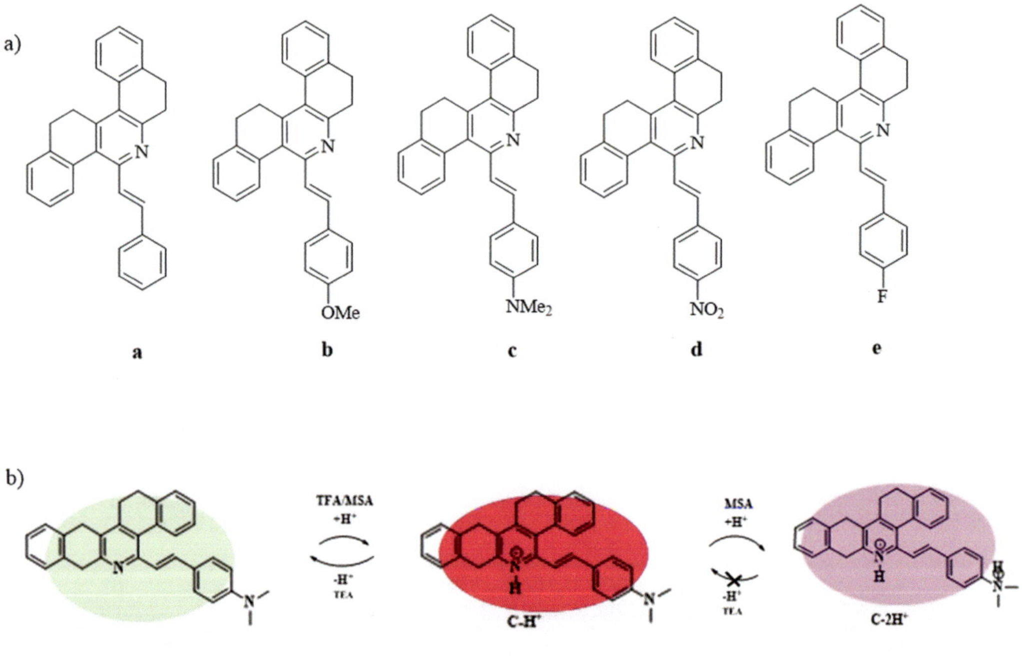

FIGURE 10.10 (a) Structure of the 5-styryl substituted terahydrodibenzo[a,i]phenanthridines. (b) Protonation-deprotonation equilibria of dye c.

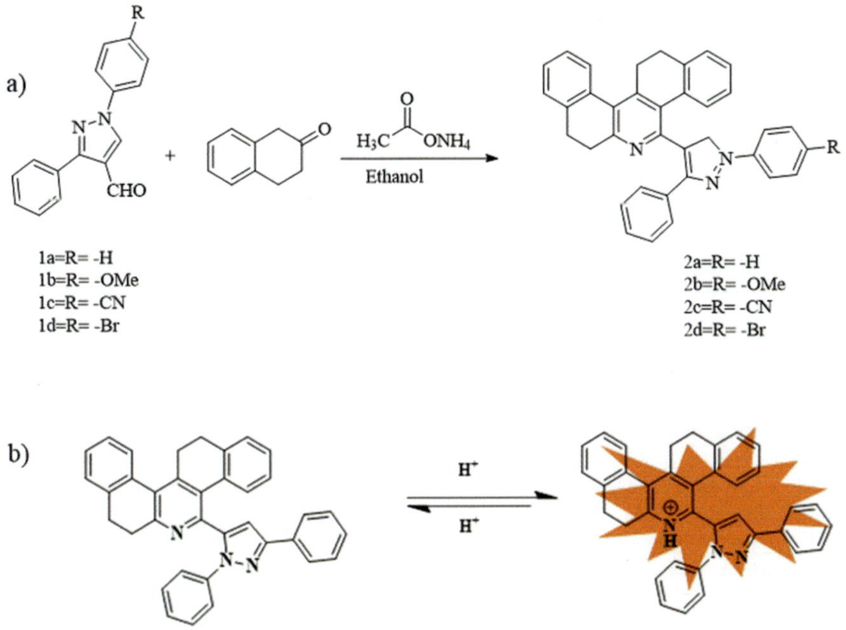

FIGURE 10.11 (a) Synthesis of probe. (b) Protonation of probe 2.

cells. Acid-sensitive molecules change their optical properties upon the addition of protons due to electron delocalization in molecules. The fluorophores containing N atom allows the protonation in acidic medium which can provide the acidochromic property to the probe . A new class of fused pyrazole-phenanthridine probes was synthesized having proton sensing ability due to the induced intramolecular charge transfer mechanism. Ratiometric fluorescence was shown by probes 2a and 2b upon the incorporation of trifluoracetic acid. After the protonation of probes, their fluorescent property can be reversed. Probe 2a has excellent fluorescence in the pH range of 1–7 and pK_a value 3.12. The probes 2a and 2b have selectivity toward H^+ ions in the presence of other ions and molecules (Figure 10.11b)

10.4 CONCLUSION

Phenanthridine fluorophore is an emerging moiety used for the development of chemosensors for use in healthcare and environmental science. Phenanthridine acts as a donor moiety, and various electron-withdrawing groups are substituted on the phenanthridine ring for the modulation of fluorescent properties and selective applications. Phenanthridine-based fluorescent probes have been investigated for the detection of metal ions, toxic chemicals, and biological compounds through various mechanisms like ICT, ESIPT, PET, FRET *etc.*

REFERENCES

Aigner, D., Borisov, S. M., Petritsch, P., & Klimant, I. (2013). Novel near infra-red fluorescent pH sensors based on 1-aminoperylene bisimides covalently grafted onto poly(acryloylmorpholine). *Chemical Communications*, *49*(21), 2139–2141.

Akbarinejad, A., Ghoorchian, A., Kamalabadi, M., & Alizadeh, N. (2016). Electrospun soluble conductive polypyrrole nanoparticles for fabrication of highly selective n-butylamine gas sensor. *Sensors and Actuators, B: Chemical*, *236*, 99–108.

Antonijević, M. M., Dimitrijević, M. D., Šerbula, S. M., Dimitrijević, V. L. J., Bogdanović, G. D., & Milić, S. M. (2005). Influence of inorganic anions on electrochemical behaviour of pyrite. *Electrochimica Acta*, *50*(20), 4160–4167.

Balijapalli, U., Manickam, S., Thirumoorthy, K., Natesan Sundaramurthy, K., & Sathiyanarayanan, K. I. (2019). (Tetrahydrodibenzo[a, i]phenanthridin-5-yl)phenol as a fluorescent probe for the detection of aniline. *Journal of Organic Chemistry*, *84*(18), 11513–11523.

Balijapalli, U., Manickam, S., Thiyagarajan, M. D., & Iyer, S. K. (2016). Highly emissive naked-eye solvatochromic probe based on styryl tetrahydrodibenzo[: A, i] phenanthridine for acidochromic applications. *RSC Advances*, *6*(63), 58549–58560.

Beer, P. D., & Gale, P. A. (2001). Anion recognition and sensing: The state of the art and future perspectives. *Angewandte Chemie International Edition, 40*, 486–516.

Bhattacharjee, Y. (2008). New efforts to detect explosives require advances on many fronts. *Science, 320,* 1416–1417.

Bowman, W. R., Lyon, J. E., & Pritchard, G. J. (2012). Palladium and radical routes to phenanthridines. *Arkivoc*, *2012*(7), 210–227.

Brugnara, C. (2003). Iron deficiency and erythropoiesis: New diagnostic approaches. *Clinical chemistry,* *49*(10), 1573–1578

Budén, M. E., Dorn, V. B., Gamba, M., Pierini, A. B., & Rossi, R. A. (2010). Electron-transfer-mediated synthesis of phenanthridines by intramolecular arylation of anions from n-(ortho-Halobenzyl)arylamines: regiochemical and mechanistic analysis. *Journal of Organic Chemistry*, *75*(7), 2206–2218.

Chung, S. Y., Nam, S. W., Lim, J., Park, S., & Yoon, J. (2009). A highly selective cyanide sensing in water via fluorescence change and its application to in vivo imaging. *Chemical Communications*, *20*, 2866–2868.

Enbanathan, S., Manickam, S., Munusamy, S., Jothi, D., Manoj Kumar, S., & Kulathu Iyer, S. (2022a). A phenanthridine-based probe for selective detection of hypochlorite ions. *New Journal of Chemistry*, *46*(14), 6570–6576.

Enbanathan, S., & Iyer, S. K. (2022b). A novel phenanthridine and terpyridine based D-π-A fluorescent probe for the ratiometric detection of Cd^{2+} in environmental water samples and living cells. Ecotoxicology and Environmental Safety, 247, 114272.

Faye, D., Zhang, H., Lefevre, J. P., Bell, J., Delaire, J. A., & Leray, I. (2012). Mercury detection in a microfluidic device by using a molecular sensor soluble in organoaqueous solvent. *Photochemical and Photobiological Sciences*, *11*(11), 1737–1743.

Gao, Y., Hong, Y. L., Yin, L. C., Wu, Z., Yang, Z., Chen, M. L., Liu, Z., Ma, T., Sun, D. M., Ni, Z., Ma, X. L., Cheng, H. M., & Ren, W. (2017). Ultrafast Growth of High-Quality Monolayer WSe_2 on Au. *Advanced Materials*, *29*(29), 1700990

Garg, P., Rawat, K. S., Bhattacharyya, G., Kumar, S., & Pathak, B. (2019). Hexagonal CuCl monolayer for water splitting: A DFT study. *ACS Applied Nano Materials*, *2*(12), 4238–4246.

Gilbert Morgan, B. T., & Percy Walls, L. (1931). Besearches in the phenanthridine series. Part I. A new synthesis of phenunthridine hornologues and derivatives. *Chemical Society*, 2447–2456

Gimenez, L. E., Noblin, T. A., Williams, S. Y., Mullick Bagchi, S., Ji, R.-L., Tao, Y.-X., Jeppesen, C. B., Conde-Frieboes, K. W., Sawyer, T. K., Grieco, P., & Cone, R. D. (2022). Demonstration of a common DPhe7 to DNal(2')7 peptide ligand antagonist switch for the melanocortin-3 and melanocortin-4 receptors identifies systematic mischaracterization of the pharmacological properties of melanocortin peptides. *Journal of Medicinal Chemistry*, *65*(8), 5990–6000.

Górski, Ł., Malinowska, E., Parzuchowski, P., Zhang, W., & Meyerhoff, M. E. (2003). Recognition of anions using metalloporphyrin-based ion-selective membranes: State-of-the-art. *Electroanalysis*, *15*(15–16), 1229–1235.

Huang, Z., Yu, S., Wen, K., Yu, X., & Pu, L. (2014). Zn(ii) promoted dramatic enhancement in the enantioselective fluorescent recognition of functional chiral amines by a chiral aldehyde. *Chemical Science*, *5*(9), 3457–3462.

Jia, X., Yang, Y., He, Y., Ma, Q., & Liu, Y. (2019). Theoretical study on the sensing mechanism of a fluorescence chemosensor for the cyanide anion. *Spectrochimica Acta - Part A: Molecular and Biomolecular Spectroscopy*, *216*, 258–264.

Jiang, T., He, P., Wang, G., Shen, Y., Nan, C. W., & Fan, L. Z. (2020). Solvent-free synthesis of thin, flexible, nonflammable garnet-based composite solid electrolyte for all-solid-state lithium batteries. *Advanced Energy Materials*, *10*(12), 1903376.

Jin, M. U., Jin-Lan, Y. A., Zhang, D. W., Qiong, J. I. (2021). Progress in preparation of metal nanoclusters and their application in detection of environmental pollutants. *Chinese Journal of Analytical Chemistry*, *49*(3), 319–329. Chinese Academy of Sciences.

Katritzky, A. R., Du, W., Matsukawa, Y., Ghiviriga, I., Denisenko, S. N. (1999). Polycyclic fused phenanthridine: An alternated approach from benzotriazole. *Journal of Heterocyclic Chemistry*, *36*, 927–932.

Katritzky, A. R., & Yang, B. (1996). Radical mediated synthesis of 6-arylphenanhtridine via benzotriazole ring opening. *Journal of Heterocyclic Chemistry*, *33*, 607–610.

Krishnaveni, K., Iniya, M., Jeyanthi, D., Siva, A., & Chellappa, D. (2018). A new multifunctional benzimidazole tagged coumarin as ratiometric fluorophore for the detection of Cd2+/F– ions and imaging in live cells. *Spectrochimica Acta - Part A: Molecular and Biomolecular Spectroscopy*, *205*, 557–567.

Lakowicz, J. R. (Eds.)(1999). Introduction to fluorescence. In *Principles of Fluorescence Spectroscopy* (pp. 1–23). Springer.

Landete, J. M., Ferrer, S., Polo, L., & Pardo, I. (2005). Biogenic amines in wines from three Spanish regions. *Journal of Agricultural and Food Chemistry*, *53*(4), 1119–1124.

Leardini, R., Tundo, A., Zanardi, G., & Pedulli, G. F. (1985). A new and convenient synthesis of phenanthridines. *G. F. Synthesis*, 107–110. doi:10.1055/s-1985-31129

Li, M., Xiong, T., Du, J., Tian, R., Xiao, M., Guo, L., Long, S., Fan, J., Sun, W., Shao, K., Song, X., Foley, J. W., & Peng, X. (2019). Superoxide radical photogenerator with amplification effect: Surmounting the Achilles' heels of photodynamic oncotherapy. *Journal of the American Chemical Society*, *141*(6), 2695–2702.

Li, Z., Fang, M., LaGasse, M. K., Askim, J. R., & Suslick, K. S. (2017). Colorimetric recognition of aldehydes and ketones. *Angewandte Chemie - International Edition*, *56*(33), 9860–9863.

Lin, S., Xu, L., Chi Wang, A., & Wang, Z. L. (2020). Quantifying electron-transfer in liquid-solid contact electrification and the formation of electric double-layer. *Nature Communications*, *11*(1), 399.

Lysén, M., Kristensen, J. L., Vedsø, P., & Begtrup, M. (2002). Convergent synthesis of 6-substituted phenanthridines via anionic ring closure. *Organic Letters*, *4*(2), 257–259.

Mamalis, B. P., & Petrow, V. (1950). Mamalis and petrow: Some heterocyclic N-oxides. *Journal of the Chemical Society*, *142*, 703–711.

Manickam, S., & Iyer, S. K. (2020). Highly sensitive turn-off fluorescent detection of cyanide in aqueous medium using dicyanovinyl-substituted phenanthridine fluorophore. *RSC Advances*, *10*(20), 11791–11799.

Ponram, M., Balijapalli, U., Sambath, B., Iyer, S. K., Cingaram, R., & Sundaramurthy, K. N. (2018). Development of paper-based chemosensor for the detection of mercury ions using mono-and tetra-sulfur bearing phenanthridines. *New Journal of Chemistry*, *42*(11), 8530-8536

Matić, J., Tandarić, T., Stojković, M. R., Šupljika, F., Vianello, R., & Tumir, L.-M. (2019). Phenanthridine-pyrene conjugates as fluorescent probes for DNA/RNA and an inactive mutant of dipeptidyl peptidase enzyme. *International Journal of Biological Macromolecules*, *134*, 422–434.

McBurney, R. T., Slawin, A. M. Z., Smart, L. A., Yu, Y., & Walton, J. C. (2011). UV promoted phenanthridine syntheses from oxime carbonate derived iminyl radicals. *Chemical Communications*, *47*(28), 7974–7976.

McDonagh, C., Burke, C. S., & MacCraith, B. D. (2008). Optical chemical sensors. *Chemical Reviews*, *108*(2), 400–422.

Meng, D., Yang, S., Sun, D., Zeng, Y., Sun, J., Li, Y., Yan, S., Huang, Y., Bielawski, C. W., & Geng, J. (2014). A dual-fluorescent composite of graphene oxide and poly(3-hexylthiophene) enables the ratiometric detection of amines. *Chemical Science*, *5*(8), 3130–3134.

Montañez, R., Sánchez-Jiménez, F., Aldana-Montes, J. F., & Medina, M. Á. (2007). Polyamines: Metabolism to systems biology and beyond. *Amino Acids*, *33*(2), 283–289.

Muthukumar, V., Munusamy, S., Thirumoorthy, K., Sawminathan, S., & Kulathuiyer, S. (2019). Fused pyrazole-phenanthridine based dyads: Synthesis, photo-physical and theoretical studies, and live cell pH imaging. *RSC Advances*, *9*(66), 38687–38696.

Na Kim, H., Xiu Ren, W., Seung Kim, J., & Yoon, J. (2012). Fluorescent and colorimetric sensors for detection of lead, cadmium, and mercury ions. *Chemical Society Reviews, 41*(8), 3210–3244.

Newson, T., Dyer, Tom, Adam, C., & Sharp, S. (2006). Effect of structure on the geotechnical properties of bauxite residue. *Journal of Geotechnical and Geoenvironmental Engineering, 132,* 143–151.

Pacquit, A., Frisby, J., Diamond, D., Lau, K. T., Farrell, A., Quilty, B., & Diamond, D. (2007). Development of a smart packaging for the monitoring of fish spoilage. *Food Chemistry, 102*(2), 466–470.

Peng, Z., Cheng, Y., Tan, B. C. M., Kang, L., Tian, Z., Zhu, Y., Zhang, W., Liang, Y., Hu, X., Tan, X., Guo, J., Dong, Z., Liang, Y., Bao, L., & Wang, J. (2012). Comprehensive analysis of RNA-Seq data reveals extensive RNA editing in a human transcriptome. *Nature Biotechnology, 30*(3), 253–260.

Petruci, J. F. D. S., & Cardoso, A. A. (2016). Portable and disposable paper-based fluorescent sensor for in situ gaseous hydrogen sulfide determination in near real-time. *Analytical Chemistry, 88*(23), 11714–11719.

Pictet, A., Ankersmit, H. J. Justus Liebigs. (1891). Ueber das phenanthridin. *Annalen der Chemie und Pharmacie, 266,* 138–153.

Pinto, S. M. A., Calvete, M. J. F., Ghica, M. E., Soler, S., Gallardo, I., Pallier, A., Laranjo, M. B., Cardoso, A. M. S., Castro, M. M. C. A., Brett, C. M. A., Pereira, M. M., Tóth, É., & Geraldes, C. F. G. C. (2019). A biocompatible redox MRI probe based on a Mn(ii)/Mn(iii) porphyrin. *Dalton Transactions, 48*(10), 3249–3262.

Rakow, N. A., Sen, A., Janzen, M. C., Ponder, J. B., & Suslick, K. S. (2005). Molecular recognition and discrimination of amines with a colorimetric array. *Angewandte Chemie - International Edition, 44*(29), 4528–4532.

Reddy, S. S., Sree, V. G., Gunasekar, K., Cho, W., Gal, Y. S., Song, M., Kang, J. W., & Jin, S. H. (2016). Highly efficient bipolar deep-blue fluorescent emitters for solution-processed non-doped organic light-emitting diodes based on 9,9-Dimethyl-9,10-dihydroacridine/Phenanthroimadazole derivatives. *Advanced Optical Materials, 4*(8), 1236–1246.

Sawminathan, S., & Iyer, S. K. (2021). A new imidazole-based phenanthridine probe for ratiometric fluorescence monitoring of methanol in biodiesel. *New Journal of Chemistry, 45*(13), 6033–6041.

Sawminathan, S., Munusamy, S., Jothi, D., & Iyer, S. K. (2021). Phenanthridine-based donor/acceptor fluorescent dyes: Synthesis, photophysical properties and fluorometric sensing of biogenic primary amines. *Chemistry Select, 6*(4), 858–864.

Seenan, S., Manickam, S., & Kulathu Iyer, S. (2021). Fluorescent chemosensors for Hg2+ions based on a pyridine-attached phenanthridine probe. *New Journal of Chemistry, 45*(37), 17667–17673.

Sekhar, A. R., Kaloo, M. A., & Sankar, J. (2014). Aliphatic amine discrimination by pentafluorophenyl dibromo BODIPY. *Chemistry - An Asian Journal, 9*(9), 2422–2426.

Simenhoff, M. L., Burke, J. F., Saukkonen, J. J., Ordinario, A. T., & Doty, R. (1977) Biochemical profile or uremic breath. *New England Journal of Medicine, 297,* 132–135.

Singh, H., Tiwari, K., Tiwari, R., Pramanik, S. K., & Das, A. (2019). Small molecule as fluorescent probes for monitoring intracellular enzymatic transformations. *ACS. Chemical Reviews, 119*(22), 11718–11760.

Singh, M., & Kushwaha, M. (2021). IJA-Nutrient content, uptake and NUE of oats. *Indian Journal of Agronomy, 66*(4), 466–473.

Soh, N. (2018). DNA-based fluorescent probe. *Trends in Analytical Chemistry, 34,* 515.

Song, C., Yang, B., Yang, Y., & Wang, L. (2016). SERS-based mercury ion detections: Principles, strategies and recent advances. *Science China Chemistry, 59*(1), 16–29.

Stojković, M. R., Škugor, M., Dudek, Ł., Grolik, J. A., Eilmes, J., & Piantanida, I. (2014). Molecular recognition of AT-DNA sequences by the induced CD pattern of dibenzotetraaza[14] annulene (DBTAA)-adenine derivatives. *Beilstein Journal of Organic Chemistry, 10,* 2175–2185.

Tephlyl, T. R. (1991). The toxicity of methanol. *Life Sciences, 48*(11), 1031–1041.

Wang, J., Liu, W., Luo, G., Li, Z., Zhao, C., Zhang, H., Zhu, M., Xu, Q., Wang, X., Zhao, C., Qu, Y., Yang, Z., Yao, T., Li, Y., Lin, Y., Wu, Y., & Li, Y. (2018). Synergistic effect of well-defined dual sites boosting the oxygen reduction reaction. *Energy and Environmental Science, 11*(12), 3375–3379.

Wang, P., Duan, L., & Liao, Y. (2019). A retrievable and highly selective peptide-based fluorescent probe for detection of Cd^{2+} and Cys in aqueous solutions and live cells. *Microchemical Journal, 146,* 818–827.

WHO guideline 1996, Vol-II, ISBN 9241544805, International Programme on Chemical Safety (2nd edition) .

Wu, D., Sedgwick, A. C., Gunnlaugsson, T., Akkaya, E. U., Yoon, J., & James, T. D. (2017). Fluorescent chemosensors: The past, present and future. *Chemical Society Reviews, 46*(23), 7105–7123.

Wu, X., Yue, H., Zhang, Y., Gao, X., Li, X., Wang, L., Cao, Y., Hou, M., An, H., Zhang, L., Li, S., Ma, J., Lin, H., Fu, Y., Gu, H., Lou, W., Wei, W., Zare, R. N., & Ge, J. (2019). Packaging and delivering enzymes by amorphous metal-organic frameworks. *Nature Communications, 10*(1), 5165.

Xiang, M., Li, Y., Yang, J., Lei, K., Li, Y., Li, F., Zheng, D., Fang, X., & Cao, Y. (2021). Heavy metal contamination risk assessment and correlation analysis of heavy metal contents in soil and crops. *Environmental Pollution, 278,* 116911.

Yuan, Y., Wang, D., Long, W., Deng, F., Yu, S., Tian, J., Ouyang, H., Lin, S., Zhang, X., & Wei, Y. (2021). Ratiometric fluorescent detection of hypochlorite in aqueous solution and living cells using an ionic probe with aggregation-induced emission feature. *Sensors and Actuators, B: Chemical, 330,* 129324.

Zhang, P., Shi, B., You, X., Zhang, Y., Lin, Q., Yao, H., & Wei, T. (2014). A highly selective and sensitive chemosensor for instant detection cyanide via different channels in aqueous solution. *Tetrahedron, 70*(10), 1889–1894.

Zhang, Y., Deng, Y., Wang, C., Li, L., Xu, L., Yu, Y., & Su, X. (2019). Probing and regulating the activity of cellular enzymes by using DNA tetrahedron nanostructures. *Chemical Science, 10*(23), 5959–5966.

Zhang, Y., Li, D., Li, Y., & Yu, J. (2014). Solvatochromic AIE luminogens as supersensitive water detectors in organic solvents and highly efficient cyanide chemosensors in water. *Chemical Science, 5*(7), 2710–2716.

Zhang, Y., Peng, C., Ma, X., Che, Y., & Zhao, J. (2015). Fluorescent and photoconductive nanoribbons as a dual-mode sensor for selective discrimination of alkyl amines versus aromatic amines. *Chemical Communications, 51*(81), 15004–15007.

Zhang, Y. Y., Chen, X. Z., Liu, X. Y., Wang, M., Liu, J. J., Gao, G., Zhang, X. Y., Sun, R. Z., Hou, S. C., & Wang, H. M. (2018). A highly sensitive multifunctional sensor based on phenylene-acetylene for colorimetric detection of Fe^{2+} and ratiometric fluorescent detection of Cd^{2+} and Zn^{2+}. *Sensors and Actuators, B: Chemical, 273,* 1077–1084.

11 Benzothiazole-Based Fluorescent Probes for Various Applications

Rohan Bhatia, Preeti Ashokkumar Chaudhran and Abha Sharma
National Institute of Pharmaceutical Education and Research,
Raebareli, Lucknow, UP, India

LIST OF ABBREVIATIONS

ESIPT	Excited-state intramolecular proton transfer
AIE	Aggregation-induced emission
LOD	Limit of detection
CHEF	Chelation-enhanced fluorescence
BTT	3-(benzo[d]thiazol-2-yl)-5-bromosalicylaldehyde-4 N-phenyl thiosemicarbazone
BZDM	2-amino-3-((E)-3-(benzo[d]thiazol-2-yl)-2-hydroxy-5-methylbenzylidene)amino)maleonitrile
DMSO	Dimethyl sulfoxide
THF	Tetrahydrofuran
ICT	Intramolecular charge transfer
DFT	Density functional theory
EMEA	European Medicines Evaluation Agency
PET	Photoinduced electron transfer
Cys	Cysteine
Hcy	Homocysteine
GSH	Glutathione
NIR	Near-infrared
BMP	2-(benzo[d]thiazol-2-yl)-4-methyl-6-((morpholinoimino)methyl)phenol
SPECT	Single-photon emission computed tomography
AD	Alzheimer's disease
HOMO	Highest occupied molecular orbital
LUMO	Lowest unoccupied molecular orbital
IC_{50}	Half maximal inhibitory concentration
DNA	Deoxyribonucleic acid

11.1 INTRODUCTION

Benzothiazole, with the chemical formula C_7H_5NS, is a heterocyclic aromatic moiety comprising a 1,3-thiazole ring fused to benzene (Quin, 1998). It has a flavor reminiscent of meat, a sulfurous smell, and is slightly viscous in nature (Bozec & Moody, 2009). Benzothiazole derivatives are prepared by reacting 2-aminothiophenol with corresponding acids using molecular iodine under the solvent-free condition (Gupta & Moorthy, 2007), with various aldehydes in DMF; acetic acid promoted condensation of aldehyde in air, iron promoted via nitroarene ortho-C-H sulfuration with elemental sulfur or decarboxylative redox cyclization reaction of o-chloronitroarenes and aryl acetic acids via elemental sulfur, copper catalyzed synthesis from 2-iodoanilines, benzyl chlorides, and elemental sulfur, visible light-induced cyclization of ortho-halothiobenzanilides, group transfer cyclization via iron-catalyzed oxidative C-H functionalization and by use of one-pot tandem approach from benzyl halides. Benzothiazole-based fluorescent probe (Thioflavin-T) is used to stain clumps of Aβ and α-synuclein (Razus et al., 2010; Watanabe et al., 2017; Henriksen et al., 2007; Klunk et al., 2001). Moreover, benzothiazole congeners are employed as fluorescent probes for analysis of several substances, including arsenic, cysteine, Zn^{2+}, and Al^{3+}(Chauhan et al., 2017; Wu et al., 2021; Tian et al., 2019). The benzothiazole fluorescent probes exhibit selective interaction with the analyte based on excited-state intramolecular proton transfer (ESIPT), aggregation-induced emission (AIE), intramolecular charge transfer (ICT), and photoinduced electron transfer (PET) (Yan et al., 2020).

Benzothiazole probes that target mitochondria can be used to visualize and track viscosity in lung tissues, live cells, and animals. The identification of intracellular viscosity using fluorescent probes is crucial since abnormalities in cellular viscosity are mostly linked to illnesses and cellular malfunctions. Probe HBTD-V responds sensitively to variations in biological systems' viscosity (Tang et al., 2021). Using two-photon imaging, a benzothiazole-based near-infrared fluorescent probe (Triph-SZ) was utilized to detect SO_2 derivatives and to generate a response to a change in lysosome viscosity. When HSO_3^- is added, Triph-SZ exhibits a highly sensitive and selective fluorescence response. Fluorescence increase was observed with this probe when the viscosity of the MeOH-Gly solution varied (Li et al., 2021).

These days heavy metal contamination is a major concern. Even very small amounts of heavy metals can be harmful to human health. This means that different fluorescent probes based on benzothiazoles can identify Hg^{2+}, Pb^{2+}, and Cd^{2+} with greater specificity. There are four types of benzothiazole probes: ratiometric, reaction-based, turn-on, and turn-off. Various design methodologies are employed in the creation of benzothiazole probes (Chen et al., 2021). The endoplasmic reticulum, Golgi apparatus, lysosomes, nucleus, membranes, mitochondria, and other organelles like peroxisomes, autophagosomes, and melanosomes were

 DOI: 10.1201/9781032706245-11

studied with differently substituted benzothiazole derivatives for sensing (Choi et al., 2021). Additionally, benzothiazole and 8-hydroxyquinoline may combine to successfully image Cd^{2+} (Lu et al., 2019). Organic chemists and biologists are paying more and more attention to the development of benzothiazole-appended fluorescent probes for various applications.

11.2 APPLICATION

11.2.1 Metal Ion Detection

11.2.1.1 Bioimaging and Detection of Copper

A benzothiazole-based probe BTP-LU has fluorescence signals to selectively recognize Cu^{2+}. The recovery of the ESIPT process is linked to the ratio of fluorescence signals at low concentrations of Cu^{2+} that are produced by hydrolysis of 2-picolinic formate in BTP-LU. The chelation of BTP with Cu^{2+} (Figure 11.1i) causes fluorescence emission intensity to decrease gradually (Figure 11.1ii). Moreover, it is capable of detecting Cu^{2+} in bovine and calf serum (Li *et al.*, 2020). Another probe (BTPPA) is a straightforward derivative based on benzothiazoles that exhibits strong Cu^{2+} ion selectivity with a 1:1 binding stoichiometry and LOD of 5.06×10^{-7} M (Figure 11.1iii). Additionally, it is capable of detecting Cu^{2+} in meat, fish, crab, and tea (Tang et al., 2020).

Another probe named BT-1 was synthesized for Cu^{2+} detection and live cell imaging. It is built on diarylacetylene with a benzothiazole moiety. Job's plot indicates that the binding stoichiometry of BT-1 to Cu^{2+} is 1:1. There was a noticeable increase in fluorescence on adding Cu^{2+} in probe BT-1 solution. Here, N and alkyne were used to chelate Cu^{2+} to BT-1 rather than S and alkyne (Figure 11.2i). This probe has a LOD of 3.2×10^{-9} M. Large Stokes shift and excellent copper ion selectivity characterize this probe (Figure 11.2ii). The BT-1 probe can be a chemosensor for Cu^{2+} ions in living cells (Zhao et al., 2019).

The basis of BSRH, a ratiometric chemosensor for Cu^{2+} detection, is a spiro-lactamized benzothiazole-substituted diethylrhodol probe. This probe can detect Cu^{2+} from its ability to block ESIPT (Figure 11.2iii) and generate highly sensitive and selective delocalized xanthene. This probe has a LOD of approximately 0.11–0.18 µM (Figure 11.2iv). This probe is a helpful tool for identifying Cu^{2+} in viable hepatocytes (Jin et al., 2021).

11.2.1.2 Bioimaging and Detection of Al^{3+}

After oxygen and silicon, aluminum is the third most prevalent element in the earth's crust. When high amounts of Al^{3+} are detected in the nervous system, it has been associated with neurological diseases. BTZ-SF is an active keto-enol (Figure 11.3i) fluorescent probe that detects Al^{3+} in aqueous medium and living cells ratiometrically via AIE and excited-state intramolecular proton transfer (ESIPT). BTZ-SF was checked with different metal ions and found to respond with the color change from orange to blue for Al^{3+} ion only (Figure 11.3ii). The BTZ-SF can detect Al^{3+} with exceptional selectivity, and there is a strong linear connection ($R^2 = 0.9911$) between the fluorescence intensity ratio (I_{476} nm/I_{568} nm) and Al^{3+} concentration (0–100 µM). It was calculated that the detection threshold might be as low as 2.2 µM. BTZ-SF chelated Al^{3+} through the O atoms in phenolic hydroxyl groups, the N atoms in -C=N groups, the carbonyl O, and three H_2O molecules, according to the hexa-coordinated structure of BTZ-SF-Al in a single crystal. Furthermore, BTZ-SF may be used to follow Al^{3+} in vivo, as demonstrated by confocal fluorescent

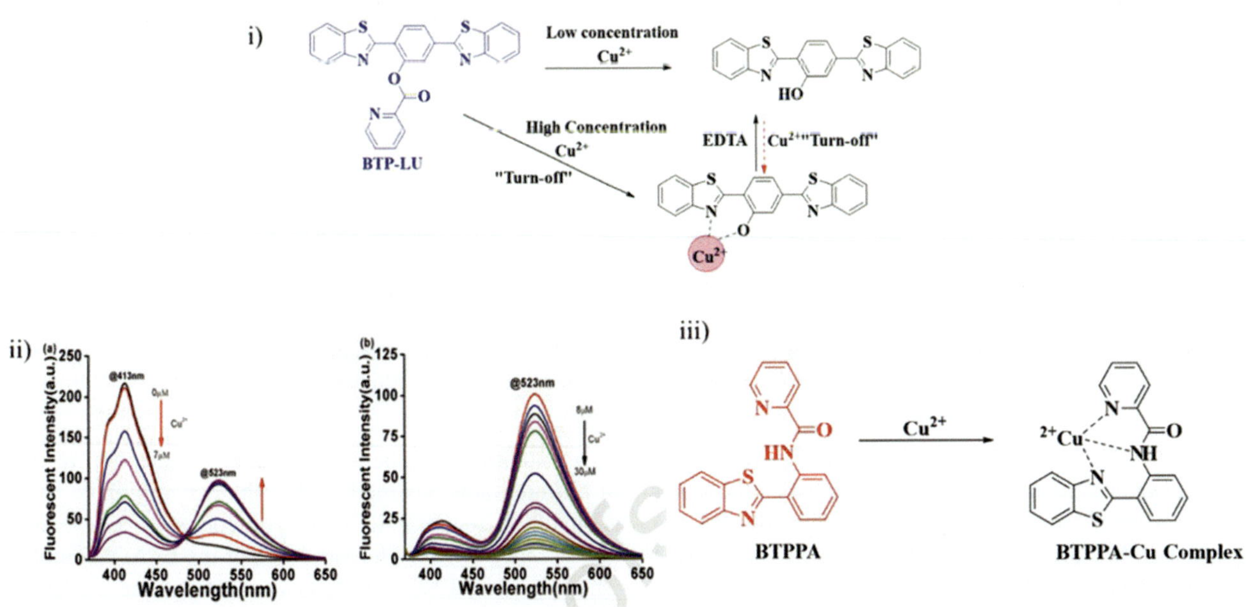

FIGURE 11.1 (i) Response mechanism of probe BTP-LU and copper. (ii) Fluorescence spectra of probe BTP-LU (1 µM) in the presence of copper ions (0–7 µM [left], 8–30 µM [right]). (iii) Response mechanism of probe BTPPA and copper. Reproduced with permission Li et al. and Tang et al. Copyright (2020), Elsevier.

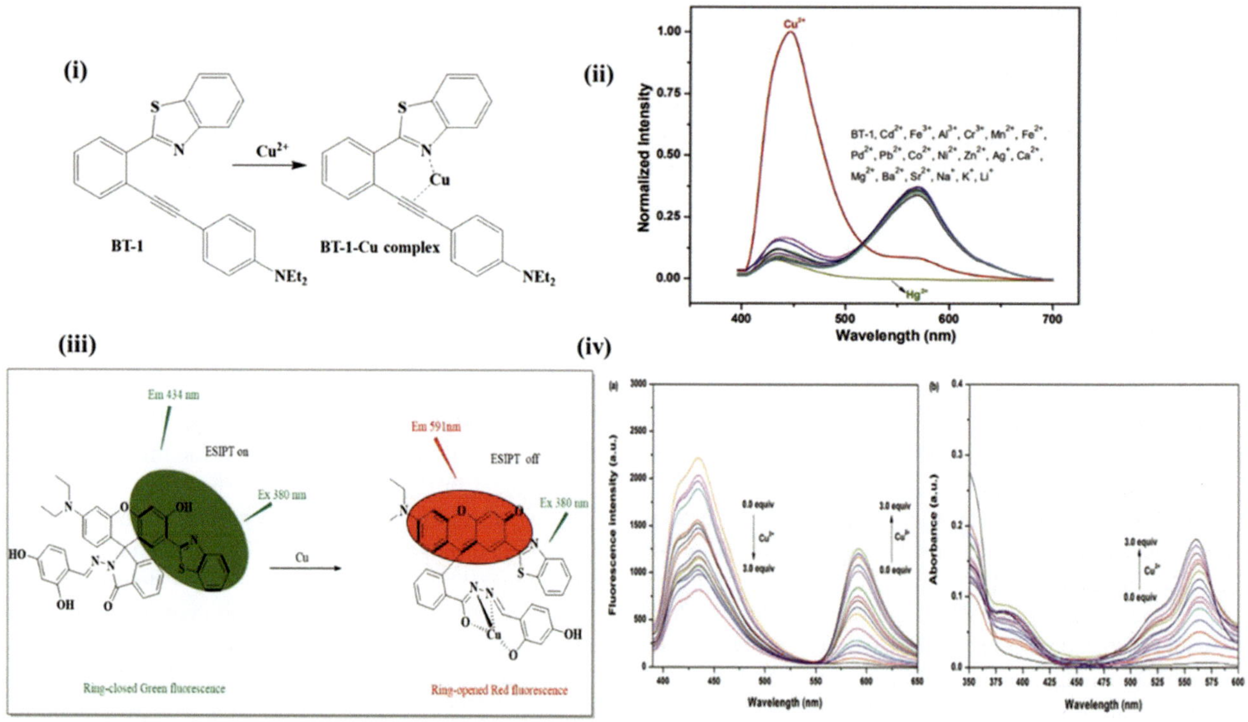

FIGURE 11.2 (i) Response mechanism of BT-1 in presence of copper. (ii) Fluorescence spectra of BT-1 in presence and absence of different metal ions. (iii) Response mechanism of probe BSRH. (iv) Fluorescence spectra (left) and UV-vis spectra (right) of BSRH in different concentration of copper ions. Reproduced with permission, Zhao et al., Copyright (2019), Elsevier and Jin et al., Copyright (2021), Elsevier.

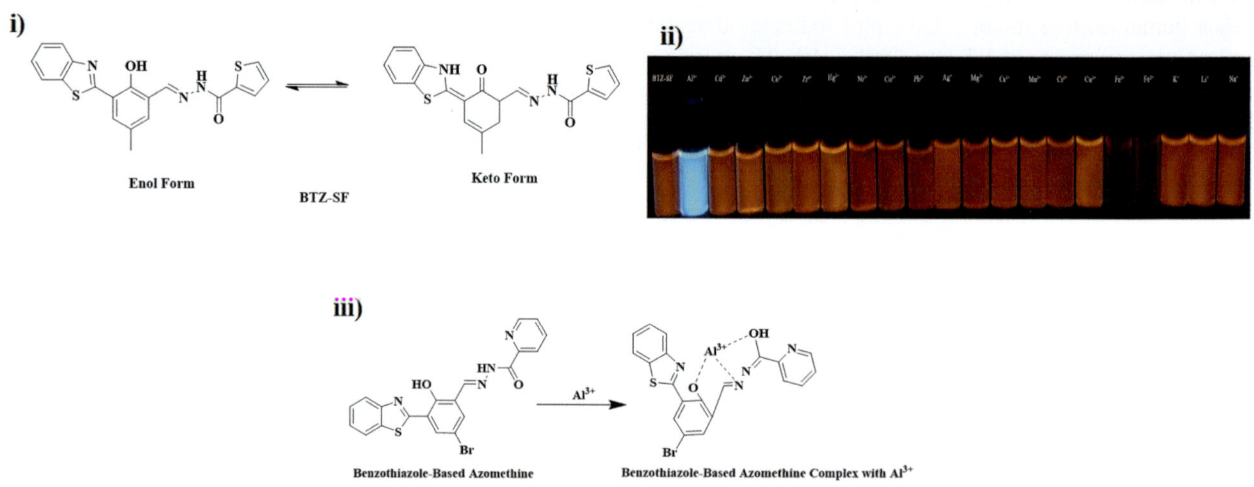

FIGURE 11.3 (i) Keto-enol form of BTZ-SF probe. (ii) Picture of BTZ-SF (10 μM) with 200 equiv. of metal ion analytes seen under UV lamp in HEPES buffer solution. (iii) Response mechanism of benzothiazol-based azomethine receptor L against Al^{3+}. Reproduced with permission, Chen et al., Copyright (2017), American Chemical Society.

images of HeLa cells. The red fluorescence emission in the THF/water system happens at 582 nm when the water percentage rises from 0 to 90%, which is caused by the exceptional AIE property of the free BTZ-SF. When Al^{3+} was introduced, there was a noticeable shift in ratiometric fluorescence at 476 nm. BTZ-SF will be a promising tool to respond to Al^{3+} in biological and environmental systems (Chen et al., 2017).

An additional probe for the selective detection of Al^{3+} in buffered aqueous methanol is the benzothiazole-based azomethine receptor L, which functions as a reversible fluorescence sensor. The formation of the L–Al^{3+} complex, which inhibits ESIPT, C-N isomerization, and chelation-enhanced fluorescence (CHEF), may be responsible for the fluorescence and absorption enhancement response of L toward Al^{3+} along with a significant blue shift

(Figure 11.3iii). Under a fluorescent microscope, the application of the probe was evaluated in BEL-7402 human hepatoma cancer cells. In the presence of Al^{3+} ions, turn-on, intense bluish-green fluorescence was detected. Under a UV lamp, noticeable changes in fluorescence can be seen. The Job's plot analysis, ^{1}H NMR, HRMS, IR, and DFT computing were used to suggest the chelation mechanism and the binding mode of L–Al^{3+} in a 1:1 stoichiometric ratio. Therefore, it may be used to find Al^{3+} in biological systems (Wang et al., 2019).

2-amino-3-((E)-3-(benzo[d]thiazol-2-yl)-2-hydroxy-5-methylbenzylidene)amino)maleonitrile is a benzothiazole-based fluorescence chemosensor (BZDM) that has been developed and synthesized to detect Al^{3+} in aqueous medium. The sensing behavior showed that BZDM had a noticeable fluorescence enhancement in DMSO buffer solution along with an outstanding selectivity toward Al^{3+} by chelation as shown in Figure 11.4i. Other significant metal ions hardly interfered with Al^{3+} detection. The LOD value for Al^{3+} under ideal circumstances was discovered to be 7.06 μM (Figure 11.4ii). Notably, the probe was used to detect Al^{3+} in living HeLa cells and was found to have good biological compatibility. When exposed to UV radiation, the fluorescence color changed from red to yellow, which was caused by the formation of the BZDM-Al^{3+} complex (Li et al., 2021).

Tian et al. developed BHM that showed a strong color change perceptible to the naked eye and a clear blue shift in the fluorescence spectra with Al^{3+} was observed, whereas remain intact with other metal ions. BHM forms chelate with Al^{3+} as shown in Figure 11.4iii. The Job's plot showed that probe BHM binding ratio to Al^{3+} was found to be 1:1 (Tian et al., 2019).

11.2.1.3 Bioimaging and Detection of Zn^{2+} Ions

Zinc is a trace element that is required in all human cells and is crucial for numerous biological processes, such as DNA synthesis, immunological response, gene transcription, brain signal transmission, apoptosis, and others. Zinc is also a cofactor for more than 100 enzymes and takes part in all major metabolic processes. Problems with zinc metabolism may be the root cause of many chronic illnesses. While an excess of zinc in cells can cause major neurological disorders including Alzheimer's, Parkinson's, and epilepsy, a deficit in zinc can lead to growth retardation, immunological dysfunction, eye and skin diseases, and diabetes in humans. 3-(benzo[d]thiazol-2-yl)-5-bromosalicylaldehyde-4-N-phenyl thiosemicarbazone (BTT) was investigated for ratiometric detection of Zn^{2+} ions in methanol/HEPES buffer solution (3:2, pH = 7.4). In the presence of Zn^{2+} ions, BTT emission peak blue-shifted dramatically, from 570 to 488 nm, and fluorescence color changes from orange to green. When BTT binds 1:1 with Zn^{2+} (Figure 11.5), it exhibits a clear color shift that is visible to the naked eye as well as a notable blue shift (82 nm) in its emission maxima. BTT shows solvatochromic behavior in different polar solvents. BTT was also checked against other metal ions where the response was only observed for Zn^{2+}, for which the LOD calculated as 37.7 nM. The cytotoxicity of the probe was evaluated on living HeLa cells using the MTT test, indicating the minimal cytotoxicity of the probe. A plausible sensing mechanism is the obstruction of ESIPT and CHEF as a result

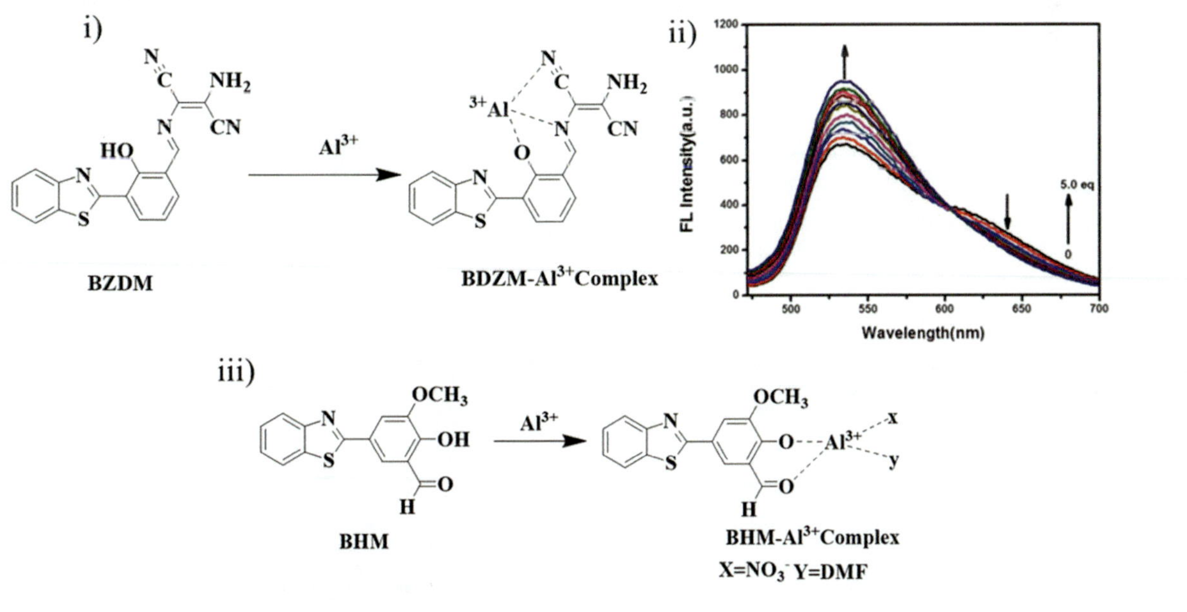

FIGURE 11.4 (i) Response mechanism of probe BZDM. (ii) Emission spectra of probe BZDM with addition of Al^{3+} in DMSO buffer solution at rt. (iii) Response mechanism of probe BHM. (iv) Job's plot of BHM with Al^{3+}. Reproduced with permission, Li et al., Copyright (2021), Elsevier.

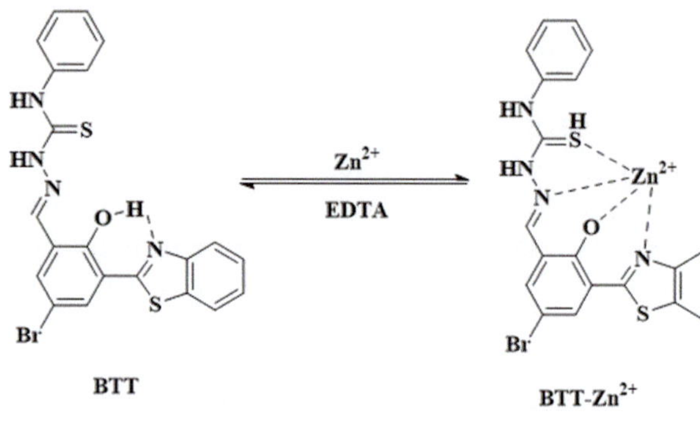

FIGURE 11.5 Response mechanism of probe BTT.

of the formation of the probe-Zn^{2+} complex. Furthermore, the probe BTT can scan both exogenous Zn^{2+} ions injected into HeLa cells and the endogenous Zn^{2+} distribution in SH-SY5Y neuroblastoma cell. Because of its large Stokes shift, quick ratiometric emission response at physiological pH, low detection limit, and minimal cytotoxicity, the probe could be useful for tracking and in vitro visualization of Zn^{2+} (Wu et al., 2021).

11.2.1.4 Bioimaging and Detection of Palladium

Palladium (Pd) is widely employed as a crucial material in many scientific fields, such as fuel cells, electronics, jewelry, dental alloys, medical devices, and catalytic converters. One prominent area of study is Pd-catalyzed reactions, which include transformations for the manufacture of medicines and functional compounds. There is a lot of leftover palladium in soil and water systems as a result of its widespread use. Palladium is a concern to human health because it can create complexes with proteins, amino acids that include thiols, DNA, RNA, and vitamins that can alter the physiological processes of the human life cycle and cause other major health issues. The European Medicines Evaluation Agency (EMEA) has established the threshold level of palladium discovered in pharmaceuticals as 5–10 ppm.

Chen et al. created two fluorescent probes based on the 2-(2'-hydroxy-)benzothiazole moiety to detect Pd^{2+} (Figure 11.6i). When Pd^{2+} was added, probe 1 produced green fluorescence with a Stokes shift of roughly 163 nm, but probe 2 showed red fluorescence under the same circumstances with a significant Stokes shift of roughly 217 nm (Figure 11.6ii). The detection limits of probe 1 and probe 2 to Pd^{2+} were found to be 285 nM and 14.6 nM, respectively. This outcome demonstrated that probe 2 is more Pd^{2+} sensitive than probe 1. Pd^{2+}-induced release of the 2-(2'-hydroxyphenyl)benzothiazole unit caused the two probes, 1 and 2, to show significant Stokes shifts with reduced excitation interference (Chen et al., 2017).

Li et al. developed the ESIPT ratiometric fluorescent probe (BTP-Pd) for detection of Pd(0). When it reacts with Pd(0), the ESIPT process is recovered, which accounts for the notable red shift (Figure 11.6iii). The free probe BTP-Pd (1 μM) showed prominent absorption bands at around 350 nm. As Pd(0) (5 μM) was added, the absorption bands showed a red shift. The free probe BTP-Pd(0) displayed a significant emission band at 411 nm upon excitation at 350 nm. The band at 523 nm increased as a result of the subsequent sharp decline in the fluorescence emission intensity of 411 nm with the addition of increasing Pd(0) (0–10 μM) (Figure 11.6iv). The emission peak moved up by 112 nm, from 411 nm to 523 nm. Probe BTP-Pd fluorescence

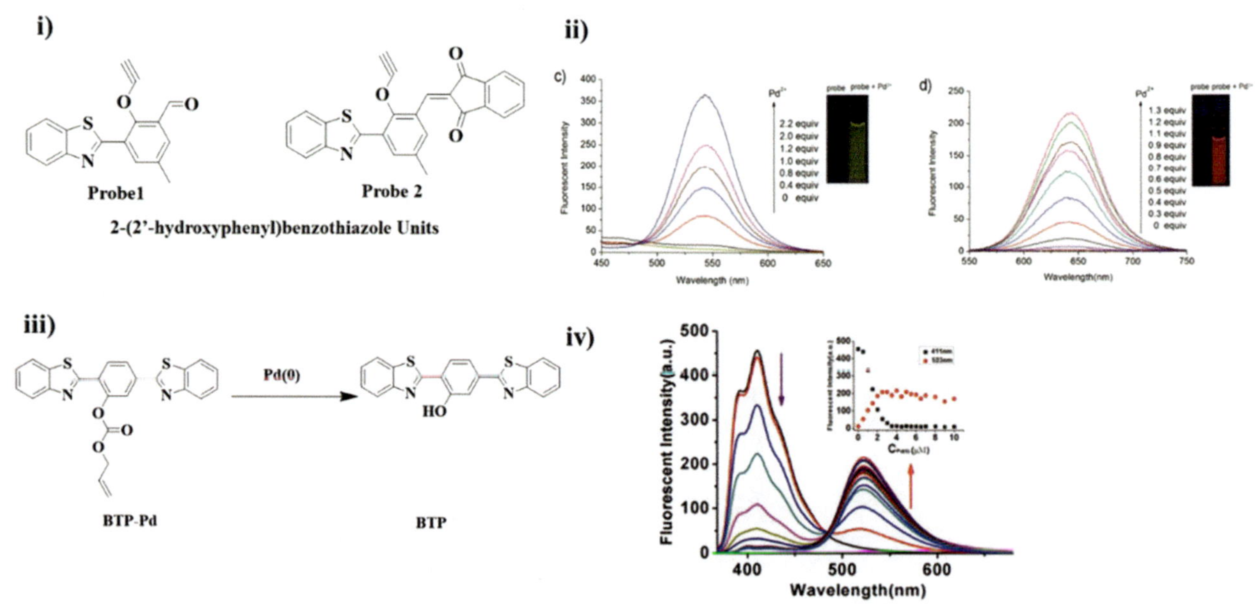

FIGURE 11.6 (i) Chemical structure of probe 1 and probe 2 (2-(2'-hydroxy)benzothiazole units). (ii) (c) Fluorescence spectra of probe 1 with different concentrations of Pd^{2+}; (d) fluorescence spectra of probe 2 with different concentrations of Pd^{2+}. (iii) Probe BTP-Pd reacts with Pd(0). (iv) Fluorescence spectra of BTP-Pd(0) in absence and presence of Pd(0) (0–10 μM). Reproduced with permission, Chen et al., Copyright (2017), Elsevier and Li et al., Copyright (2019), Elsevier.

emission intensity remained constant until Pd(0) concentration approached approximately 4 equivalent. The fluorescence intensity ratio demonstrated a noteworthy shift from 0.025 14.8 when Pd(0) (Li et al., 2019).

11.2.1.5 Bioimaging and Detection of Iron

The body needs iron, a vital mineral, for the development of human cells. An essential component of hemoglobin's ability to carry oxygen is iron. In muscle, iron is also involved in the uptake and storage of oxygen. Additionally, iron is essential for the synthesis of enzymes. An imbalance of iron in the body can lead to a number of illnesses. Inadequate iron levels in the body can impede liver growth, lower immunity, and result in iron deficiency anemia. Environmental pollution is the primary cause of high blood iron levels, which can result in hemorrhagic ironosis, diabetes, and other conditions.

Liu et al. developed a 2-(2-hydroxyphenyl)benzothiazole (HBT) probe (Figure 11.7i) with a significant red shift and fluorescence enhancement to detect Fe^{3+} with exceptional accuracy and selectivity over other metal ions. HBT displayed two fluorescence emissions in acetonitrile at 360 and 377 nm, respectively, with an excitation at 319 nm. It should be mentioned that there was a noticeable red shift of the emission, ranging from 370 nm to 420 nm, and that the emission strength gradually increased upon Fe^{3+} addition and a blue color fluorescence appeared. When the Fe^{3+} molar fraction was greater than 0.5, the fluorescence intensity at 420 nm increased somewhat, which is compatible with the binding stoichiometry shown by UV-vis absorption titration (Liu et al., 2013). The LOD calculated for the probe was 6.04×10^{-8} M, and the dissociation constant K_d was 1.13×10^{-11} M. According to experiments using UV-vis and fluorescence titrations, the binding stoichiometry of the probe-Fe^{3+} complex was 2:1.

Padghan and colleagues created a pyrene-based probe with benzothiazole ionophore (Py-BTZ, Figure 11.7ii) that can be used to detect Fe^{3+} and Fe^{2+} ions in CH_3CN:H_2O (1:1, v/v). Py-BTZ was created in a single step by condensing 2-amino-6-methoxy benzothiazole and pyrene aldehyde with a catalytic quantity of H_2SO_4, which produced 72% of Py-BTZ. The UV-vis absorption spectra showed distinct pyrene absorption bands at 297 and 438 nm. Py-BTZ exhibited a notable increase in fluorescence emission in addition to optical and colorimetric alterations toward Fe^{3+} and Fe^{2+} ions in comparison to other metal ions. Py-BTZ was discovered to have a LOD of 2.61 μM for Fe^{3+} ions and 2.06 μM for Fe^{2+} ions. Py-BTZ demonstrates features of aggregation-induced emission enhancement (AIEE) in mixtures of polar solvents, such as CH_3CN:H_2O (1:1, v/v). In Py-BTZ, the conjugated aza-bridge is responsible for the lowest energy absorption band at 438 nm. Compared to Fe^{3+} ions, the Py-BTZ color change for Fe^{2+} ions (at 3 equivalent) is less pronounced. With a low quantum yield ($\phi = 0.003$), the Py-BTZ fluorescence spectrum show a very weak emission band at 456 nm ($\lambda_{ex} = 400$ nm) (Padghan et al., 2017).

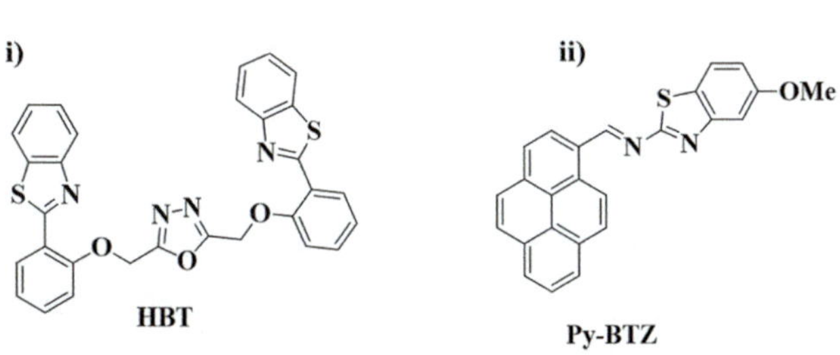

FIGURE 11.7 (i) Chemical structure of probe HBT. (ii) Chemical structure of probe Py-BTZ.

Guan et al. created the luminophore (Rh-F) that comprises rhodol and 2-hydroxy benzothiazole. Excellent fluorescence properties were demonstrated by Rh-F, including ESIPT feature and AIE. The blue-green fluorescence produced by Fe^{2+} was caused by an irreversible Fe^{2+} chelate that involves the N atom of the benzothiazole moiety and the oxygen atom of the hydroxyl group of rhodol (Figure 11.8i). Rh-F fluorescence was quenched as a result of blocking the ESIPT process. A THF/H_2O (7:3, v/v, pH = 7.4) solution was used to test the Rh-F probe fluorescence characteristics toward metal ions. Separate additions of different metal ion solutions (100 μM) were made to the 10 μM Rh-F analytical solution. The fluorescence spectra demonstrate that upon excitation at 350 nm, Rh-F solution (10 μM) exhibited a strong fluorescence intensity. The addition of Na^+, K^+, Pb^{2+}, Mg^{2+}, Ca^{2+}, Cu^{2+}, Zn^{2+}, Co^{2+}, Ag^{2+}, Cd^{2+}, Hg^{2+}, Mn^{2+}, Fe^{3+}, Cr^{3+}, and Al^{3+} ions results in nearly no discernible change in fluorescence intensity. The color of the fluorescent sensor Rh-F solution was observed to change from colorless to yellow after adding Fe^{2+} with LOD of 115.2 nM (Figure 11.8ii) (Guan et al., 2019).

Gong et al. created two new benzothiazole-based probes (sensors 1 and 2, Figure 11.8iii). In DMF solution, 5-formylsalicylic acid and o-aminothiophenol reacted to generate sensor 1. The sensor 1 was then esterified with ethanol and refluxed with tris to produce sensor 2. Both probes showed strong anti-interference capabilities and a selective fluorescence quenching response when exposed to Fe^{3+} ions in aqueous solutions. The LOD values of 8.43 μM and 5.86 μM for Fe^{3+} were observed by sensor 1 and sensor 2, respectively (Figure 11.8iv). The formation of 1:1 complex between sensors and Fe^{3+} ions was demonstrated by Job's plot, mass spectrometry, ^{1}H NMR titration, and DFT calculations. Selective response is generally influenced by solvent polarity, protophilia, solubility, and pH value. To maximize the response selectivity, the solvent systems C_2H_5OH:H_2O (1:1 and 4:1, v:v), CH_3CN:H_2O (1:1 and 4:1, v:v), DMF:H_2O (1:1, v:v), and pure water solvent at pH = 7.4 were chosen. Since sensors 1 and 2 shared the same conjugated unit of benzene-benzothiazole, their fluorescence spectra were identical.

Strong fluorescence emission was seen in both sensor 1 (1×10^{-5} M) and sensor 2 (5×10^{-6} M), with the center of emission at 411 nm and 421 nm, respectively. The selective complexing effect of sensor 1 and sensor 2 with Fe^{3+} ions

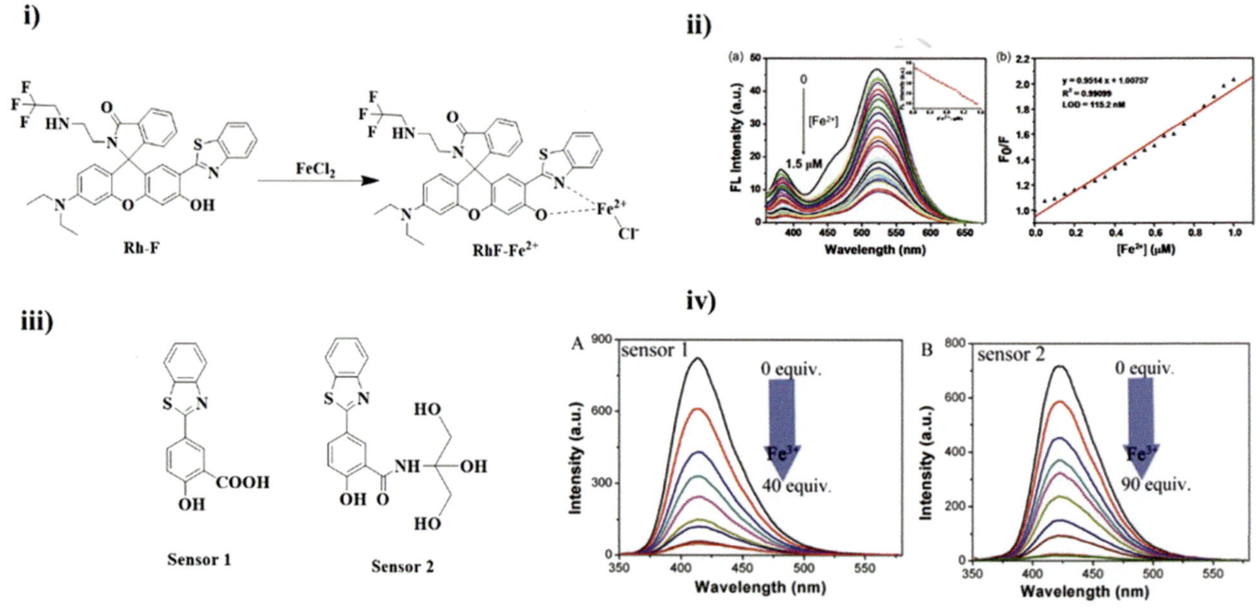

FIGURE 11.8 (i) Response mechanism of probe Rh-F. (ii) Fluorescence spectra of Rh-F with gradual increase in the concentration of Fe^{2+} (left); plot of fluorescence intensity (right). (iii) Chemical structures of sensor 1 and sensor 2. (iv) Fluorescence spectral changes of sensor 1 (left); fluorescence spectral changes of sensor 2 (right) upon addition of Fe^{3+}ions. Reproduced with permission, Guan et al. and Gong et al., Copyright (2019), Elsevier.

resulted in the reduction of their fluorescence. Finally, sensor 1 and sensor 2 might be employed to track the presence of Fe^{3+} ions in living cells due to their minimal cell toxicity (Gong et al., 2020).

11.2.1.6 Bioimaging and Detection of Chromium

Toxic heavy metal ions in particular are of great concern due to their adverse effect on the environment and living beings (Mitra et al., 2022). Lv et al. created the benzothiazole-functionalized Schiff-base derivative L by the condensation reaction of 2-formylbenzothiazole (1) with thiosemicarbazide. The sensor L selective sensing toward Cr^{3+} in an aqueous solution at pH 7.2 was determined with an association constant 5.73×10^4 M^{-1} and a LOD of 2.1×10^{-8} M. The L and L-Cr^{3+} combination was optimized using a B3LYP/6-31G (d,p) basis set. The fluorescence emission steadily rises with the progressive addition of Cr^{3+} (0–2 equiv.) to the sensor L solution, reaching saturation at 1 equiv. of Cr^{3+}. The high selectivity of sensor L points to a possible use of the chemosensor for biological tracking and monitoring of Cr^{3+}. Its response mechanism is shown in Figure 11.9i (Lv et al., 2017).

Erdemir et al. also produced a fluorescent probe based on anthracene that contains a benzothiazole group (BFA, Figure 11.9ii). The combination of restricted C=N isomerization, a unique anthracenyl static excimer production, and the inhibition of a highly efficient PET pathway led to a considerable boost in emission intensity upon the interaction of BFA with Cr^{3+}. For Cr^{3+}, the BFA detection limit was 0.46 μM. In the absence of Cr^{3+}, absorption peaks at 369 nm dominate the UV-vis absorption spectra of BFA. A new absorption band was seen at 442 nm following the addition of 0–2.0 equiv. of Cr^{3+} ion to the BFA (20 μM) solution. This indicates that a 2:1 complex between BFA and Cr^{3+} was formed. Finally, confocal fluorescence microscopy can be used to investigate the use of BFA as a bioimaging fluorescent probe to detect Cr^{3+} in human prostate cancer cell lines (Erdemir & Kocyigit, 2016).

11.2.2 Amino Acid Detection

11.2.2.1 Bioimaging and Detection of Cysteine

In many biological systems, cysteine (Cys), a by-product of the metabolism of homocysteine (Hcy) and a precursor of the antioxidant glutathione (GSH), is receiving more and more interest as a possible biomarker for several disorders. Many disorders, including Parkinson's, Alzheimer's, cardiovascular diseases, cancer, liver damage, poor growth in children, rheumatoid arthritis, neurotoxicity, edema or skin lesions, and so forth, are intimately linked to aberrant levels of Cys (30–200 μM) in the body. Therefore, knowing the physiological function and clinical indicators of Cys is very crucial for medical professionals.

HBTMP is a fluorescent probe based on benzothiazoles that can detect cysteine with high sensitivity and selectivity. Benefits of this probe include a low detection limit of 1.2×10^{-7} M and a large Stokes shift. With a 210 nm Stokes shift, there was an increase in emission at 607 nm when cysteine was added to the probe solution. Compared to homocysteine and glutathione, cysteine reacts with HBTMP at a quicker rate. The cyclization of the HBTMP-homocysteine intermediate is hindered by intramolecular electrostatic interaction, but Cys exhibits sensitivity to this interaction (Figure 11.10i). This Cys detection was carried out in aqueous solution, and potentially imagine Cys in living MCF-7 cells, targeting mitochondria (Tang et al., 2019). BPBC is

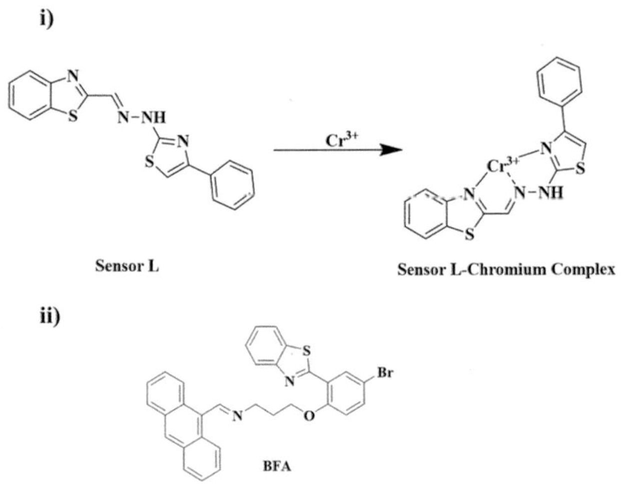

FIGURE 11.9 (i) Response mechanism of sensor L. (ii) Chemical structure of probe BFA.

an ESIPT-based, benzothiazole-based turn-on fluorescence probe for Cys detection. In essence, the two chloroacetate groups on this probe serve as Cys reaction sites. Compared to other amino acids like homocysteine and glutathione, Cys is the only amino acid for which this probe is highly selective and sensitive (Figure 11.10ii). With the benefit of low cytotoxicity (<50 μM), this probe may successfully bio-image intracellular Cys in living cells. It can be synthesized with good yield from inexpensive, commercially accessible ingredients (Yu et al., 2018).

Cys in aqueous solution can be quickly and selectively detected using the fluorescence probe Doc-Ac, which is based on benzothiazole and quinoline. This probe consists of a positively charged benzothiazole unit that withdraws electrons and an electron-donating acrylate quinoline derivative that connects to an unsaturated olefinic bond to form a "push-pull" conjugated dye structure that has an ICT effect. The acrylate group of the probe, which is employed

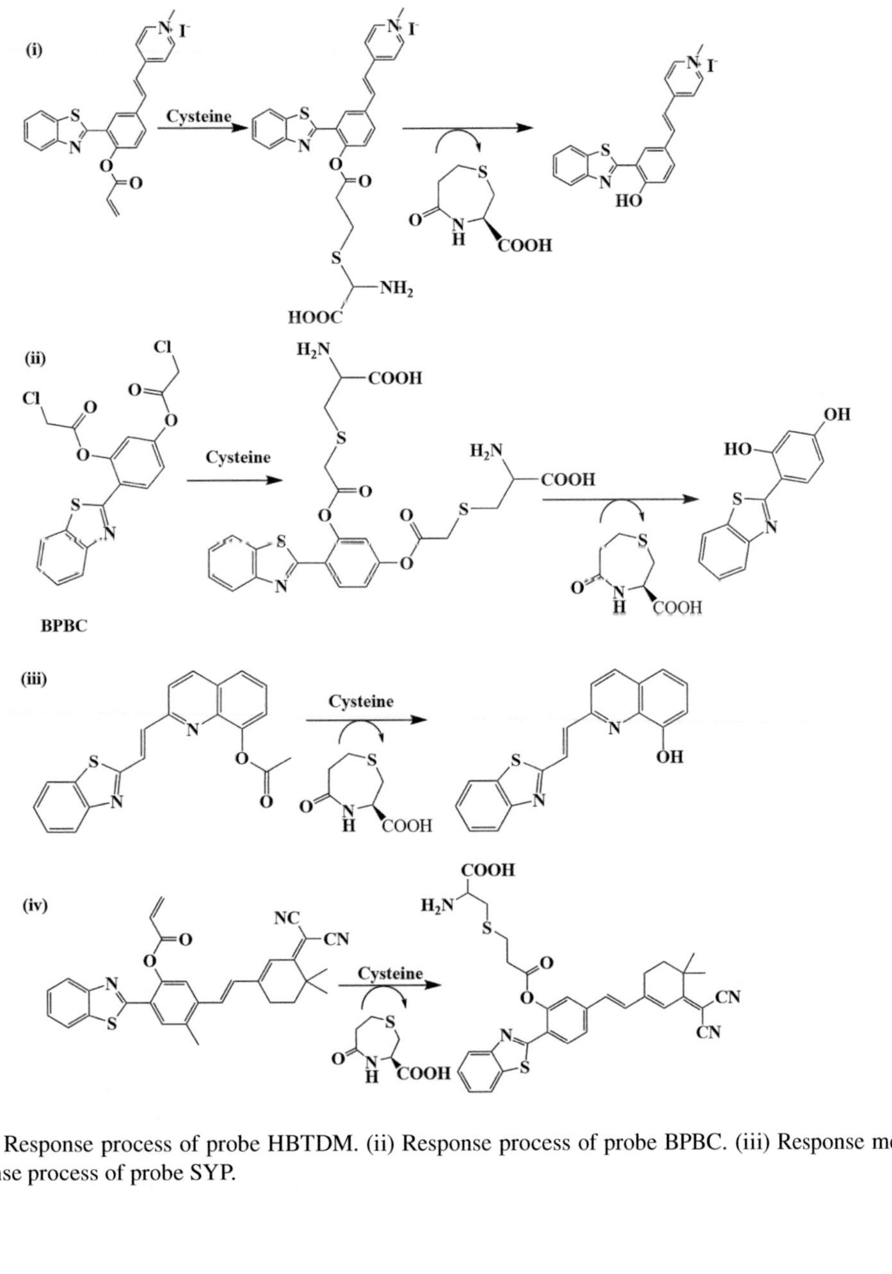

FIGURE 11.10 (i) Response process of probe HBTDM. (ii) Response process of probe BPBC. (iii) Response mechanism of probe Doc-Ac. (iv) Response process of probe SYP.

to detect Cys, is responsible for the sensing mechanism. The free probe displays the acrylate group's proton signals at 6.75, 6.70, and 6.34 ppm. The proton signals move toward a higher field upon the addition of Cys, signifying the creation of a cyclized adduct. When the concentration of the probe was gradually increased, the peak at 377 nm gradually decreased and the peak at 290 nm shifted toward 280 nm. The free probe exhibits significant absorption at 377 nm and weak absorption at 290 nm. In mice and live cells, this probe can also distinguish between indigenous and exogenous Cys. The response process of the probe is shown in Figure 11.10iii (Mei et al., 2021).

Another probe for the selective detection of Cys and bioimaging in cells and mice is the NIR-ESIPT-based fluorescent probe (SYP), which has substantial Stokes shift. Dicyanoisophorone and benzothiazole are coupled with toluene, which has an acryloyl group acting as a nucleophilic addition and leaving site, to create this probe. The nucleophilic addition of Cys in the SYP probe releases hydroxy toluene, which generates a turn-on NIR emission based on the ESIPT action of benzothiazole (Figure 11.10iv). Additionally, this probe exhibits excellent selectivity for Cys without any hindrance. This fluorophore exhibits two types of emission: keto-form emission and enol-form emission. The application of the MTT analysis on HeLa cells demonstrates its applicability to biological tissues. Moreover, it works well for the detection of Cys, which has a low detection limit of 0.20 μM or 0.98 μM at pH 8 or 7.4 (Ren et al., 2020).

11.2.2.2 Bioimaging and Detection of Glutathione

Glutathione (GSH) performs many functions in reversible redox reactions and is necessary for live cells to survive. Notably, abnormal GSH concentrations increase the risk of developing AIDS, cancer, and Alzheimer's disease. Benzothiazole-based fluorescent probes have been created as a result of their unique advantages, which include high sensitivity, strong specificity, real-time determination, and ease of use. One of the probes is 2'-hydroxy-3'-ethoxyphenyl benzothiazole based on the ESIPT fluorescence mechanism (Figure 11.11i). It can differentiate GSH from other thiols in an aqueous solution containing CTAB micelles in less than 30 minutes. It has also been effectively utilized to monitor GSH in living cells. The probe shows the fluorescence enhancement (Figure 11.11ii) at 485 nm due to the release of 2-(2'-hydroxy-3'-ethoxyphenyl) benzothiazole via the nucleophilic substitution process.

The introduction of CTAB micelles as a catalyst to quicken the reaction may have been motivated by the synchronous occurrence of hydrophobic and electrostatic interactions in the reaction system. Because of its negative charge, GSH was able to create a strong connection and quickly penetrate positively charged CTAB micelles with a hydrophobic substrate. The probe's greater selectivity to GSH over Cys and Hcy was also believed to be significantly influenced by the quantity of carboxyl groups in thiols. Compared to Cys and Hcy, GSH with two carboxyl groups was more easily able to have a considerable attraction to positively charged CTAB micelles, which boosted the ability to generate a negative charge. GSH was therefore probably going to attack the probe substrate that was inside the micellar aggregates, which would release the fluorophore. When GSH was added to the probe solution, the detection limit was improved by 15 times and can be as low as 0.81 M. The extra carboxyl group in GSH accounts for its enhanced selectivity over Cys and Hcy, and this probe can be applied to detect glutathione in living cells (Ren et al., 2016).

11.2.3 Detection of Anions

11.2.3.1 Bioimaging and Detection of Peroxynitrite

Within biological systems, superoxide and nitric oxide (NO) interact instantaneously to generate peroxynitrite, a highly reactive molecule. Overexposure to $ONOO^-$ may damage several molecular components of cells, including DNA and proteins, which can impair crucial cell components and functions. This could lead to the emergence of illnesses such as Alzheimer's, cancer, inflammatory conditions, and more. Furthermore, $ONOO^-$ was difficult to quantify due to its short half-life (about 1 second) and steady-state concentration in biological systems in the nanomolar range. Two simple fluorescent probes (ADB and ANB, Figure 11.12i) with the phenyl boronate moiety recognition receptor were developed for $ONOO^-$ detection. The effect of the

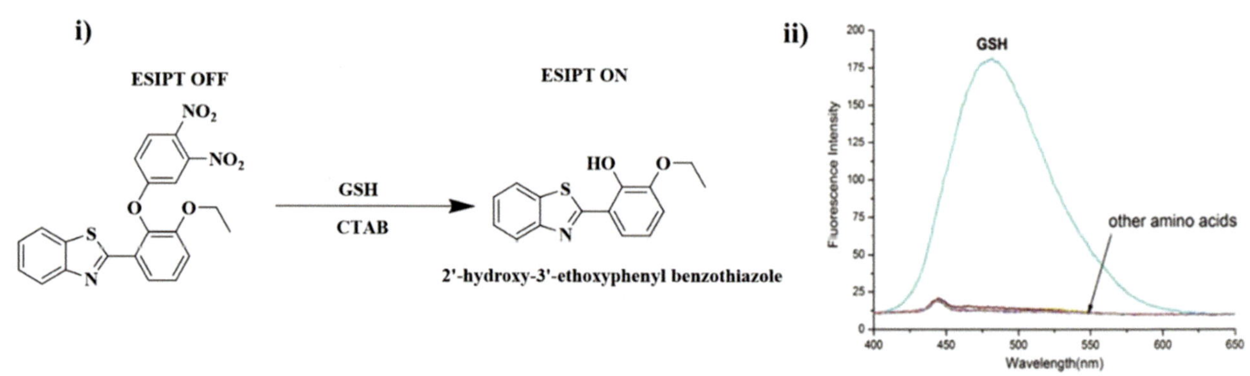

FIGURE 11.11 (i) Response process of probe 2'-hydroxy-3'-ethoxyphenyl benzothiazole. (ii) Fluorescence spectra of probe in presence and absence of different amino acids. Reproduced with permission, Ren et al., Copyright (2016), Elsevier.

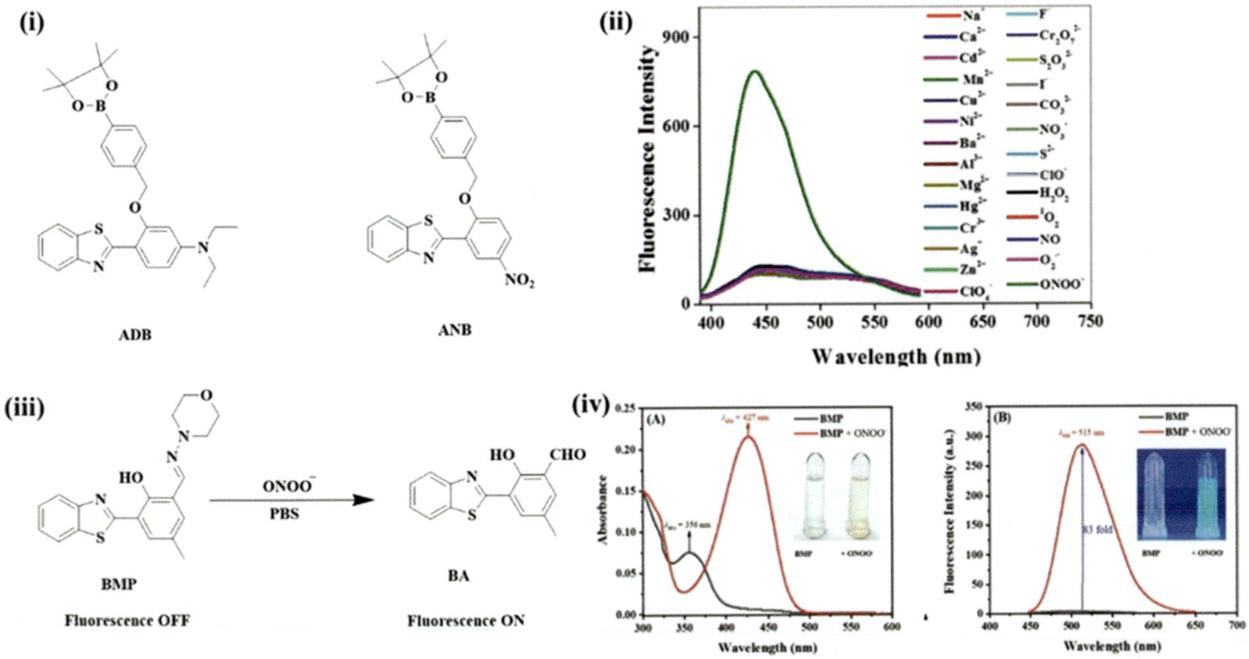

FIGURE 11.12 (i) Chemical structure of probe ADB and ANB. (ii) Fluorescence spectra of ADB in the presence and absence of different cations and anions. (iii) Response process of BMP. (iv) UV absorption spectra of BMP with or without $ONOO^-$ (left). Fluorescence spectra of BMP with or without $ONOO^-$ (right). Reproduced with permission, Li et al., and Gu et al., Copyright (2021), Elsevier.

substituents (electron donor or acceptor) on shifting in the emission spectrum was not uniform. ADB using the AIE technique showed quantitative $ONOO^-$ detection capability in response to $ONOO^-$ concentrations in less than 20 seconds, along with good sensitivity for $ONOO^-$ with a LOD of 75 nM. In an aqueous solution, it was demonstrated that ANB had a low quantum yield and sensitivity toward $ONOO^-$. It is noteworthy that ADB also identified exogenous $ONOO^-$ in the RAW 264.7 cells. The opening fluorescence response of the arlyboronate-based $ONOO^-$ fluorescence probe was moderate, but the sensitivity of $ONOO^-$ was significantly increased by the borate-masked probe. The main ADB absorption band at 362 nm disappeared and was replaced by a new band at 378 nm when $ONOO^-$ was introduced one after the other. The ANB absorption spectra at 363 nm become visible after adding $ONOO^-$ one at a time. The ADB and ANB fluorescence spectra in a PBS buffer solution (20 mM, pH 7.4, 1% DMSO) with increasing $ONOO^-$ amounts were recorded. Following incubation with $ONOO^-$ at different doses (0–5.7 M), a dose-dependent rise in fluorescence was observed with a quantum yield of 0.10. The presence of $ONOO^-$ increased the fluorescence of ADB by 7.6 times (Figure 11.12ii), and the intensity at 440 nm showed a good linear correlation ($R^2 = 0.9921$) with the added ONOO concentration (0.15–3.9 M). Through a series of oxidative hydrolysis processes, ADB was successfully employed to detect cellular $ONOO^-$ and showed great promise for the selective detection of exogenous $ONOO^-$ in living cells (Li et al., 2021).

The lysosome targetable fluorescent probe (E)-2-(benzo[d]thiazol-2-yl)-4-methyl-6-((morpholinoimino)methyl) phenol (BMP) is an additional tool for $ONOO^-$ detection in biological systems (Figure 11.12iii). It employs a benzothiazole derivative as a fluorophore and a reactive morpholino hydrazone as a novel $ONOO^-$ response site. An 83-fold increase in fluorescence after reacting with $ONOO^-$ as well as dual colorimetric and fluorescence changes were observed. The probe could scan $ONOO^-$ in live cell lysosomes and exhibited significant cell penetration and low cytotoxicity. Under identical conditions, the fluorescence spectra of BMP containing many analytes were analyzed to elucidate the selectivity of BMP for $ONOO^-$. Reactive oxygen species, cations, and anions that are relevant to biology made up the majority of the examined species. When $ONOO^-$ (10 mM) was added to the BMP (10 mM) solutions, the detection system produced a significant fluorescence enhancement signal (Figure 11.12iv); however, no discernible changes in fluorescence were seen when other analytes at higher concentrations (100 mM) were added. BMP excellent $ONOO^-$ selectivity enables it to meet the requirements for selective $ONOO^-$ detection in complex biosystems. It is especially remarkable that BMP is markedly insensitive to higher concentrations of $ONOO^-$ (10 times $ONOO^-$) and exhibits substantial activity and selectivity for $ONOO^-$ over ClO^-. This is expected to be the case since $ONOO^-$ oxidizes more readily than ClO^- to the hydrazone moiety. Compared to a few previously published $ONOO^-$ fluorescent probes, BMP offers some remarkable advantages, such as a low

LOD of 6 nM and fast response. It was observed that $ONOO^-$ oxidized the morpholino hydrazone of the BMP to create the aldehyde group. This probe was able to image $ONOO^-$ in HeLa cells due to its efficient biocompatibility and lysosome targeting (Gu et al., 2021).

11.2.3.2 Bioimaging and Detection of Bisulfite

Bisulfite is one of the primary forms of sulfur dioxide in the physiological system. Sulfur dioxide, long recognized as a significant air pollutant, is principally produced by burning fuels containing sulfur and using sulfur-containing industrial raw materials. Sulfite and bisulfite derivatives are the main forms of sulfur dioxide that dissolve in water. Sulfite and bisulfite are commonly used as enzyme inhibitors and antiseptic agents in food, alcohol, and pharmaceutical products because they can lower oxidation and bacterial growth as well as both enzymatic and nonenzymatic browning. HNa and HNb, 2-(2'-hydroxyphenyl)-benzothiazole (HBT)-derived fluorescent probes (Figure 11.13i), exhibit a fluorescence turn-on response that is specific to the presence of bisulfite, based on the ICT and ESIPT mechanisms. The conjugate addition reaction between the probes' nitroolefin moiety and bisulfite produced a blue shift in the absorption spectra (Figure 11.13ii). HNa and HNb demonstrated exceptional anti-interference properties, great selectivity, and strong sensitivity, suggesting that the probes are ideally adapted for biological systems with LOD values of 13 nM and 89 nM (Wang et al., 2018).

1-hydroxy-2-(benzothiazol-2-yl)naphthalene (HBTN) is a probe for sensing bisulfite through the Michael addition process (Figure 11.13iii). While adding bisulfite in a probe, a decrease in the intensity of emission at 655 nm was accompanied by a new emission at 475 nm in the fluorescence spectrum (Figure 11.13iv). The structure of the probe can be understood as a D-A-A type, where the electron-donating groups are hydroxyl groups, the electron-withdrawing

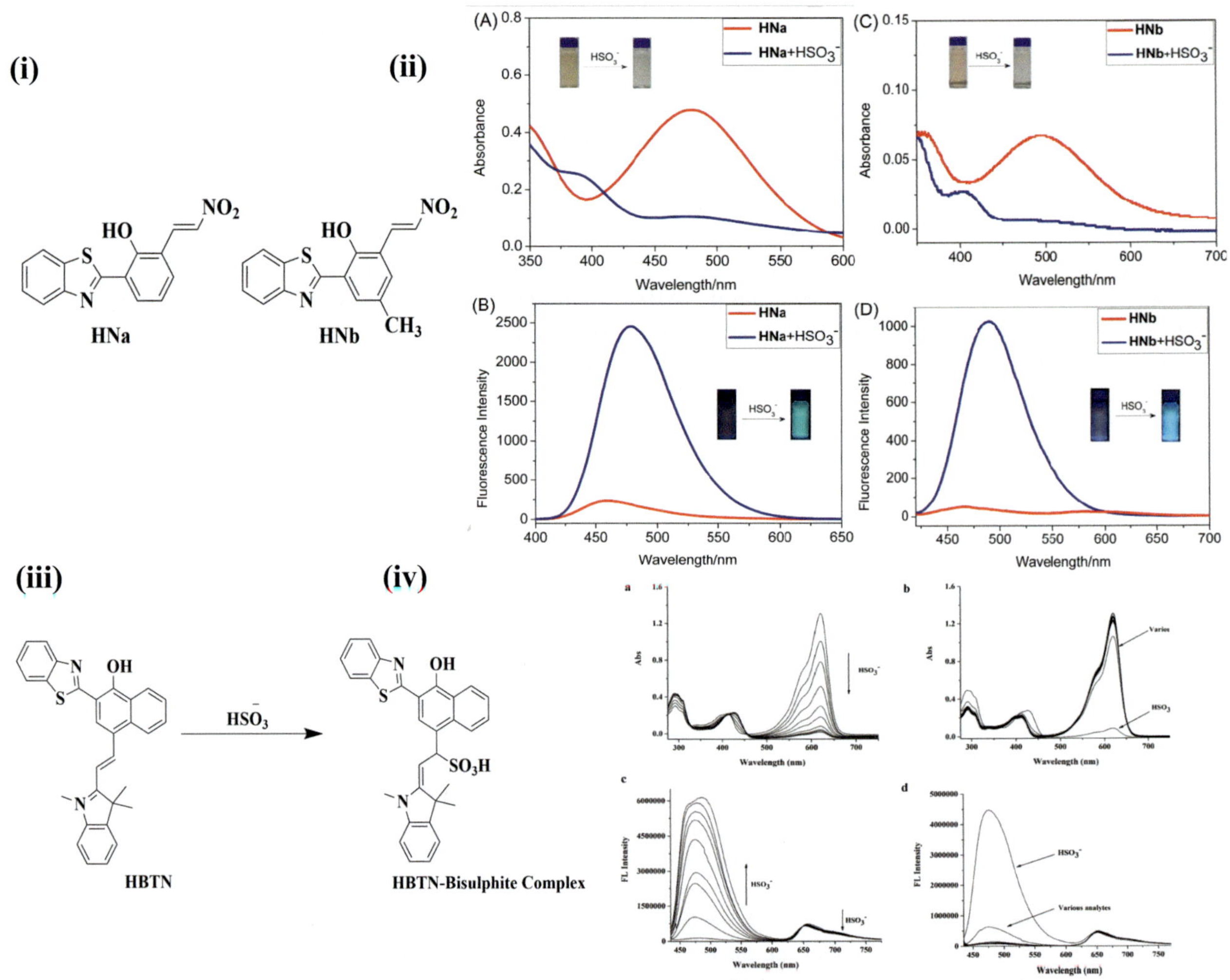

FIGURE 11.13 (i) Chemical structure of probes HNa and HNb. (ii) (A and B) Absorption spectra and fluorescence spectra of HNa before and after reacting with bisulfite in PBS buffer solution, respectively. (C and D) Absorption spectra and fluorescence spectra of HNb before and after reacting with bisulfite in PBS buffer solution. (iii) Response process of probe HBTN. (iv) (A and B) Changes in UV-vis absorption of probe with increase in bisulfite and with various ions, respectively. (C and D) Fluorescence responses of probe with increase in bisulfite concentration and with various ions, respectively. Reproduced with permission, Wang et al., Copyright (2018), Wiley-VCH and Qin et al., Copyright (2022), American Chemical Society.

groups are benzothiazole and indolium iodide, and the bridge is made up of ethylene and naphthalene. The probe, consisting of a naphthyl ring functioning as a donor, a hydroxyl moiety serving as a bridge, and benzothiazole/hemicyanine functioning as an acceptor, exhibits excellent selectivity and high sensitivity to bisulfite in aqueous solution. Due to its exceptional two-photon fluorescence properties, the probe is successfully used for fluorescence imaging of bisulfite in MCF-7 cells and rat liver tissues (Qin et al., 2022).

11.2.3.3 Bioimaging and Detection of Hypochlorite

During pathogen invasion, hypochlorous acid (HClO), which is created biologically from hydrogen peroxide (H_2O_2) and chloride ion (Cl^-) by MPO (myeloperoxidase), functions as a potent microbicide agent in the innate immune systems. HClO is a very strong oxidant; it can also react with proteins, nucleosides, amino acids, and cholesterol to block a variety of biomolecular processes. According to reports, a number of disorders, including cardiovascular diseases, inflammatory diseases, certain malignancies, and neurological diseases, are intimately linked to aberrant HClO synthesis. Numerous fluorescent probes have been developed by researchers to locate HClO. However, several of the developed sensors are not well suited to a healthy pH range and require large amounts of organic solvents.

The HBT-HBZ probe that consists of 3-(benzo[d]thiazol-2-yl)2-hydroxy-5-methylbenzaldehyde and 2-hydrazino benzothiazole was studied for the detection of HClO (Figure 11.14i). The LOD of 24 nM suggests that the sensor is extremely sensitive to HClO. HBT-HBZ can be used practically to find HClO deposits in drinking water. Moreover, studies employing confocal laser imaging have demonstrated that HBT-HBZ can detect HClO in HeLa cells selectively. Ratiometric fluorescent probes typically perform well in quantitative detection and provide more precise data because they measure the changes in intensity of two emission peaks. Additionally, fluorescent probes may become increasingly appealing due to their successful usage in the detection of living cells and a few other beneficial real-world applications. When HClO is added to a PBS solution containing HBT-HBZ, it produces a mild orange fluorescence. This is because unbridged C=N becomes bridged C=N due to the intramolecular cyclization process between HClO and HBT-HBZ. The HBT-HBZ probe is a useful tool for analyzing HClO residues in tap water and HeLa cells (Wang *et al.*, 2020).

Another benzothiazole-based fluorescent probe with a substantial Stokes shift (190 nm) was produced by Nguyen et al.; the probe exhibits a low detection limit of 0.08 μM and a rapid fluorescence response for HClO exhibiting remarkable selectivity over other reactive oxygen species. The mechanism suggests that the oxime moiety of the probe undergoes an oxidation to nitrile oxide when exposed to HClO (Figure 11.14ii). Both UV-vis and fluorescence titration were used to examine the spectrum responses of the probe toward ClO^- in an H_2O-EtOH solution (9:1, v/v, 10 mM phosphate buffer, pH 7.4). The primary absorption band of the probe is visible at about 350 nm, with the addition of ClO^-, the initial absorption at 350 nm was slightly reduced and a weak shoulder absorption band at 410 nm eventually appeared. Given that the nitrile oxide that is produced is a stronger electron-withdrawing group than the unreacted hydroxylamine, this bathochromic shift can be explained by the accelerated intramolecular charge transfer (ICT) activity within the oxidized product of the probe by ClO^-. Because of the C=N isomerization-related decay process, the probe shows modest fluorescence at an excitation of 350 nm when HClO was not present. The addition of HClO to the probe solution resulted in a noticeable increase in fluorescence at 540 nm, which is explained by ClO^- converting the hydroxylamine quenching group. The probe can be used to find HClO in living cells because it is permeable to cells (Nguyen et al., 2018).

11.2.3.4 Bioimaging and Detection of Fluoride

Fluoride is the most electronegative atom among the anions and the smallest one with a high charge density. It is used in many chemical, biological, and medicinal processes. Osteoporosis and dental issues can be avoided and treated

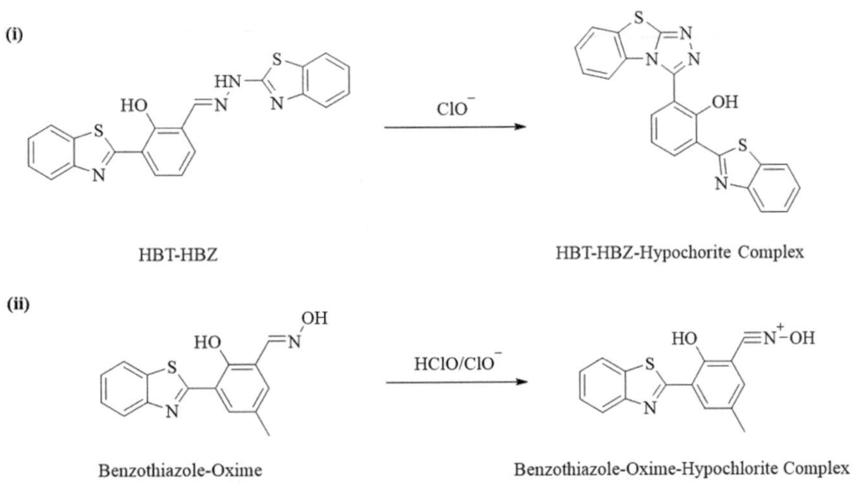

FIGURE 11.14 (i) Response process of probe HBT-HBZ. (ii) Response process of probe benzothiazole-oxime.

with appropriate fluoride intake. Fluoride must therefore be added to drinking water and toothpaste because the body absorbs fluoride readily and eliminates it gradually. Too much fluoride can cause bone diseases, tooth mottling, liver, thyroid, and other organ abnormalities, as well as dental and skeletal fluorosis. Additionally, because of the growing importance of phosphate fertilizers and the massive increase in industrial production of aluminum fluoride, there is a higher hazard to human health and the environment. Therefore, detection of fluoride anion is very important in terms of both.

Erdemir and Kocyigit prepared a novel Schiff base derivative of benzothiazole-bisphenol A (BBA, Figure 11.15i) that exhibited selective chromogenic and fluorescent sensor properties toward fluoride. In the absence of anions, absorption bands at 334 and 384 nm dominate the UV-vis absorption spectra of BBA. The intensities of absorption at 334 and 384 nm dropped as the concentration of fluoride ions rose, whereas a new absorption band at 510 nm rapidly formed (Figure 11.15ii (right)). No UV-vis alterations could be induced by other anions (Figure 11.15ii (left)). The LOD for the fluoride anion was calculated as 9.13 ± 0.21 μM (Erdemir & Kocyigit, 2015). Du et al. synthesized PBT after preparing a probe, 2-(2'-hydroxyphenyl)benzothiazole. The ESIPT of the HBT moiety is prevented by the design of the probe (Figure 11.15iii). In the dynamic range of 0.5–10 mM, the probe's F^- detection limit was found to be 3.8 nM. The results of the experiment indicate that PBT has the potential to be used as a fluorescent light-up sensor in an aggregation state by producing a highly emissive HBT product by cleaving the dimethylphosphinothionyl group (Du et al., 2018).

11.2.3.5 Bioimaging and Detection of Cyanide

The most well-known hazardous anions that can impact a body physiological process, such as the neurological system, metabolic system, endocrine system, cardiovascular system, and even the process of cellular respiration, are cyanide anions. Therefore, it is crucial to recognize it to stop contamination. The 3TBN sensor is designed for both fluorescent "turn-on" and visual detection of cyanide. This sensor has a LOD of 0.46 μM, and the cyanide concentration ranges from 0 to 40 μM. The nucleophilic addition reaction of the sensor with cyanide on the cyanovinyl electrophilic group forms the basis of the suggested sensing mechanism. Furthermore, 3TBN is an inexpensive, selective fluorescence sensor. It works well for both bioimaging in living HeLa cells and the detection of cyanide in water. Its response process is shown in Figure 11.16i (Niu et al., *2018*). The ICT concept is used by benzothiazole and diaminomaleonitrile-based fluorescent probes to detect cyanide. Red emission is produced when cyanide is present because it deprotonates the functional moieties of NH_2 and OH. It exhibits excellent fluorescence, responding in less than a minute to cyanide at 604 nm by enhancing emission by 124 times. Excellent features of this probe include 1.32 μM LOD, quick response time, and high selectivity and sensitivity (Malkondu et al., 2020).

A colorimetric and fluorogenic chemodosimeter based on carbazole-benzothiazole ICT is used to detect cyanide ions. Following the addition of cyanide to the probe, there was a small red shift and an increase in fluorescence intensity. It is discovered that the carbazole-benzothiazole probe has a LOD of 3.37 nM. The disappearance of vinylic protons in the resonance spectra of the carbazole-benzothiazole and

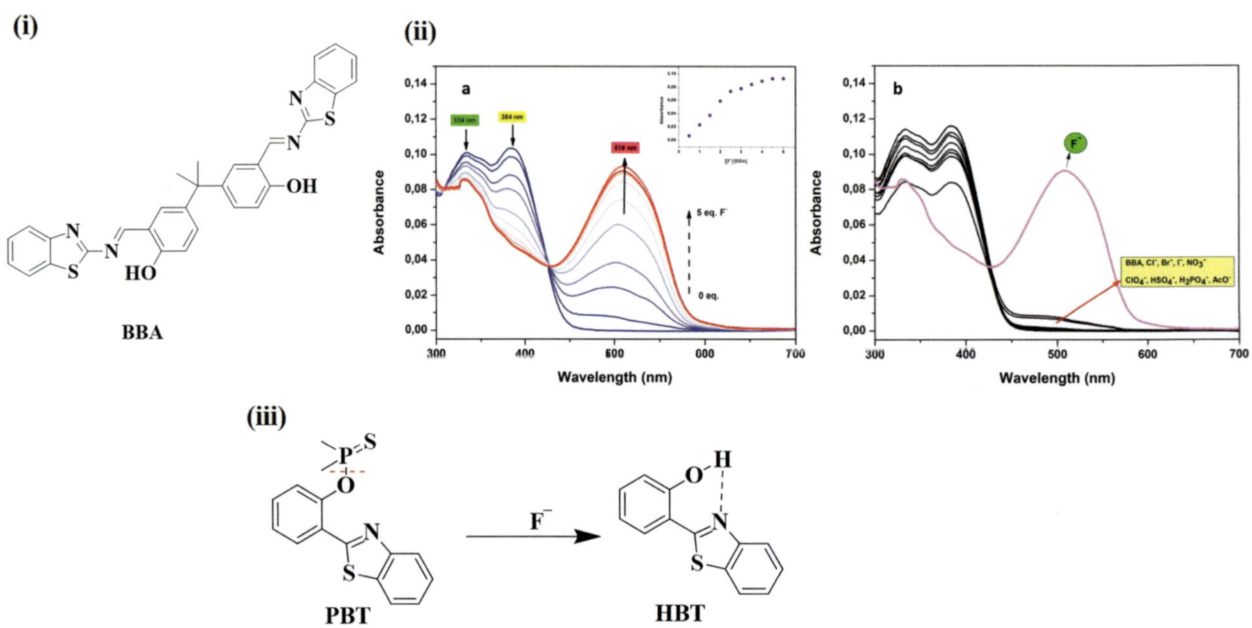

FIGURE 11.15 (i) Chemical structure of probe BBA. (ii) UV-visible absorption response of BBA in presence of various amounts of fluoride ion in ACN (left) and in the presence of various anions (right). (iii) Response process of probe PBT (conversion of PBT to HBT). Reproduced with permission, Erdemir and Kocyigit et al., Copyright (2015), Elsevier.

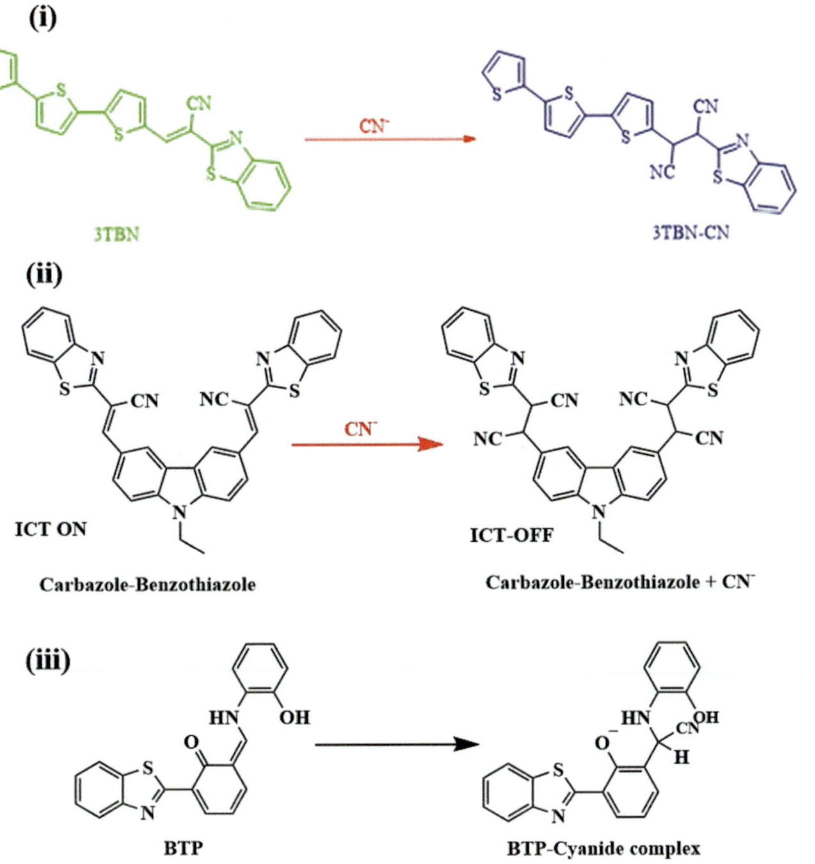

FIGURE 11.16 (i) Response process of probe 3TBN. (ii) Response process of probe carbazole-benzothiazole (CB). (iii) Response process of probe BTP.

cyanide complex indicates the likely mechanism of sensing. This probe UV-vis spectrum was also examined for anions such as fluoride, chloride, bromide, iodide, sulfate, and thiocyanate, but no appreciable alterations were found. The proposed mechanism of sensing is shown in Figure 11.16ii (Patra et al., 2018).

Another probe for cyanide detection is enaminone BTP, which produces fluorescence and colorimetric dual-channel alterations and is very selective to cyanide among all other anions. When cyanide was applied to BTP, it exhibited blue-shifted fluorescence amplification and changed from orange to yellow-green in appearance. When cyanide was added, BTP produced strong absorption at 480 nm, but a new band emerged at 426 nm. The addition of cyanide ions to the β-olefinic carbon atom of enaminone BTP was discovered to be the proposed mechanism of BTP (Figure 11.16iii). It can be detected over a broad pH range, especially in neutral environments (Tang et al., 2016).

11.2.4 Biological Sensing

11.2.4.1 Bioimaging and Detection of Amyloid-Beta

Alzheimer's disease (AD) is a chronic neurological illness marked by behavioral abnormalities, language impairment, memory loss, and decreased cognitive function. The accumulation of amyloid-beta peptide ($A\beta$) in the brain is one of the pathophysiological markers of AD. Amyloid-beta ($A\beta$) peptide synthesis and accumulation are considered to be the primary causes of AD. Therefore, fluorescent probes that interact with $A\beta$ aggregates can be a promising tool for its identification. RMs molecules are similar to the ThT architecture because they consist of the benzothiazole moiety with electron donor and electron acceptor end groups in their structure. RMs, in contrast to ThT, feature a C-C bond in their structure and an extended double bond conjugation system. An emission shifts to the red in this system due to an intramolecular charge transfer (ICT) transition. Moreover, RMs have a molecular weight of less than 5 kDa and a reasonable log P. The emission maxima of these probes show significant solvent dependence and longer wavelength emission maxima with increasing solvent polarity. This is compatible with the ongoing ICT transition, which has a greater dipole moment in the excited state compared to the ground state, with their projected HOMO–LUMO energy gaps (Mallesh et al., 2022).

Ono and group have developed two probes named PP-BTA-1 and PP-BTA-2 (Figure 11.17i). The absorption/emission maxima of PP-BTA-1 and PP-BTA-2 in EtOH appeared at 540 nm and 634 nm, respectively. PP-BTA-2 showed a shorter wavelength than PP-BTA-1 even though it had a longer π-conjugation (Figure 11.17ii). However, the wavelength of PP-BTA-1 is so close to the near-infrared

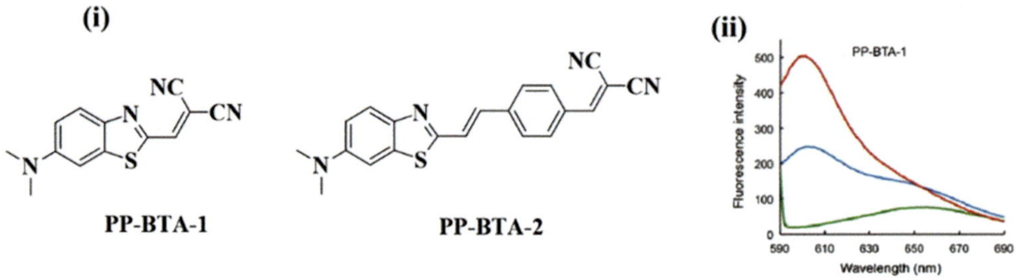

FIGURE 11.17 (i) Chemical structure of probes PP-BTA-1 and PP-BTA-2. (ii) Fluorescence spectra of PP-BTA-1 (right) and PP-BTA-2 (left) with and without A$\beta_{(1-42)}$ aggregates. Reproduced with permission, Ono et al., Copyright (2009), Elsevier.

spectrum that a small modification should yield a wavelength appropriate for in vivo imaging. Furthermore, the fluorescence intensity of PP-BTA-1 and PP-BTA-2 rose in proportion to the concentration of A$\beta_{(1-42)}$ aggregates in the solution, indicating an affinity for Aβ aggregates. Push-pull benzothiazole derivatives' affinity for Aβ plaques was assessed using inhibition assays on the binding to A$\beta_{(1-42)}$ aggregates using thioflavin T as a competitive ligand. PP-BTA-1 and PP-BTA-2 showed a dose-dependent displacement of thioflavin T, indicating their affinity for A$\beta_{(1-42)}$ aggregates. This result also suggests that PP-BTA-1 and PP-BTA-2 might occupy a binding position on Aβ aggregates similar to the binding site of thioflavin T. PP-BTA-1 and PP-BTA-2 have lower IC_{50} values than PIB, a drug commonly used in clinical trials. This implies that A$\beta_{(1-42)}$ aggregates are more favored by PP-BTA-1 and PP-BTA-2. Despite having a different phenyl group in the phenylbenzothiazole structure than PIB, PP-BTA-1 showed more binding to the Aβ aggregates. In vitro binding experiments revealed a significant affinity of

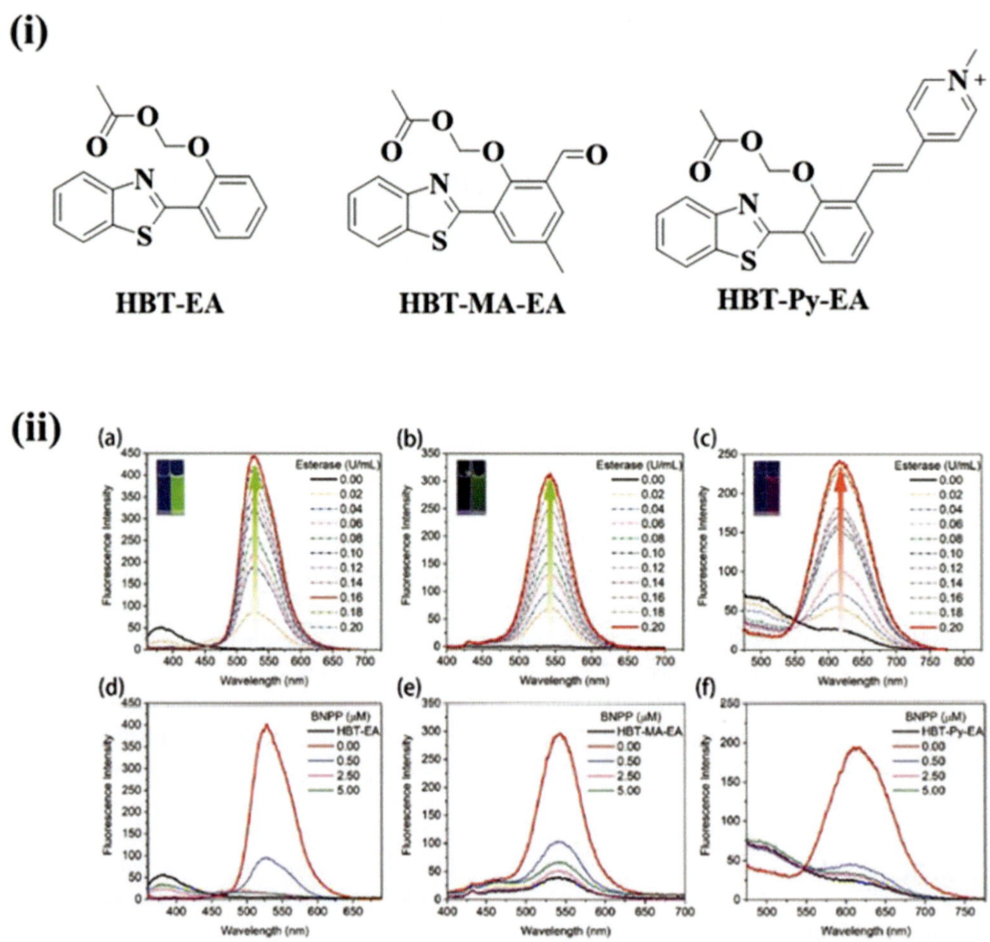

FIGURE 11.18 (i) Chemical structure of probes HBT-EA, HBT-MA-EA, HBT-Py-EA. (ii) Fluorescence spectral changes of (a) HBT-EA, (b) HBT-MA-EA, and (c) HBT-Py-EA with various esterase concentration. Fluorescence spectral changes of (d) HBT-EA, (e) HBT-MA-EA, and (f) HBT-Py-EA in different concentrations of BNPP and esterase in PBS buffer solution at rt. Reproduced with permission, Kong et al., Copyright (2021), Elsevier.

these benzothiazole compounds for $A\beta_{(1-42)}$ aggregates. PP-BTA-1 and PP-BTA-2 highly stained the mouse and human brains, respectively, indicating their affinity for $A\beta$ clumps. These findings suggest that in addition to optical imaging, other structural alterations to the benzothiazole backbone might be made to potential probes for PET and SPECT imaging (Ono et al., 2009).

11.2.4.2 Bioimaging and Detection of Enzymes Like Esterases

Esterase has a crucial role in regulating many biological processes in organisms. The measurement of esterase levels has applications in the treatment of several disorders. The fluorescence signal from benzothiazole probes was significantly increased in the presence of esterase. Many fluorescent probes have been made using the ESIPT fluorophore, 2-(2'-hydroxyphenyl)-benzothiazole (HBT, Figure 11.18i), because of its large Stokes shift, high fluorescence quantum yield, and good photostability. Benzothiazole derivatives also have outstanding optical characteristics because of their vast, conjugated rigid plane and delocalized bonds. However, at large concentrations, fluorescence rises as in an aggregate or solid form due to their AIE features (Figure 11.18ii). Additionally, in biological environments, this class of probes is better adapted for imaging and detecting endogenous esterase. Furthermore, multicolor esterase fluorescent probes provide an efficient way to identify esterase on-site. Benzothiazole probes have great potential in biological research and materials science (Kong et al., 2021).

11.3 CONCLUSION

Benzothiazole is known to serve as an important class of pharmaceutical drugs, also used in different dye preparations. Along with this, the fluorescent properties of the scaffold were explored exclusively, where some highly potential probes were generated for applications, such as metal sensing, amino acid detection, anion sensing, bioimaging of some enzymes, biomarkers of neurodegenerative disease, etc. For such diverse applications, various photophysical mechanisms, such as ICT, FRET, ESIPT, AIE, etc., play a vital role that allows the fluorescent probes to change their luminescent feature to respond against various analytes that in turn affords the detection of the analytes. Also, the benzothiazole fluorescent probes showed merits of large Stokes shift, excellent color change, and high quantum yield.

REFERENCES

Chauhan, K., Singh, P., Kumari, B., & Singhal, R. K. (2017). Synthesis of new benzothiazole Schiff base as selective and sensitive colorimetric sensor for arsenic on-site detection at ppb level. *Analytical Methods*, *9*(11), 1779–1785.

Chen, S.-Y., Li, Z., Li, K., & Yu, X.-Q. (2021). Small molecular fluorescent probes for the detection of lead, cadmium and mercury ions. *Coordination Chemistry Reviews*, *429*, 213691.

Chen, T., Wei, T., Zhang, Z., Chen, Y., Qiang, J., Wang, F., & Chen, X. (2017). Highly sensitive and selective ESIPT-based fluorescent probes for detection of Pd2+ with large Stocks shifts. *Dyes and Pigments*, *140*, 392–398.

Chen, Y., Wei, T., Zhang, Z., Chen, T., Li, J., Qiang, J., Lv, J., Wang, F., & Chen, X. (2017). A benzothiazole-based fluorescent probe for ratiometric detection of Al3+ in aqueous medium and living cells. *Industrial & Engineering Chemistry Research*, *56*(43), 12267–12275.

Choi, N.-E., Lee, J.-Y., Park, E.-C., Lee, J.-H., & Lee, J. (2021). Recent advances in organelle-targeted fluorescent probes. *Molecules (Basel, Switzerland)*, *26*(1), 217.

Du, M., Huo, B., Li, M., Shen, A., Bai, X., Lai, Y., Liu, J., & Yang, Y. (2018). A "Turn-On" fluorescent probe for sensitive and selective detection of fluoride ions based on aggregation-induced emission. *RSC Advances*, *8*(57), 32497–32505.

Erdemir, S., & Kocyigit, O. (2015). Reversible "OFF–ON" fluorescent and colorimetric sensor based benzothiazole-bisphenol A for fluoride in MeCN. *Sensors and Actuators B: Chemical*, *221*, 900–905.

Erdemir, S., & Kocyigit, O. (2016). Anthracene excimer-based "turn on" fluorescent sensor for Cr3+ and Fe3+ ions: Its application to living cells. *Talanta*, *158*, 63–69.

Gong, X., Ding, X., Jiang, N., Zhong, T., & Wang, G. (2020). Benzothiazole-based fluorescence chemosensors for rapid recognition and "turn-off" fluorescence detection of Fe3+ ions in aqueous solution and in living cells. *Microchemical Journal*, *152*, 104351.

Gu, B., Wu, C., Zhang, C., He, S., Tang, S., Li, H., & Shen, Y. (2021). A morpholino hydrazone-based lysosome-targeting fluorescent probe with fast response and high sensitivity for imaging peroxynitrite in living cells. *Spectrochimica Acta Part A: Molecular and Biomolecular Spectroscopy*, *262*, 120100.

Guan, J., Tu, Q., Chen, L., Yuan, M.-S., & Wang, J. (2019). A benzothiazole-rhodol based luminophor: ESIPT-induced AIE and an application for detecting Fe2+ ion. *Spectrochimica Acta Part A: Molecular and Biomolecular Spectroscopy*, *220*, 117114.

Gupta, S. D., Singh, H. P., & Moorthy, N. S. H. N. (2007). Iodine catalyzed, one-pot, solid-phase synthesis of benzothiazole derivatives. *Synthetic Communications*, *37*(24), 4327–4329.

Henriksen, G., Hauser, A. I., Westwell, A. D., Yousefi, B. H., Schwaiger, M., Drzezga, A., & Wester, H.-J. (2007). Metabolically stabilized benzothiazoles for imaging of amyloid plaques. *Journal of Medicinal Chemistry*, *50*(6), 1087–1089.

Jin, M., Jin, D., Wang, H., & Yin, W. (2021). Ratiometric fluorescence imaging of Cu2+ based on spirolactamized benzothiazole-substituted N,N-diethylrhodol probe. *Journal of Molecular Structure*, *1226*, 129360.

Klunk, W. E., Wang, Y., Huang, G. F., Debnath, M. L., Holt, D. P., & Mathis, C. A. (2001). Uncharged thioflavin-T derivatives bind to amyloid-beta protein with high affinity and readily enter the brain. *Life Sciences*, *69*(13), 1471–1484.

Kong, Q., Wang, J., Chen, Y., Zheng, S., Chen, X., Wang, Y., & Wang, F. (2021). The visualized fluorescent probes based on benzothiazole used to detect esterase. *Dyes and Pigments*, *191*, 109349.

Le Bozec, L., & Moody, C. J. (2009). Naturally occurring NitrogenSulfur compounds. The Benzothiazole Alkaloids. *Australian Journal of Chemistry*, *62*(7), 639–647.

Li, M., Chen, H., Liu, X., Zhang, N., Sun, Q., & Zheng, K. (2020). A selective and sensitive sequential ratio/"turn-off" dual mode fluorescent chemosensor for detection of copper ions in aqueous solution and serum. *Inorganica Chimica Acta*, *511*, 119825.

Li, M., Chen, H., Zheng, K., Liu, X., Xiao, S., & Zhang, N. (2019). Benzothiazole derived ratiometric fluorescent probe for selective detection of Pd(0) based on Tsuji-Trost reaction. *Inorganica Chimica Acta*, *495*, 119000.

Li, M., Han, H., Song, S., Shuang, S., & Dong, C. (2021). AIE-based fluorescent boronate probe and its application in peroxynitrite imaging. *Spectrochimica Acta Part A: Molecular and Biomolecular Spectroscopy*, *261*, 120044.

Li, Y., Zhu, Y., Cai, X., Guo, J., Yao, C., Pan, Q., Wang, X., & Wang, K.-N. (2021). A benzothiazole-based near-infrared fluorescent probe for sensing SO2 derivatives and viscosity in HeLa cells. *Spectrochimica Acta Part A: Molecular and Biomolecular Spectroscopy*, *251*, 119457.

Li, Z., Wang, J., Xiao, L., Wang, J., & Yan, H. (2021). A dual-response fluorescent probe for Al3+ and Zn2+ in aqueous medium based on benzothiazole and its application in living cells. *Inorganica Chimica Acta*, *516*, 120147.

Liu, S., Zhang, L., & Liu, X. (2013). A highly sensitive and selective fluorescent probe for Fe3+ based on 2-(2-hydroxyphenyl)benzothiazole. *New Journal of Chemistry*, *37*(3), 821–826.

Lu, Z.-N., Wang, L., Zhang, X., & Zhu, Z.-J. (2019). A selective fluorescent chemosensor for Cd2+ based on 8-hydroxylquinoline-benzothiazole conjugate and imaging application. *Spectrochimica Acta Part A: Molecular and Biomolecular Spectroscopy*, *213*, 57–63.

Lv, R.-G., Chen, S.-W., & Gao, Y. (2017). A selective fluorescence probe based on benzothiazole for the detection of Cr3+. *Heterocyclic Communications*, *23*(5), 389–394.

Malkondu, S., Erdemir, S., & Karakurt, S. (2020). Red and blue emitting fluorescent probe for cyanide and hypochlorite ions: Biological sensing and environmental analysis. *Dyes and Pigments*, *174*, 108019.

Mallesh, R., Khan, J., Pradhan, K., Roy, R., Jana, N. R., Jaisankar, P., & Ghosh, S. (2022). Design and development of benzothiazole-based fluorescent probes for selective detection of Aβ aggregates in alzheimer's disease. *ACS Chemical Neuroscience*, *13*(16), 2503–2516.

Mei, H., Gu, X., Wang, M., Cai, Y., & Xu, K. (2021). A novel cysteine fluorescent probe based on benzothiazole and quinoline with a large stokes shift and application in living cell and mice. *Journal of Photochemistry and Photobiology A: Chemistry*, *418*, 113335.

Mitra, S., Chakraborty, A. J., Tareq, A. M., Emran, T. B., Nainu, F., Khusro, A., Idris, A. M., Khandaker, M. U., Osman, H., Alhumaydhi, F. A., & Simal-Gandara, J. (2022). Impact of heavy metals on the environment and human health: Novel therapeutic insights to counter the toxicity. *Journal of King Saud University-Science*, 34(3), 101865.

Nguyen, K. H., Hao, Y., Zeng, K., Fan, S., Li, F., Yuan, S., Ding, X., Xu, M., & Liu, Y.-N. (2018). A benzothiazole-based fluorescent probe for hypochlorous acid detection and imaging in living cells. *Spectrochimica Acta Part A: Molecular and Biomolecular Spectroscopy*, *199*, 189–193.

Niu, Q., Lan, L., Li, T., Guo, Z., Jiang, T., Zhao, Z., Feng, Z., & Xi, J. (2018). A highly selective turn-on fluorescent and naked-eye colorimetric sensor for cyanide detection in food samples and its application in imaging of living cells. *Sensors and Actuators B: Chemical*, *276*, 13–22.

Ono, M., Hayashi, S., Kimura, H., Kawashima, H., Nakayama, M., & Saji, H. (2009). Push–pull benzothiazole derivatives as probes for detecting β-amyloid plaques in Alzheimer's brains. *Bioorganic & Medicinal Chemistry*, *17*(19), 7002–7007.

Padghan, S. D., Puyad, A. L., Bhosale, R. S., Bhosale, S. V, & Bhosale, S. V. (2017). A pyrene based fluorescent turn-on chemosensor: aggregation-induced emission enhancement and application towards Fe3+ and Fe2+ recognition. *Photochemical & Photobiological Sciences*, *16*(11), 1591–1595.

Patra, L., Aich, K., Gharami, S., & Mondal, T. K. (2018). A new carbazole-benzothiazole based chemodosimeter for chromogenic and fluorogenic detection of CN–. *Journal of Luminescence*, *201*, 419–426.

Qin, J., Kong, F., Guo, Y., Wang, D., Zhang, C., & Li, Y. (2022). Rational construction of a two-photon NIR ratiometric fluorescent probe for the detection of bisulfite in live cells, tissues, and foods. *Journal of Agricultural and Food Chemistry*, *70*(23), 7314–7320.

Quin, L. D. (1998). Heterocyclic Chemistry. Third Edition By Thomas L. Gilchrist (University of Liverpool). Addison Wesley Longman: Essex. 1997. xvii + 414 pp. £22.99. ISBN 0-582-27843–0. *Journal of the American Chemical Society*, *120*(17), 4257–4258.

Razus, A. C., Bîrzan, L., Cristea, M., Ungureanu, E. M., & Cretu, M.-S. (2010). Investigation on the correlation between the structure of azulen-1-yl-benzothiazol-2-yl diazenes and their properties. Acidity and electrochemical redox potentials. *Dyes and Pigments*, *86*, 1–5.

Ren, H., Huo, F., Zhang, Y., Zhao, S., & Yin, C. (2020). An NIR ESIPT-based fluorescent probe with large stokes shift for specific detection of Cys and its bioimaging in cells and mice. *Sensors and Actuators B: Chemical*, *319*, 128248.

Ren, X., Wang, F., Lv, J., Wei, T., Zhang, W., Wang, Y., & Chen, X. (2016). An ESIPT-based fluorescent probe for highly selective detection of glutathione in aqueous solution and living cells. *Dyes and Pigments*, *129*, 156–162.

Tang, L., Xia, J., Zhong, K., Tang, Y., Gao, X., & Li, J. (2020). A simple AIE-active fluorogen for relay recognition of Cu2+ and pyrophosphate through aggregation-switching strategy. *Dyes and Pigments*, *178*, 108379.

Tang, L., Xu, D., Tian, M., & Yan, X. (2019). A mitochondria-targetable far-red emissive fluorescence probe for highly selective detection of cysteine with a large Stokes shift. *Journal of Luminescence*, *208*, 502–508.

Tang, L., Zhou, L., Liu, A., Yan, X., Zhong, K., Liu, X., Gao, X., & Li, J. (2021). A new cascade reaction-based colorimetric and fluorescence "turn on" dual-function probe for cyanide and hydrazine detection. *Dyes and Pigments, 186*, 109034.

Tang, L., Zou, Y., Zhong, K., & Bian, Y. (2016). A novel benzothiazole-based enaminone as a fluorescent probe for highly selective and sensitive detection of CN. *RSC Advances*, *6*(54), 48351–48356.

Tian, Z.-N., Wu, D.-Q., Sun, X.-J., Liu, T.-T., & Xing, Z.-Y. (2019). A benzothiazole-based fluorescent probe for ratiometric detection of Al(3+) and its application in water samples and cell imaging. *International Journal of Molecular Sciences*, *20*(23), 5993.

Wang, J., Feng, L., Chao, J., Wang, Y., & Shuang, S. (2019). A new 'turn-on' and reversible fluorescent sensor for Al3+ detection and live cell imaging. *Analytical Methods*, *11*(43), 5598–5606. https://doi.org/10.1039/C9AY01438A

Wang, K., Xi, D., Liu, C., Chen, Y., Gu, H., Jiang, L., Chen, X., & Wang, F. (2020). A ratiometric benzothiazole-based fluorescence probe for selectively recognizing HClO and its practical applications. *Chinese Chemical Letters*, *31*(11), 2955–2959.

Wang, S.-J., Li, J., Gao, Y., & Guo, Y. (2018). Highly selective and sensitive 2-(2′-Hydroxyphenyl)benzothiazole-based turn-on fluorescent probes for detecting and imaging bisulfite in living cells. *Asian Journal of Organic Chemistry*, *7*(3), 563–569.

Watanabe, H., Ono, M., Ariyoshi, T., Katayanagi, R., & Saji, H. (2017). Novel benzothiazole derivatives as fluorescent probes for detection of β-Amyloid and α-Synuclein aggregates. *ACS Chemical Neuroscience*, *8*(8), 1656–1662.

Wu, Q., Feng, L., Chao, J. Bin, Wang, Y., & Shuang, S. (2021). Ratiometric sensing of Zn2+ with a new benzothiazole-based fluorescent sensor and living cell imaging. *Analyst*, *146*(13), 4348–4356.

Yan, F., Sun, J., Zang, Y., Sun, Z., Zhang, H., & Wang, X. (2020). Benzothiazole applications as fluorescent probes for analyte detection. *Journal of the Iranian Chemical Society*, *17*(12), 3179–3203.

Yu, Y., Xu, H., Zhang, W., Wang, B., & Jiang, Y. (2018). A novel benzothiazole-based fluorescent probe for cysteine detection and its application on test paper and in living cells. *Talanta*, *176*, 151–155.

Zhao, Y.-H., Luo, Y., Wang, H., Wei, H., Guo, T., Tan, H., Yuan, L., & Zhang, X.-B. (2019). A novel ratiometric and reversible fluorescence probe with a large Stokes shift for Cu2+ based on a new clamp-on unit. *Analytica Chimica Acta*, *1065*, 134–141.

12 Coumarin-Based Molecules
Design, Synthesis, and their Applications

Garima, Esther Rani Motamarri and Abha Sharma
National Institute of Pharmaceutical Education and Research,
Raebareli, Lucknow, UP, India

LIST OF ABBREVIATIONS

ICT	Intramolecular charge transfer
PET	Photoinduced electron transfer
CSD	Cambridge structural database
4-AMU	4-azidomethyl umbeliferone
TrxRs	Thioredoxin reductases
Xn	Xanthohumol
VIC	Vinylbenzylaminoimidazolylmethyl coumarin
AIC	Acrylamidoimidazolylmethyl coumarin
SIC	Styryl imidazolylmethyl coumarin
CS-P	7-diethylamino-3-[2-(4-piperidin-1-yl-phenyl)-vinyl]-chromen-2-one
LOD	Limit of detection
BENZPYR	(E)-3-(1-(2-(benzo[d]thiazol-2-yl)hydrazono)ethyl)-7-(diethylamino)-2H-chromen-2-one
CHEQ	Chelation-enhanced quenching
DMF	Dimethyl formamide
DFT	Density functional theory
GSH	Glutathione
NAC	N-acetylcysteine
CHEF	Chelation-enhanced fluorescence effect
BSA	Bovine serum albumin
PPi	Pyrophosphate
EDTA	Ethylenediaminetetraacetic acid
DMSO	Dimethyl sulfoxide
CNB	Coumarinnitrobenzene conjugate
D-PET	Donor-excited photoinduced electron transfer
ROS	Reactive oxygen species
LD	Lipid droplet
AIEgens	Aggregation-induced emission luminogens
5-MOS	5-methoxyseselin
6-MOS	6-methoxyseselin
RIM	Restriction of the intramolecular motion
MMP	Mitochondrial membrane potential

12.1 INTRODUCTION

Coumarin is a naturally occurring organic compound, and the first parent coumarin was discovered from tonka bean by A. Vogel. The majority of the natural coumarins currently known have been extracted from higher plants and microorganisms (e.g., novobiocin and coumermycin from *streptomycin* and aflatoxins from *Aspergillus* species). Coumarins are naturally occurring secondary metabolites present in many plant roots, flowers, leaves, peels, seeds, and fruits. Bai et al. identified more than 20 coumarins from the roots of *Angelica dahurica* with antioxidant activity. Different *Ferulago* species also contain many coumarins mostly in the roots (Bai et al., 2016). *Zanthoxylum schinifolium* and *Zanthoxylum avicennae* leaves were used to extract compounds with anti-inflammatory properties (Nguyen et al., 2016; Cho et al., 2012). Classic synthetic routes for synthesizing coumarin derivatives include Perkin condensation, Knoevenagel condensation, Pechmann condensation, and metal-catalyzed cyclizations (Perkin et al., 1868; Vekariya and Patel, 2014; Heravi et al., 2014). As a result, new approaches such as microwave and ultrasound chemistry, utilization of novel catalysts, and greener solvents have been generated in recent years to produce coumarins with much ease and in a environmentally friendly fashion (J. W. Jung et al., 2018). Coumarin derivatives are used as anticancer, antiproliferative, antioxidant, antifungal, anticoagulant, anti-inflammatory, antiviral, antibacterial, anti-neurodegenerative, anti-HIV20–22, and as bioimaging agents (Emami et al., 2015; An et al., 2018; Kathiravan et al., 2012; Satish, 2016; Grover et al., 2015; Hassan et al., 2016; Jameel et al., 2016;Yu et al., 2003). Thus, the use of coumarin fluorescent probes is widespread for various purposes, such as medicine and environmental protection.

12.2 CLASSIFICATION

Coumarins are divided into following categories: simple coumarins, dihydrofuranocoumarins, furanocoumarins, pyranocoumarins (linear and angular), phenylcoumarins, and bicoumarins (Figure 12.1). Simple coumarins, such as coumarin, 7-hydroxycoumarin, and 6,7 dihydroxycoumarin, are the parent compounds of hydroxylated, alkoxylated, and alkylated derivatives. Furanocoumarins consist of five furan rings that are attached to the coumarin nucleus; these rings can be classified as linear or angular, and they have substituents at either or both benzoid positions. Pyranocoumarins are analogous to furanocoumarins, which have a six-membered ring. 4-hydroxycoumarin is

DOI: 10.1201/9781032706245-12

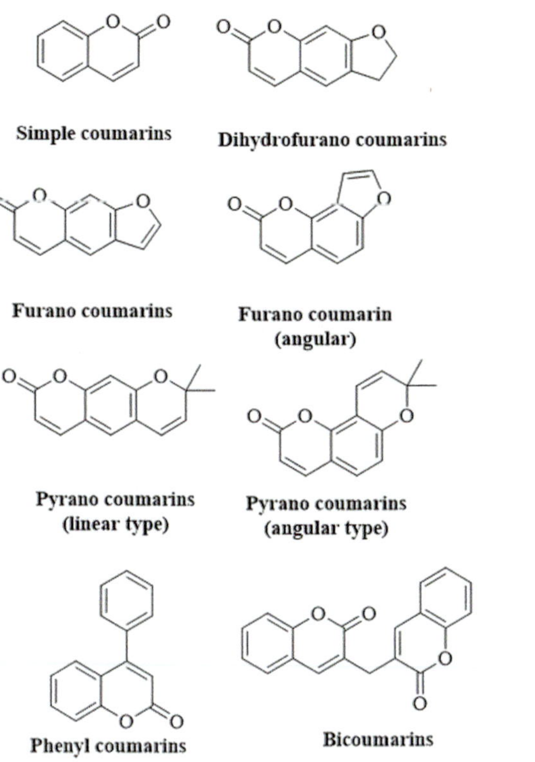

FIGURE 12.1 Structure of various types of naturally occurring coumarins.

coumarin that has been replaced in the pyrone ring (Matos et al., 2015).

12.3 APPLICATIONS

12.3.1 Detection and Bioimaging of Cancer Cells

Thasnim and Bahulayan reported a series of triazoyl-coumarin peptidomimetics that act as fluorescent click-on inhibitors of human cancer cells, MCF-7. According to quantum mechanical calculations, the decrease in electron density and increase in the space between the excited coumarin unit and the methyl group upon triazole formation slow down the process of acceptor PET that causes click-on fluorescence, which is usually observed when azidomethyl coumarin precursors undergo click cycloaddition reactions with various alkynes. A series of 4-AMU (Figure 12.2(a)) derivatives also act as probes that are CuAAC-activatable. 4-AMU showed fluorescence ($\varphi_f = 0.14$) before the click reaction, and a strong fluorescence is created with remarkable improvement of fluorescence quantum yield upon triazole synthesis with diverse alkynes. An acceptor PET pathway was inhibited during the CuAAC reaction thereby causing increase in fluorescence. The evaluation results of the cytotoxicity suggest that these compounds may be useful as inhibitors of the human breast cancer cell line, MCF-7 (Thasnim and Bahulayan , 2017).

Wang and his team reported that selenoprotein TrxRs are a popular prospective target as anticancer medications. Development of several coumarin-chalcone hybrids as TrxR inhibitors was reported where majority of them showed stronger anticancer activity compared to Xn. Representative Xn-2 (Figure 12.2 (b)) (IC_{50} = 3.6 M) is a fluorescent chemical that makes it simple to monitor drug uptake in living cells using red fluorescence imaging. The maximal excitation and emission peaks of Xn-2 are observed at 490 nm and 595 nm, respectively. A large Stokes shift of 105 nm of Xn-2 made it appropriate for bioimaging (Zhang et al., 2017, 2019). The probe superior sensing qualities were examined for its bioimaging capabilities using a cell uptake assay on cancer cells causing apoptosis. When the probe was built up inside cancer cells, it effectively caused the production of intracellular ROS, which in turn prompted cancer cell demise. This method of incorporating a small-molecule fluorescent dye into a naturally occurring active chemical could be a flexible way to create theragnostic probes. It is interesting to note that Xn containing an unsaturated ketone with red emission and might be utilized both as a TrxR inhibitor and a TrxR probe. Additionally, Xn 2 prevented cancer cell metastasis and eliminated the capacity for cancer cells to form colonies. Potential theragnostic TrxR for treating human cancer may be Xn-2 (Wang et al., 2020).

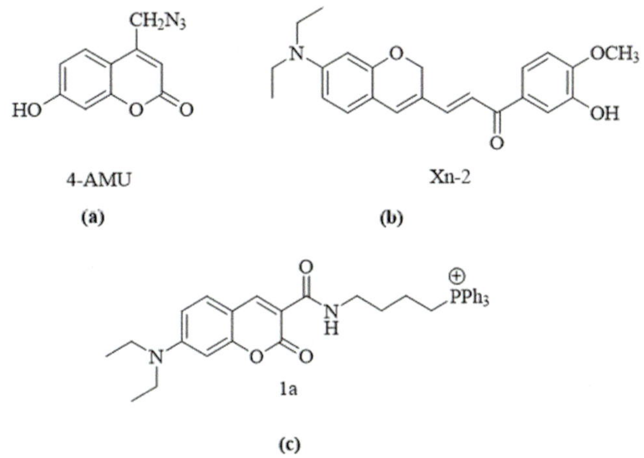

FIGURE 12.2 Coumarin-based fluorescence probes for bio-imaging of cancer lines.

Zhang et al. reported theragnostic probes that are mitochondrial targeting and simultaneously report and trigger mitochondrial malfunction in cancer cells, which is very intriguing. By combining the widely used traditional Chinese drug artemisinin with the mitochondrial dye coumarin-3-carboximide, three fluorescent mitochondrial-targeting theragnostic probes have been produced to destroy cancer cells. The intended coumarin-artemisinin conjugates are primarily found in the mitochondrial region; fluorescence shows greater anticancer activity than artemisinin. Strong ability to accumulate in mitochondria was connected with high cytotoxicity against cancer cells, which could effectively raise intracellular ROS levels and trigger cell apoptosis (Shen et al., 2013). An emission wavelength (λ_{em}) of around 450 nm and a maximum absorption wavelength (λ_{abs}) of approximately 400 nm are the characteristic peaks of the courmarin-3-carboxamide fluorophore. In these methods of drug-mediated fluorophore delivery, the approach of using a small-molecule fluorescent dye to selectively deliver natural products to mitochondria with enhanced cytotoxicity is one of the versatile methods for the design of theragnostic probes. Many efforts are underway to use near-infrared fluorophores in the construction of mitochondrial-targeting theragnostic probes that enable in vivo cancer detection and treatment (Zhang et al., 2015).

Li and team reported that it is possible to eliminate cancer cells while sparing their healthy counterparts by specifically targeting mitochondrial abnormalities. Derivatives of coumarin have shown that potential as anti-breast cancer agents. Mitocoumarins were made and tested for the anti-breast cancer properties by combining coumarin-3-carboxamide derivatives with mitochondria carrier triphenylphosphonium. With an IC_{50} range of 3.0–4.1 μM, **1a** (Figure 12.2(c)) is an amide-based medication that effectively inhibited the development of MCF-7, MDA-231, and SK-BR-3 breast cancer cells while having a lesser cytotoxicity toward normal MCF-10A cells (IC_{50} = 45.30 ± 2.45 μM). A mechanistic analysis of **1a** in MDA-MB-231 cells revealed that it might localize in the mitochondria. As a result of an irreversible alkylation process, it can also slightly block the mitochondrial enzyme TrxR2 and deplete GSH, exacerbating cellular oxidative stress. The compounds showed fluorescence with λ_{ex} around 420 nm and λ_{em} around 475 nm. This proved that **1a** might be helpful for developing an anti-triple negative breast cancer medication that targets the mitochondria (Li et al., 2021).

12.3.2 pH-Sensitive Probe

Nguyen and his team reported new pH indicators VIC, AIC, and SIC based on polymerizable coumarins. The absorption spectra of VIC and AIC in neutral aqueous solution revealed that each dye has a single major absorption band in the UV region, at 370 nm and 330 nm, respectively. The SIC spectrum exhibited two primary peaks, a weaker one at 266 nm and a stronger one at 334 nm. Only one band was observed in the 415–470 nm range in the emission spectra of all the compounds recorded in the same solvent with excitation at the absorbance maxima. AIC and SIC are absorbed and emitted at shorter wavelengths than VIC, although VIC fluorescence quantum yield is substantially lower than SIC and more like AIC. All of the dyes exhibited the same fluorescence characteristics in aqueous solution in the neutral and nearly neutral pH regions, and as the pH increases in the alkaline region, the fluorescence intensity decreases. This was explained by the equilibrium between the imidazolyl group protonated and deprotonated forms. The fluorescent form that is protonated is significantly more intense than the deprotonated form. As a result, with increasing pH values, the dye's fluorescence intensities decrease. The fluorescence intensity of all coumarins significantly decreased in the strongly alkaline pH. The materials can be used in optical pH sensors to measure pH in a variety of alkaline medium since they have similar pH sensitivity (Nguyen et al., 2020) Figure 12.3 (**I**)).

Liu and team reported a novel pH-sensitive probe CS-P, developed by adding piperidine to coumarin (Figure 12.3 (**II**)). In vitro cell imaging has shown that fluorescence intensity is significantly correlated with the cellular pH changes generated by chloroquine or dexamethasone. It showed high pH sensitivity, ranging from pH 5.9 to pH 3.0 (pKa = 4.55). The probe is useful to demonstrate the intracellular pH sensitivity ranging from neutral to acidic conditions, and it has shown good cell membrane permeability. In the case of CS-P, pH was reduced from 7.6 to 3.0, and there was a notable increase in absorbance at 443 nm as changing color of the solution from dark yellow to bright yellow-green, suggesting that CS-P might be used as a visual indicator for the presence of H^+. When pH levels decreased from 5.9 to 3.0, the intensity of the fluorescence increased. A pKa of 4.55 analyzed how emission intensities varied with pH at 515 nm. The pKa value was nearly identical to the normal pH of acidic organelles, such as lysosomes and endosomes, suggesting that CS-P might be useful for in vitro pH monitoring of changes in acidic environments at pH 4–6 (Liu et al., 2015).

Long and team used pyridine as the receptor and 7-diethylamino-coumarin as the fluorophore to design two new ratiometric sensors for acidic pH, 1a and 1b (Figure 12.3). Sensor 1a in pH from 8.35 to 2.36 showed a red shift in the fluorescence emission spectra from 529 to 616 nm. In the pH range of 4.0–6.5, the emission ratio also showed good linearity, useful for quantitative pH determination in this acidic pH. NMR studies and theoretical computations revealed that H^+ interaction with pyridine N and the resulting increase in the ICT process is responsible for the sensor's ratiometric response to an acidic pH. Both sensors 1a and 1b showed fluorescence ratiometric response in acidic pH. The sensors have been successfully used to measure biological fluid pH. The sensor's ability to identify acidic pH variations in living cells was further proved by the intracellular pH imaging studies (Long et al., 2013)

12.3.3 Detection and Bioimaging of Metal Ions

Applications for metal ion sensing fluorescent probes are useful in many areas, which include chemistry, biology, environmental protection, etc. Thus, there is a lot of research value in developing a heavy metal ion fluorescence probe to detect environmental contamination. The ionic forms of some metal ions concentration in the body can reflect some chemical reaction process, and as a basis for the diagnosis and indication of the disease development process, which is due to many metal ions participating in important biochemical reactions in the body, or in the formation of coenzymes, that act as signal molecules to affect normal physiological functions. These crucial roles make the study of metal ion fluorescence probes extremely valuable. Since coumarin derivatives have very variable sizes

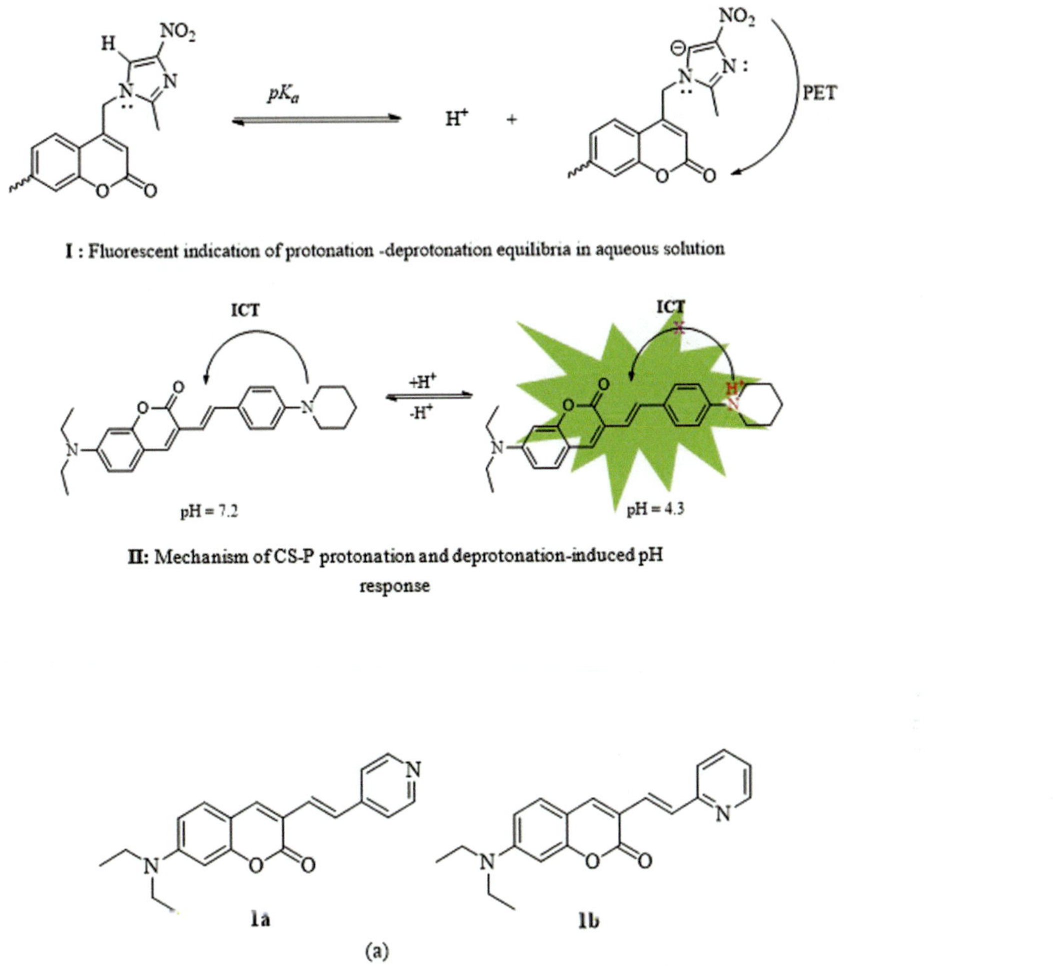

FIGURE 12.3 Coumarin-based pH-sensitive probes.

and are hydrophobic and chelate, they can be employed as fluorescent probes for metal ions.

(a) **Copper**

Mukherjee and his team synthesized a probe containing coumarin in a metal-free environment that was able to detect copper (II) in an acetonitrile solvent. This synthetic probe demonstrated fluorescence quenching by preferentially attaching to the metal ion Cu^{2+} over other metal ions. Condensation of aryl gyloxals, aryl amines, and 4-aminocoumarin in the presence of Bronsted acid-surfactant-combined reagent PEG-OSO_3H under solvent-free conditions, a convenient methodology for the one-pot synthesis of novel fluorescent turn-off chemosensors based on pyrrole-fused coumarin derivatives, has been done. The synthesized pyrrolocoumarin derivatives displayed "turn-off" sensing for Cu^{2+} ions and fluorescence activity with excellent quantum yield ($\varphi_f = 0.4$). It is a simple, inexpensive, and effective technique for monitoring Cu^{2+} contamination since it can detect Cu^{2+} at low levels (LOD = 11.3 nM), which is much lower than the United States Environmental Protection Agency maximum allowable contaminant level for Cu (~1.3 ppm). When excitation was λ_{ex} = 325 nm, a sharp fluorescence response was seen in the emission spectra in the absence of Cu^{2+} at λ_{em} = 468 nm. As the quantity of Cu^{2+} ions increased, there was a decrease in the intensity of the characteristic fluorescence peak that appeared at 468 nm. In metal to ligand, the ratio was 1:1, according to Job's plot. Additionally, a preliminary biological investigation revealed that the pyrrolocoumarin compounds were cell permeable and demonstrated fluorescence "turn-off" in the presence of Cu^{2+} ions in fluorescent imaging of human cervical carcinoma (HeLa) cells (Mukherjee et al., 2018).

Tan and his team reported a colorimetric fluorescent probe called 3-benzothiazolyl-coumarin-7-yl picolinate (**P1**) to detect Cu^{2+} in a CH_3CN-water (1:1,v/v) solution, based on the coumarin derivative. Through a selective hydrolysis reaction of the picolinate moiety, the probe generated a strongly fluorescent 3-benzothiazolyl-7-hydroxycoumarin (**2**), exhibiting a highly selective and sensitive fluorescence enhancement in response to Cu^{2+}. In contrast, other competing metal cations (Na^+, K^+, Mg^{2+}, Ag^+, Fe^{3+}, Pb^{2+}, Mn^{2+}, Ca^{2+}, Ni^{2+}, Cr^{3+}, Zn^{2+}, Sn^{2+}, Hg^{2+}, Al^{3+}, Pd^{2+}, Pt^+, Co^{2+}, Au^{3+}) did not exhibit this property. As Cu^{2+} was added, it caused a shift in **P1** from colorless to yellow-green and produced an emission at about 470 nm (Tan et al., 2016) Figure 12.4 (**I**)).

Wu and his team reported a brand-new fluorescent probe based on coumarin for the sensitive detection of copper (II). Probe 1 operation depends on copper (II) hydrolyzing the pyridine carboxylate site. Under a 365-nm UV lamp, probe 1 fluorescence intensity fluctuations at the highest permitted concentrations of copper (II) in wine and water may

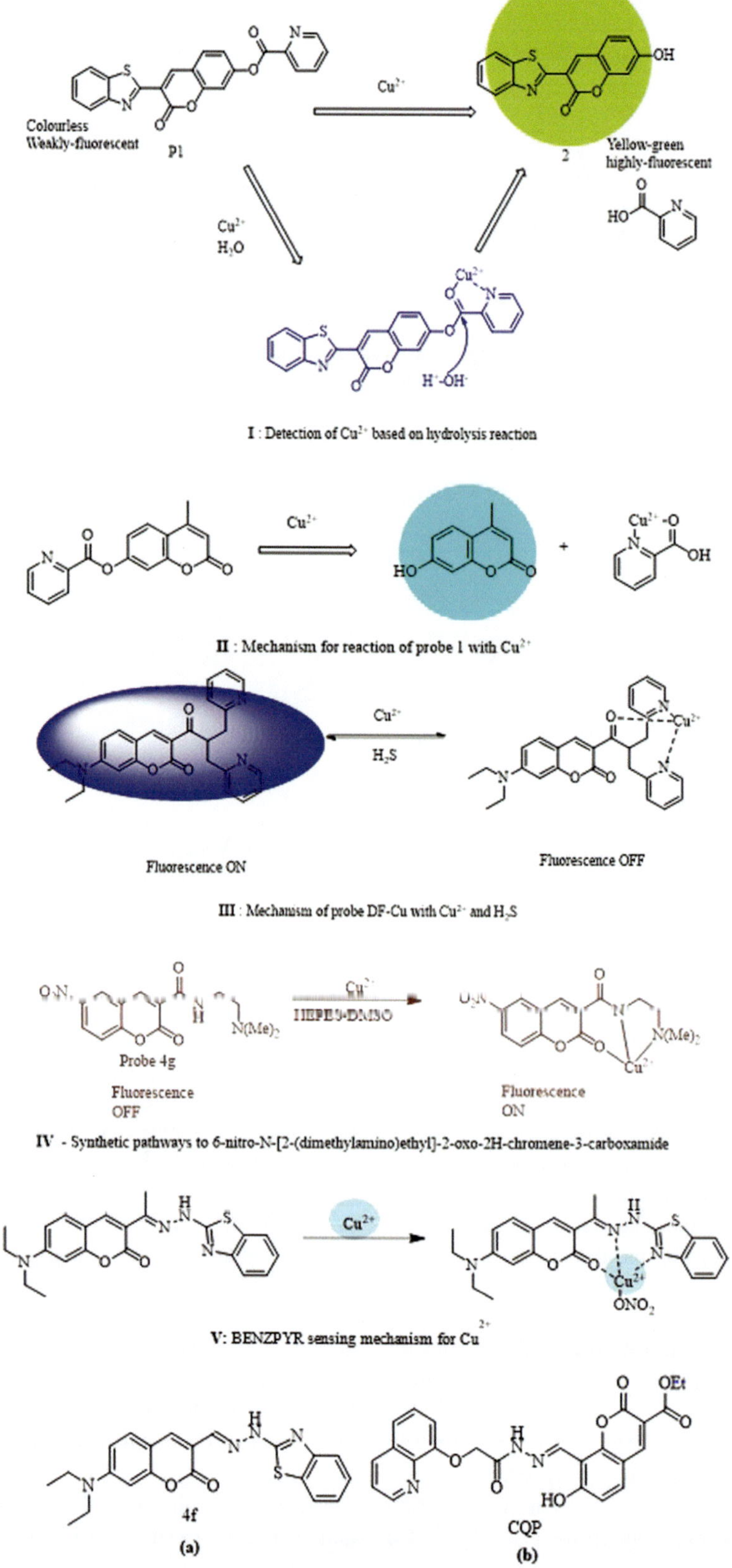

FIGURE 12.4 Coumarin-based fluorescence probe for the detection of copper.

be seen with the naked eye within the range of 0–16 μM (0–1.02 mg/L); there is a strong linear correlation between the fluorescence intensity of probe 1 and the concentration of copper (II). Copper (II) levels in wine and beer were successfully measured using probe 1 in a qualitative, semi-quantitative, and quantitative manner (Wu et al., 2019) (Figure 12.4 (**II**)).

Hien and his team synthesized a new coumarin derivative 4f that was developed and employed as a colorimetric and fluorescent probe for the selective detection of Cu^{2+} ions even in the presence of other metal ions. 4f+Cu^{2+} showed 1:1 stoichiometry that caused a red shift in the absorption maximum from 460 to 510 nm and nearly total quenching of probe 4f fluorescence intensity at 536 nm. Probe 4f may also be used to detect Cu^{2+} ions in living cells because of its wide pH range and suitability for physiological conditions. Using the PBE0/6-31+G(d) theory, the stable structures of probe 4f and its 1:1 complex with Cu^{2+} ion were optimized. The complex formation caused a substantial transfer of electron density from probe 4f, which was acting as a ligand, to the Cu^{2+} ion. Breaking the π- conjugated system, caused the 4f-Cu^{2+} complex to change colour followed by quenching of fluorescence (Hien et al., 2020) (Figure 12.4).

Nadeem Ahmed and his team reported a ICT-based fluorescent ON-OFF-ON probe for detecting H_2S and Cu^{2+}. The sensitivity and selectivity of the DF-Cu probe were verified in an aqueous tris-HCl buffer (10 mM, pH 7.4, 90% ACN), which proved that high selectivity of the probe DF-Cu over other analytes. A linear relationship (R_2 = 0.9919) between the degree of quenching of fluorescence and the concentration of Cu^{2+} was observed. For Cu^{2+}, the LOD of 6.4 μM was calculated using the 3σ/slope method. The probe operates in nearly any pH. The probe responds to Cu^{2+} within 10 seconds. H_2S can reverse the probes' reaction toward the copper ions. One possible use of the Cu^{2+}-probe complex is the detection of H_2S. A 400 nm centered absorption peak was observed by the DF-Cu probe. The absorption peak moved to 410 nm after 100 μM of Cu^{2+} was added, and it remained unchanged even when additional Cu^{2+} ions were added. Selectivity of the DF-Cu probe was proved using UV-based Cu^{2+} detection. Except Cu^{2+}, the fluorescence intensity of the DF-Cu probe at 465 nm was unaffected by the addition of different common ions. Due to the complex formation between the DF-Cu probe and Cu^{2+}, the fluorescence changed. It indicates that this probe showed high fluorescence selectivity towards Cu^{2+} . DF-Cu exhibits emission at 430–510 nm and excitation at 400 nm. Moreover, Cu^{2+}-DF probe complex was used to identify hydrogen sulfide ions in MCF-7 cells (Ahmed et al., 2020) (Figure 12.4 (**III**)).

Bekhradnia and his team reported a facile synthesis of a coumarin derivative, nitro-3-carboxamide, a novel fluorescent chemosensor, using microwave irradiation. This substance has selectivity over other metal ions in aqueous solution, making it useful as a fluorescent probe for Cu^{2+}. In comparison to the other cations examined, 6-nitro-N-[2-(dimethylamino)ethyl]-2-oxo-2H-chromene-3-carboxamide (probe 4g) exhibits the strongest fluorescence when Cu^{2+} is present, with an excitation at λ = 320 nm and emission at 420 nm. 4g shows selectivity towards Cu^{2+} over other metal ions in aqueous solution; this probe can be employed for complexation of heavy hazardous metals and exhibits enhanced fluorescence when Cu^{2+} is present (Bekhradnia et al., 2016) (Figure 12.4 (**IV**)).

Mani and his team reported a coumarin-based fluorescent probe BENZPYR that showed ICT process due to condensation of N,N-diethylamino-3-acetyl coumarin with 2-hydrazinobenzothiazole. This probe can selectively identify Cu^{2+} ions over other metal ions, and LOD of 40 nM in the nanomolar range is found. The BENZPYR turn-off sensor exhibits CHEQ and ICT processes and is a great tool for fluorescent detection of Cu^{2+} ions in the DMF medium. Before the addition of Cu^{2+} ions, the emission of BENZPYR was 536 nm and excitation wavelength was 420 nm. Addition of Cu^{2+} ions to BENZPYR caused emission intensity at 572 nm to slowly reduce, and red shift from 536 nm to 572 nm was observed. The color of the solution changed from yellow to wine upon the addition of Cu^{2+} ions to BENZPYR, which is due to metal-to-ligand energy transfer and d-d transition of metal ions. The probe BENZPYR exhibits strong cell membrane permeability and can be used for sensing Cu^{2+} in living cells (Mani et al., 2019) (Figure 12.4 (**V**)).

Duan reported a unique fluorescent probe CQP, which was synthesized using 2,4-dihydroxybenzaldehyde and 8-hydroxyquinoline to detect Cu^{2+} with naked eye. For detecting Cu^{2+}, a Schiff-base fluorescent probe based on coumarin was created using the PET mechanism. It displayed great selectivity against other metal ions and robust fluorescence responses toward Cu^{2+}. In ethanol, free CQP exhibits absorption at 275 nm and 385 nm, and as Cu^{2+} was gradually added, a 50 nm and 20 nm hypochromatic shift emerged, causing the color change from colorless to yellow, which was not seen when it comes to other metal cations. The Job's plot suggested 1:1 binding stoichiometry . Also, the Benesi–Hildebrand equation was used to determine the binding constant between CQP and Cu^{2+} (Duan et al., 2014) (Figure 12.4 (a and b)).

(b) **Thiol**

Li and his team reported an ICT-based, off-on fluorescence probe 4h by utilizing 2,4-dinitrobenzene-1-sulfonamide as the recognition unit and 3-amino-7-diethylaminocoumarin as the fluorophore. The created probe identified thiophenols from aliphatic thiols with good selectivity and sensitivity. It exhibited >280-fold increase in fluorescence, and a significant Stokes shift of 145 nm was observed. The probe effectively assessed thiophenols in water samples with good recovery (>90%), indicating that it has significant potential for usage in thiophenol detection for environmental and biological purposes. Furthermore, the probe demonstrated good cell permeability, making it useful for detecting thiophenol in HEK293 cells. It showed emission at 515 nm and excitation at 370 nm. The LODs for thiophenol and

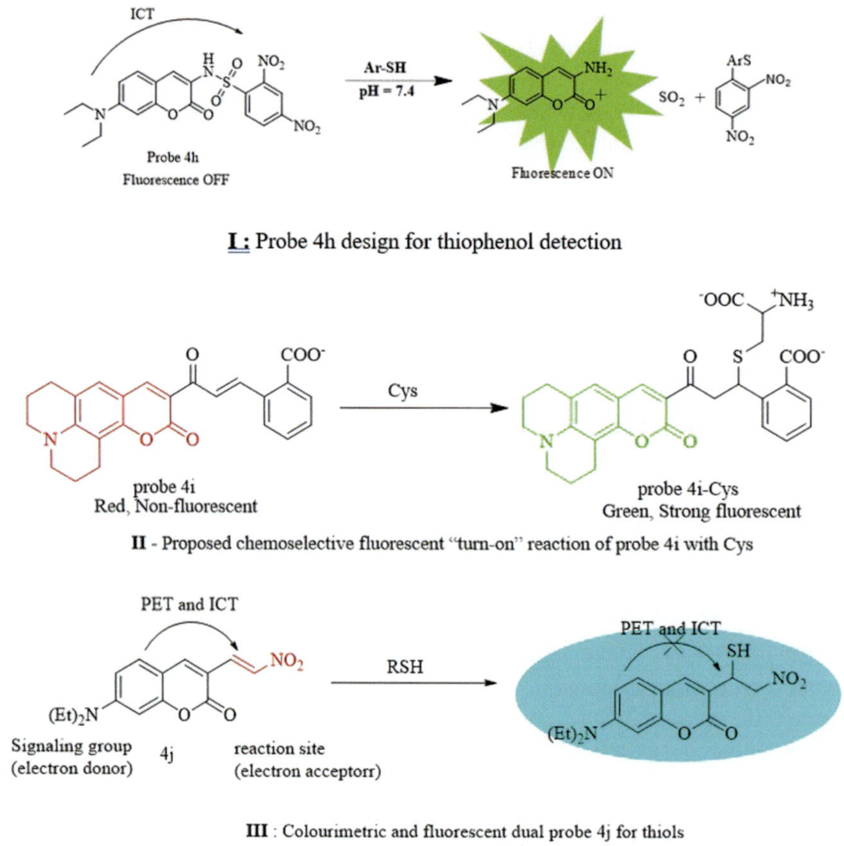

FIGURE 12.5 Coumarin-based fluorescence probe for the detection of thiol.

p-methoxy-thiophenol of the probe 4h were around 30 nM and 18 nM, respectively (Li et al., 2014) (Figure 12.5 (**I**)).

Jung and his team reported a group of coumarins that react Michael type with substances containing thiols to produce fluorescence in a turn-on fashion. The carboxyl group on a coumarin benzene ring moiety with the double-bond conjugated coumarin is the only characteristic that separates the coumarins. Fluorescence spectroscopy, DFT calculations, and kinetic studies demonstrated their selectivity for Cys, GSH, and Hcy, as well as the related fluorogenic process. Second-order kinetics shows a fold change from 48.6 to 111 fold (probe 4i). All isomers favor Cys over GSH. Given that probe 4i favors a less negatively charged nucleophile, the ortho carboxyl group on its benzene ring may be the cause of its strong selectivity toward Cys. The modest fluorescence emission of free probe 4i (φf = 0.01) can be attributed to an ICT that involves the conjugated benzoic acid and the coumarin molecule. Cys, on the other hand, causes a strong emission maximum at 513 nm and a 16 nm hypsochromic shift of the absorption maximum to 488 nm ($\varepsilon = 2.7 \times 10^4\ M^{-1}\ cm^{-1}$) in the solution. As the concentration of Cys was increased to 0.35 mM, an proportionate elevation in the fluorescence intensity at $\lambda_{em} = 513$ nm ($\varphi_f = 0.39$) was seen. An extensive level of GSH in the HepG2 cells was considered a potential source of interference during the intracellular Cys detection process employing probe 4i. L-buthionine sulfoximine (BSO), an inhibitor of γ-glutamylcysteine synthetase, was used to treat the cells. This created an experimental environment in which the cells were unable to produce GSH from Cys or other species. Next, during BSO-H_2O_2 treatment, the fluorescence intensity of probe 4i in HepG2 cells showed a significant increase in NAC, a precursor of Cys, suggesting that intracellular Cys changes rather than intracellular GSH changes are the primary cause of the fluorescence signal from the cells (Jung et al., 2012) (Figure 12.5 (**II**)).

Sun and his team reported that a coumarin-based probe's quick response time and excellent selectivity for biothiols compared to other naturally occurring amino acids could be attributed to the Michael addition of a thiol to the probe nitroolefin unit. It is anticipated that probe 4j exhibits weak fluorescence in solution as a result of the PET quenching because it contains a nitroolefin moiety, an electron acceptor responsible for the PET process, and a diethylaminocoumarin, which serves as a signaling group. The nitroolefin moiety will no longer be able to quench the diethylaminocoumarin fluorophore when thiols are added one to four times to it. This will cause the nitroolefin double bond to break, which will increase the probe 4j fluorescence. On the other hand, when thiols are added Michael-type to nitroolefin of 4j, it will also prevent the ICT occurring from the electron-rich diethylamino moiety to the electron-poor nitroolefin moiety. Cys dramatically changed the color from orange to yellow, causing an apparent blue shift from 483 to 400 nm. Cys with an intensity increase in absorbance of more than 30 times was found to have good selectivity when the absorbance ratio at 400 and 483 nm were examined. This implies that, in comparison to other amino acids, the system showed a greater degree of selectivity toward Cys, making it useful for faster detection of Cys (Sun et al., 2011) (Figure 12.5 (**III**)).

(c) **Zinc**

Su and his team reported a fluorescent compound 4k that could be used for the detection of Zn^{2+} based on coumarin. Compound 4k has lower background fluorescence because of PET that occurs within the molecules. But when probe 4k is mixed with Zn^{2+} in 30% (v/v) aqueous ethanol, there is a "turn-on" fluorescence emission. Compound 4k has an absorption maximum at 416 nm but when Zn^{2+} is bound it experiences a hypsochromic shift to 407 nm. In the absence of Zn^{2+}, the fluorescence emission maximum of 4k is 483 nm and a blue shift of λ_{em} to 457 nm due to addition of Zn^{2+} from 0 to 10 equivalent was observed. The fluorescence emission at 457 nm increases with increasing Zn^{2+} concentration. Zn^{2+}-induced emission enhancement is caused by the inhibition of the PET process from the semicarbazide amine to the coumarin ring, which is brought about by Zn^{2+} coordination. With LOD of 0.29 µmol L^{-1}, the fluorescence emission rises linearly with Zn^{2+} concentration in the range of 0.5–10 µmol L^{-1} (Su et al., 2010) (Figure 12.6 (**I**)).

Qin reported 7-(4'-(diethylamino)-2'-hydroxybenzylideneimino)-4-methyl coumarin developed one-photon and two-photon Zn^{2+} probe. The selectivity of the sensor for Zn^{2+} ions is excellent. When Zn^{2+} (λ_{ex} = 380 nm) is present, the sensor exhibits fluorescence intensity elevation at around 500 nm. This could be related to the formation of a 2:1 stoichiometry with the Zn complex, which results in the restricted C=N isomerization and the suppressed PET process (Qin et al., 2016) (Figure 12.6 (**II**)).

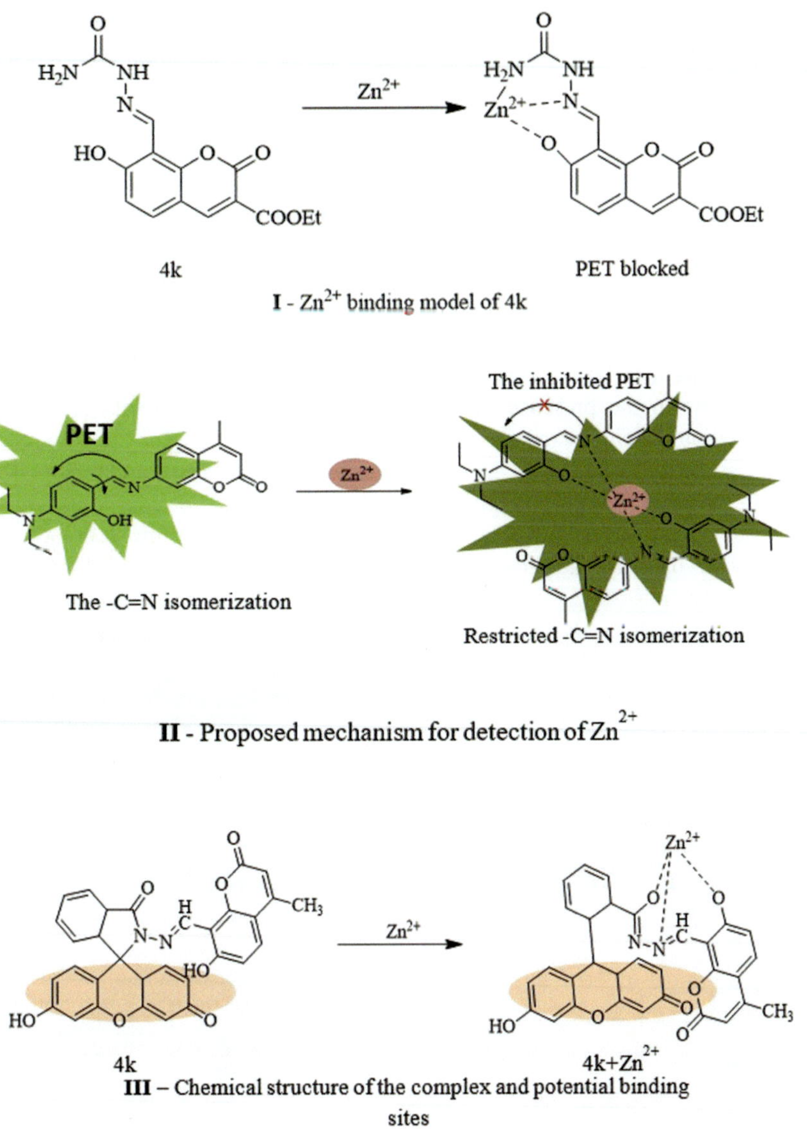

FIGURE 12.6 Coumarin-based fluorescence probe for the detection of zinc.

An and his team reported a novel 7-hydroxy-4-methylco umarin-8-carbaldehyde-(fluorescein) hydrazone fluorescent sensor to selectively recognize Zn^{2+} in HEPES buffer media with a pH of 7.4. A wavelength of 404 nm is recorded for excitation, while a wavelength of 501 nm is recorded for emission. Through the observation of alterations in the UV/vis and fluorescence spectra used as a Zn^{2+} probe, it was known that this sensor exhibited a high degree of selectivity, sensitivity, and color change for Zn^{2+} in comparison to other transition and earth metal ions. This is mostly because Zn^{2+} has the ability to open spiro-lactam rings. The dosimeter fluorescence emission intensity increased 33 times at λ_{ex} = 404 nm and λ_{em} = 501 nm after Zn^{2+} was added, and it reached saturation at 1.0 eq. of Zn^{2+}. The fluorescein moiety is responsible for the fluorescence amplification observed at 501 nm. Zn^{2+} causes spirocyclic opening of 4k by coordination. When compound 4k was used as a sensor to analyze Zn^{2+} ions, its detection limit for Zn^{2+} was found to be 6.54 ppb, showing a high level of sensitivity and fluorescence quantum yield of 4k in the free form and Zn^{2+} bound form were 0.03 and 0.28, respectively (An et al., 2013) (Figure 12.6 (**III**)).

(d) Cadmium

Zehra and team reported a novel Schiff base ligand (3)–(2–mercapto–2,7a–dihydro–1H–benzo[d]imidazol–6–yl) imino) methyl)–4H–chromen–4–one from coumarins that has selective detection of cadmium ions in the presence of other metal ions. The limited C=N isomerization that results from CHEF is thought to be the cause of the "turn on" fluorescence process in the presence of Cd^{2+}. Using Job's plot, the 1:1 binding stoichiometry was calculated, yielding an association constant of 3.3×10^5 M^{-1}. The probe fluorescence spectra by itself showed no significant emission after excitation at 300 nm wavelength. The emission intensity significantly rise upon the addition of Cd^{2+} compared to other metal ions, such as Fe^{2+}, Zn^{2+}, Cu^{2+}, Co^{2+}, Mn^{2+}, and Ni^2 but except for Mn^{2+}. The observed "turn-on" response is thought to be caused by Cd^{2+} attaching to the imine bond N atom (–C=N) and the carbonyl oxygen atom, which prevents C=N isomerization because of the CHEF process. Absorbance centered at 237 nm and 277 nm was observed in the ligand probe electronic spectra (1.0×10^{-5}), and it was determined to be the result of π–π^* transitions of the azomethine chromophore found in the ligand system. When Cd^{2+} ions were added, a red shift of around ~ 5 nm was seen with a notable increase in absorbance and there was no change in the color of probe solution. Through EDTA titration, the probe's reversible nature was examined, indicating that it can be utilized again. The sensors' practical application for the quantitative assessment of Cd^{2+} concentration in blood plasma was demonstrated by interaction studies with BSA. The sensors' practical use could involve detecting lower amounts of Cd^{2+} in biological and environmental samples due to its lower LOD of 0.114 μM (Zehra et al., 2019) (Figure 12.7 (**I**)).

Shaily and her team reported a chalcone-based naked-eye chemical sensor with colorimetry, (E)-4-hydroxy-3-(3-(4-methoxyphenyl)acryloyl)-2H-chromen-2-one, for the sensitive and selective detection of Cd^{2+} in a mixture of aqueous organic solutions. The chemosensor 4l dual colorimetric and fluorometric responses to Cd^{2+} are explained by the development of a 1:1 [4l + Cd^{2+}] complex resulting in the modification of sensors' optical features. In neutral and basic media, the chemosensor 4l with Cd^{2+} showed distinct fluorescence characteristics, including an excitation band at 387 nm and an emission band at 495 nm. Using the Benesi–Hildebrand plot and nonlinear least square fitting, the association constant for Cd^{2+} toward receptor 4l was determined to be 9.56×105 M^{-1} and $(1.34 \pm 0.87) \times 106$ M^{-1}, respectively, with LOD of 5.84×10^{-8} M. There was a noticeable color shift from yellow to colorless upon addition of Cd^{2+}. At pH 7.0, chemosensor 4l complexed with Cd^{2+} showed

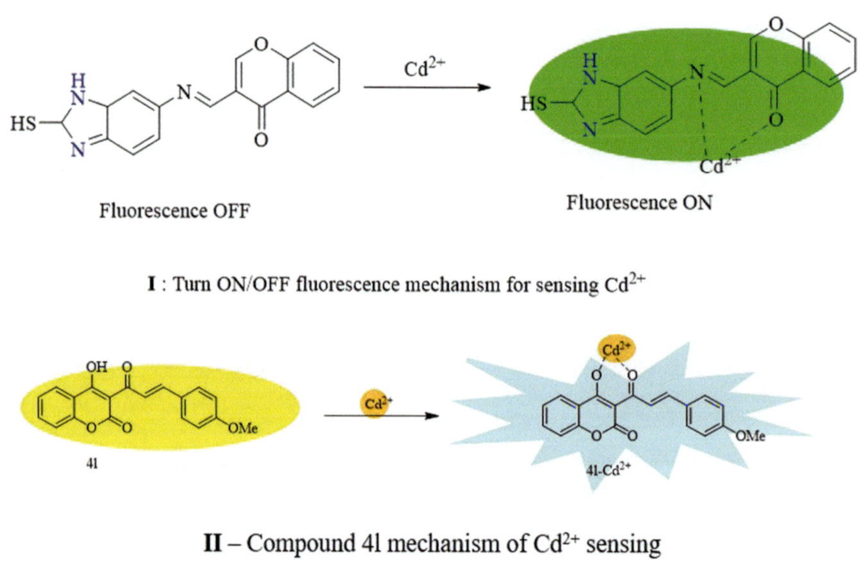

FIGURE 12.7 Coumarin-based fluorescence probe for the detection of cadmium.

an increase in fluorescence because of the CHEF process (Shaily et al., 2017) (Figure 12.7 (**II**)).

(e) **Ferric**

Wang and his team reported a chemosensor based on coumarin FB to detect Fe^{3+} and PPi. FB showed strong affinity for Fe^{3+} even when there were competing cations present. The obtained Fe^{3+} displacement method-based FB-Fe^{3+} compound showed strong sensitivity to PPi. The LODs of FB to Fe^{3+} and FB-Fe^{3+} complex to PPi were estimated to be 8.73 x 10^{-8} M and 1.25 x 10^{-8} M, respectively. Fe^{3+} and PPi identification procedures are quick and reversible. Through the development of a 1:1 complex, specific binding of FB to Fe^{3+} resulted in significant fluorescence quenching. Fe^{3+} λ_{em} = 335 nm is added after FB emission intensities at 447 nm. The confocal microscope ability to examine the fluorescent response for Fe^{3+} and PPi in living cells is made possible by the high degree of biocompatibility of FB. A basic set of B3LYP/6-31G(d,p) was utilized to optimize the FB and FB-Fe^{3+} complexes (Wang et al., 2018) (Figure 12.8 (I)).

Zhang and his team reported a new type of pyrazoline fluorescence probe based on coumarin, probe A. It shows how the inclusion of Fe^{3+} changes the probe fluorescence emission spectrum. When exposed to Fe^{3+}, an analyzable sample showed a linear fluorescent response with a detection limit of 1.01 × 10^{-7} M. Additionally, it exhibits good reversibility when EDTA is added. At 523 nm, probe A showed a prominent fluorescence emission band. The probe A system fluorescence intensity dropped to 45% upon the addition of Fe^{3+}. Through DFT simulations, the fluorescent probe structure was optimized. Due to its exceptional cell penetration and minimal toxicity, probe A set the widespread use of fluorescence detection in everyday applications (Zhang et al., 2021) (Figure 12.8 (II)).

Ge and team reported a novel ratiometric fluorescent probe with great sensitivity and selectivity for monitoring Fe^{3+} that is based on coumarin and rhodamine B (probe L). Fe^{3+} was added to the probe aqueous solution, resulting in the appearance of two fluorescence peaks at 460 and 580 nm, respectively, that correspond to coumarin and rhodamine B. This is a novel design for a ratiometric Fe^{3+} probe because the PET process was inhibited concurrently

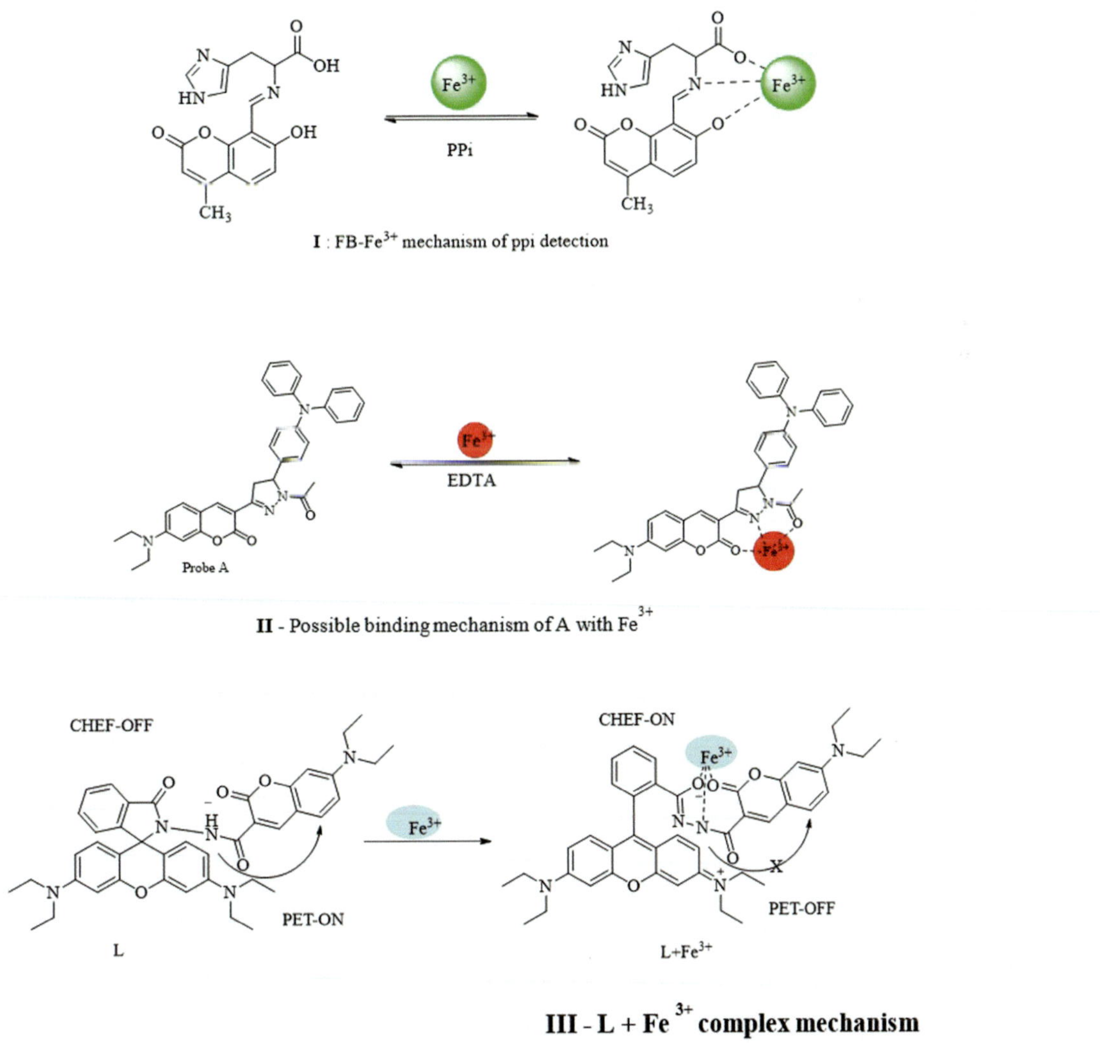

FIGURE 12.8 Coumarin-based fluorescence probe for the detection of ferric.

with the generation of the CHEF process. Five equivalents of Fe^{3+} added to an aqueous solution increased the fluorescence intensity at 580 nm by 120 fold (Ge et al., 2013) Figure 12.8 (III).

(f) **Cyanide**

Chemchem et al. reported colorimetric and fluorescent in nature, chemosensor 2 is to detect cyanide anions in aqueous solutions. Chemosensor 2's ability to enhance pink fluorescence in DMSO and change from yellow to red allowed it to optically identify fluoride ions from other anions. However, when CN^- was added to a 6:4 v/v DMSO/water mixture, its fluorescence color was vivid green. Thus, the chemosensor can be utilized to selectively detect CN^- along with other interfering anions (F^-, AcO^-, and $H_2PO_4^-$) in aqueous solution. Moreover, chemosensor 2 was able to identify CN^- at concentrations as low as 0.32 mM, which is lower than the cyanide level that the WHO recommends (2.7 mM). Its emission maxima are between 468 and 485 nm, and its maximum absorption occurs in the visible range, from 398 to 418 nm. The Stokes shifts were found in the range of 56–84 nm. Chemosensor 2 can identify and measure a low concentration of CN^- (1.08 mM). Moreover, absorbance and fluorescence spectra of the interactions between chemosensor 2 and duplex DNA show the intercalative binding of calf thymus DNA to chemosensor 2. The compound is a greenish-brown, nonfluorescent that only exhibits single absorbance and emission band in each of the used solvents (Chem chem et al., 2018) (Figure 12.9 (**I**)).

Sun et al. reported CNB a basic ICT fluorescence probe that detects cyanide ion (CN-) in a buffered aqueous solution. When cyanide is added to the electron-deficient alkene bridge, it disturbs the π*conjugation in CNB, thereby hindering the ICT process resulting in a color change shift from yellow to colorless at 410 nm. When the CNB spectrum was formed at 340 nm excitation, its emission spectrum showed a very weak emission band at 410 nm because of a strong ICT process in the excited state of CNB. Significant dual shifts in both the emission (turn-on) and absorption (blue-shift) bands for CN^- were seen by CNB that were perceptible to the naked eye. Using Job's plot analysis, a 1:1 stoichiometry was determined for the interaction between the probe (II) and cyanide anion and LOD was found to be 0.14 mM. Additionally, the probe was used to create CN^- test kits for its detection (Sun et al., 2015) Figure 12.9 (II).

A novel ratiometric fluorescent and colorimetric probe, coum-1 (Figure 12.10) for the detection of cyanide was developed by Cheng et al. which contained the characteristic nucleophilic attachment of cyanide to the α-position of the dicyano-vinyl group. It showed several benefits, including a successful bioimaging application, good sensitivity and high selective for CN^-, relative good solubility in aqueous conditions, and good ratiometric response.

When exposed to cyanide ions, the coum-1 emission moved from 585 nm to 495 nm, with an isoemissive point at 562 nm; this was backed up with DFT studies where a conjugated bridge (–C=C–) between the coumarin (donor) and dicyano groups (acceptor) had a dihedral angle of only 28° but in the presence of CN^-, the conjugated bridge (–C=C–) in coum-1 became a saturated bridge (–C-C–), resulting in a dihedral angle of up to 77°. NMR investigation identified the vinylic proton (Ha) that is responsible for the resonance

I : Chemosensor 2 structural characteristics and deprotonation in the presence of CN^-

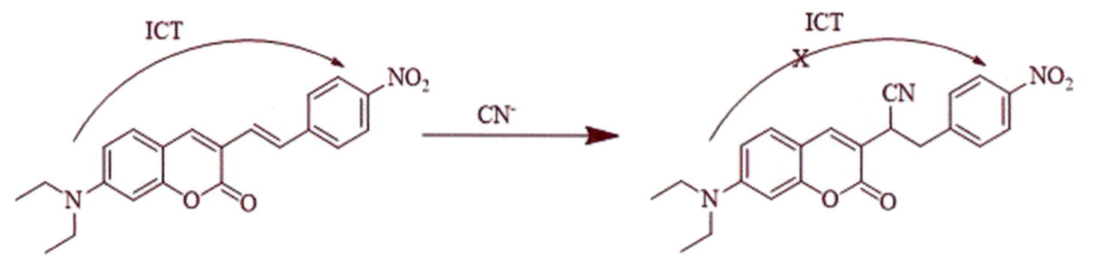

II - Possibility of cyanide interaction with CNB

FIGURE 12.9 Coumarin-based fluorescence probe for the detection of cyanide ion.

signal at δ = 9.85 ppm and introducing 0.5 equiv. of CN^- resulted in five new signals with similar intensity proportional to the original signals where a new signal at 5.51 ppm that corresponded to the α-proton (Hb) was observed proving that half of the dicyano-vinyl moiety in coum-1 had interacted with CN^-. Cell imaging investigations revealed that the probe showed intracellular fluorescence at ~590 nm (the red channel), implying that coum-1 may cross the cell membrane, thereby useful for imaging CN^- in living cells. When cells were treated with NaCN (cyanide ion source) after which the coum-1 was administered showed emission intensity at ~590 nm (red channel) dropped while the intensity at ~520 nm (green channel) was elevated (Cheng et al., 2012).

12.3.4 Bioimaging

(a) Cell Imaging

Zhang and his team reported for the H_2O_2 detection-based D-PET technique a fluorescent probe 4j through the addition of benzil, which serves as a quenching and detecting moiety. In the presence of H_2O_2, benzil might undergo a Baeyer–Villiger-type reaction that would provide benzoic anhydride. Benzoic acid would then be produced by hydrolyzing benzoic anhydride, releasing the fluorophore. To prove that the benzil was a fluorescence quencher, a DFT calculation was done. At λ_{abs} = 475 nm, probe 4j showed its highest absorption. The absorption at 475 nm decreased with 10 equiv. of H_2O_2 at 37°C, whereas a new absorption peak was visible at 360 nm with emission at 505 nm and quick increase in fluorescence intensity. After being tested in both one- and two-photon modes, the probe showed excellent sensitivity with LOD of 0.09 μM and strong selectivity toward H_2O_2 over other ROS. Additionally, the probe was effectively used for one- and two-photon microscopy cell imaging of intracellular H_2O_2 levels (Zhang et al., 2014) (Figure 12.10 (I)).

Jana et al. reported coumarin-pyridine hetrocycles based on the push-pull framework for detecting lipid-rich intracellular organelles, LDs, which control a variety of physiological activities. Solvatochromic lipophilic dyes with ultra-low concentrations (200 nM) enable selective live cell imaging of LDs in Cos-7 and McA-RH7777 cells. The hydrophobic core of the LD contributes to the observed intensity augmentation, which is demonstrated by the fluorescent probes' extraordinary increase in fluorescence intensity over time. The compounds absorb in the visible range; **J-1** absorbs at 498 nm, and **C-1** absorbs between 477 nm and 488 nm. The julolidine moiety's enhanced capacity to donate electrons accounts for the greater absorption maxima. The structural design, substantial Stokes shift (>80 nm), red emission, and good biocompatibility emphasized their value as an adaptable imaging tool for LD research (Jana et al., 2020) (Figure 12.11(a)).

(b) Mitrochondrial Imaging

Chen and his team reported that AIEgens play a major role in the domains of biomedicine and optoelectronics. 5-MOS and 6-MOS, two nontraditional rotor-free AIEgens, were found. The coumarin isomers exhibit entirely different fluorescence characteristics when they aggregate in aqueous conditions due to a slight structural variation. Additional mechanism research shows that 5-MOS forms various aggregates at varying degrees with the help of protonic solvents. This results in the transfer of electrons and energy of 5-MOS causing a distinct AIE property, which is decreased emission in aqueous medium but increased emission in crystal. Meanwhile, the AIE feature of 6-MOS is using the traditional RIM method. More interestingly, 5-MOS's

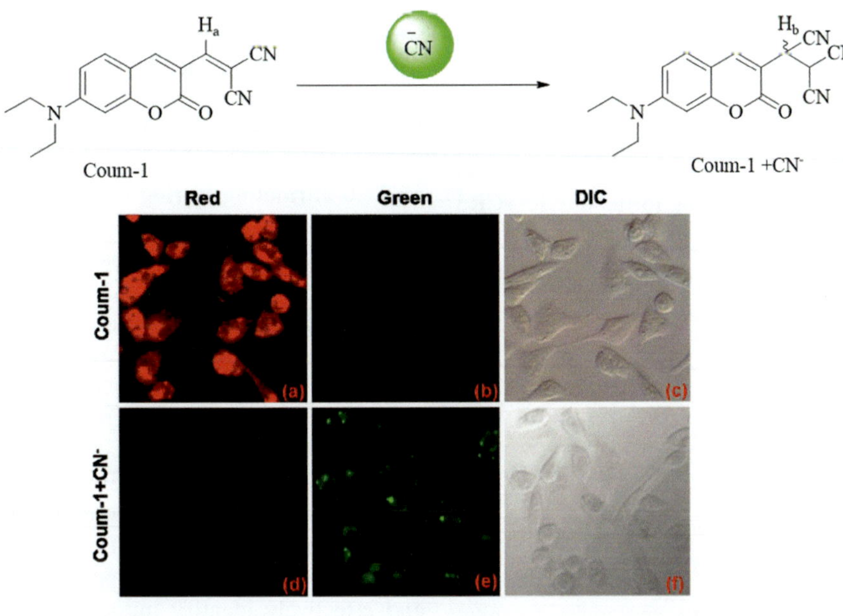

FIGURE 12.10 Mode of CN^- detection by coum-1 and cell imaging studies thereof in the absence and presence of CN^-. Reprinted (adapted) with permission from Cheng et al. (2012). Copyright: American Chemical Society, 2012.

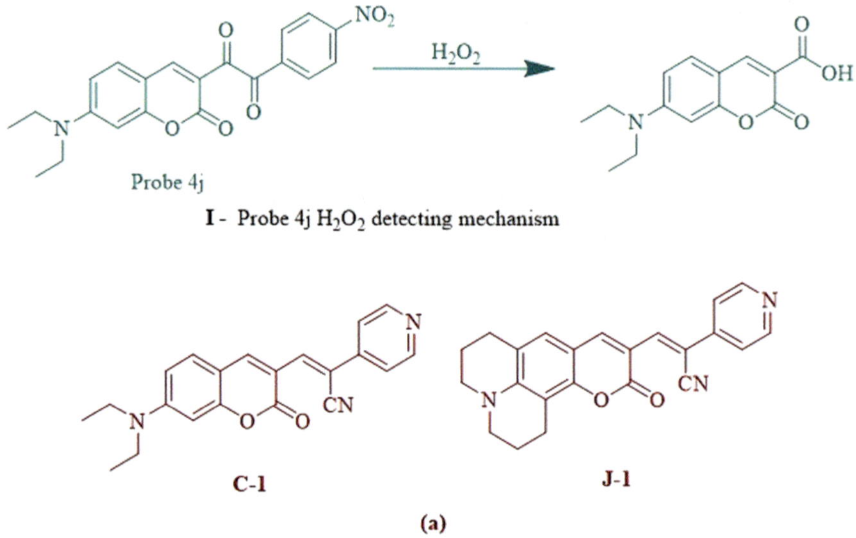

FIGURE 12.11 Coumarin-based fluorescence probe for cell imaging.

distinct water-sensitive fluorescence characteristic makes it possible for successful wash-free mitochondria imaging applications. In DMSO solution, 5-MOS emitted brightly with emission maxima of 473 nm, whereas the fluorescence emission of 6-MOS solution was very weak, with emission maxima at 428 nm. Not only does this work show a strategy for finding new AIEgens in naturally fluorescent organisms, but it also advances the design of next-generation AIEgens structures and applications (Chen et al., 2023) (Figure 12.12 (a)).

Chen et al., reported a series of deep red fluorescent dyes (CR1–CR3) (see Figure 12.13) that contained a coumarin moiety and rhodamine in the chemical skeleton, which contained characteristic features that attribute to coumarin and rhodamine derivatives, having emission wavelength that was extended to the deep red area (>650 nm) by continuously increasing the fused-ring conjugate skeleton. To demonstrate its utility, they devised and quickly manufactured a series of novel deep red bioimaging dyes (CR1E–CR3E) via esterification of CR1–CR3, which might preferentially stain mitochondria. The absorption of CR1–CR3 dropped dramatically between pH 2 and 4 but remained steady at pH > 4 due to the presence of their spirocyclic forms. As a result, the spirocyclic and open-ring forms of CR1–CR3

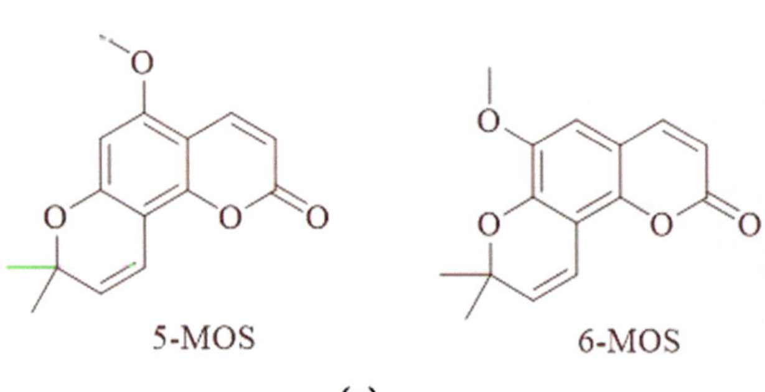

FIGURE 12.12 Coumarin-based fluorescence probes for mitochondrial imaging.

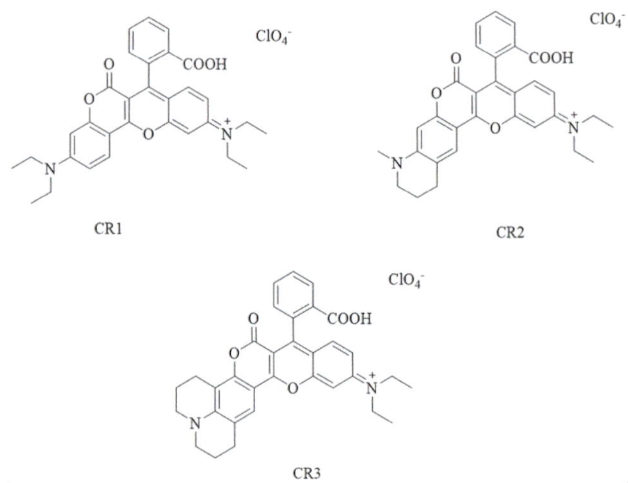

FIGURE 12.13 Structures of CR1-CR3.

were agreeable with the features of other rhodamine derivatives. CR1E–CR3E lack rhodamines' distinctive spirocyclic/open-ring mechanism; their absorption from CR1E to CR3E was unaffected in pH range from 2 to 8. The absorption of CR1E-CR3E significantly reduced in the pH range of 8–10 attributing to ring opening of coumarin moieties in higher pH. Thus, compounds CR1E–CR3E proved to be effective deep red fluorescence imaging reagents in physiological conditions. CR1E–CR3E molecules have a positive charge and are pertaining to the fluorescent dyes MTR and Rhodamine 123 targeting mitochondria that may selectively aggregate in mitochondria due to electrostatic interactions. CR1E–CR3E were tested on living A549 cells. Dual-channel laser scanning confocal microscopy was used to validate the localization of dye accumulation during the costaining studies. The observed substantial overlap with MitoTracker Green FM (MTG) demonstrated that these deep red dyes could easily penetrate the cell membrane and effectively identify mitochondria (see Figure 12.14).

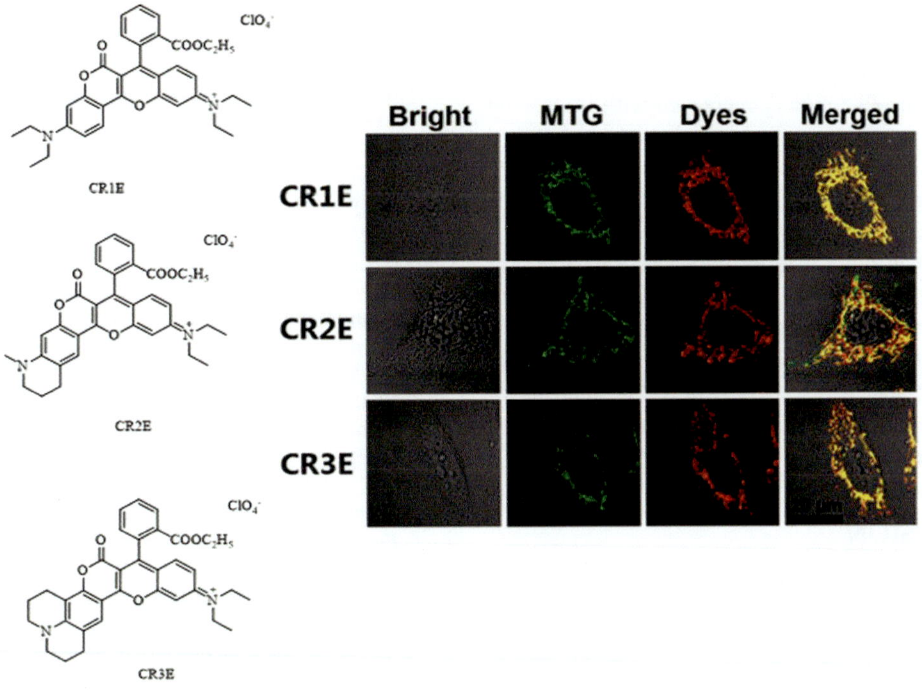

FIGURE 12.14 Structures of CR1E TO CR3E and Confocal fluorescence microscopy images of A549 cells incubated with MTG and CR1E-CR3E. Reprinted (adapted) with permission from ref. Chen et al (2013). Copyright, American Chemical Society, 2013.

To determine the method of accumulation in mitochondria, carbonyl cyanide 3-chlorophenylhydrazone (CCCP) was employed to modify mitochondrial membrane potential (MMP) prior to staining. MMP was lost when CCCP was added due to fast acidification of mitochondria and ATP synthase malfunction. Administration of CCCP reduced fluorescence intensity in mitochondria due to decreased MMP and cationic dye accumulation (CR1E–CR3E). The cytotoxicities of MTG, MTR, and MTDR were used as standards. After 24 hours of treatment with A549 cells, compounds CR1E–CR3E showed little cytotoxicity at doses up to 100 nM compared to the standards that showed cytotoxicity at 10 nM.

The absorption spectra of CR1S revealed a characteristic peak of the coumarin moiety of the spirocyclic form at 381 nm. Upon administration of Hg^{2+}, rapid reduction of peak at 381 nm was observed with red shift with a new peak at 596 nm characteristic of absorption of the rhodamine-like moiety followed by color change from colorless to blue, and colorimetric performance was readily observed with the naked

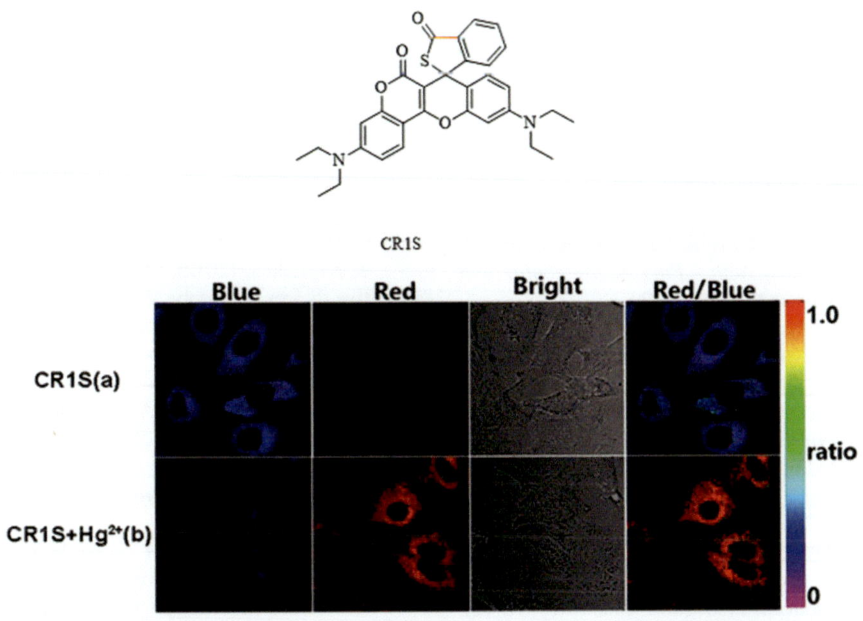

FIGURE 12.15 Structure of CR1S and Confocal fluorescence imaging showing effect of presence of Hg2+ in A549 cells with (a) CR1S (b) CR1S and Hg2+. Reprinted (adapted) with permission from ref. Chen et al (2013). Copyright, American Chemical Society, 2013.

eye for immediate visual. The fluorescence spectra show similar ratio variations where intensity at 430 nm that was characteristic of coumarin reduced, while that at 653 nm from the rhodamine-like moiety increased with increasing Hg^{2+} concentration. This coumarin- and rhodamine-fused π-conjugated system displayed spectrum shifts of up to 223 nm, substantially bigger than those of TBET- or FRET-based rhodamine systems.

A549 cells when administered with CR1S (2 μM) for 10 minutes produced obvious blue fluorescence and comparatively faint red fluorescence in the red channel indicating CR1S was initially in the spiro-lactam form. After a 10-minute treatment with Hg^{2+} (20 μm), the intensity of blue fluorescence reduced while red fluorescence increased dramatically (Chen et al., 2013) (see Figure 12.15).

12.4 CONCLUSION

Coumarin-based fluorescent probes are valuable tools in scientific research due to their sensitivity, versatility, biocompatibility, and specificity. Their stable fluorescence properties and ease of synthesis make them attractive choice of wide range of applications. The coumarin-based fluorescent probe is reported for various applications like detection of cancer, as pH-sensitive probe, for the detection of metal ions and bioimaging, etc. It is an emerging moiety in environmental and medical science.

REFERENCES

Ahmed, N., Zareen, W., Zhang, D., Yang, X., & Ye, Y. (2020). Coumarin-based reversible fluorescent probe for selective detection of Cu^{2+} in living cells. *Journal of Fluorescence*, *30*(5), 1171–1179.

An, J. M., Yan, M. H., Yang, Z. Y., Li, T. R., & Zhou, Q. X. (2013). A turn-on fluorescent sensor for Zn(II) based on fluorescein-coumarin conjugate. *Dyes and Pigments*, *99*(1), 1–5.

An, R., Hou, Z., Li, J. T., Yu, H. N., Mou, Y. H., & Guo, C. (2018). Design, synthesis and biological evaluation of novel 4-substituted coumarin derivatives as antitumor agents. *Molecules*, *23*(9), 2281.

Bai, Yan, et al. (2016). Coumarins from the roots of Angelica dahurica with antioxidant and antiproliferative activities. *Journal of Functional Foods*, *20*, 453–462.

Bekhradnia, A., Domehri, E., & Khosravi, M. (2016). Novel coumarin-based fluorescent probe for selective detection of Cu(II). *Spectrochimica Acta - Part A: Molecular and Biomolecular Spectroscopy*, *152*, 18–22.

Chemchem, M., Yahaya, I., Aydıner, B., Seferoğlu, N., Doluca, O., Merabet, N., & Seferoğlu, Z. (2018). A novel and synthetically facile coumarin-thiophene-derived Schiff base for selective fluorescent detection of cyanide anions in aqueous solution: Synthesis, anion interactions, theoretical study and DNA-binding properties. *Tetrahedron*, *74*(48), 6897–6906.

Chen, J., Liu, W., Zhou, B., Niu, G., Zhang, H., Wu, J., ... & Wang, P. (2013). Coumarin-and rhodamine-fused deep red fluorescent dyes: synthesis, photophysical properties, and bioimaging in vitro. *The Journal of Organic Chemistry*, *78*(12), 6121–6130.

Chen, S. S., Wang, H., Wu, B., Li, Q., Gong, J., Zhao, Y. L., Zhao, Y., Xiao, X., Lam, J. W. Y., Zhao, Z., Luo, X. D., & Tang, B. Z. (2023). Natural coumarin isomers with dramatically different AIE properties: Mechanism and application. *ACS Central Science*, *9*(5), 883–891.

Cheng, X., Tang, R., Jia, H., Feng, J., Qin, J., & Li, Z. (2012). New fluorescent and colorimetric probe for cyanide: direct reactivity, high selectivity, and bioimaging application. *ACS Applied Materials & Interfaces*, *4*(8), 4387–4392.

Cho, J. Y., Hwang, T. L., Chang, T. H., Lim, Y. P., Sung, P. J., Lee, T. H., & Chen, J. J. (2012). New coumarins and anti-inflammatory constituents from Zanthoxylum avicennae. *Food Chemistry*, *135*(1), 17–23.

Duan, Y. W., Tang, H. Y., Guo, Y., Song, Z. K., Peng, M. J., & Yan, Y. (2014). The synthesis and study of the fluorescent probe for sensing Cu^{2+} based on a novel coumarin Schiff-base. *Chinese Chemical Letters*, *25*(7), 1082–1086.

Emami, S., & Dadashpour, S. (2015). Current developments of coumarin-based anti-cancer agents in medicinal chemistry. *European Journal of Medicinal Chemistry*, *102*, 611–630. Elsevier Masson SAS.

Ge, F., Ye, H., Zhang, H., & Zhao, B. X. (2013). A novel ratiometric probe based on rhodamine B and coumarin for selective recognition of Fe (III) in aqueous solution. *Dyes and Pigments*, *99*(3), 661–665.

Grover, J., &Jachak, S. M. (2015). Coumarins as privileged scaffold for anti-inflammatory drug development. *RSC Advances*, *5*(49), 38892–38905). Royal Society of Chemistry. https://doi.org/10.1039/c5ra05643h

Hassan, M. Z., Osman, H., Ali, M. A., & Ahsan, M. J. (2016). Therapeutic potential of coumarins as antiviral agents. *European Journal of Medicinal Chemistry*, *123*, 236–255. Elsevier Masson SAS.

Heravi, M. M., Khaghaninejad, S., & Mostofi, M. (2014). Pechmann reaction in the synthesis of coumarin derivatives. *Advances in Heterocyclic Chemistry*, *112*, 1–50. Academic Press Inc.

Hien, N. K., Bay, M. Van, Bao, N. C., Vo, Q. V., Cuong, N. D., Thien, T. V., Ai Nhung, N. T., Van, D. U., Nam, P. C., & Quang, D. T. (2020). Coumarin-based dual chemosensor for colorimetric and fluorescent detection of Cu2+ in water media. *ACS Omega*, *5*(33), 21241–21249.

Jameel, E., Umar, T., Kumar, J., & Hoda, N. (2016). Coumarin: A privileged scaffold for the design and development of antineurodegenerative agents. *Chemical Biology and Drug Design*, *87*(1), 21–38. Blackwell Publishing Ltd.

Jana, P., Siva, A., Soppina, V., & Kanvah, S. (2020). Live-cell imaging of lipid droplets using solvatochromic coumarin derivatives. *Organic and Biomolecular Chemistry*, *18*(29), 5608–5616.

Jung, H. S., Pradhan, T., Han, J. H., Heo, K. J., Lee, J. H., Kang, C., & Kim, J. S. (2012). Molecular modulated cysteine-selective fluorescent probe. *Biomaterials*, *33*(33), 8495–8502.

Jung, J. W., Kim, N. J., Yun, H., & Han, Y. T. (2018). Recent advances in synthesis of 4-arylcoumarins. *Molecules*, *23*(10), 2417.

Kathiravan, M. K., Salake, A. B., Chothe, A. S., Dudhe, P. B., Watode, R. P., Mukta, M. S., & Gadhwe, S. (2012). The biology and chemistry of antifungal agents: A review. In *Bioorganic and Medicinal Chemistry* (Vol. 20, Issue 19, pp. 5678–5698).

Li, J., Zhang, C. F., Yang, S. H., Yang, W. C., & Yang, G. F. (2014). A coumarin-based fluorescent probe for selective and sensitive detection of thiophenols and its application. *Analytical Chemistry*, *86*(6), 3037–3042.

Li, Y., Tang, Q., Xie, Y., He, D., Yang, K., & Zheng, L. (2021). Synthesis of mitochondria-targeted coumarin-3-carboxamide fluorescent derivatives: Inhibiting mitochondrial TrxR2 and cell proliferation on breast cancer cells. *Bioorganic and Medicinal Chemistry Letters*, *33*, 127750.

Liu, M., Hu, M., Jiang, Q., Lu, Z., Huang, Y., Tan, Y., & Jiang, Q. (2015). A novel coumarin derivative as a sensitive probe for tracing intracellular pH changes. *RSC Advances*, *5*(21), 15778–15783.

Long, L., Li, X., Zhang, D., Meng, S., Zhang, J., Sun, X., Zhang, C., Zhou, L., & Wang, L. (2013). Amino-coumarin based fluorescence ratiometric sensors for acidic pH and their application for living cells imaging. *RSC Advances*, *3*(30), 12204–12209.

Mani, K. S., Rajamanikandan, R., Murugesapandian, B., Shankar, R., Sivaraman, G., Ilanchelian, M., & Rajendran, S. P. (2019). Coumarin based hydrazone as an ICT-based fluorescence chemosensor for the detection of Cu 2+ ions and the application in HeLa cells. *Spectrochimica Acta - Part A: Molecular and Biomolecular Spectroscopy*, *214*, 170–176.

Matos, M. J., Santana, L., Uriarte, E., Abreu, O. A., Molina, E., & Yordi, E. G. (2015). Coumarins—an important class of phytochemicals. *Phytochemicals-isolation, Characterisation and Role in Human Health*, *25*, 533–538.

Mukherjee, S., Hazra, S., Chowdhury, S., Sarkar, S., Chattopadhyay, K., & Pramanik, A. (2018). A novel pyrrole fused coumarin based highly sensitive and selective fluorescence chemosensor for detection of Cu2+ ions and applications towards live cell imaging. *Journal of Photochemistry and Photobiology A: Chemistry*, *364*, 635–644.

Nguyen, T. H., Sun, T., & Grattan, K. T. (2020). Novel coumarin-based pH sensitive fluorescent probes for the highly alkaline pH region. *Dyes and Pigments*, *177*, 108312.

Nguyen, P. H., Zhao, B. T., Kim, O., Lee, J. H., Choi, J. S., Min, B. S., & Woo, M. H. (2016). Anti-inflammatory terpenylated coumarins from the leaves of Zanthoxylum schinifolium with α-glucosidase inhibitory activity. *Journal of Natural Medicines*, *70*, 276–281.

Perkin, W. H. (1868). On the artificial production of coumarin and formation of its homologues. *Journal of the Chemical Society*, *21*, 53–63.

Qin, J. C., Fan, L., & Yang, Z. Y. (2016). A small-molecule and resumable two-photon fluorescent probe for Zn2+ based on a coumarin Schiff-base. *Sensors and Actuators, B: Chemical*, *228*, 156–161.

Satish, G. (2016). Anticoagulant agents. In *Advances in Structure and Activity Relationship of Coumarin Derivatives* (pp. 151–159). Academic Press.

Shaily, Kumar, A., & Ahmed, N. (2017). A coumarin-chalcone hybrid used as a selective and sensitive colorimetric and turn-on fluorometric sensor for Cd^{2+} detection. *New Journal of Chemistry*, *41*(23), 14746–14753.

Shen, B., He, P. J., & Shao, C. L. (2013). Norcantharidin induced DU145 cell apoptosis through ROS-mediated mitochondrial dysfunction and energy depletion. *PloS One*, *8*(12), e84610.

Su, Z., Chen, K., Guo, Y., Qi, H., Yang, X. F., & Zhao, M. (2010). A coumarin-based fluorescent chemosensor for Zn^{2+} in aqueous ethanol media. *Journal of Fluorescence*, *20*(4), 851–856.

Sun, X., Wang, Y., Zhang, X., Zhang, S., & Zhang, Z. (2015). A new coumarin based chromo-fluorogenic probe for selective recognition of cyanide ions in an aqueous medium. *RSC Advances*, *5*(117), 96905–96910.

Sun, Y. Q., Chen, M., Liu, J., Lv, X., Li, J. F., & Guo, W. (2011). Nitroolefin-based coumarin as a colorimetric and fluorescent dual probe for biothiols. *Chemical Communications*, *47*(39), 11029–11031.

Tan, W., Leng, T., Lai, G., Li, Z., Wang, K., Shen, Y., & Wang, C. (2016). A novel coumarin-based fluorescence enhancement and colorimetric probe for Cu^{2+} via selective hydrolysis reaction. *Journal of Photochemistry and Photobiology A: Chemistry*, *324*, 81–86.

Thasnim, P., & Bahulayan, D. (2017). Click-on fluorescent triazolyl coumarin peptidomimetics as inhibitors of human breast cancer cell line MCF-7. *New Journal of Chemistry*, *41*(22), 13483–13489.

Vekariya, R. H., & Patel, H. D. (2014). Recent advances in the synthesis of coumarin derivatives via Knoevenagel condensation: A review. *Synthetic Communications*, *44*(19), 2756–2788. Taylor and Francis Inc.

Wang, W., Wu, J., Liu, Q., Gao, Y., Liu, H., & Zhao, B. (2018). A highly selective coumarin-based chemosensor for the sequential detection of Fe^{3+} and pyrophosphate and its application in living cell imaging. *Tetrahedron Letters*, *59*(19), 1860–1865.

Wang, Y., Zhang, W., Dong, J., & Gao, J. (2020). Design, synthesis and bioactivity evaluation of coumarin-chalcone hybrids as potential anticancer agents. *Bioorganic Chemistry*, *95*, 103530.

Wu, X., Wang, H., Yang, S., Tian, H., Liu, Y., & Sun, B. (2019). A novel coumarin-based fluorescent probe for sensitive detection of copper (II) in wine. *Food Chemistry*, *284*, 23–27.

Yu, D., Suzuki, M., Xie, L., Morris-Natschke, S. L., & Lee, K. H. (2003). Recent progress in the development of coumarin derivatives as potent anti-HIV agents. *Medicinal Research Reviews*, *23*(3), 322–345.

Zehra, S., Khan, R. A., Alsalme, A., & Tabassum, S. (2019). Coumarin derived "turn on" fluorescent sensor for selective detection of cadmium (II) Ion: Spectroscopic studies and validation of sensing mechanism by DFT calculations. *Journal of Fluorescence*, *29*(4), 1029–1037.

Zhang, K. M., Dou, W., Li, P. X., Shen, R., Ru, J. X., Liu, W., Cui, Y. M., Chen, C. Y., Liu, W. S., & Bai, D. C. (2014). A coumarin-based two-photon probe for hydrogen peroxide. *Biosensors and Bioelectronics*, *64*, 542–546.

Zhang, W., Liu, F., Zhang, C., Luo, J. G., Luo, J., Yu, W., & Kong, L. (2017). Near-infrared fluorescent probe with remarkable large stokes shift and favorable water solubility for real-time tracking leucine aminopeptidase in living cells and in Vivo. *Analytical Chemistry*, *89*(22), 12319–12326.

Zhang, W., Wang, Y., Dong, J., Zhang, Y., Zhu, J., & Gao, J. (2019). Rational design of stable near-infrared cyanine-based probe with remarkable large stokes shift for monitoring Carbon monoxide in living cells and in vivo. *Dyes and Pigments*, *171*, 107753.

Zhang, X., Ba, Q., Gu, Z., Guo, D., Zhou, Y., Xu, Y., Wang, H., Ye, D., & Liu, H. (2015). Fluorescent coumarin-artemisinin conjugates as mitochondria-targeting theranostic probes for enhanced anticancer activities. *Chemistry - A European Journal*, *21*(48), 17415–17421.

Zhang, Y. P., Teng, Q., Yang, Y. S., Guo, H. C., & Xue, J. J. (2021). A novel coumarin-based pyrazoline fluorescent probe for detection of Fe^{3+} and its application in cells. *Inorganica Chimica Acta*, *525*, 120469.

13 Carbazole-Based Fluorescent Molecules

Design and Applications

Somidi Srinu, Mareechika Gaddam and Abha Sharma
National Institute of Pharmaceutical Education and Research,
Raebareli, Lucknow, UP, India

LIST OF ABBREVIATIONS

AD	Alzheimer's disease
AIE	Aggregation-induced emission
AIEE	Aggregation-induced enhanced emission
APTESMS	3-aminopropyl-triethoxysilane (APTES)-modified silica
ATP	Adenosine triphosphate
CD	Circular dichroism
CE	Capillary Electrophoresis
CMFS	Carbazole modified fluorescent silica
DFT	Density functional theory
DNA	Deoxyribonucleic acid
EI-MS	Electron ionization mass spectrometry
ESI-MS	Electrospray ionization mass spectrometry
ESIPT	Excited-state intramolecular proton transfer
FONs	Fluorescent organic nanoparticles
GFP	Green fluorescent protein
HPLC-FLD	High-performance liquid chromatography with fluorescence detection
HPLC-UV	High-Performance liquid chromatography with ultraviolet detection
HPTLC	High-performance thin-layer chromatography
HRP	Horseradish peroxidase
ICT	Intramolecular charge transfer
IMP	Imino-phenol
LC	Liquid chromatography
LDs	Lipid droplets
LOD	Limit of detection
MIPs	Molecularly imprinted polymers
MTG	MitoTracker Green
MTR	MitoTracker Red
NIR	Near-infrared
NMR	Nuclear magnetic resonance
NPs	Nanoparticles
NUV-LEDs	Near-ultraviolet light-emitting diodes
OLEDs	Organic light-emitting diodes
PDT	Photodynamic therapy
PET	Photoinduced electron transfer
pHi	Intracellular pH
RNA	Ribonucleic acid
RNS	Reactive nitrogen species
ROS	Reactive oxygen species
STED	Stimulated emission depletion
TD-DFT	Time-dependent density functional theory
TLC	Thin-layer chromatography
TPA	Two-photon absorption
TPF	Two-photon excitation fluorescence
TPM	Two-photon microscopy
WHO	World Health Organization

13.1 INTRODUCTION

A tricyclic nitrogen heterocyclic compound was first isolated in 1872 by Graebe and Glaser from coal tar and named 9H-carbazole (Graebe, 1872). The carbazole moiety has characteristics like rich optical properties, good planarity, rigid ring structure, and electronic donor (Figure 13.1). To address the limitations associated with the carbazole moiety, including its short excitation and emission properties, as well as its low water solubility, researchers have introduced a range of substituents and functional groups at various positions on the carbazole structure. This approach has led to the creation of innovative carbazole derivatives with improved properties (Knölker & Reddy, 2002). Carbazole has wide applications including cell organelle imaging (Kong et al., 2018). Cell organelles like mitochondria, endoplasmic reticulum, lysosomes, and ribosomes are functional units that control the biological processes (Mallick et al., 2020; Zhang et al., 2016; Zhu et al., 2016). The abnormal functioning of these functional units causes disturbance in homeostasis leading to various disorders. For example, mitochondria are responsible for the generation of energy in the form of ATP. During this process, reactive oxygen species (ROS) are produced; if they are produced at an abnormal level, they may cause oxidative stress, which can result in diseases like Alzheimer's disease (AD) (Hao et al., 2018). So, the monitoring of biological processes controlled by cell organelles is helpful to diagnose the disorders. The carbazole fluorescent probes selectively bind and provide accurate information about the pathophysiological condition.

DOI: 10.1201/9781032706245-13

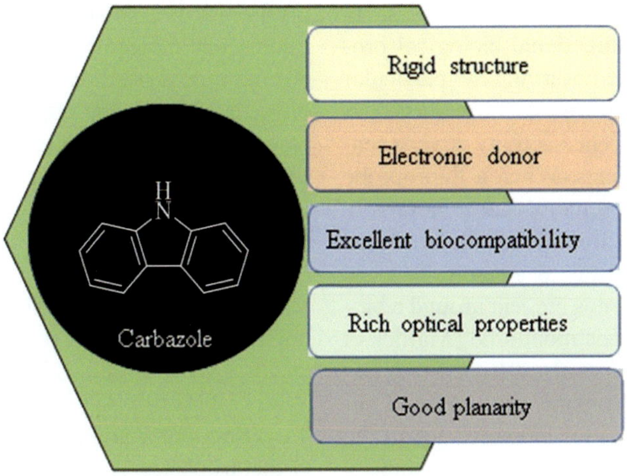

FIGURE 13.1 The structure and characteristics of 9H-carbazole.

Carbazole has great importance in monitoring and tracking nucleic acid physiology; many fluorescent probes are developed to track and monitor DNA and RNA, especially G-quadruplex tracking fluorescent probes as G-quadruplex has significant importance in maintaining the stability of chromosomes (Feng et al., 2019; Chang et al., 2003). RNA binding probes are also made from carbazole to detect RNA in living cells. RNA acts as a genetic carrier in translation including the transfer of genetic information from genes to proteins that are made from amino acids (Liu et al., 2016). Carbazole probes can detect the ions in the biological system, which maintain homeostasis by osmotic balance. The abnormal levels of the ions may cause a disturbance in normal physiology like cell death and neurodegeneration. Recently, a novel carbazole fluorescent probe has been developed as a chemosensor to monitor the D-glucose levels in the blood; the increased glucose levels cause the disorder diabetes (Oloub et al., 2022). Carbazole fluorescent probes are used in organic light-emitting diodes (OLEDs) because of their electron donor characteristic; carbazole analogs are mostly used as fluorescent probes in OLEDs (Konidena et al., 2020; Qiu et al., 2020).

13.2 BIOMOLECULAR IMAGING

Fluorescent probes are emerging tools in the realm of molecular biology and biotechnology by enabling the visualization of biomolecules and cellular processes with high precision and sensitivity. Among these probes, carbazole derivatives have emerged as a fascinating class of fluorescent molecules due to their unique photophysical properties and versatile applications.

13.2.1 Fluorescent Probes for Nucleic Acids

This section provides a comprehensive overview of the role of carbazole derivative-based fluorescent probes in nucleic acid-related research. We will delve into the structural characteristics that make carbazole derivatives suitable for DNA/RNA binding and discuss their mechanisms of interaction. Furthermore, we will explore the diverse applications of these probes, ranging from DNA sequence detection and DNA structural analysis to real-time monitoring of DNA-related processes within living cells. The discussion will encompass both the design strategies employed to enhance the DNA-binding affinity, fluorescence properties of carbazole derivatives, and the analytical techniques used to exploit their capabilities effectively (Mallick et al., 2020). Additionally, we will highlight key studies and advancements in this field, showcasing the potential of carbazole derivatives in revolutionizing DNA visualization and manipulation.

13.2.1.1 Deoxyribonucleic Acid

Carbazole derivatives possess a distinctive molecular structure characterized by a tricyclic aromatic system with remarkable photostability, strong fluorescence emission, and significant binding affinity to deoxyribonucleic acid (DNA) molecules (Hao et al., 2018). In recent years, the development and utilization of carbazole derivative-based fluorescent probes for DNA studies have garnered substantial attention from researchers. These probes offer a wide array of benefits, such as their ability to selectively bind to DNA through various modes of interaction, including intercalation, groove binding, and electrostatic interactions. This selectivity is crucial for minimizing background signals and ensuring specific detection in complex biological environments (Feng et al., 2019). The utilization of carbazole derivative-based fluorescent probes represents a cutting-edge approach to the study of DNA, offering researchers an invaluable toolset to unravel the intricate dynamics of genetic material. By shedding light on the synthesis, binding mechanisms, and applications of these probes, this chapter aims to contribute to a deeper understanding of their significance and potential in advancing molecular and cellular biology research (Dahm, 2008). Many large biological molecules, such as nucleic acids (DNA/RNA), proteins, and enzymes, play crucial roles in governing biological processes within living organisms. Disease in living things is caused by a dysfunctional biological process, which may be researched by observing the particular biomacromolecule involved (Swierczewska et al., 2012). A fluorophore (fluoroprobe) fluorescence activity is very sensitive to its immediate surroundings. When a fluoroprobe interacts with a particular biomacromolecule (or biotarget), it causes a disturbance in the area around it. Its fluorescence behavior changes in a distinctive way as a result. To learn more about the physicochemical environment around a biotarget's active site, this shift may be monitored (Mallick et al., 2020; Raibaut et al., 2017; Zhu et al., 2016).

Mallick et al. (2020) studied organically modified silica nanoparticles (NPs) as sensors for biomolecules in the realm of biomedical diagnostics. However, the majority of instances involve the attachment of synthetic fluorescent organic dyes onto the surface of silica NPs. Using fluorescent compounds based on the structural framework of natural products, such as carbazole alkaloids, remains uncommon. In their study, efforts have been directed toward introducing appropriate functionalities into the original carbazole structure. This alteration enables these fluorescent molecules to be affixed onto the surface of Amino propyl triethoxy silane nano particles (APTESMS) NPs, resulting in the creation of carbazole-modified fluorescent silica (CMFS) NPs (Figure 13.2). The integration of these biofriendly carbazole units onto the NPs' surface contributes to an enhanced level of biocompatibility for CMFS NPs when compared to bare silica NPs. Furthermore, CMFS NPs demonstrate a specific ability to sense fluorescence in response to biomacromolecules like DNA and horseradish peroxidase (HRP).

13.2.1.1.1 Fluorescent Probes for G-Quadruplex DNA

Targeting noncanonical DNA secondary structures like G-quadruplexes is now regarded as an attractive opportunity for clinical diagnosis and the development of new chemotherapeutic agents, particularly because mounting evidence suggests that G-quadruplexes exist in vivo (Rhodes & Lipps, 2015). The understanding of their biological processes would greatly benefit from the selective detection of G-quadruplex structures. Fluorescence probes,

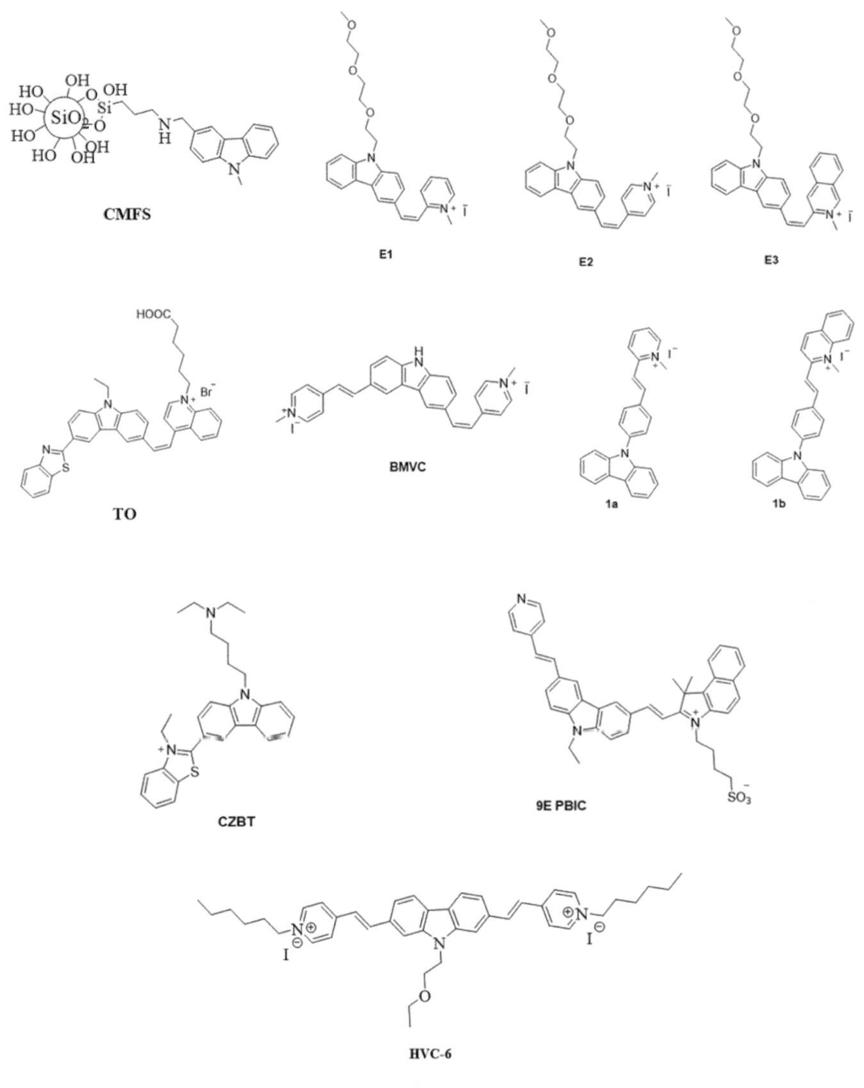

FIGURE 13.2 Nucleic acid imaging carbazole-based fluorescent probes.

based on small organic molecules, have significantly contributed because of their numerous benefits, including rapidity, high selectivity, sensitivity, low cost, noninvasive detection, and real-time detectability (Wang et al., 2018). Recently, numerous classes of small-molecule fluorescent probes with significant G-quadruplex over duplex DNA selectivity have been established (Dumat et al., 2011; Yu & Wang, 2020). Due to their attractive structural characteristics and high binding affinities, carbazole derivatives are the most popular fluorescent compounds for G-quadruplex DNA targeting (Chang et al., 2003; Henderson et al., 2014).

Yu and Wang (2020) have developed effective fluorescent probes for G-quadruplex DNA sensing. The new fluorescent probes (E1, E2, and E3) (Figure 13.2) were developed based on the combination of a triethylene glycol mono methyl side chain in the center of a carbazole scaffold and various side groups that resemble styrene at its 3-position. On various DNA forms, their fluorescence performance was examined. The resulting probes exhibited turn-on fluorescent signal enhancements toward G-quadruplex DNAs, which displayed notable structural, positional, and side group-type differentiation. More significantly, probe E1 can enter living cells with low cytotoxicity and showed almost no fluorescence response to different duplex DNAs. Through both experimental and modeling studies, the precise binding characteristics for G-quadruplex DNA were thoroughly evaluated.

Gu et al. (2018) reported carbazole TO (Figure 13.2). It is highly selective for bcl-2 2345 G4. Fluorescence studies revealed that when the probe carbazole TO selectively bound to bcl-2 2345 G4, the fluorescence amplification took place as compared to other DNA forms. The binding of carbazole TO to the bcl-2 2345 G-quadruplex occurs through an end-stacking model, which has been confirmed through various methods including CD spectrum analysis, NMR studies, and docking simulations. The high selectivity to the bcl-2 2345 G4 of the probe has applicable use in diagnostics, therapeutics, and biosensors.

Chang et al. (2003) studied the interaction of BMVC (Figure 13.2) with DNA resulting in a considerable increase in fluorescence, up to 90 fold with Hum24 at excitation 460 nm. In addition, after interacting with different DNA, visible fluorescence bands in electrophoresis gels were found, ranging from green to red. This discovery might help with high-throughput screening using fluorescence imaging. It is possible to distinguish between duplex and quadruplex DNAs using the fluorescence of BMVC. Additionally, some specific quadruplex structures may be seen in prestained gels containing 0.1 M BMVC. Three different kinds of quadruplexes can be identified by combining them with circular dichroism (CD) spectra. The great sensitivity of the BMVC to a particular DNA suggests that the BMVC is not only beneficial as a quadruplex stabilizer and telomerase inhibitor but also serves as a sensitive biosensor.

Wang et al. (2018) conducted a study to design and investigate the specific recognition ability of carbazole derivatives with various styrene-like substituents for G-quadruplex structures. In pursuit of this objective, they synthesized and characterized two probes, 1a and 1b (Figure 13.2). The findings revealed that the introduction of different side groups to the same core scaffold yielded distinct binding properties toward G-quadruplex structures. More specifically, compound 1a, featuring a pyridine side group, has shown the capability to distinguish G-quadruplexes from duplex and single-stranded DNA by displaying a substantial increase in fluorescence intensity in the presence of G-quadruplexes. On the other hand, 1b, which featured a quinolinium group, exhibited poor selectivity. These results underscore the significance of side groups in the G-quadruplex binding process, suggesting the potential to develop probes with high selectivity. Additionally, the study provided a comprehensive understanding of the precise interactions between 1a and G-quadruplex DNAs through a series of experiments and molecular docking simulations. In solution, 1a has extreme sensitivity toward G-quadruplex DNAs, achieving impressively low limit of detection (LOD) values of less than 1 μM. Furthermore, CD experiments provided evidence that 1a could detect G-quadruplex structures without causing any changes to their topologies. Subsequent studies discovered that 1a intercalated within the groove section of G-quadruplex structures and demonstrated a higher affinity for anti-parallel G-quadruplex formations. These findings offer valuable insights that can significantly contribute to the development of efficient probes for various applications involving G-quadruplex DNA.

Gao et al. (2019) reported the development of a new fluorescent probe named CZ-BT, which is derived from carbazole-benzothiazole (Figure 13.2). This probe was designed and synthesized with the specific aim of detecting G4 DNA, both in laboratory settings (in vitro) and within living organisms (in vivo). The researchers confirmed the selectivity of CZ-BT for G4 DNA through a series of experiments, including UV-vis, fluorescence emission, and lifetime measurements. Additionally, molecular docking calculations provided further evidence, demonstrating that CZ-BT indeed binds to G4 DNA through π-π stacking interactions. Importantly, the researchers utilized two-photon confocal fluorescent imaging and conducted cell cycle experiments to showcase CZ-BT's specific affinity for G4 DNA within living cells. These cumulative results strongly indicate that CZ-BT has substantial promise for advancing research in the domain of G4 DNA.

Lin et al. (2015) reported the development of a new cyanine dye, 9E PBIC (Figure 13.2), designed for the detection of c-myc G-quadruplex structures. Dye is synthesized by combining the benzindole and carbazole moieties. The selective binding capability of dye with c-myc 2345 G-quadruplex was studied. Under aqueous conditions, no fluorescence was observed with 9E PBIC. However, when it interacted with c-myc G-quadruplex, there was a remarkable 100-fold increase in fluorescence intensity. For other types of G-quadruplexes, 9E PBIC exhibited fluorescence enhancements ranging from 10 to 30 folds. The cyanine dye with the carbazole moiety is an effective fluorescent dye for studying the c-myc G-quadruplexes.

13.2.1.2 Ribonucleic Acid

Ribonucleic acid (RNA) is formed from nucleotide monomers. It is crucial in transforming genetic data from genes into the protein's amino acid sequences (Johnston & Bose, 1972). Friedrich Miescher discovered RNA in 1869. In 1984, for the first time, dyes were used to identify RNA (Melton et al., 1984). In recent times, multiple varieties of molecular probes have been developed to investigate RNA within living cells. SYTO RNA-Select, a commercially available RNA probe, is used to detect RNA in the cells (Johnston & Bose, 1972).

A groundbreaking moment occurred when researchers successfully designed a novel two-photon fluorescent probe named HVC-6 (Figure 13.2) through a rational approach for RNA sensing. The RNA was monitored at 488 nm and 800 nm with the help of a probe having a detection limit of 0.2 μM. The probe's unique properties make it suitable for use in both one- and two-photon imaging of endogenous RNA within living systems. Probe HVC-6 has shown excellent sensitivity and stability when compared to the commercially available RNA sensing probes (Liu et al., 2016).

13.2.2 Fluorescent Probe for Cellular Imaging

The progress in creating fluorescent probes for medical uses such as diagnosis, imaging, and sensing has garnered significant attention. Fluorescent imaging offers several benefits, including straightforward procedures, easily controllable tools, and a wide array of affordable imaging substances. Additionally, fluorescence imaging delivers heightened sensitivity and exceptional spatial and temporal resolution images, showcasing intricate details even at the cellular level. Over the past few decades, there has been a notable focus on fluorescent NPs due to their improved compatibility with living systems and impressive optical characteristics. These characteristics encompass broader absorption ranges, adjustable emission wavelengths based on size, larger differences between absorption and emission wavelengths, and enhanced stability under light exposure, all of which surpass traditional organic dyes. Fluorescent NPs have attracted significant research interest over the past two decades (Zhang et al., 2016; Zhu et al., 2016).

Research carried by Zhang et al. (2016) has created high-performance fluorescent organic nanoparticles (FONs) in various colors using three carbazole derivatives with donor-acceptor structures. These FONs exhibit strong fluorescence when excited by both one- and two-photon processes (Figure 13.3d). Cell imaging experiments conducted on the A549 cell line using both one- and two-photon excitation techniques demonstrated promising outcomes. The study highlights the benefits of two-photon imaging, offering improved visualization of cellular structures and reduced interference from background autofluorescence (Figure 13.3c). The research suggests that self-assembled nanostructures composed of these carbazole derivatives, featuring donor-acceptor molecular structures, have significant potential as highly effective fluorescence probes for bioimaging applications.

13.2.3 Fluorescent Probe for Nucleus Imaging

The nucleus is the primary location for DNA, the genetic information reporter. DNA is also present in the mitochondria and chloroplasts, although to a far smaller amount. For a variety of species, it is a crucial factor in controlling the cell cycle. The combination of a two-photon fluorescent DNA probe with LCSM (stimulated by NIR) allows for an in-depth analysis of morphological changes in living cells. In the past, bioimaging of live cells and tissues has seldom made use of commercially available nucleic acid fluorescent markers partly because they are stimulated by ultraviolet-visible light. For long-wavelength laser scanning microscopy, a fluorophore with a high cross-section for two-photon absorption (TPA) is essential (Kapuscinski et al., 1978; Zhu et al., 2016).

Feng et al. (2019) developed a novel compound with quinolinium as the terminal moiety and carbazole (Figure 13.3). The analogue with the quinolinium moiety shows the best improvement in the fluorescence and quantum yield. While being exposed to NIR laser radiation, they exhibit enhanced red fluorescence in the presence of DNA. The novel dyes can penetrate through the cell membrane and they stain the nucleus with a high degree of consistency to Hoechst 33342 while also having less cytotoxicity and higher photostability. To stain the nuclei using a confocal laser scanning microscope, the novel probes may have a potential use.

13.2.4 Fluorescent Probe for Plasma Membrane

Plasma membranes change constantly because they are the way for information and materials to move between cells and their surroundings. Endocytosis, exocytosis, signaling, and other processes depend on the ability of cells to instantly change their shapes and sizes. The shape of plasma membranes is a clear sign of the health of a cell. For example, if a drug causes apoptosis, the burst or partial eating of a plasma membrane is strong proof. Also, since plasma membranes are the first line of defense for cells, they will act when they feel damage from outside poisons (like heavy metal ions). So, monitoring the plasma membrane on video in real time is useful for cell biology, medicine, toxicology, and other fields. Bright-field imaging has been used to look at membranes, but the sharpness and color mode aren't very good, therefore these results aren't very useful. Fluorescence photography is a stronger and more accurate way that gives clear images of the information. However, this method is very dependent on how well light probes work (Schlessinger et al., 1977).

Mem-SQAC (Figure 13.3), a novel bipolar marker discovered by Zhang et al. (2014) demonstrates a long retention duration of over 30 minutes. Like other members of the BODIPY dye family, mem-SQAC has remarkable

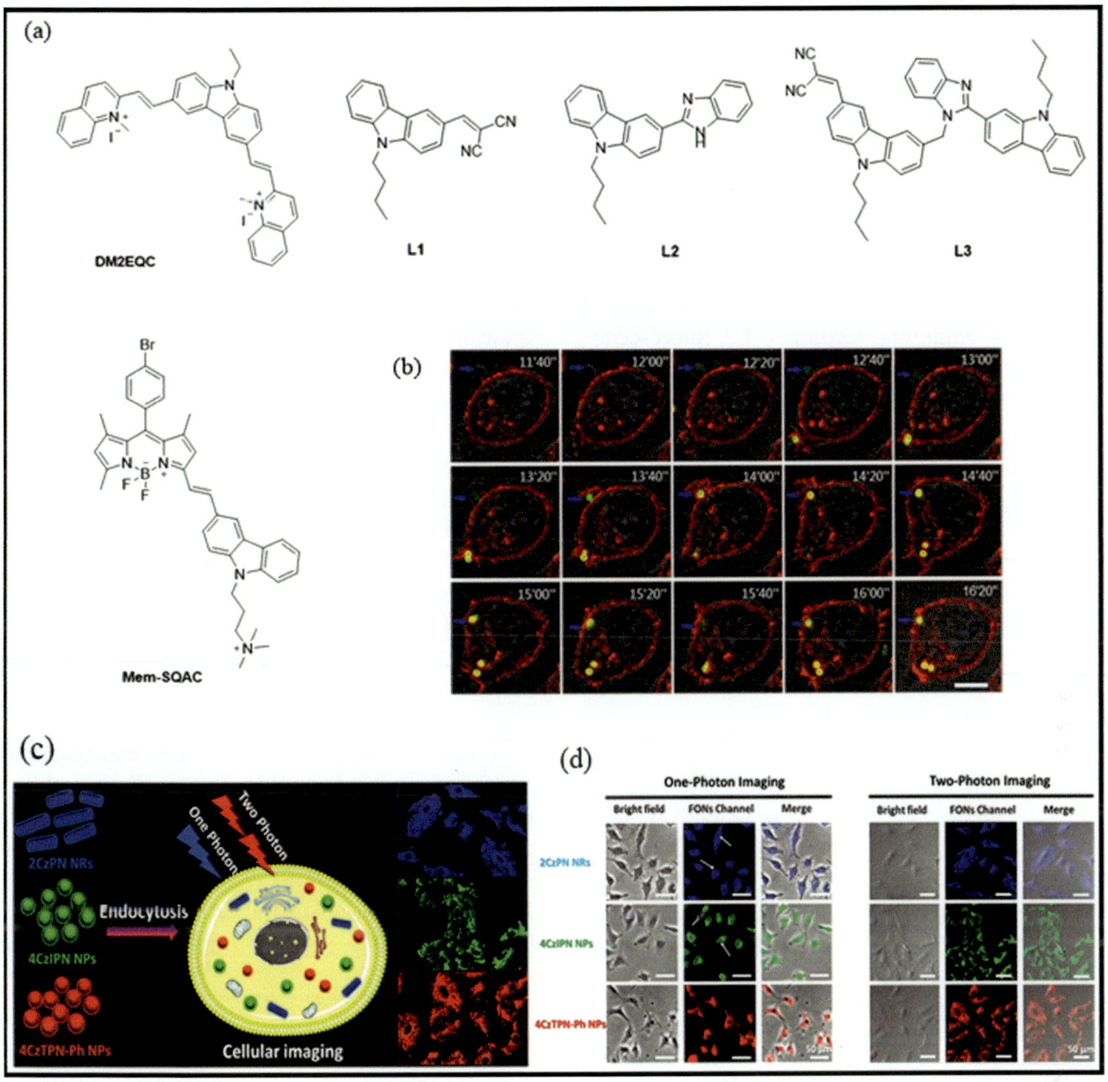

FIGURE 13.3 (a) AIEE probes. (b) Dynamic fluorescent images depict RAW 264.7 cells engulfing *S. aureus* (expressing GFP). Mem-SQAC is represented by pseudo red color with an excitation wavelength (λex) of 559 nm and an emission wavelength (λem) of 570–630 nm. GFP is shown in pseudo green color with λex at 488 nm and λem at 500–560 nm. The scale bar corresponds to 10 μM. (c) a schematic illustration of the process for preparing blue/green/red fluorescent organic nanoparticles (FONs) using the reprecipitation method, specifically for use in one- and two-photon cellular imaging. (d) One-photon and two-photon cellular imaging were performed to visualize and localize the three FONs. One-photon imaging was monitored using a fluorescence microscope with an excitation wavelength range of 380–420 nm, while two-photon imaging was carried out using a laser scanning confocal fluorescence microscope with an excitation wavelength of 800 nm. These imaging experiments were conducted in A549 cells at a final FON concentration of 8 μg/mL. Reprinted (adapted) with permission from Zhang et al., American Chemical Society, Copyright 2014. Reprinted (adapted) with permission from Zhang, et al., American Chemical Society, Copyright 2016.

photostability. Mem-SQAC's advantage as a membrane marker is reinforced by its notable attributes, including substantial two-photon absorption cross-sections and extended wavelength fluorescence emissions. Using Mem-SQAC, they observed substantial alterations in the morphology of plasma membranes in MCF-7 cells subjected to heavy metal poisoning and drug-induced apoptosis; the change tendencies are so dissimilar that they can serve as indicators of distinguishing between the various cell injuries. Additionally, RAW 264.7 cell endocytosis toward *Staphylococcus aureus* and *Escherichia coli* has been dynamically monitored. It turns out that the plasma membrane reacts quite differently to both bacteria, a fact that was previously unknown.

13.2.5 Fluorescent Probe for Imaging of Endoplasmic Reticulum

Organic materials with aggregation-induced emission (AIE) or aggregation-induced enhanced emission (AIEE) characteristics have recently been extensively studied due to the enhanced fluorescence emission in an aggregated state, particularly in aqueous media, which is the main constituent of biological systems. Designing new fluorescent probes with AIE/AIEE properties is essential in this regard, as it involves the understanding of the of relationship between structure and property for their potential biological applications, such as fluorescent bioimaging (Kruusma et al., 2006).

Kong et al. developed three probes L1, L2, and L3 (Figure 13.3) with AIEE properties. When the molecular configuration is disturbed, it may disrupt the π-π interactions between nearby molecules, which would then improve the fluorescence emission at the aggregated state and the AIEE property, which is advantageous for bioimaging applications. In HepG2 cells and plant tissues, probes L1–L3 had functional fluorescent cell imaging applications. While L2 was specifically targeted to the lysosome, L1 and L3 were primarily stained at the endoplasmic reticulum of HepG2 cells. In *Arabidopsis* seedlings, L3 fluorescence emission was more intense than L1 or L2 fluorescence emission (Kong et al. 2018).

13.2.6 Fluorescent Probe for Mitochondria Visualization

Mitochondria, ranging in size from 0.5 to 10 μm, are essential organelles enclosed within membranes, acting as cell power centers. They play a critical role in various cellular processes like differentiation, apoptosis, signal transmission, and the generation of reactive oxygen species. Their primary function is to facilitate respiration, providing energy crucial for cell survival and growth. Consequently, most eukaryotic cells contain plentiful mitochondria, and their size, quantity, and shape change in response to the cell's metabolic state, cycle, development stages, differentiation, and disease conditions. Scientists have observed a significant link between mitochondrial structure, number, and diseases like cancer and Alzheimer's disease. Hence, exploring mitochondrial structure and function is essential for understanding mitochondrial-related physiological processes and advancing diagnostics and therapies for mitochondrial diseases.

A recent research study introduced two new pyridine cation derivatives, CAI and CAEI (Figure 13.4), which exhibit impressive fluorescence properties when excited by two-photon excitation (TPF). These compounds can selectively stain mitochondria in living SiHa cells within a short span of 30 minutes. Notably, CAI, emitting a green light,

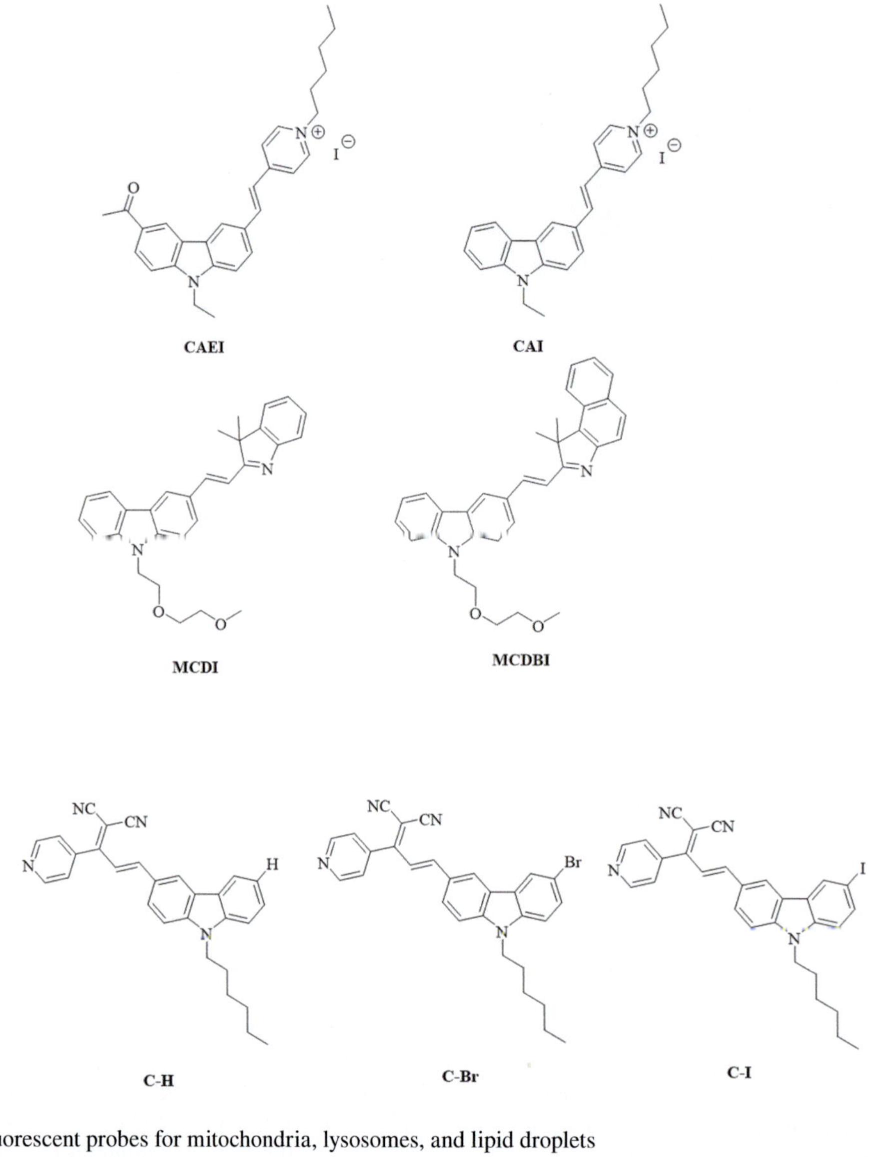

FIGURE 13.4 Fluorescent probes for mitochondria, lysosomes, and lipid droplets

displayed a significantly larger two-photon excitation fluorescence action absorption cross-section (δ * Φ) of 328 g at 860 nm compared to the commercial MitoTracker Green (MTG) with a maximum δ * Φ value of 2.18 g at 850 nm. In the field of two-photon microscopy (TPM), the parameter δ * Φ is crucial for obtaining high-quality microscopic images in living cells. Additionally, colocalization experiments conducted in SiHa cells demonstrated a strong association (co-localization coefficient of 0.95) between CAI and conventional MitoTracker Red (MTR), confirming CAI's specific effectiveness in staining mitochondria. Both CAI and CAEI served as excellent biosensors, possessing advantageous features such as significant δ * Φ and Stokes shifts, good membrane permeability, extended retention time, high resistance to photobleaching, and seamless compatibility with various other biosensors such as Hoechst 33342 and D 307. These qualities position them as highly effective TPEF probes for mitochondria. Further research into their localization revealed a dependence on the mitochondrial membrane potential in living cells, opening up new avenues for exploring mitochondrial functions in biological contexts (Miao et al., 2014).

13.2.7 Fluorescent Probe for Lysosome

Maintaining the right pH levels inside cells is crucial for many biological processes, including cell growth, programmed cell death (apoptosis), resistance to multiple drugs, and movement of ions. Different parts within cells have varying pH values; for example, lysosomes and endosomes have slightly acidic interiors, while active mitochondria and the cytosol have slightly basic pH levels. Lysosomes are particularly vital organelles; they contain enzymes that break down various large molecules. Any abnormal changes in lysosomal pH can disrupt the functioning of enzymes and proteins inside, as they can only work properly in an acidic environment. Irregular lysosomal pH levels are linked to diseases like cancer, AD, and lysosome storage disorders. Monitoring lysosomal pH in real time is crucial for diagnosing these diseases and gaining a deeper understanding of the natural processes within lysosomes.

Two innovative fluorescent pH sensors, MCDBI and MCDI (Figure 13.4), both based on carbazole, have been created specifically to target lysosomes. These probes undergo a distinct color change from yellow to red and orange in response to a shift in pH from alkaline to acidic conditions, with respective pKa values of 4.26 and 4.51. Notably, MCDBI and MCDI offer significant advantages, including a substantial Stokes shift, high sensitivity, and the ability to prevent interference from natural cellular fluorescence. Additionally, they feature rapid response times, strong selectivity for protons, exceptional photostability, excellent reversibility, and low toxicity. These qualities make the probes well-suited for imaging intracellular pH levels. Furthermore, confocal fluorescent microscopy successfully captured the distribution of MCDBI and MCDI within cells, along with their colocalization with LysoTracker Green DND-26 in live HepG2 cells. This accomplishment underscores the probes' effective cell membrane permeability and their potential for practical applications in studying physiological and pathological processes associated with lysosomes (Ge et al., 2019).

13.2.8 Fluorescent Probe for Lipid Droplets

Lipid droplets (LDs) are subcellular compartments that store different neutral lipids like trioctanoic acid and cholesterol esters. These LDs are surrounded by a single layer of phospholipids and play a key role in regulating cellular energy balance. Numerous studies have indicated that abnormal lipid storage can lead to various metabolic disorders such as obesity, fatty liver, cardiovascular diseases, and diabetes. Hence, it is essential to develop reliable methods to monitor LDs for biomedical research purposes.

A series of fluorescent probes (Figure 13.4) with carbazole units were created with different carbon-hydrogen (C–H), carbon-bromine (C–Br), and carbon-iodine (C–I) bonds. The research highlighted the iodine-modified carbazole derivative, C–I, which displayed exceptional targeting capabilities for lipid droplets (LDs) due to its lipophilicity (log P = 1.62). Significantly, C–I was effectively utilized in both stimulated emission depletion (STED) and two-photon fluorescence imaging, providing high-resolution visualization of lipid droplets. This efficacy was attributed to its large Stokes shift (~120 nm). Additionally, exploiting the heavy atom effect, C–I efficiently generated singlet oxygen ($^1O_2$) with a quantum yield of 0.59, leading to cancer cell apoptosis under near-infrared (NIR) laser irradiation at 880 nm. Furthermore, in vitro simulations illustrated the transfer of reactive oxygen species from LDs to the cytoplasmic matrix, laying the groundwork for employing LD-targeting probes in photodynamic therapy (PDT) research. In summary, they introduced a promising method for two-photon imaging of lipid droplets and triggering two-photon-excited reactive oxygen species generation (Du et al., 2022).

13.3 METAL ION DETECTION

The essential trace element ions, which include cations and anions, are important for maintaining the osmotic balance. But by upsetting the balance in cellular processes, too many ions can cause cell death and neurodegenerative illnesses. The monitoring of ions in biological systems holds paramount significance in the realm of biomedical research.

13.3.1 Fluorescent Detection of Fe^{3+} and Cr^{3+}

An ongoing and captivating field of investigation within chemistry and environmental science revolves around the development of chemosensors designed for the detection of transition and heavy metal ions. There have been several reports on chemosensors for metal ions including Zn^{2+}, Hg^{2+}, Cu^{2+}, and Ag^+. It's worth noting that the chemosensors described, which exhibit selectivity for Fe^{3+} and

Cr^{3+} ions, are quite rare and not commonly encountered in scientific research. One of the most critical nutrients for maintaining a person's body's regular physiological functions is iron, which is crucial for numerous cellular functions such as oxygen transport, DNA and RNA synthesis, and energy production. Iron deficiency can potentially lead to various health conditions, such as anemia, liver damage, Parkinson's disease, a decrease in intelligence quotient (IQ), diabetes, and even cancer. However, consuming more iron than what the body needs may lead to harmful illnesses including thalassemia and Friedreich's ataxia. A healthy diet for people and animals must have 50–200 mg of chromium daily. A chromium deficit may raise the risk factors for diabetes and cardiovascular disease, but an excess of the metal can have genotoxic consequences. Furthermore, one of the most well-known environmental contaminants is chromium. Therefore, it is crucial to prioritize the development of specialized chemosensors for metal ions in environmental and biological systems (Saluja et al., 2012).

Zhu et al. (2015) have documented the synthesis of two carbazole-based Schiff bases, designated as L1 and L2 (Figure 13.5). Notably, L1 functions as a colorimetric chemosensor for detecting Fe^{3+} ions and also serves as a fluorescent chemosensor, detecting both Fe^{3+} and Cr^{3+} ions. The fluorescence method was employed to confirm the binding stoichiometry as 2:1 (L1–Fe^{3+}) and (L1–Cr^{3+}). The increase in fluorescence intensity observed upon the binding of Fe^{3+} and Cr^{3+} ions can be attributed to excited-state intramolecular proton transfer (ESIPT) and the suppression of C=N isomerization. However, the lack of a phenolic hydroxyl (-OH) group in the ortho position of L2 leads to limited selectivity in recognizing the tested metal ions. This suggests that the oxygen atom within the hydroxyl group (-OH) plays a crucial role as the primary binding site for compounds containing metal ions (Figure 13.5). The results suggest that improving the coordination binding site can significantly increase the selectivity of chemosensors. This study introduces a novel approach to the creation of chemosensor materials.

13.3.2 Fluorescent Probe for Al^{3+} Detection

The consumption of aluminum ions (Al^{3+}) is not essential for biological processes and should be carefully regulated. As stated in a report by the World Health Organization (WHO), the typical daily intake of Al^{3+} for the human body is around 30 mg. Consuming excessive amounts of Al^{3+} could result in various health conditions such as headaches, AD (Walton, 2006), and a specific type of anemia called microcytic hypochromic anemia. As a result, there is a continuous effort to enhance methods for detecting Al^{3+} due to its significant role in biological contexts (Diao et al., 2016).

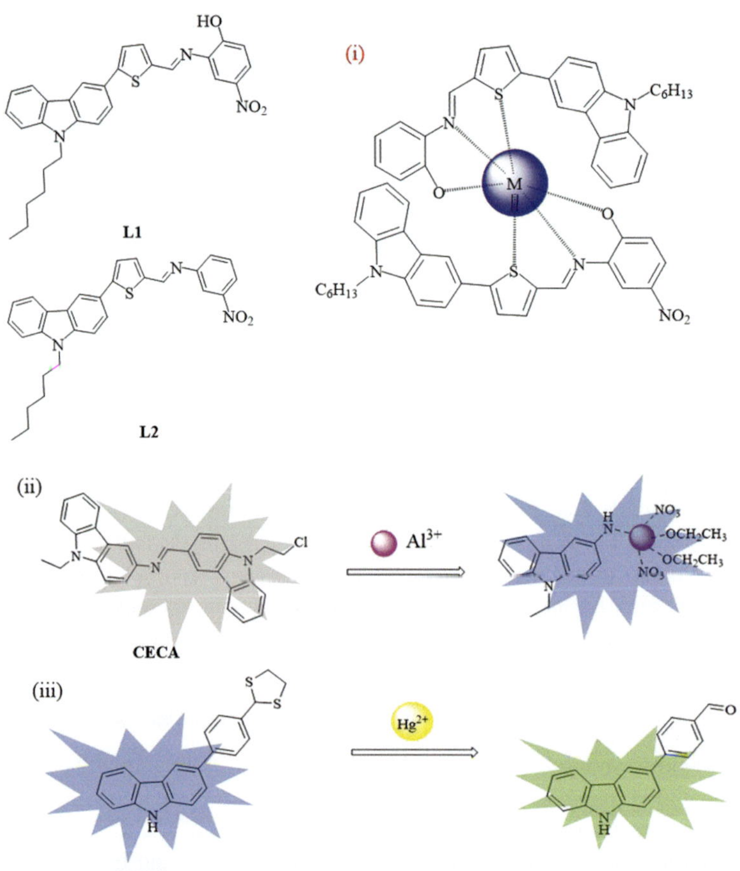

FIGURE 13.5 Probes for metal ion detection. (i) Binding mode of L1 and M (M stands for Fe^{3+} and Cr^{3+}). (ii) Schematic illustration of the reaction between CECA and Al^{3+} ions. (iii) Schematic illustration of the reaction between DTCB and Hg^{2+} ions.

Gao et al. (2019) reported a novel probe, known as CECA (Figure 13.5), which utilizes colorimetric, ultraviolet absorption, and fluorescence signals for detection purposes. The CECA method demonstrated the ability to qualitatively and quantitatively detect Al^{3+} ions with exceptional performance, characterized by high sensitivity and selectivity. The investigation focused on elucidating the mechanism by which CECA decomposes to produce carbazole aldehyde and carbazole amine complex in the presence of Al^{3+} ions. This decomposition process effectively cancels out the photoinduced electron transfer (PET) process, resulting in a noticeable increase in fluorescence intensity. The investigation employed various analytical techniques, including Job's plot analysis, ^{1}H NMR spectroscopy, electron ionization mass spectrometry (EI-MS), and fluorescence measurements. Furthermore, the ability of CECA to accurately detect the presence of Al^{3+} in HeLa cell was successfully shown by CECA, thus highlighting the potential use of CECA in the study of the effects of Al^{3+} in biological systems. Additionally, it is important to highlight that these three techniques for Al^{3+} detection, using a common probe, not only address the needs of various populations but also enable reciprocal modification of test results to improve accuracy. These research approaches provide a systematic strategy for creating several signal probes to identify particular targets (Gao et al., 2019).

13.3.3 Fluorescent Probe for Hg^{2+} Ions

Using the well-known deprotection process of thioacetal started by Hg^{2+}, scientists have created a new fluorescent detector called DTCB (3-(4-(1,3-dithiolan-2-yl)phenyl)-9 H-carbazole), which demonstrated a robust fluorescence response to Hg^{2+} when utilized in a mixed THF-H_2O solution (with a water fraction of 5%) (Figure 13.5). The introduction of Hg^{2+} significantly enhances the probe's fluorescence intensity, resulting in a shift from dark blue to green fluorescence. Over a wide concentration range of Hg^{2+} (ranging from 0 to 6×10^{-5} mol·L^{-1}), the fluorescence intensity at 480 nm exhibits a strong linear relationship with the concentration of Hg^{2+} ($R_2 = 0.993$). The detection limit of the probe for Hg^{2+} is impressively low, at less than 2.05×10^{-7} mol·L^{-1}. This indicates the probe's high sensitivity for detecting Hg^{2+} ions. The DTCB probe demonstrates excellent selectivity and resistance to interference from various metal ions and anions. Many other metal ions and anions have minimal to no effect on their detection performance, highlighting their specificity and reliability in distinguishing the presence of Hg^{2+} ions from other substances. The sensing mechanism of probe DTCB has been validated through theoretical calculations based on the density functional theory (DFT) and time-dependent DFT (TD-DFT). These calculations have revealed that the significant fluorescence change observed in DTCB is primarily attributed to intramolecular charge transfer (ICT). Additionally, the DTCB probe has been successfully employed for the detection of Hg^{2+} in real water samples. It is noteworthy that when Hg^{2+} is introduced, the DTCB-based test strip exhibits strong fluorescence, and this is the first instance where such a test strip has revealed significant fluorescence from solid-state 3-(4-formylphenyl)-9H-carbazole (FCB) upon the addition of Hg^{2+}. This demonstrates the practical utility and versatility of the DTCB probe for real-world applications in detecting Hg^{2+} ions (Pan et al., 2021).

13.4 DETECTION OF REACTIVE OXYGEN SPECIES (ROS)

13.4.1 Imaging of Hypochlorite in Living Cells

Reactive oxygen species (ROS) are essential intermediates involved in various biological and pathological processes, originating from oxygen during electron transfer events. They serve to prevent uncontrolled chain reactions caused by foreign organisms. Excessive ROS accumulation leads to heightened oxidative stress, contributing to conditions like Alzheimer's disease, neuronal degeneration, cancer, kidney disorders, and cardiovascular ailments. In the immune system, myeloperoxidase enables the reaction between chloride ions (Cl^-) and hydrogen peroxide (H_2O_2), forming hypochlorite (ClO^-), a vital active oxygen species and disinfectant in everyday life. There is a need to develop fluorescent probes for detecting ClO^- (Tang et al., 2018).

Wang et al. (2020) developed a new fluorescent sensor named FCZ (Figure 13.6) to detect ClO^- in aqueous solutions at physiological pH levels. The sensing mechanism underlying ClO^- detection and the luminescent attributes of the sensor were scrutinized through a combination of methods including ESI-MS, TLC, NMR, and DFT/TDDFT calculations. The outcomes of fluorescence titration revealed that the sensor exhibited a robust linear correlation ($r = 0.9981$) and demonstrated a low limit of detection (LOD) of 0.056 mmol/L (three times the standard deviation over the slope). Furthermore, the sensor was effectively applied for fluorescence imaging of both exogenously introduced and naturally occurring ClO^- in RAW264.7 cells. Notably, the sensor showcased minimal cytotoxicity and superior cell permeability, enabling its successful entry into cells. In a zebra fish experiment, the FCZ sensor exhibited favorable biocompatibility, thereby suggesting its potential applicability as a tool for real-time ClO^- monitoring within living organisms.

13.4.2 Detection of Hydrogen Peroxide in Living Cells

Aerobic organisms produce ROS using molecular oxygen. Even though ROS are important species for many physiological processes, their excessive production would have negative effects on living systems. Therefore, it is very important to create sensitive tools that can track specific ROS in organisms (Lambeth, 2007). The research team led by Kai Zhang developed and produced a novel probe with two-photon fluorescence capabilities, designed specifically

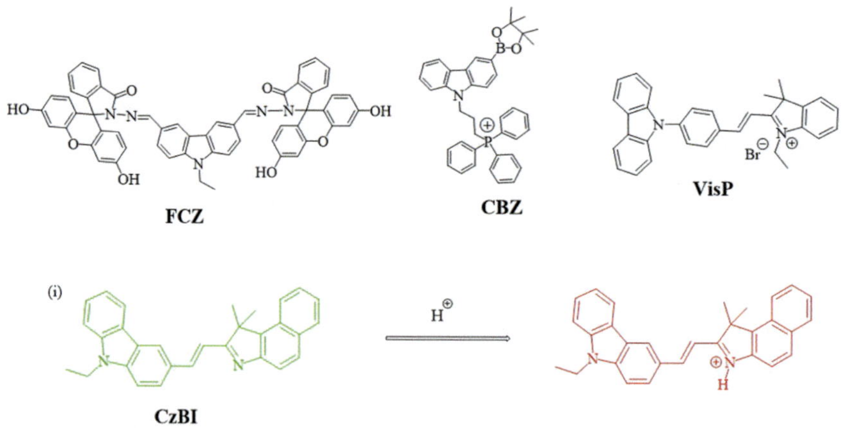

FIGURE 13.6 ROS probes and viscosity probes. (i) Schematic illustration of intracellular acidity sensing.

for the detection of mitochondrial H_2O_2. The probe is made up of a triphenylphosphonium moiety that acts as a localizer that targets mitochondria, a carbazole derivative that serves as a fluorophore, and a boronate ester functionality that serves as a sensing site. The observed characteristics of CBZ (Figure 13.6) include rapid response time and strong specificity in comparison to other ROS, reactive nitrogen species (RNS), and associated biomolecules. Additionally, it demonstrates a low detection limit of 1.96 μM. The results of colocalization experiments provide clear evidence that CBZ-H_2O_2 exhibits selective localization and accumulation within mitochondria. CBZ-H_2O_2 exhibits sensitivity toward both exogenous and endogenous H_2O_2. The utilization of two-photon fluorescence microscopy enables the detection of H_2O_2 within living cells and tissues at a considerable depth of 105 μm, while maintaining exceptional imaging quality over an extended duration. This experimental tool will facilitate a comprehensive examination of the various roles and functions of H_2O_2 within a sophisticated biological system (Zhang et al., 2016).

13.5 INTRACELLULAR ENVIRONMENT IMAGING

13.5.1 In Situ Tracking of Viscosity

The transmission of signals and the interactions between biomolecules depend heavily on the intracellular microenvironment. Viscosity is a crucial element in regulating intracellular transport, intermolecular interaction, and membrane fluidity (Kuimova et al., 2009). Viscosity must be normal and stable for biological interactions and biochemical processes to occur. Atherosclerosis, hypertension, diabetes, inflammation, Alzheimer's disease, and cell malignancy are only a few of the aberrant physiological processes and disorders that are brought on when the viscosity equilibrium of the microenvironment is disturbed (Liu et al., 2022). It is crucial to reveal variations in physiological viscosity to fully comprehend the biological impacts of viscosity performance as well as to monitor the emergence and progression of illness (Rynearson, 2017). For this reason, identifying viscosity variation is important for biological study and medical diagnosis.

To detect viscosity in biological systems, Liu et al. (2022) synthesized the fluorescent probe, VisP (Figure 13.6). VisP specifically detects viscosity because of its traditional molecular rotor design, which exhibits high sensitivity to the viscosity of the environment. The fluorescent probe VisP was able to detect the viscosity changes induced by monensin and nystatin. Additionally, the variation in viscosity in inflamed cells treated with dexamethasone sodium phosphate has been observed. The investigation of diseases related to viscosity alteration will be greatly aided by VisP.

13.5.2 Fluorescent Probe for Imaging Extreme Acidity

Cellular metabolic processes, such as apoptosis, proliferation, adhesion, cell cycle regulation, and endocytosis, are intricately regulated by intracellular pH (pHi) (Han & Burgess, 2010). The pHi of the cells can vary significantly, ranging from highly acidic to basic, depending on the specific bacterial species or subcellular compartments within eukaryotic cells (Krulwich et al., 2011). In diseases related to the heart and nervous system, such as cancer and Alzheimer's disease, abnormal pHi levels are frequently associated with cellular dysfunction and may play a role in the development and progression of these conditions (Chesler, 2003). Therefore, it is extremely important to monitor pH variations inside living cells to investigate cellular activities and comprehend physiological and pathological processes.

Fan et al. (2015) reported a ratiometric emission pH probe, CzBI (Figure 13.6). The ICT effect of this fluorophore is stronger when the N atom of benzoindole is protonated. This causes a red change in both the absorption (105 nm) and emission (82 nm) spectra. With a pKa of 2.44, CzBI has a very high emission ratio (F605 nm/F523 nm) when the pH drops from 7.00 to 0.50. Interestingly, the probe reacts quickly and smoothly to small changes in pH within the very acidic range of 1.05–3.60. CzBI also exhibits several favorable characteristics, including a substantial Stokes shift of 100 nm under highly acidic conditions. Additionally, it demonstrates high sensitivity, excellent reversibility, strong selectivity, impressive photostability, and minimal cytotoxicity, making it an excellent candidate for measuring pHi. The ability of CzBI to effectively function within live HeLa cells underscores its capability to permeate the cell membrane. This enables the visualization of alterations in acidic pHi levels with minimal interference from autofluorescence. Crucially, CzBI represents a novel pH fluorescent probe with a ratiometric emission feature, making it valuable for detecting whether *E. coli* cells possess an extremely acidic intracellular environment. The CzBI probe can be used in the health and biological fields to track changes in pH in real time, especially in situations where the pH is very low.

13.6 ORGANIC LIGHT-EMITTING DIODES

Extensive research efforts have been directed toward the exploration of wide-bandgap (WBG) organic semiconducting materials. This is because of their broad range of potential applications, which include laser diodes, detectors, and near-ultraviolet light-emitting devices (NUV-LEDs) (Nakamura, 1998). In particular, the organic semiconductive materials developed by the World Bank Group have the potential to function as host materials for a range of dopants in doped organic light-emitting diodes (OLEDs). To counteract the adverse impacts of triplet-triplet annihilation and collision-induced exciton quenching, a common approach is to introduce phosphors and thermally activated delayed fluorescence (TADF) emitters into suitable host materials (Hu et al., 2012; Meng et al., 2019). Theoretically, the doped OLEDs have the potential to achieve an internal quantum efficiency of almost 100% by effectively utilizing both singlet and triplet excitons concurrently (Holmes et al., 2003). The host materials play a significant role in determining the performance and stability of OLED devices, as they constitute the majority of emissive layers. Consequently, it is of utmost importance to carefully design appropriate host molecules to achieve efficient OLEDs.

Qiu et al. (2020) have documented the creation of two innovative categories of type D–π –A host materials (Figure 13.7). These materials are formed by bonding an electron acceptor formate to an electron donor carbazole unit, joined together by a connecting component called a styryl group. The incorporation of the formate moiety leads to

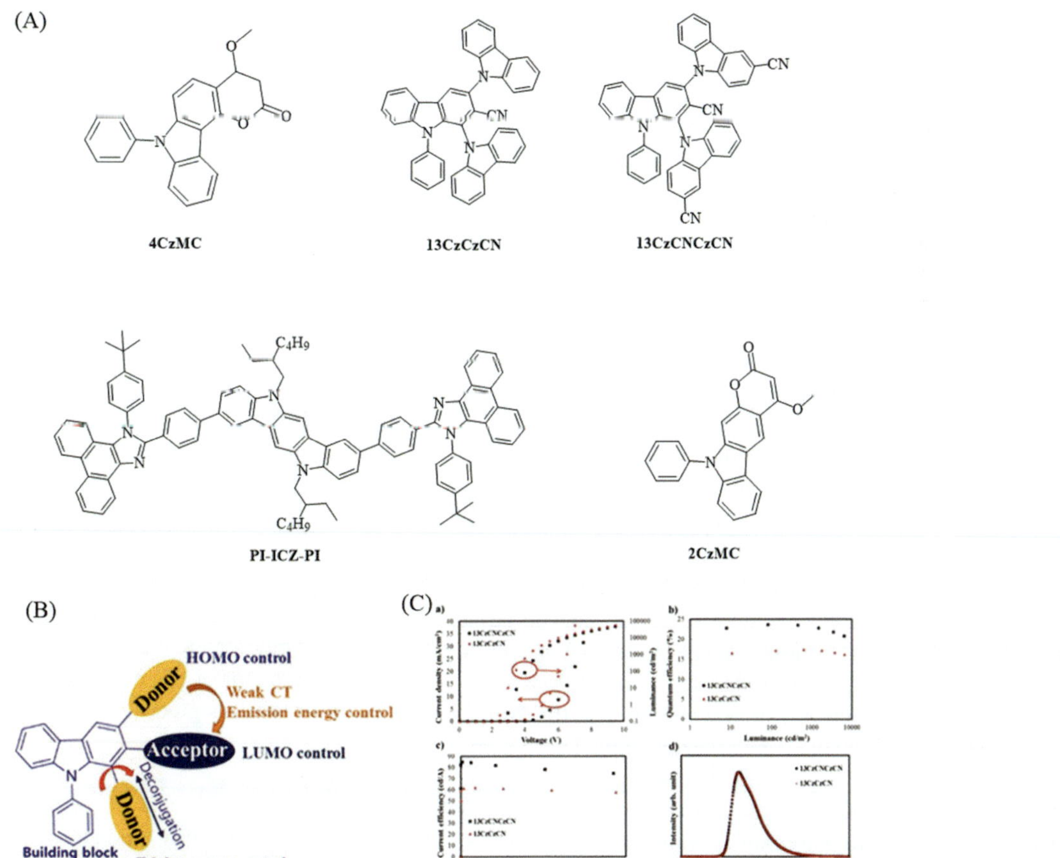

FIGURE 13.7 (A) OLED probes. (B) Molecular design concept of C1-, C2-, and C3-modified carbazole-based hosts. (C) (a) *I–V–L*, (b) EQE versus *L*, (c) CE versus L, and (d) EL plots of the green devices. Reprinted (adapted) with permission from Konidena et al., American Chemical Society Copyright 2020.

the emergence of intermolecular hydrogen bonding interactions, resulting in an elevated glass transition temperature (Tg) for these substances. The presence of a shortened conjugation pathway and a relatively weak intramolecular charge transfer (ICT) state imparts high electron transfer (ET) energies of 2.83 eV for 2CzMC and 2.90 eV for 4CzMC. These ET values are pivotal characteristics for these materials to serve as versatile host materials in phosphorescent organic light-emitting diodes (PhOLEDs) spanning blue, green, and red emissions. Due to the higher ET values, blue devices found on 4CzMC exhibit more efficient performance compared to those based on 2CzMC with maximum external quantum efficiency, (EQEmax) reaching 18.9% for 4CzMC and 12% for 2CzMC. This study convincingly illustrates the potential of formate-carbazole-based derivatives as a promising foundation for the design of host materials featuring high ET values, attributable to their moderately balanced electron-accepting and electron-donating properties.

Konidena et al. (2020) developed two novel bipolar host materials, 13CzCzCN and 13CzCNCzCN (Figure 13.7). These materials were synthesized by modifying a carbazole unit with cyanocarbazole or carbazole at specific positions (C1/C3 and C2) and incorporating a cyano acceptor group. In comparison to 13CzCzCN, the compound 13CzCNCzCN demonstrated elevated triplet energy, enhanced thermal stability, and a well-balanced charge transport performance. Certainly, the green device hosted by 13CzCNCzCN exhibited exceptional performance, achieving an EQEmax of 23.6% and a CE (current efficiency) of 84.9 cdA–1. Notably, it maintained an impressively low-efficiency roll-off of only 2.1% even at high brightness levels (1,000 cd/m^2). This outstanding performance can be attributed to the high triplet energy and well-balanced charge transport characteristics of 13CzCNCzCN. They believed that these findings provide valuable insights into the potential of the carbazole building block for creating innovative molecular designs aimed at developing efficient bipolar host materials for OLEDs.

Yang et al. (2021) reported the synthesis of four new derivatives of indolo[3,2-*b*]carbazole. These compounds incorporate imidazole-based components and are named Py-ICZ-Py, PI-ICZ-PI, BI-ICZ-BI, and IM-ICZ-IM. These derivatives were specifically designed for application in blue OLEDs. The design strategy involved incorporating two imidazole-based groups, namely pyreno[4,5-*d*] imidazole, phenanthrol [9,10-*d*]imidazole, benzo[*d*]imidazole, and imidazole, as electron-withdrawing units at the 2- and 8-positions of the indolo[3,2-*b*]carbazole core. These compounds emitted blue light and demonstrated significant thermal stability with a decomposition temperature (Td) surpassing 460°C. Remarkably, among these derivatives, PI-ICZ-PI (Figure 13.7) exhibited superior photoluminescence and electroluminescence (EL) properties. Devices incorporating PI-ICZ-PI as a dopant demonstrated exceptional EL performance, achieving a peak luminance of 3,973 cd/m^2, a maximum current efficiency of 3.67 cd/A, and a peak external quantum efficiency of 2.64%.

13.7 FLUORESCENT AND COLORIMETRIC CHEMOSENSORS

13.7.1 Sensor Based on Carbazole for Glucose

The existence of carbohydrates is essential to life. Although monosaccharides are necessary for the operation of living systems, they can also lead to serious illnesses. Glucose serves as the primary and most abundant source of energy in tissues and cells. However, diabetes mellitus can lead to a range of complications, including heart disease, kidney failure, and blindness (Qasemi & Ghaemy, 2020). This condition is a significant global health concern arising from impaired glucose regulation. The establishment of a proficient glucose detection system is a valuable tool for the effective management of diabetes. D-glucose monitoring techniques have been used, such as enzyme-based and electrochemical sensors. To detect carbohydrates, synthetic chemical sensors, on the other hand, have proven to be much more reliable and popular. Since boronic acids can generate cyclic boronic esters with carbohydrate cis-1,2- or 1,3-diols that are reversible, scientists have recently become very interested in making boronic acid fluorescent sensors (Xu et al., 2020).

Oloub et al. (2022) reported water-soluble fluorescent sensor by employing carbazole pyridinium boronic acid (CPBA) as its foundational compound (Figure 13.8). Extensive fluorescence investigations conducted on the fabricated CPBA chemosensor unveiled a distinct and specific fluorescent response to glucose when compared to various other monosaccharides. The results reveal that CPBA serves as a pH-sensitive sensor, exhibiting distinct behavior in the pH range of 4–10, with notable differences in the absence and presence of glucose. Through an amalgamation of pH assessments, dynamic light scattering analyses, Job's plot evaluations, UV-visible spectroscopy, and fluorescence titration studies, it was established that CPBA's preference for glucose emanates from the phenomenon of aggregation-induced emission. The fluorescence emission intensity of CPBA undergoes a remarkable change of more than 2,100 fold when glucose is introduced, whereas the change is only two fold for fructose. The derived binding constant value of CPBA for glucose ($K = 2.3 \times 10^6\ M^{-1}$) surpasses that for fructose by a factor of 85, underscoring the strong affinity of the sensor towards glucose.

13.7.2 Acid and Base Sensing

Researchers have recently paid close attention to the stimulus response to fluorescent materials due to their potential applications in the analysis of trace contaminants in the environment, biosensing, and the detection of toxins in food, among other fields (Moon et al., 2015). Numerous

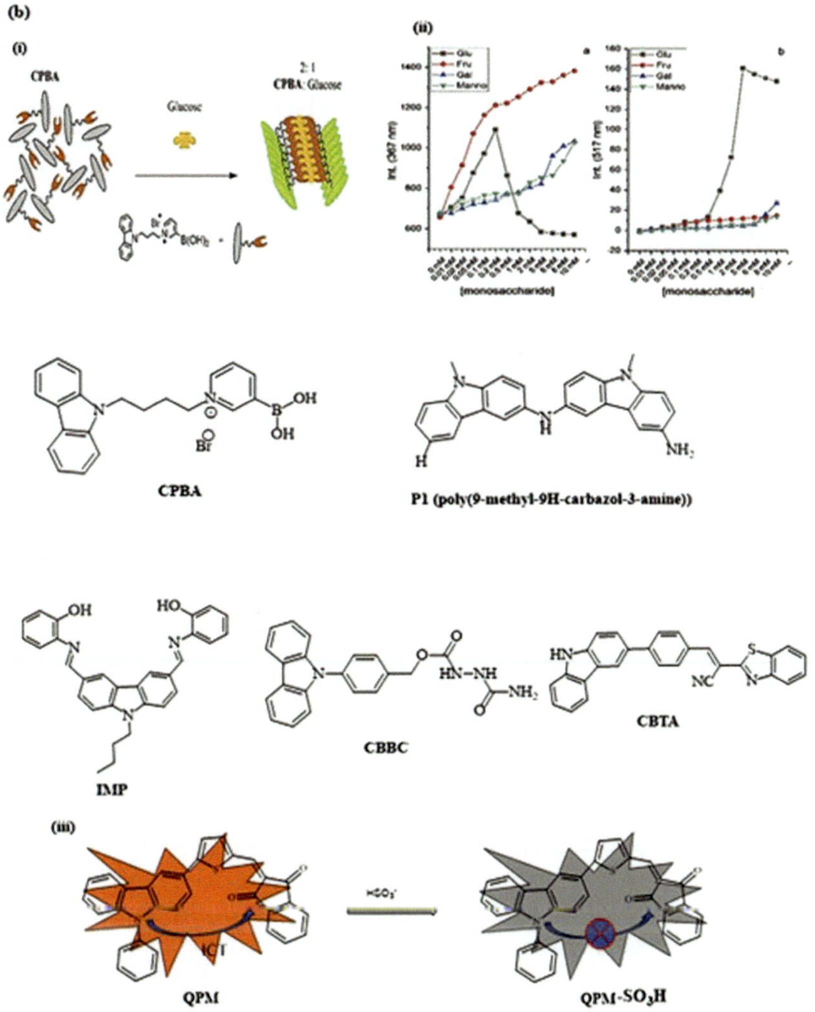

FIGURE 13.8 (i) Schematic representation of **CPBA** interaction with glucose. (ii) The fluorescent emission of CPBA at wavelengths 367 nm and 517 nm was measured in the presence of varying concentrations (ranging from 0 to 10 mM) of glucose, fructose, galactose, and mannose, with CPBA concentration at 0.05 mM. These measurements were conducted in a carbonate buffer with a pH of 9.5, and the excitation wavelength (λ_{ex}) used was 260 nm. The measurements were taken after 25 minutes. (iii) The proposed sensing mechanism of **QPM** toward HSO_3^-. Reprinted (adapted) with permission from Oloub et al., Royal Society of Chemistry Copyright 2022.

studies have been conducted on the variables that affect fluorescence, including temperature, scattered light, pH, and solvent types (Yáñez-Sedeño et al., 2017).

Qian et al. (2019) reported two variations of polyaniline to act as chemical sensors for identifying acids and amines. The interaction involving the π-π* transition of the inflexible fluorene and carbazole components within the structure of these connected arrangements is advantageous in augmenting the sensing capabilities. This distinct arrangement empowers these conductive polymers to exhibit rapid and highly perceptive responsiveness. Moreover, it has been observed that P1 (poly(9-methyl-9H-carbazol-3-amine)) showcases more proficient sensing characteristics compared to P2 (poly(9,9-dihexyl-9H-fluoren-2-amine)) through a comprehensive analysis of their structures, electrochemical attributes, and fluorescence properties. Notably, P1 (Figure 13.8) possesses the remarkable capability of fluorescence-based detection for both acids and amines. This property extends to both liquid and gas-phase fluorescence detection applications. The detection threshold of P1 is as low as 3.3 μM. Exceptional fluorescence sensing qualities of P1 are anticipated that these responsive fluorescent derivatives of polyaniline hold promise in the advancement of highly sensitive chemical sensors.

13.7.3 Caffeine Quantification

Caffeine, an alkaloid drug commonly consumed worldwide, can be found in a variety of sources including coffee, black tea, guarana paste, cola nuts, and cocoa beans. It is also a component of cola beverages and energy drinks. While caffeine is generally considered safe for adults when consumed in moderation, there are concerns about its potential negative effects on young children and pregnant women. These concerns are primarily related to the potential risks it may pose to fetal health during pregnancy. Given the widespread and varied use of caffeine, there is a need for advanced analytical tools to ensure product safety. Various established chromatographic methods are applied for quantifying caffeine, including HPTLC (high-performance thin-layer chromatography), HPLC-UV (high-performance liquid chromatography with ultraviolet detection), LC (liquid chromatography), MIPs (molecularly imprinted polymers), and CE (capillary electrophoresis). Nevertheless, it's worth noting that these methods can often be time-consuming for analysis. To address this, the field of chemosensors has emerged as a dynamic area within supramolecular chemistry. Various research teams have successfully made synthetic receptors specifically designed for caffeine. Although hydrogen-bond receptors have primarily been studied in organic solvents, the goal of achieving significant selectivity and high affinity for caffeine in aqueous environments is of particular interest. This is important for potential applications in the detection of caffeine in water-based solutions.

Mahapatra et al. (2012) documented the use of iminophenol (IMP) (Figure 13.8), which is based on carbazole, as a highly effective fluorescent probe for the detection of xanthine alkaloids, specifically caffeine. IMP demonstrates a remarkable degree of sensitivity and specificity in its ability to enhance fluorescence specifically for caffeine, as compared to other alkaloids related to xanthine, when present in an aqueous buffer solution at pH 7.2. This particular fluorescent probe is notable for its ability to selectively detect caffeine in water through a "turn-ON" mechanism, making it one of the few probes with this capability. Indeed, the successful expression of IMP in cells serves as compelling evidence of its utility as a probe in biological systems.

13.7.4 Nitrofuran Quantification in Foodstuffs

Nitrofurans (NFs) are a class of compounds characterized by a molecular structure containing a 5-nitrofuran ring group. Examples of NFs include furazolidone (FZD), nitrofurazone (NFZ), nitrofurantoin (NFT), and furaltadone (FTD). These compounds possess broad-spectrum antibacterial properties and have found extensive application in veterinary medicine and as feed additives for food-producing animals (Vass et al., 2008). The primary use is in the treatment of gastrointestinal infections. NFs are widely recognized as the most frequently employed antibacterial agents in the context of food production and preservation (Picó & Barceló, 2008). NFs undergo rapid metabolism within the human body, typically occurring within a few hours after ingestion. This metabolic process results in a substantial reduction in the concentration of NFs in the plasma (Leitner et al., 2001). Nevertheless, the residual metabolites have the potential to persist in the body for extended periods, potentially lasting for weeks or even months, due to their binding to proteins (Niessen, 1998). In addition, certain metabolite compounds derived from them have been scientifically demonstrated to exhibit mutagenic and carcinogenic properties (Conneely et al., 2003).

Yu et al. (2019) introduced a straightforward and highly responsive technique using high-performance liquid chromatography with fluorescence detection (HPLC-FLD). The method employed a pre-column derivatization strategy that utilized a new fluorescence-labeling reagent called CBBC (Figure 13.8). This approach was designed to quantify the remaining NFs metabolite compounds found in food products. The process of fluorescence labeling with CBBC for NFs metabolite compounds occurred rapidly, completing in just 5 minutes at room temperature. The resulting derivatives of NFs metabolites, labeled with CBBC, exhibited exceptional sensitivity to fluorescence, displaying peak excitation and emission wavelengths at 375 nm and 410 nm, respectively. The method was established as a straightforward, cost-effective, sensitive, and dependable means for analyzing NFs metabolite compounds within food products. Moreover, this highly sensitive and dependable approach is likely to find broader applications in quantifying NFs metabolite compounds across various sample types.

13.7.5 Detection of Sulfur and Hydrazine Derivatives

Hydrazine and sulfur dioxide (SO_2) compounds both pose risks to human health and the environment. Numerous studies are being conducted on the use of fluorescent probes to detect hydrazine and sulfur. Few fluorescent probes can detect both sulfur and hydrazine, whereas many can only detect a single kind of analyte. When compared to a single fluorescent analyte, which can only detect a single analyte, a multifunctional fluorescence probe has greater applications and detection efficiency and is also consistent with the idea of atomic economy.

Li et al. (2020) have successfully developed a novel fluorescent probe, CBTA (Figure 13.8), utilizing carbazole and benzothiazole as its constituent components. This bifunctional probe exhibits the capability to detect both SO_2 derivatives and hydrazine. The probe exhibits rapid changes in fluorescence when exposed to SO_2 or hydrazine, resulting in two distinct emission channels. This ratiometric detection method enables the simultaneous detection of SO_2 and hydrazine. Furthermore, the probes designed for SO_2 and hydrazine exhibit remarkably low detection limits, measuring as low as 5.6 nmol and 39 nmol, respectively. These probes demonstrate a high level of selectivity toward SO_2 derivatives and hydrazine in comparison to other analytes. The probe exhibits a low level of toxicity, making it suitable for detecting SO_2 derivatives and hydrazine in an in vitro setting. Additionally, it can serve as a visual aid for detecting SO_2 derivatives within cells.

13.7.6 Detection of Bisulfite

Bisulfite (HSO_3^-) is an important food additive with antibacterial, antioxidant, and bleaching capabilities that are widely utilized in pharmaceutical products and a variety of foods. HSO_3^- inhibits enzymatic processes to stop food spoiling brought on by oxidation and microbial growth. It also stops epinephrine medication from becoming inactive

(Duan et al., 2019). Research has shown that HSO_3^- is involved in a variety of physiological and pathological processes in living things, including controlling heart function, preventing atherosclerosis, preserving intracellular redox homeostasis, reducing blood pressure, and more. Researchers are especially interested in fluorescent probes for HSO_3^- because of their unique advantages such as high sensitivity, minimum harm to detection targets, real-time detection, and simple operation. Several fluorescent probes for HSO_3^- have been developed in recent years (Yang et al., 2019).

Shang et al. (2022) reported the development of a near-infrared-emitting fluorescence probe named QPM (Figure 13.8). This probe is based on carbazole and was synthesized to detect HSO_3^- in real food samples and live organisms. The underlying sensing mechanism was confirmed to be associated with the disruption of QPM's donor-π-acceptor (D-π-A) structure, which resulted from the Michael addition reaction between HSO_3^- and C=C bond bridge within the probe. This interaction interrupted the intramolecular charge transfer (ICT) effect from carbazole to indanedione in QPM. Notably, the QPM probe offers several advantages including a substantial Stokes shift (220 nm), a prolonged emission wavelength (710 nm), rapid response time (45 seconds), excellent sensitivity (58 nM), and specific responsiveness to HSO_3^-. Most significantly, QPM was effectively employed to detect HSO_3^- in living organisms as well as two authentic food samples. As a result, the QPM probe holds great potential as a robust instrument for detecting and monitoring HSO_3^- in the realms of food chemistry and biological systems.

13.8 CONCLUSION

The exploration of carbazole derivatives as fluorescent probes has revealed a wide array of applications in various fields. In the realm of biomolecular imaging, these probes have been utilized for visualizing nucleic acids and cellular components such as the nucleus, plasma membrane, endoplasmic reticulum, mitochondria, lysosomes, and lipid droplets. They have also played a crucial role in metal ion detection, enabling the fluorescence detection of ions such as Fe3+, Cr3+, A13+, and Hg2+. Furthermore, these probes have been instrumental in imaging the intracellular environment, detecting factors like viscosity and acidity. In the context of ROS, they have been employed for the fluorescent detection of hypochlorite and hydrogen peroxide in living cells. Additionally, carbazole derivatives have found application in organic light-emitting diodes (OLEDs), particularly in blue OLEDs. Lastly, they have been pivotal in the development of fluorescent and colorimetric chemosensors for substances like glucose, acids, bases, caffeine, nitrofuran, bisulfite, sulfur, and hydrazine derivatives. The versatile nature of carbazole derivatives as fluorescent probes highlights their significant impact on diverse scientific and technological domains, paving the way for innovative research and practical applications.

REFERENCES

Chang, C. C., Wu, J. Y., Chien, C. W., Wu, W. S., Liu, H., Kang, C. C., Yu, L. J., & Chang, T. C. (2003). A fluorescent carbazole derivative: High sensitivity for quadruplex DNA. *Analytical Chemistry*, *75*(22), 6177–6183.

Chesler, M. (2003). Regulation and modulation of pH in the brain. *Physiological Reviews, 83,* 1183 1221.

Conneely, A., Nugent, A., O'Keeffe, M., Mulder, P. P. J., Van Rhijn, J. A., Kovacsics, L., Fodor, A., McCracken, R. J., & Kennedy, D. G. (2003). Isolation of bound residues of nitrofuran drugs from tissue by solid-phase extraction with determination by liquid chromatography with UV and tandem mass spectrometric detection. *Analytica Chimica Acta*, *483*(1–2), 91–98.

Dahm, R. (2008). Discovering DNA: Friedrich miescher and the early years of nucleic acid research. *Human Genetics*, *122*(6), 565–581.

Diao, Q., Ma, P., Lv, L., Li, T., Sun, Y., Wang, X., & Song, D. (2016). A water-soluble and reversible fluorescent probe for A13+ and F- in living cells. *Sensors and Actuators, B: Chemical*, *229*, 138–144.

Du, W., Lu, X., Yuan, T., Sun, Z., Li, X., Li, S., Zhang, Q., Tian, X., Li, D., & Tian, Y. (2022). Halogen-modified carbazole derivatives for lipid droplet-specific bioimaging and two-photon photodynamic therapy. *Analyst*, *147*(1), 66–71.

Duan, C., Zhang, J. F., Hu, Y., Zeng, L., Su, D., & Bao, G. M. (2019). A distinctive near-infrared fluorescence turn-on probe for rapid, sensitive and chromogenic detection of sulfite in food. *Dyes and Pigments*, *162*, 459–465.

Dumat, B., Bordeau, G., Faurel-Paul, E., Mahuteau-Betzer, F., Saettel, N., Bombled, M., Metgé, G., Charra, F., Fiorini-Debuisschert, C., & Teulade-Fichou, M. P. (2011). N-phenyl-carbazole-based two-photon fluorescent probes: Strong sequence dependence of the duplex vs quadruplex selectivity. *Biochimie*, *93*(8), 1209–1218.

Fan, L., Gao, S. Q., Li, Z. B., Niu, W. F., Zhang, W. J., Shuang, S. M., & Dong, C. (2015). An indole-carbazole-based ratiometric emission pH fluorescent probe for imaging extreme acidity. *Sensors and Actuators, B: Chemical*, *221*, 1069–1076.

Feng, Y., Chi, S., Gao, G., Zhao, Y., & Wu, Y. (2019). Mono-, bi-quinolinium branches carbazole as groove binding mode of two-photon excited fluorescent probe interacting with DNA and nucleus imaging. *Sensors and Actuators, B: Chemical*, *282*, 16–26.

Gao, F., Cao, S., Sun, W., Long, S., Fan, J., & Peng, X. (2019). Development of a two-photon carbazole derivative probe for fluorescent visualization of G-quadruplex DNA in cells. *Dyes and Pigments*, *171*, 107749.

Gao, Z. Y., Zhang, C. J., Zhang, X., Xing, S., Yao, J. shui, Qiao, C. de, & Liu, Q. ze. (2019). A colorimetric, ultraviolet absorption and fluorescence three-signal probe based on bis-carbazole for Al 3+ detection and the application in cell imaging. *Journal of Molecular Structure*, *1188*, 14–22.

Ge, J., Zhang, K., Fan, L., Wang, X., Zhang, C., Dong, C., Wong, M. S., & Shuang, S. (2019). Novel long-wavelength emissive lysosome-targeting ratiometric fluorescent probes for imaging in live cells. *Analyst*, *144*(14), 4288–4294.

Graebe, C. (1872). Zur Abwehr. *Berichte Der Deutschen Chemischen Gesellschaft*, *5*(2), 968–969.

Gu, Y., Lin, D., Tang, Y., Fei, X., Wang, C., Zhang, B., & Zhou, J. (2018). A light-up probe targeting for Bcl-2 2345 G-quadruplex DNA with carbazole TO. *Spectrochimica Acta - Part A: Molecular and Biomolecular Spectroscopy*, *191*, 180–188.

Han, J., & Burgess, K. (2010). Fluorescent indicators for intracellular pH. *Chemical Reviews*, *110*(5), 2709–2728.

Hao, G., Sun, J., & Wei, C. (2018). Studies on interactions of carbazole derivatives with DNA, cell image, and cytotoxicity. *Bioorganic and Medicinal Chemistry*, *26*(1), 285–294.

Henderson, A., Wu, Y., Huang, Y. C., Chavez, E. A., Platt, J., Johnson, F. B., Brosh, R. M., Sen, D., & Lansdorp, P. M. (2014). Detection of G-quadruplex DNA in mammalian cells. *Nucleic Acids Research*, *42*(2), 860–869.

Holmes, R. J., Forrest, S. R., Tung, Y.-J., Kwong, R. C., Brown, J. J., Garon, S., & Thompson, M. E. (2003). Blue organic electrophosphorescence using exothermic host–guest energy transfer. *Applied Physics Letters*, *82*(15), 2422–2424.

Hu, D., Shen, F., Liu, H., Lu, P., Lv, Y., Liu, D., & Ma, Y. (2012). Separation of electrical and optical energy gaps for constructing bipolar organic wide bandgap materials. *Chemical Communications*, *48*(24), 3015.

Johnston, R. E., & Bose, H. R. (1972). Correlation of messenger RNA function with adenylate-rich segments in the genomes of single-stranded RNA viruses. *Proceedings of the National Academy of Sciences*, *69*(6), 1514–1516.

Kapuscinski, J., Skoczylas, B., & Nencki, M. (1978). Fluorescent complexes of DNA with DAPI 4',6-diamidine-2-phenyl indole.2HCl or DCI 4\6-dicarboxyamide-2-pbenyl indole. *Nucleic Acids Research, 5*(10), 3775–3800.

Knölker, H. J., & Reddy, K. R. (2002). Isolation and synthesis of biologically active carbazole alkaloids. *Chemical Reviews*, *102*(11), 4303–4427.

Kong, L., Yang, L., Zhang, G. bin, Chen, Q. yu, Wang, H., Gan, X. ping, Li, H., Zhou, H. ping, Yang, J. xiang, & Tian, Y. peng. (2018). High fluorescence of carbazole derivatives for bioimaging of endoplasmic reticulum and lysosome: The influence of molecular structure on optical properties and cell uptake. *Journal of Materials Science*, *53*(2), 921–936.

Konidena, R. K., Chung, W. J., & Lee, J. Y. (2020). C1-, C2-, and C3-modified carbazole derivatives as promising host materials for phosphorescent organic light-emitting diodes. *Organic Letters*, *22*(7), 2786–2790.

Krulwich, T. A., Sachs, G., & Padan, E. (2011). Molecular aspects of bacterial pH sensing and homeostasis. In *Nature Reviews Microbiology*, *9*(5), 330–343.

Kruusma, J., Benham, A. M., Williams, J. A. G., & Kataky, R. (2006). An introduction to thiol redox proteins in the endoplasmic reticulum and a review of current electrochemical methods of detection of thiols. *Analyst 131*(4), 459–473.

Kuimova, M. K., Botchway, S. W., Parker, A. W., Balaz, M., Collins, H. A., Anderson, H. L., Suhling, K., & Ogilby, P. R. (2009). Imaging intracellular viscosity of a single cell during photoinduced cell death. *Nature Chemistry*, *1*(1), 69–73.

Lambeth, J. D. (2007). Nox enzymes, ROS, and chronic disease: An example of antagonistic pleiotropy. In *Free Radical Biology and Medicine*, *43*(3), 332–347.

Leitner, A., Zöllner, P., & Lindner, W. (2001). Determination of the metabolites of nitrofuran antibiotics in animal tissue by high-performance liquid chromatography–tandem mass spectrometry. *Journal of Chromatography A*, *939*(1–2), 49–58.

Li, W., Hu, Y., Song, Y., Gu, Y., & Yang, W. (2020). New carbazole fluorescent sensor for ultrasensitive and ratiometric sensing of SO2 derivatives and hydrazine. *Journal of Photochemistry and Photobiology A: Chemistry*, *389*, 112269.

Lin, D., Fei, X., Gu, Y., Wang, C., Tang, Y., Li, R., & Zhou, J. (2015). A benzindole substituted carbazole cyanine dye: A novel targeting fluorescent probe for parallel c-myc G-quadruplexes. *Analyst*, *140*(16), 5772–5780.

Liu, S., Yue, Y., Hao, S., Huang, S., Lin, X., & Yin, J. (2022). A fluorogenic sensor for *in situ* tracking of viscosity in response to cellular inflammatory stress. *New Journal of Chemistry*, *46*(41), 20022–20027.

Liu, Y., Meng, F., He, L., Yu, X., & Lin, W. (2016). Fluorescence behavior of a unique two-photon fluorescent probe in aggregate and solution states and highly sensitive detection of RNA in water solution and living systems. *Chemical Communications*, *52*(57), 8838–8841.

Mahapatra, A. K., Roy, J., Sahoo, P., Mukhopadhyay, S. K., Mukhopadhyay, A. R., & Mandal, D. (2012). Fluorescence sensing of caffeine in aqueous solution with carbazole-based probe and imaging application in live cells. *Bioorganic and Medicinal Chemistry Letters*, *22*(17), 5379–5383.

Mallick, T., Karmakar, A., Bag, J., Sahu, S., Mishra, M., & Begum, N. A. (2020). Carbazole analog anchored fluorescent silica nanoparticle showing enhanced biocompatibility and selective sensing ability towards biomacromolecule. *Dyes and Pigments*, *173*, 107994.

Melton, D. A., Krieg, P. A., Rebagliati, M. R., Maniatis, T., Zinn, K., & Green, M. R. (1984). Efficient *in vitro* synthesis of biologically active RNA and RNA hybridization probes from plasmids containing a bacteriophage SP6 promoter. *Nucleic Acids Research*, *12*(18), 7035–7056.

Meng, G., Chen, X., Wang, X., Wang, N., Peng, T., & Wang, S. (2019). TADF emitters: Isomeric bright sky-blue TADF emitters based on bisacridine decorated DBNA: Impact of donor locations on luminescent and electroluminescent properties (advanced optical materials 11/2019). *Advanced Optical Materials*, *7*(11), 1970043.

Miao, F., Zhang, W., Sun, Y., Zhang, R., Liu, Y., Guo, F., Song, G., Tian, M., & Yu, X. (2014). Novel fluorescent probes for highly selective two-photon imaging of mitochondria in living cells. *Biosensors and Bioelectronics*, *55*, 423–429.

Moon, I. K., Yoon, S., Chun, K. Y., & Oh, J. (2015). Highly elastic and conductive n-doped monolithic graphene aerogels for multifunctional applications. *Advanced Functional Materials*, *25*(45), 6976–6984.

Nakamura, S. (1998). The roles of structural imperfections in InGaN-based blue light-emitting diodes and laser diodes. *Science*, *281*(5379), 956–961.

Niessen, W. M. A. (1998). Analysis of antibiotics by liquid chromatography–mass spectrometry. *Journal of Chromatography A*, *812*(1–2), 53–75.

Oloub, M., Hosseinzadeh, R., Tajbakhsh, M., & Mohadjerani, M. (2022). A new fluorescent boronic acid sensor based on carbazole for glucose sensing via aggregation-induced emission. *RSC Advances*, *12*(40), 26201–26205.

Pan, J., Ma, J., Liu, L., Li, D., Huo, Y., & Liu, H. (2021). A novel carbazole-based highly sensitive and selective turn-on fluorescent probe for mercury (II) ions in aqueous THF. *Journal of Photochemistry and Photobiology A: Chemistry*, *416*, 113322.

Picó, Y., & Barceló, D. (2008). The expanding role of LC-MS in analyzing metabolites and degradation products of food contaminants. *TrAC Trends in Analytical Chemistry*, *27*(10), 821–835.

Qasemi, S., & Ghaemy, M. (2020). Highly sensitive and strongly fluorescent gum tragacanth based superabsorbent hydrogel as a new biosensor for glucose optical detection. *Journal of Materials Chemistry C*, *8*(12), 4148–4156.

Qian, J., Zhang, Y., Liu, X., & Xia, J. (2019). Carbazole and fluorene polyaniline derivatives: Synthesis, properties and application as multiple stimuli-responsive fluorescent chemosensor. *Talanta*, *204*, 592–601.

Qiu, X., Ying, S., Yao, J., Zhou, J., Wang, C., Wang, B., Li, Y., Xu, Y., Jiang, Q., Zhao, R., Hu, D., Ma, D., & Ma, Y. (2020). Universal host materials based on carbazole-formate derivatives for blue, green and red phosphorescent organic light-emitting diodes. *Dyes and Pigments*, *174*, 108045.

Raibaut, L., Vasseur, W., Shimberg, G. D., Saint-Pierre, C., Ravanat, J. L., Michel, S. L. J., & Sénèque, O. (2017). Design of a synthetic luminescent probe from a biomolecule binding domain: Selective detection of AU-rich mRNA sequences. *Chemical Science*, *8*(2), 1658–1664.

Rhodes, D., & Lipps, H. J. (2015). Survey and summary G-quadruplexes and their regulatory roles in biology. *Nucleic Acids Research*, *43*(18), 8627–8637.

Rynearson, T. A. (2017). Navigating in a sea of genes. *Science*, 358(6367), 1129–1130.

Saluja, P., Sharma, H., Kaur, N., Singh, N., & Jang, D. O. (2012). Benzimidazole-based imine-linked chemosensor: Chromogenic sensor for Mg2+ and fluorescent sensor for Cr3+. *Tetrahedron*, *68*(10), 2289–2293.

Schlessinger, J., Axelrod, D., Koppel, D. E., &Elson, E. L. (1977). Lateral transport of a lipid probe and labeled proteins on a cell membrane. *Science*, *195*(4275), 307–309.

Shang, Z., Liu, J., Meng, Q., Jia, H., Gao, Y., Zhang, C., Zhang, R., & Zhang, Z. (2022). Carbazole-based near-infrared-emitting fluorescence probe for the detection of bisulfite in live animals and real food samples. *New Journal of Chemistry*, *46*(43), 20737–20744.

Swierczewska, M., Liu, G., Lee, S., & Chen, X. (2012). High-sensitivity nanosensors for biomarker detection. *Chemical Society Reviews*, *41*(7), 2641–2655.

Tang, X., Zhu, Z., Wang, Y., Han, J., Ni, L., Zhang, H., Li, J., & Yanli, M. (2018). A cyanobiphenyl based fluorescent probe for rapid and specific detection of hypochlorite and its bio-imaging applications. *Sensors and Actuators, B: Chemical*, *262*, 57–63.

Vass, M., Hruska, K., & Franek, M. (2008). Nitrofuran antibiotics: a review on the application, prohibition and residual analysis. *Veterinární Medicína*, *53*(9), 469–500.

Walton, J. R. (2006). Aluminum in hippocampal neurons from humans with Alzheimer's disease. *NeuroToxicology*, *27*(3), 385–394.

Wang, M. Q., Ren, G. Y., Zhao, S., Lian, G. C., Chen, T. T., Ci, Y., & Li, H. Y. (2018). Development of a carbazole-based fluorescence probe for G-quadruplex DNA: The importance of side-group effect on binding specificity. *Spectrochimica Acta - Part A: Molecular and Biomolecular Spectroscopy*, *199*, 441–447.

Wang, N., Xu, W., Song, D., & Ma, P. (2020). A fluorescein-carbazole-based fluorescent probe for imaging of endogenous hypochlorite in living cells and zebrafish. *Spectrochimica Acta - Part A: Molecular and Biomolecular Spectroscopy*, *227*, 117692.

Xu, S., Sedgwick, A. C., Elfeky, S. A., Chen, W., Jones, A. S., Williams, G. T., Jenkins, A. T. A., Bull, S. D., Fossey, J. S., & James, T. D. (2020). A boronic acid-based fluorescent hydrogel for monosaccharide detection. *Frontiers of Chemical Science and Engineering*, *14*(1), 112–116.

Yáñez-Sedeño, P., Campuzano, S., & Pingarrón, J. M. (2017). Electrochemical sensors based on magnetic molecularly imprinted polymers: A review. *Analytica Chimica Acta*, *960*, 1–17.

Yang, B., Xu, J., & Zhu, H. L. (2019). Recent progress in the small-molecule fluorescent probes for the detection of sulfur dioxide derivatives. *Free Radical Biology and Medicine*, *145*, 42–60.

Yang, H., Li, Y., Zhao, Y., Yu, S., Ma, H., Qian, L., Wang, R., Yu, T., & Su, W. (2021). Four new bipolar Indolo[3,2-b]carbazole derivatives for blue OLEDs. *Dyes and Pigments*, *187*, 109096.

Yu, Q. Q., & Wang, M. Q. (2020). Carbazole-based fluorescent probes for G-quadruplex DNA targeting with superior selectivity and low cytotoxicity. *Bioorganic and Medicinal Chemistry*, *28*(17), 115641.

Yu, Y., Li, N., Jin, Q., Ji, Z., Sun, Z., Li, G., Zhang, S., & You, J. (2019). Novel fluorescence labeling reagent 4-(carbazole-9-yl)-benzyl chloroformate and its application in the determination of nitrofuran metabolites compounds in foodstuffs by high performance liquid chromatography with fluorescence detection. *Microchemical Journal*, *145*, 9–17.

Zhang, J., Chen, W., Kalytchuk, S., Li, K. F., Chen, R., Adachi, C., Chen, Z., Rogach, A. L., Zhu, G., Yu, P. K. N., Zhang, W., Cheah, K. W., Zhang, X., & Lee, C. S. (2016). Self-assembly of electron donor-acceptor-based carbazole derivatives: Novel fluorescent organic nanoprobes for both one- and two-photon cellular imaging. *ACS Applied Materials and Interfaces*, *8*(18), 11355–11365. https://doi.org/10.1021/acsami.6b03259

Zhang, K., Wu, W., Li, Y., Sun, M., Yu, H., & Wong, M. S. (2016). Carbazole-based two-photon fluorescent probe for selective imaging of mitochondrial hydrogen peroxide in living cells and tissues. *RSC Advances*, *6*(116), 115298–115302.

Zhang, X., Wang, C., Jin, L., Han, Z., & Xiao, Y. (2014). Photostable bipolar fluorescent probe for video tracking plasma membranes related cellular processes. *ACS Applied Materials and Interfaces*, *6*(15), 12372–12379.

Zhu, H., Fan, J., Du, J., & Peng, X. (2016). Fluorescent probes for sensing and imaging within specific cellular organelles. *Accounts of Chemical Research*, *49*(10), 2115–2126.

Zhu, W., Yang, L., Fang, M., Wu, Z., Zhang, Q., Yin, F., Huang, Q., & Li, C. (2015). New carbazole-based Schiff base: Colorimetric chemosensor for Fe^{3+} and fluorescent turn-on chemosensor for Fe^{3+} and Cr^{3+}. *Journal of Luminescence*, *158*, 38–43.

Index

A

B

C

D

F